Handbook of Combustion Engineering

Handbook of Combustion Engineering

Edited by **Alden Whitley**

NY RESEARCH
P R E S S

New York

Published by NY Research Press,
23 West, 55th Street, Suite 816,
New York, NY 10019, USA
www.nyresearchpress.com

Handbook of Combustion Engineering
Edited by Alden Whitley

International Standard Book Number: 978-1-63238-516-1 (Hardback)

Printed in the United States of America.

Contents

Permissions

List of Contributors

Preface

Over the recent decade, advancements and applications have progressed exponentially. This has led to the increased interest in this field and projects are being conducted to enhance knowledge. The main objective of this book is to present some of the critical challenges and provide insights into possible solutions. This book will answer the varied questions that arise in the field and also provide an increased scope for furthering studies.

Combustion engineering applies the concept of using fuel to produce heat energy. It has applications in diverse areas such as home heating systems, car engines and manufacturing, etc. This discipline deals with evaluation of energy burning systems, combustion supervision and management, heat transference, combustion equipment, etc. This book is a compilation of chapters that discuss the most vital concepts and emerging trends in the field of combustion engineering. Different approaches, evaluations, methodologies and advanced studies revolving around combustion engineering have been included in this book. It is a valuable compilation of topics, ranging from the basic to the most complex technological progress in this area. It is an essential guide for researchers, academicians, students and anyone else who wishes to pursue this discipline further.

I hope that this book, with its visionary approach, will be a valuable addition and will promote interest among readers. Each of the authors has provided their extraordinary competence in their specific fields by providing different perspectives as they come from diverse nations and regions. I thank them for their contributions.

Editor

Development and Parametric Evaluation of a Tabulated Chemistry Tool for the Simulation of n-Heptane Low-Temperature Oxidation and Autoignition Phenomena

George Vourliotakis, Dionysios I. Kolaitis, and Maria A. Founti

Laboratory of Heterogeneous Mixtures and Combustion Systems, Thermal Engineering Section, School of Mechanical Engineering, National Technical University of Athens, 9 Heroon Polytechniou Street, Polytechnioupoli Zografou, 15780 Athens, Greece

Correspondence should be addressed to Dionysios I. Kolaitis; dkol@central.ntua.gr

Academic Editor: Michael Fairweather

Accurate modelling of preignition chemical phenomena requires a detailed description of the respective low-temperature oxidative reactions. Motivated by the need to simulate a diesel oil spray evaporation device operating in the "stabilized" cool flame regime, a "tabulated chemistry" tool is formulated and evaluated. The tool is constructed by performing a large number of kinetic simulations, using the perfectly stirred reactor assumption. n-Heptane is used as a surrogate fuel for diesel oil and a detailed n-heptane mechanism is utilized. Three independent parameters (temperature, fuel concentration, and residence time) are used, spanning both the low-temperature oxidation and the autoignition regimes. Simulation results for heat release rates, fuel consumption and stable or intermediate species production are used to assess the impact of the independent parameters on the system's thermochemical behaviour. Results provide the physical and chemical insight needed to evaluate the performance of the tool when incorporated in a CFD code. Multidimensional thermochemical behaviour "maps" are created, demonstrating that cool flame activity is favoured under fuel-rich conditions and that cool flame temperature boundaries are extended with increasing fuel concentration or residence time.

1. Introduction

The prevailing kinetic mechanisms governing the oxidative reactions of hydrocarbon fuels are continuously changing depending on the temperature of the mixture. In this context, it is possible to define low-, intermediate-, and high-temperature oxidative regions, in which significantly different oxidizing schemes are effective [1–3]; each oxidative region can be described by a different subgroup of elementary chemical reactions. However, for the majority of the most commonly used fuels, comprehensive chemical kinetics mechanisms, capable of describing the complete thermochemical behaviour in all three oxidative regions are currently not available. As a result, in detailed computational fluid dynamics (CFD) simulations of conventional combustion systems, it is common practice to effectively describe the high-temperature oxidative region, which is the most

important in terms of heat release and pollutant formation [3]. However, there are specific applications, where the low- and intermediate-temperature oxidative regions become important, for example, homogeneous charge compression ignition (HCCI) internal combustion engines (ICE) [4–6], lean premixed prevaporized (LPP) combustors in gas turbines [7, 8], and liquid fuel reformers for fuel cell applications [9–11] or industrial safety [12, 13]; in these cases, cool flame reactions largely determine the overall reactivity and have to be taken into account.

The term "cool flame" is conventionally used to collectively describe the low- and intermediate-temperature (approximately 500 K–900 K) oxidative preignition reactions emerging in reactive systems fed with organic fuels, such as saturated and unsaturated hydrocarbons, alcohols, aldehydes, ketones, acids, oils, ethers, and waxes [13, 14]. Cool flame reactions are modestly exothermic; they are often described as partial or intermediate oxidation reactions, since they result in the formation of a large variety of long-lived intermediate species (e.g., alcohols, acids, aldehydes, and peroxides) [14–16]. Cool flames occur preferentially under fuel-rich conditions [17] and are characterized by the appearance of a faint bluish light, which is attributed to the chemiluminescence of excited formaldehyde [18]. Cool flames appear in the form of a self-quenching temperature and pressure pulse, during the two-stage ignition, and are associated with the "ignition delay time" phenomenon, as well as "knocking" phenomena in spark-ignition ICE [15, 19].

The cool flame oxidation region is characterized by a competition between chain-termination and chain-branching reactions [15, 19]. At lower temperatures, chain-branching reactions are favoured, leading to the acceleration of the overall reaction rate. When the temperature is increased, the decomposition of chain-branching precursors is enhanced; in contrast, chain-termination reactions become more important, since their activation energy is decreasing with increasing temperature. As a result, a negative temperature coefficient (NTC) region emerges in the temperature regime where transition between the low- and intermediate-temperature oxidative mechanisms occurs. In the NTC region, the overall reaction rate is decreasing with increasing temperature; this behaviour is a distinctive characteristic of cool flame reactions. Cool flame reactions can be "stabilized" when a technical device operates in the NTC region. In this case, the heat losses through the device boundaries are fully compensated by the heat release of the exothermal cool flame reactions; the device achieves steady-state operation, which is not accompanied by thermal ignition [20]. The utilization of such innovative stabilized cool flame (SCF) reactors allows better control over the combustion process in liquid fuel premixed combustion systems; they can also be used to assist liquid fuel (e.g., diesel oil) reforming to a hydrogen-rich gas for fuel cell applications [9–11].

In order to describe in sufficient detail the degenerate chain-branching cool flame oxidative activity leading to the NTC behaviour, multistep detailed chemical kinetics mechanisms have to be used [19, 21–28]. Due to the high computational costs associated with implementing detailed chemical kinetic mechanisms in multiphase CFD tools [29], cool flames are usually neglected in detailed CFD simulations of turbulent reactive flows. In the context of numerical simulations of reactive systems, cool flame reactions are commonly described using simplified zero- or one-dimensional modelling tools [22, 23, 30, 31].

Motivated by the need to numerically simulate the turbulent, multiphase, multicomponent, and reactive flow-field developing in a non-igniting SCF technical device, the authors have developed a variety of alternative modelling techniques [24, 32–35], aiming to reduce the heavy computational load imposed by the implementation of detailed kinetic schemes into CFD solver. As a part of this ongoing work, the present study focuses on the development of a "tabulated chemistry" modelling approach, capable of simulating the thermochemical behaviour of n-heptane in the low-temperature oxidation and autoignition regimes.

The development of the "tabulated chemistry" tool focuses on accuracy and ease of implementation in multiphase CFD codes. In contrast to a previous "tabulated chemistry" approach developed by the authors [34], which focused solely on the SCF regime, the tool presented in this work is capable of describing the entire low- and intermediate-temperature oxidation regions, including autoignition phenomena. The latter are important in technical devices, where ignition must be prevented. Furthermore, the proposed tool provides information for important species concentrations, which is essential for accurate prediction of the reformate mixture composition in fuel cell reforming applications. In addition to the development of the "tabulated chemistry" tool, the obtained numerical results are used to evaluate the impact of varying significant operational parameters on the thermochemical behaviour of the system. Therefore, the main scope of this study is two-fold, that is, to develop a simulation tool (tabulated chemistry) and to thoroughly investigate the thermochemical behaviour of n-heptane, by highlighting and explaining the physical and chemical underlying phenomena (e.g., through sensitivity analysis). The latter is a critical step prior to the implementation of the developed tool in a CFD code.

2. Numerical Simulations

The developed "tabulated chemistry" tool is focused on the simulation of technical devices using diesel oil fuel, which is essentially a mixture of a large variety of organic species; the exact composition of the commercially available diesel oil fuel mixture varies significantly, depending on the raw fuel source and refinement processes. As a result, detailed chemical kinetics models describing diesel oil fuel's oxidation behaviour are currently not available; contemporary modelling approaches are focused on simulating surrogate fuels that are considered thermochemically "equivalent" to diesel oil fuel [2, 36, 37]. In the present study, n-heptane is used as a surrogate fuel for diesel oil fuel; this assumption is considered reasonable [24], especially in the case of autoignition modelling, since n-heptane's cetane number is very close to that of diesel oil [38, 39]. Additionally, it has been shown that both fuels feature a similar heat release rate for both conventional compression ignition ICE and HCCI ICE [40, 41].

There is a broad body of literature focusing on the development of detailed and reduced/skeletal chemical kinetic mechanisms for n-heptane combustion phenomena [21, 23, 38, 42–45]. However, the majority of the available reduced kinetic schemes have been developed aiming to describe the high-temperature, high-pressure conditions prevailing in an ICE cylinder; as a result, they are not capable of accurately reproducing the thermochemical characteristics of the NTC region in low-temperature, atmospheric pressure conditions. The authors have previously performed a comparative study of 7 n-heptane oxidation mechanisms (e.g., reduced, skeletal, semidetailed, and detailed), under conditions relevant to a diesel oil SCF reactor operation [24]. Only 2 mechanisms were found to yield relatively accurate results in all 3 modelled cases for which experimental measurements were available for validation; these were the detailed kinetic mechanism of

Curran et al. [21], consisting of 550 species and 2481 reactions, and the semidetailed mechanism of Tao et al. [38], involving 57 species and 290 reactions. The mechanism of Tao et al. [38] has been used in a previous effort of developing a "tabulated chemistry" tool capable of simulating diesel oil fuel's cool flame oxidation characteristics [34, 46]; in the current study, aiming to extend the applicability of the developed tool to higher temperatures, beyond autoignition, the more detailed kinetic mechanism of Curran et al. [21] is utilized. The latter mechanism is capable of thoroughly describing the cool flame regime, by utilizing 25 reaction classes of the characteristic species formed under such conditions (alkylperoxy, hydroperoxy-alkyl and hydroperoxy-alkyl-peroxy radicals, cyclic ethers, carbonyl species, ketohydroperoxydes, and so on). The mechanism has been extensively validated under a variety of reactor types (e.g., plug flow reactor, jet-stirred reactors, shock tubes, and rapid compression machines) [21, 27] and operational conditions (e.g., low-temperature cool flames and high pressure ignition) [21, 27, 28]. Additional validation relevant to the present work has been already shown in [24], while the mechanism has been also shown to be capable of accurately describing the most recent relevant experimental data [27].

Aiming to decrease the high computational cost associated with the solution of the resulting stiff system of nonlinear differential equations, when a chemical kinetics mechanism is implemented in multiphase CFD codes, a broad range of methodologies has been proposed [15, 19]. The most commonly used methodology is the "reduction" of the detailed chemical kinetics scheme, aiming to significantly decrease the number of the required chemical species and chemical reactions. However, it has been shown that the complex nature of low-temperature oxidation reactions cannot be accurately described by reduced kinetics mechanisms, for example, [24]. Thus, an alternative low-computational cost methodology is used in the current study; the "tabulated chemistry" approach allows the use of detailed chemical kinetics mechanisms, thus being able to effectively retain the essential information regarding the cool flame oxidation phenomena occurring in the low- and intermediate-temperature regimes. In this case, the time-consuming chemistry numerical calculations are *a priori* performed *off-line* and the respective results are stored in a multidimensional database ("look-up table"). During the *on-line* CFD simulations, the tabulated chemical information is accessed by means of a low computational cost retrieval algorithm (e.g., a multilinear interpolation or multivariable correlations).

In order to generate the look-up table data matrix, a large number of chemical kinetic simulations were performed using the perfectly stirred reactor (PSR) assumption. Since low-temperature oxidative phenomena are considered to be kinetically-controlled [14, 46, 47], the PSR approach is considered to be a valid assumption. In this case, each computational cell of the CFD simulations is considered to be an "infinitely fast mixing" PSR exhibiting spatially uniform temperature and mixture composition [48]; turbulence timescale variations are indirectly accounted for through the property transport calculations, whereas the completely homogeneous mixture is exclusively affected by chemical kinetics. A known limitation of the utilized PSR assumption is the need to estimate the mean residence time in each computational (CFD) cell. When each numerical grid cell is assumed to be an independent "equivalent PSR", local velocities and numerical grid length scales can be used to determine the respective residence time; in this case, the obtained results theoretically depend on the selected grid size. However, a range of grid independence studies in a variety of SCF reactors and operational conditions have revealed that the obtained results are not significantly affected by the size of the computational grid, since cool flame heat release rate predictions are monotonously increasing with increasing residence time [24, 34].

All computations were performed in an open, adiabatic, PSR environment, featuring unit volume, where the inlet temperature and mixture stoichiometry as well as the reactor's residence time are explicitly specified (independent parameters). The mass balance, species conservation, and energy equations are solved using the CHEMKIN software [49]. Particular attention has been paid in providing the initial guess of the outlet temperature and in determining the convergence criteria, since solution of the respective equations using large detailed kinetic mechanisms is mathematically stiff.

2.1. Selection of Parameters. It is well documented that the main operational parameters affecting cool flame behaviour are pressure, temperature, and fuel concentration [13, 15, 17]. Since the current study focuses on atmospheric pressure conditions, pressure was not varied. As a result, the 3 operational parameters selected were the inlet temperature (T_{in}), the inlet fuel concentration (expressed via the respective fuel mass fraction, Y_{fuel}, defined as the ratio of the fuel mass over the total inlet mixture mass), and the total residence time in the PSR (t_{res}). The variation range for each parameter was decided using relevant information obtained both in experimental [50, 51] and simulation [25, 30] studies; the limiting values were selected to extend beyond the typical values expected in the corresponding technical devices. Inlet temperature values extended from 550 K to 1000 K, thus covering the entire low- and intermediate-temperature oxidation regions, as well as autoignition phenomena. Fuel mass fraction values extended from 0.005 to 0.5; these values correspond to an equivalence ratio ranging from very lean mixtures ($\lambda = 13$) to very rich mixtures ($\lambda = 0.07$). Here, the lambda factor, λ, is defined as the ratio of the actual air-to-fuel ratio over the stoichiometric one. The majority of the selected values are distributed near stoichiometric conditions. Finally, residence time values spanned 3 orders of magnitude, ranging from 0.5 ms to 0.5 s. All in all, 34848 different chemical kinetics simulations were performed, corresponding to all triplet combinations of the respective values (Table 1).

For each triplet (T_{in}, Y_{fuel}, t_{res}) combination, values of the obtained simulation results were stored in the data matrix. The stored values regarded *thermal* quantities, such as total heat release per unit volume, HR(kJ/m^3) or outlet temperature, T_{out} (K), and *kinetic* quantities, for example, species rate of production or destruction, ω_i(kg/m^{3}s), or outlet species mole fraction, X_i. The size of the obtained database can be

TABLE 1: Variation range of selected independent parameters.

Parameter	Symbol	Units	Range	Number of nodes
Inlet temperature	T_{in}	K	550–1000	33
Fuel mass fraction	Y_{fuel}	$kg_{\text{fuel}}/kg_{\text{mixture}}$	0.005–0.5	33
Residence time	t_{res}	s	0.0005–0.5	32

customized, by excluding unneeded information, depending on the required accuracy of the performed CFD simulations. For instance, if only a rather simplistic approximation of the thermochemical behaviour of the cool flame region is needed, the size of the database may be significantly reduced so as to contain only simulation results regarding heat release rate, which is introduced to the CFD code as a local heat "source" term, and fuel consumption rate, which corresponds to a fuel concentration "sink" term in the CFD code. Despite the significant simplifying assumptions, this methodology is known to yield adequately accurate predictions in the case of multiphase CFD simulations of atmospheric pressure SCF diesel oil evaporation devices, as demonstrated in relevant validation studies, for example, [24, 34].

3. Results and Discussion

The obtained numerical results allow a systematic investigation of the effect of the examined independent parameters on the thermochemical behaviour of n-heptane, both in the low-temperature oxidation and autoignition regimes. A selection of characteristic results is presented in the following sections, aiming to highlight the most prominent observations. Furthermore, in order to obtain a more in-depth understanding of the thermochemical behaviour of the n-heptane oxidation, extensive reaction path and sensitivity analyses have been performed aiming to delineate the major chemical pathways controlling the heat release and species production/consumption processes occurring during n-heptane oxidation. Reaction path (or rate-of-production) and sensitivity analyses constitute major tools in understanding reacting flows. Rate-of-production analysis determines the contribution of each reaction to the net production or destruction rate of a species and provides a static picture of the entire reaction network. Species formation and destruction paths are revealed through such an analysis. On the other hand, sensitivity analysis gives information regarding the influence of a specific reaction to a certain species, thus providing a more dynamic understanding of chemical interactions. The sensitivity coefficient is defined as $\partial F/\partial \alpha$, where F is a target value (e.g., temperature, heat release rate, or species concentration) and α is the preexponential factor in the Arrhenius rate constant expression. The nomenclature of species naming as well as reaction numbering of the considered detailed kinetic mechanism are identical to the original publication of Curran et al. [21].

3.1. Predictions of Heat Release. A typical snapshot of the formulated database is presented in Figure 1, where predictions of heat release per unit volume (HR) are depicted as a

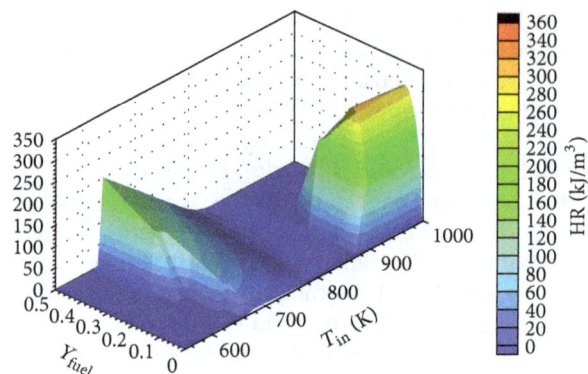

FIGURE 1: Predictions of heat release per unit volume as a function of inlet temperature and fuel concentration ($t_{\text{res}} = 5$ ms).

function of both inlet temperature and fuel mass fraction, for a characteristic residence time of 5 ms. Two main chemical activity regimes can be identified. The conventional "ignition" region emerges at temperatures higher than 880 K; the corresponding exothermal activity practically ceases at extremely fuel-rich conditions ($Y_{\text{fuel}} > 0.3$). This "quenching" behaviour is attributed to the significant reduction of essential radical species concentrations observed at high temperatures ($T_{\text{in}} > 900$ K) and fuel-rich mixtures; in fact, predictions of H and OH radicals concentrations are decreased by almost 5 orders of magnitude, when the inlet fuel concentration is quintupled, from a near stoichiometric value ($Y_{\text{fuel}} = 0.06$, $\lambda = 1.04$) to significantly fuel-rich conditions ($Y_{\text{fuel}} = 0.3$, $\lambda = 0.16$). On the other hand, heat release due to cool flame reactions is intensified with increasing fuel concentration; the "cool flame" regime is mainly observed at the low- and intermediate-temperature oxidation regions, extending approximately between 620 K and 800 K. Maximum heat release values in the cool flame regime are found to emerge at the highest fuel concentration levels considered in this study ($Y_{\text{fuel}} = 0.5$) and correspond to approximately 70% of the maximum heat release values observed at the ignition regime, observed near stoichiometric ($Y_{\text{fuel,st}} = 0.062$) conditions.

By further investigating the information stored in the formulated chemical database, taking into consideration similar multiparameter plots corresponding to all the investigated residence time values, certain observations are found to be universally valid. In general, the "cool flame" regime broadens with increasing fuel concentration and the temperature where the respective peak heat release value is observed decreases with increasing fuel concentration. On the other hand, the ignition regime appears to be rather insensitive to the variation of the main operating parameters; peak values correspond invariably to approximately stoichiometric mixture conditions, whereas a "pre-ignition" regime featuring moderate heat release values is formed at richer mixtures.

In Figure 2, predictions of heat release per unit volume are depicted as a function of inlet temperature, for 5 characteristic fuel concentration levels, at a constant residence time ($t_{\text{res}} = 30$ ms). It is evident that cool flame exothermic activity is enhanced with increasing fuel concentration; such behaviour is corroborated by similar findings in relevant

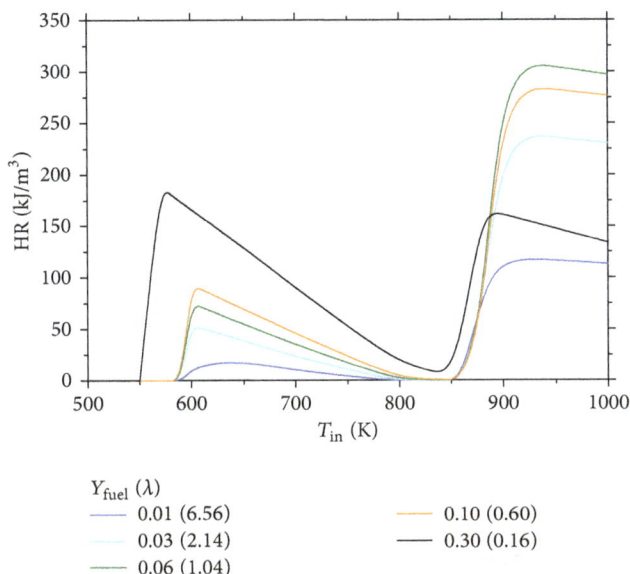

FIGURE 2: Predictions of heat release per unit volume as a function of inlet temperature for 5 characteristic values of fuel concentration (t_{res} = 30 ms).

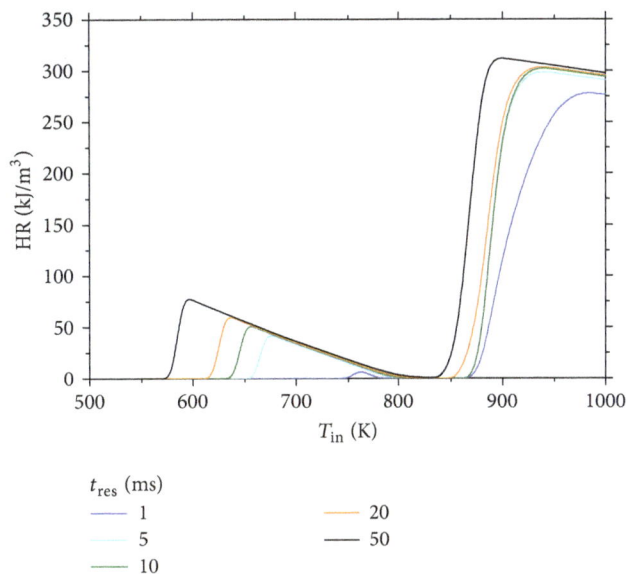

FIGURE 3: Predictions of heat release per unit volume as a function of inlet temperature for 5 characteristic values of residence time (Y_{fuel} = 0.06).

experimental studies [17, 29]. The "peak" of the cool flame heat release predictions becomes more pronounced at fuel-rich conditions, resulting in steeper curve slopes at the NTC region; the characteristic inlet temperature that corresponds to the maximum heat release value decreases with increasing fuel concentration. However, the temperature region where cool flame activity is observed remains practically constant (580 K–800 K), with the sole exception of extremely fuel-rich conditions (Y_{fuel} = 0.3), where the respective temperature limits are broadened (550 K–840 K). On the other hand, the temperature region where high-temperature oxidation is observed remains practically constant (T_{in} > 850 K) for all the considered fuel concentration values. Peak heat release predictions at the ignition regime are observed for stoichiometric conditions; the respective values decrease when fuel concentration is either increased (fuel-rich region) or decreased (fuel-lean region).

The variation of heat release per unit volume predictions as a function of inlet temperature, for 5 characteristic residence time values, at a constant fuel concentration level (Y_{fu} = 0.06), corresponding to nearly stoichiometric conditions, is depicted in Figure 3. As expected, longer residence times result in higher heat release values. The effect of residence time on the cool flame activity regime is prominent; when the residence time is increased, the cool flame temperature region is significantly extended, whereas the maximum heat release value is "shifted" towards lower temperatures. A kinetic interpretation of such behaviour is that given sufficient time, the continuous production of intermediate species (e.g., aldehydes) results in the increased production of hydroper-oxyalkyl radicals (QOOH) and hydrogen peroxide (H_2O_2), which favour chain-propagation reactions, thus leading to enhanced cool flame exothermal activity [52]. The respective effects on the ignition regime are less prominent; increasing the residence time results in a slight decrease of the lower

temperature limit of the respective region, as well as a slight increase of the observed maximum value.

Predictions of heat release per unit volume as a function of residence time, for 5 characteristic fuel concentration levels, at a constant inlet temperature (T_{in} = 740 K), corresponding to the main cool flame activity region, are depicted in Figure 4. The cool flame exothermal activity is enhanced with increasing residence time; at residence time values higher than approximately 3 ms, the predicted heat release values are practically constant. A further increase is observed at residence times higher than 0.05 s, only under extremely fuel-rich conditions (Y_{fuel} = 0.3).

3.2. Predictions of Species Production and Consumption. Information regarding the exact mixture composition is very important for a large variety of technical applications (e.g., high temperature reforming for fuel cell applications, HCCI ICE). Figure 5 presents predictions of the fuel consumption rate per unit volume as a function of both inlet temperature and fuel concentration, for a constant residence time value (t_{res} = 5 ms). Fuel consumption rates in the cool flame regime are almost twice the respective values observed in the ignition regime. Interestingly enough, modest fuel consumption rates are observed in the fuel-rich region at higher temperatures, despite the fact that there is no significant exothermic heat release observed in the same region (compare with Figure 1).

In order to further investigate the thermochemical behaviour of the fuel-rich high-temperature oxidation regime, a reaction path analysis has been performed. In the fuel-rich and high-temperature case (Y_{fuel} = 0.3, T_{in} = 950 K), n-heptane is equally consumed through the competing *endothermic* H-abstraction reactions with the HO_2 radical and the *exothermic* reactions with the OH radical (each type of reaction accounts for almost 40% of the total fuel consumption rate). Therefore, although the

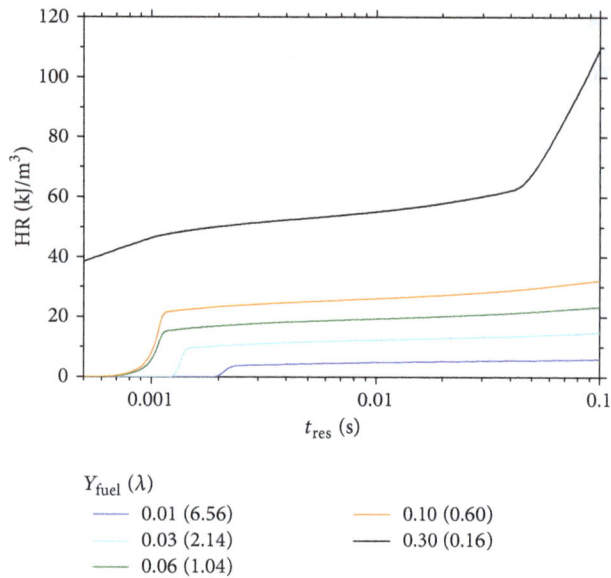

Y_{fuel} (λ)

— 0.01 (6.56)   — 0.10 (0.60)
— 0.03 (2.14)   — 0.30 (0.16)
— 0.06 (1.04)

FIGURE 4: Predictions of heat release per unit volume as a function of residence time for 5 characteristic values of fuel concentration (T_{in} = 740 K).

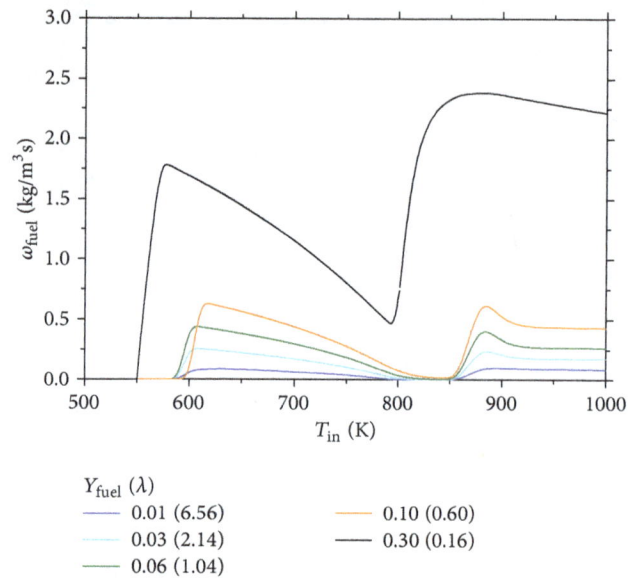

Y_{fuel} (λ)

— 0.01 (6.56)   — 0.10 (0.60)
— 0.03 (2.14)   — 0.30 (0.16)
— 0.06 (1.04)

FIGURE 6: Predictions of fuel consumption rate per unit volume as a function of inlet temperature for 5 characteristic values of fuel concentration (t_{res} = 30 ms).

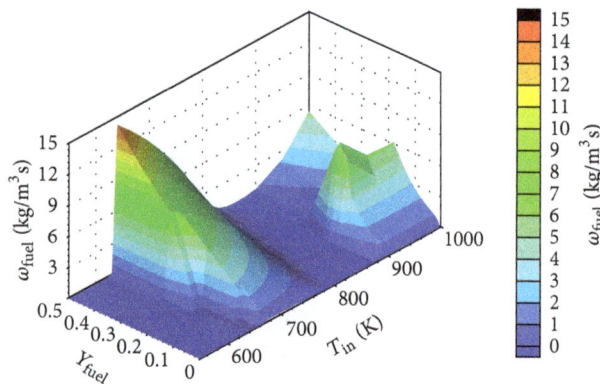

FIGURE 5: Predictions of fuel consumption rate per unit volume as a function of inlet temperature and fuel concentration (t_{res} = 5 ms).

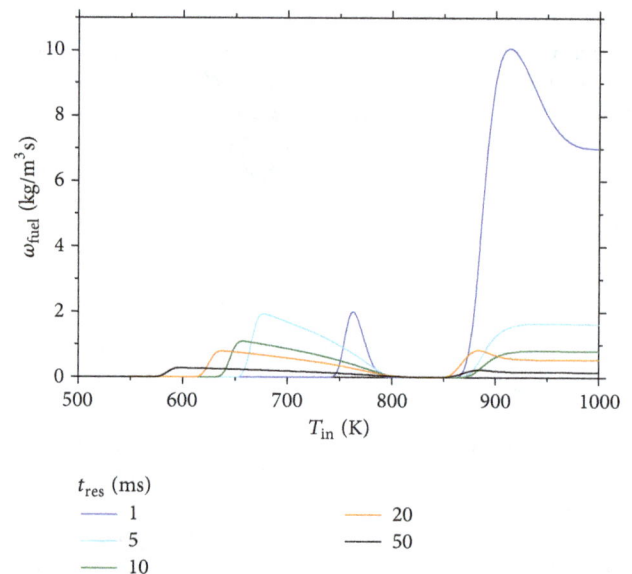

t_{res} (ms)

— 1   — 20
— 5   — 50
— 10

FIGURE 7: Predictions of fuel consumption rate per unit volume as a function of inlet temperature for 5 characteristic values of residence time (Y_{fuel} = 0.06).

fuel molecule is being consumed, the net heat production from gas-phase reactions is slightly negative. Decreasing the inlet fuel concentration (Y_{fuel} = 0.2, T_{in}=950 K) and moving towards stoichiometric conditions, the H-abstraction reactions as well as fuel decomposition reactions to C_6 + C_1 (e.g., n-C_7H_{16} <=> C_6H_{13}-1 + CH_3), C_5 + C_2 (e.g., n-C_7H_{16} <=> C_5H_{11}-1 + C_2H_5), and C_4+ C_3 species (e.g., n-C_7H_{16} <=> pC_4H_9 + n-C_3H_7) become dominant. Due to the high outlet temperature (T_{out} > 2000 K), the aforementioned fuel decomposition reactions serve as initiation steps, despite their high activation energy. Subsequently, the formed C_4 and C_5 species, promptly further decompose to C_2 and C_3 species, thus initiating the heat release producing chain, from C_2H_4 down to CO; therefore, significant exothermal activity emerges near the stoichiometric mixture conditions (compare with Figure 1).

In Figure 6, predictions of the fuel consumption rate per unit volume are depicted as a function of inlet temperature,

for 5 characteristic fuel concentration levels, at a constant residence time (t_{res} = 30 ms). The emerging "cool flame" and "ignition" regimes are once more identified; as expected, increasing initial fuel concentration results in higher fuel consumption rates. The general form of the obtained curves is similar to that of the respective curves corresponding to heat release predictions (c.f. Figure 2); however, as mentioned before, the significant fuel consumption rates observed in the ignition regime under fuel-rich conditions are not accompanied by respective important exothermic heat release.

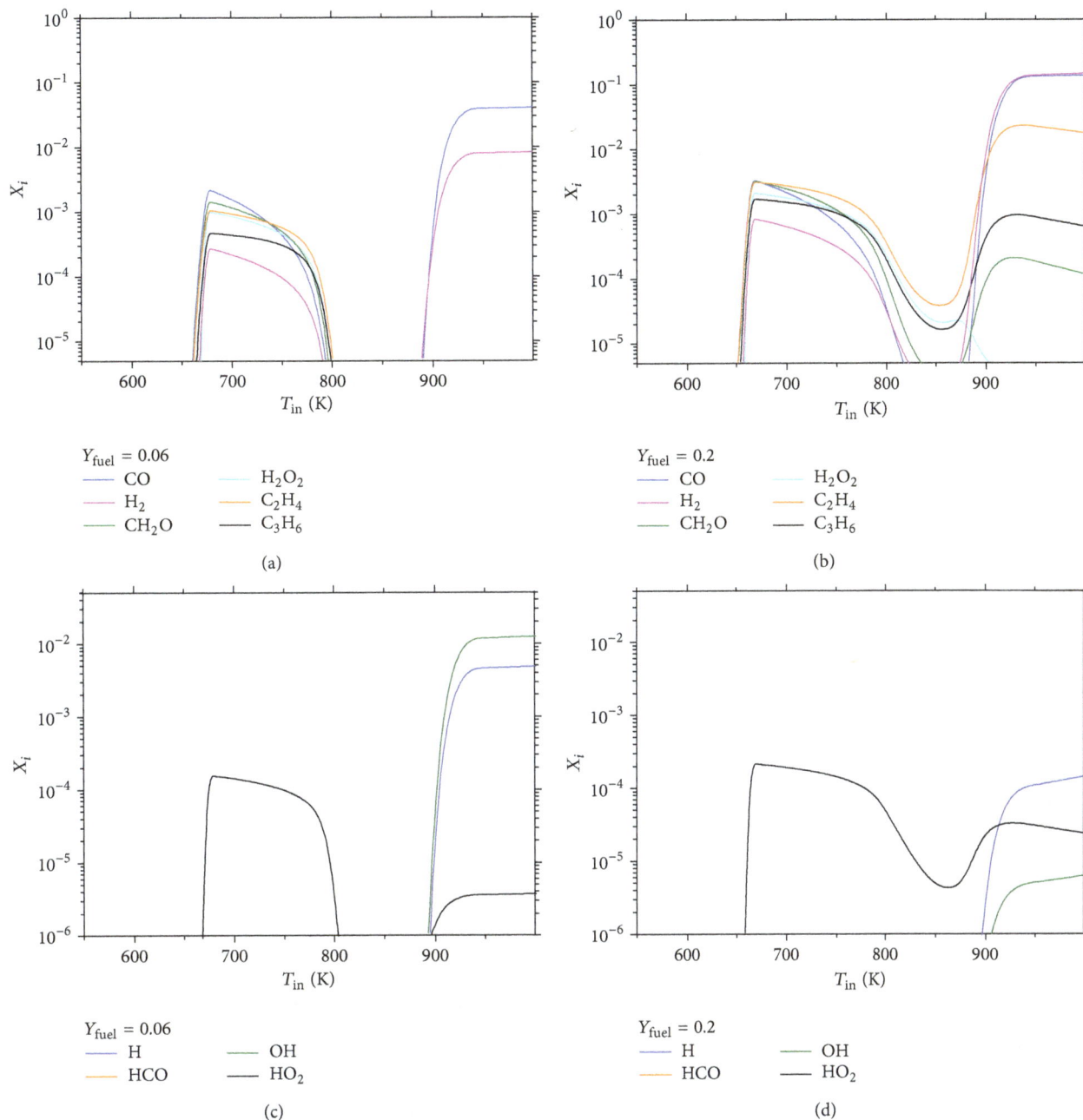

$Y_{fuel} = 0.06$

—— CO	—— H_2O_2
—— H_2	—— C_2H_4
—— CH_2O	—— C_3H_6

(a)

$Y_{fuel} = 0.2$

—— CO	—— H_2O_2
—— H_2	—— C_2H_4
—— CH_2O	—— C_3H_6

(b)

$Y_{fuel} = 0.06$

—— H	—— OH
—— HCO	—— HO_2

(c)

$Y_{fuel} = 0.2$

—— H	—— OH
—— HCO	—— HO_2

(d)

FIGURE 8: Predictions of species mole fraction as a function of inlet temperature for 2 characteristic inlet fuel concentration values ($t_{res} = 5$ ms).

The variation of the fuel mass consumption rate per unit volume as a function of inlet temperature, for 5 characteristic residence time values, at a constant (near stoichiometric) fuel concentration ($Y_{fuel} = 0.06$), is depicted in Figure 7. As expected, increasing the residence time results in decreasing fuel consumption rate values, since the latter correspond, essentially, to time gradients; this is evident in the ignition region, when the residence time is sufficiently small ($t_{res} = 1$ ms). Similar to the respective heat release predictions (c.f. Figure 3), the effect of residence time variation on the cool flame activity regime is prominent; when the residence time is increased, the cool flame regime is significantly

extended, whereas the maximum fuel consumption rate values are "shifted" towards lower temperatures. Once more, the moderate fuel consumption rates observed in the ignition region result in significantly higher exothermic heat release, compared to the respective values of the cool flame regime (c.f. Figure 3).

In order to further investigate the effect of the different kinetic paths prevailing in the cool flame and ignition regimes, predictions of several crucial major stable and radical species are presented in Figure 8. The depicted mole fraction predictions are presented as a function of inlet temperature ($t_{res} = 5$ ms), for two fuel concentration

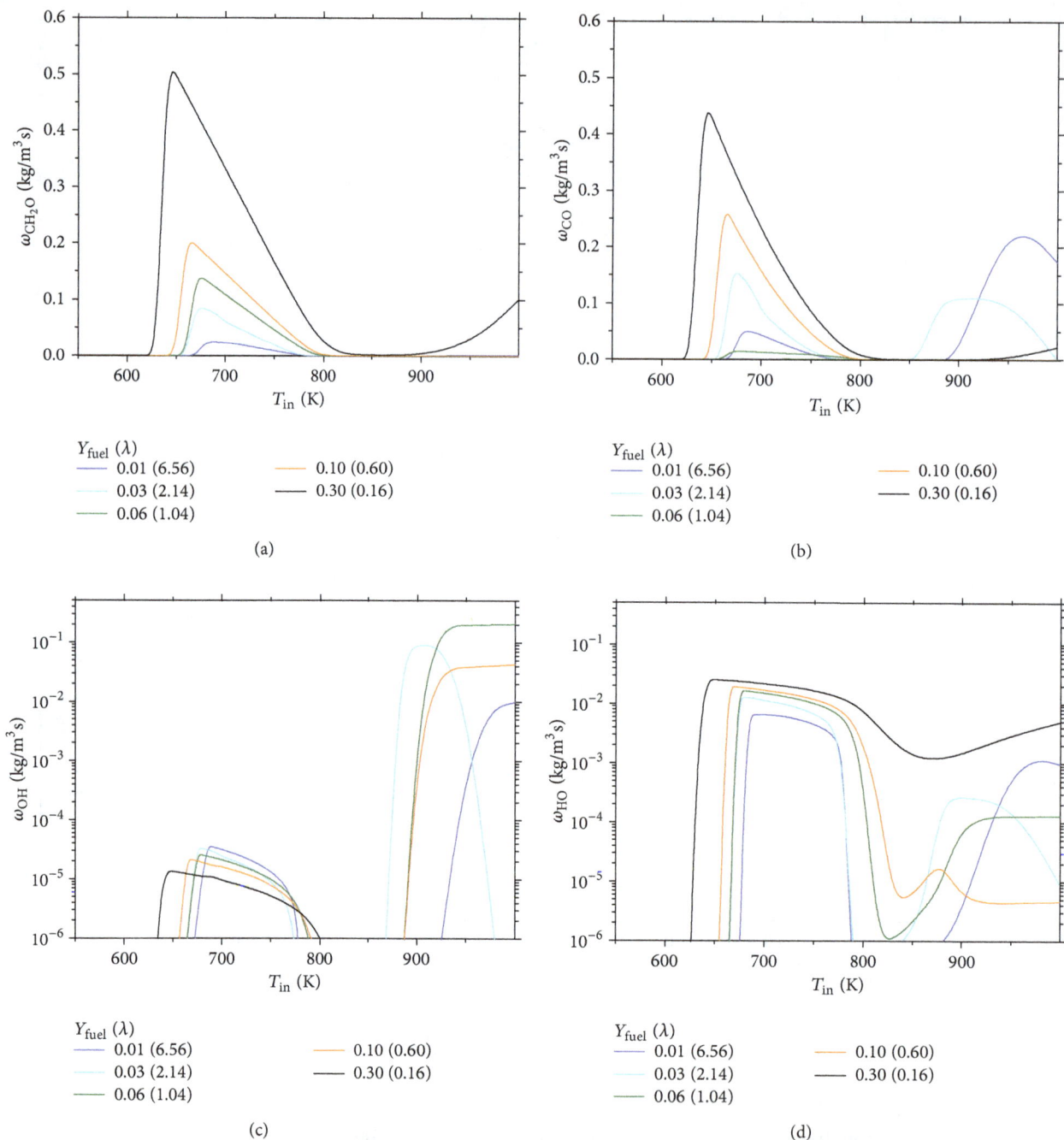

(a)

(b)

(c)

(d)

FIGURE 9: Predictions of CH_2O, CO, OH, and HO_2 production rates per unit volume as a function of inlet temperature for 5 characteristic values of fuel concentration (t_{res} = 5 ms).

levels, one corresponding to approximately stoichiometric conditions (Y_{fuel} = 0.06) and the other referring to fuel-rich conditions (Y_{fuel} = 0.2). It is evident that species concentration levels are significantly different in the low- and high-temperature oxidation regimes. In terms of radical species, the cool flame region is characterized by significant OH and H_2O_2 formation, whereas the ignition region is dominated by H and OH production. Also, while the major stable species emerging in the ignition region are CO and H_2, the cool flame regime is characterized by significant

production of a large variety of species, such as CH_2O, H_2O_2, C_2H_4, C_3H_6, and so forth. The above findings are also in agreement with recent experimental evidence [28].

The prevailing kinetic mechanisms governing the oxidative reactions of n-heptane are continuously changing depending on the temperature of the mixture; this remark is clearly elucidated by comparing predictions of the production rate per unit volume for various important stable (CH_2O, CO) and radical (HO_2, OH) species, which are depicted in Figure 9. The selected species are indicative of the difference

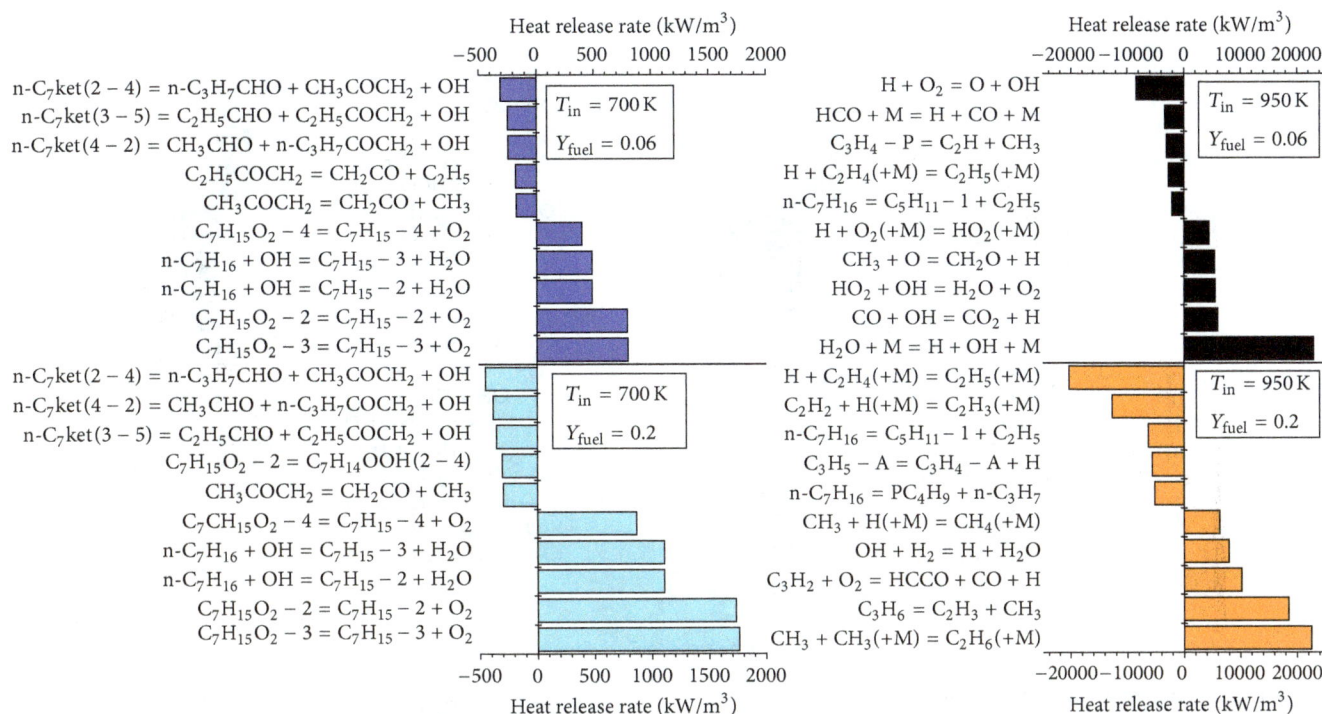

FIGURE 10: Predictions of heat release rate per unit volume for the reactions exhibiting maximum absolute endothermic and exothermic values (t_{res} = 5 ms).

in reactivity between the low- and high-temperature oxidation regimes. For instance, increasing inlet fuel concentrations results in significant production of formaldehyde in the cool flame regime; in the ignition regime, the produced quantities are rather insignificant. On the other hand, CO production rates are significant in both the cool flame and ignition regimes. Similar observations are obtained regarding the important radical species. HO_2 production is significant only in the cool flame regime (increasing with increasing fuel concentration), whereas OH is essentially produced mainly under high-temperature oxidation conditions; in the latter case, maximum values are observed near the stoichiometric conditions.

3.3. Sensitivity Analysis. As it has been already demonstrated in similar studies, for example, [45, 53], sensitivity analysis of the utilized kinetic mechanism may reveal the kinetic paths prevailing under different operating conditions. Figure 10 presents the elementary reactions exhibiting maximum absolute endothermic and exothermic heat release values, for a near stoichiometric (Y_{fuel} = 0.06) and a rich (Y_{fuel} = 0.2) n-heptane/air mixture, at two different inlet temperatures, one indicative of the cool flame regime (T_{in} = 700 K) and the other corresponding to the ignition regime (T_{in} = 950 K). Clearly, mixture stoichiometry has a less prominent effect on the system's behavior as compared to that imposed by the inlet temperature. In the cool flame regime, decomposition reactions of the ketohydroperoxides (n-C_7ket) species are the major endothermic chemical steps. The thermochemical behaviour is also controlled by the competition between isomerisation (endothermic) and decomposition (exothermic)

alkylperoxy reactions. Naturally, fuel depletion reactions through OH radical attack and HO_2 production processes largely contribute to the overall exothermicity. On the other hand, heat release processes at higher temperatures, where ignition occurs, are largely determined by the H_2/O_2 system. Note that in the case of the fuel-rich mixture (Y_{fuel} = 0.2), ketyl radical (HCCO) chemistry becomes important as well as reactions of C_3 species (see also computed C_3H_6 levels in Figure 8).

The calculated outlet temperature sensitivity coefficients for elementary reactions that exhibit the largest absolute values are shown in Figure 11, for two inlet temperature values, T_{in} = 700 K (cool flame regime) and T_{in} = 950 K (ignition regime), and two residence time values (t_{res} = 5 and 30 ms), at a constant inlet fuel concentration (Y_{fuel} = 0.06). Positive values of the sensitivity coefficient indicate that an increase in the particular reaction rate constant results in a corresponding increase in the outlet temperature value.

In the cool flame regime, residence time has a negligible impact on the temperature sensitivity; however, it does play a role in the high-temperature oxidation regime. It is evident that the elementary reactions affecting the heat release are significantly different in the two main oxidation regimes. An interesting observation is that, in the cool flame regime (T_{in} = 700 K), the alkylperoxy radical isomerisation reaction class (e.g., $C_7H_{15}O_2$ <=> $C_7H_{14}OOH$) results in both positive and negative sensitivity coefficients. However, it is arguable to suggest that a faster overall isomerization towards hydroperoxy-alkyl species formation would result in increased overall reactivity; see also [21]. The major fuel oxidation paths involving OH radical attack

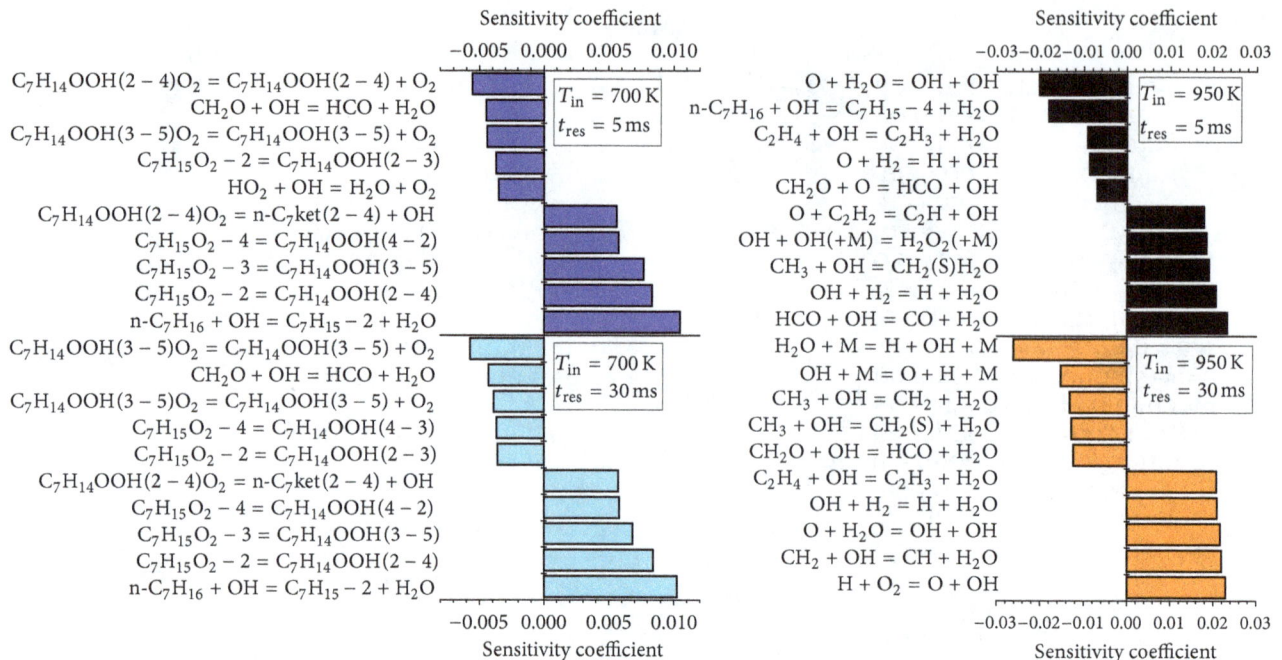

FIGURE 11: Computed outlet temperature sensitivity coefficients for two inlet temperature values, T_{in} = 700 K (cool flame regime) and T_{in} = 950 K (ignition regime), and two residence time values (t_{res} = 5 and 30 ms), at a constant inlet fuel concentration (Y_{fuel} = 0.06).

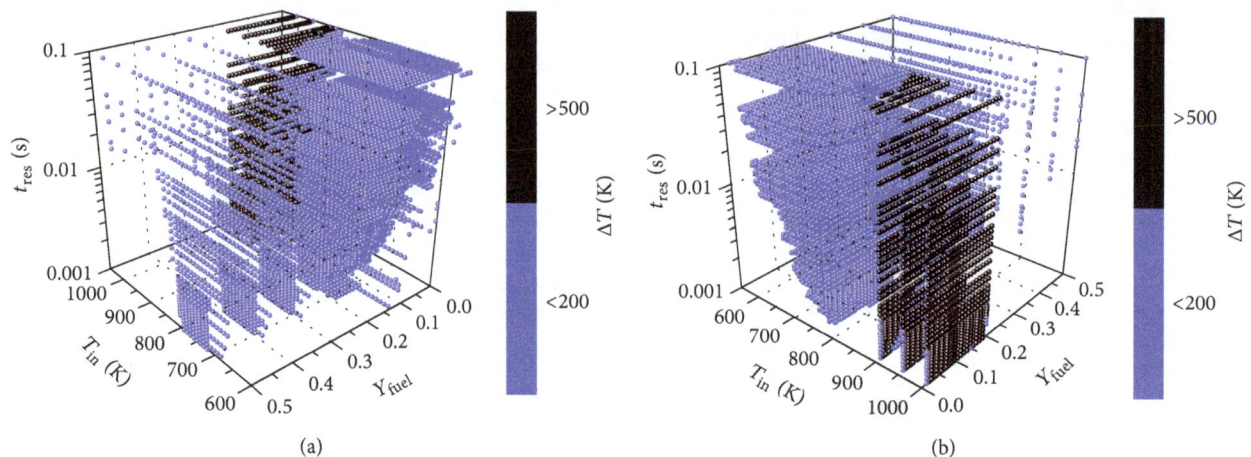

FIGURE 12: Predictions of PSR temperature increase as a function of inlet temperature, fuel concentration, and residence time.

and ketoxyhydro-peroxide (notated as n-C_7ket) species production reactions favour temperature rise. On the other hand, $O_2C_7H_{14}OOH$ decomposition reactions and the OH consuming reactions (e.g., $CH_2O + OH \Leftrightarrow HCO + H_2O$ and $HO_2 + OH \Leftrightarrow H_2O + O_2$) tend to decrease the observed outlet temperature values. In the case of the high-temperature regime, no fuel specific reactions appear to control ignition. In fact, the system's overall exothermic activity is exclusively controlled by the fate of the available radical pool, which in turn depends on the competition of chain-branching and chain-terminating reactions.

3.4. *Thermochemical Behaviour and Tool Validation.* The information obtained in the process of developing the tabulated chemistry tool can be also utilized to develop characteristic thermochemical behaviour "maps," which enable the assessment of the expected conditions emerging in technical devices when certain operational parameter values are employed. Such multidimensional maps are essentially similar to 2D "ignition diagrams," describing the thermal behaviour of technical devices operating in the low- to high-temperature oxidation regimes [9, 51]. In Figure 12, an indicative thermochemical behaviour map is depicted; the

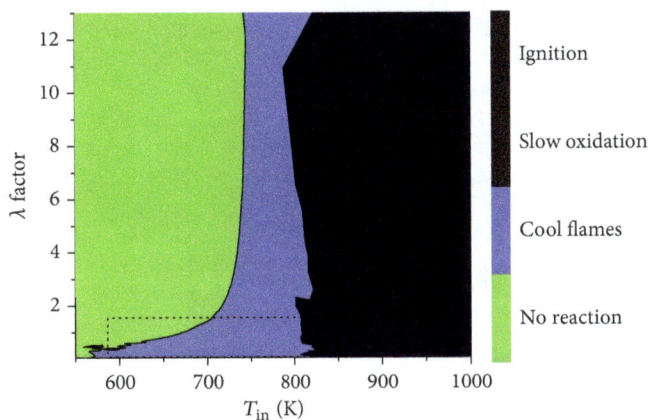

FIGURE 13: Predictions of main reactivity regions as a function of inlet temperature and lambda factors.

main operating regions, that is, cool flame and autoignition, are identified by utilizing the predicted temperature increase, between the reactor's inlet and outlet ($\Delta T = T_{out} - T_{in}$). Conventionally, a temperature increase lower than 200 K [45] corresponds to the cool flame regime, whereas a temperature increase higher than 500 K suggests that autoignition is occurring. Each point depicted in the multidimensional plot corresponds to the predicted temperature increase for each of the 34848 PSR simulations; however, for the sake of clarity, operating conditions resulting in negligible ΔT (cf. areas of near zero heat release in Figure 1) are not shown.

The multidimensional graph of T_{in}, Y_{fuel}, t_{res}, and ΔT enables a broad definition of the main chemical activity regions. It is evident that ignition is observed regardless of residence time values for temperatures higher than 900 K and inlet fuel mass fractions lower than 0.2. When sufficient time is available ($t_{res} = 0.1$ s), ignition may also occur at slightly lower inlet temperatures ($T_{in} = 860$ K). On the other hand, cool flame activity extends to a broad region, spanning a much wider range of the investigated parameter space. Cool flame activity is favoured in fuel-rich conditions; the respective temperature region is increased with increasing inlet fuel concentration. In addition, increasing the residence time results in extending the temperature region where cool flame reactions emerge.

The developed tool is validated on the basis of experimental observations in a stabilized cool flame reactor [51]. Edenhofer et al. [51] have constructed a thermochemical behaviour map of their reactor (c.f. Figure 2 of [51]) using 4 different reaction "regimes," namely, the "no reaction" regime, the "cool flame" regime (corresponding to outlet temperatures up to 740 K), the "slow oxidation" regime (outlet temperatures up to 875 K), and the "ignition" regime (outlet temperatures over 1300 K). The outlet temperature criteria used in [51] have been applied to formulate Figure 13, where predictions of the developed tool are classified into the same 4 different reaction regimes. Numerical predictions achieve, both qualitatively and quantitatively, a good level of agreement with available experimental data [51]; the dashed line in Figure 13 corresponds to the operational conditions presented in [51].

4. Conclusions

A tabulated chemistry tool, capable of simulating the thermochemical behaviour of n-heptane oxidation in the low-, intermediate-, and high-temperature regimes, has been developed. n-Heptane has been used as a surrogate fuel of diesel oil. Extensive numerical simulations have been performed, using a detailed n-heptane oxidation chemical kinetic mechanism and utilizing the perfectly stirred reactor assumption. Three independent parameters have been varied in order to construct the chemical database: inlet temperature, inlet fuel concentration, and residence time. The comprehensive parametric analysis of the obtained predictions, that is, heat release, outlet temperature, species production and consumption rates, species mole fractions, and so forth, allowed the comparative assessment of the impact of the varying variables on the thermochemical behaviour of n-heptane.

Two main chemical activity regimes appear in all cases; the conventional "ignition" region emerges at temperatures higher than 880 K, whereas the "cool flame" oxidation region extends approximately between 620 K and 800 K. Increasing the fuel concentration favours cool flame activity; the respective temperature region is extended and the peak heat release is increased. In addition, when the residence time is increased, the temperature region of the cool flame regime is significantly extended and the maximum heat release value is "shifted" towards lower temperatures; however, residence time variations slightly affect the characteristics of the ignition regime. Rate of production and sensitivity analyses have been also performed, aiming to elucidate the different chemical paths prevailing in each oxidative region. Competition between the endothermic decomposition and the exothermic isomerization reactions of C7 alkylperoxy species appear to largely determine overall system reactivity in the cool flame regime. Finally, the obtained predictions have been used to develop an indicative thermochemical behaviour "map," which assists in the estimation of the expected oxidation regime (e.g., cool flame or ignition) in technical devices.

Conflict of Interests

The authors declare that there is no conflict of interests regarding the publication of this paper.

Acknowledgments

This work has been partly performed in the framework of the CM0901 COST action "Detailed Chemical Kinetic Models for Cleaner Combustion." Financial support through the FC-District (Grant no. 260105) and the ECCO-MATE (Grant no. 607214) EU research projects is acknowledged.

References

[1] T. Lu and C. K. Law, "Toward accommodating realistic fuel chemistry in large-scale computations," *Progress in Energy and Combustion Science*, vol. 35, no. 2, pp. 192–215, 2009.

[2] F. Battin-Leclerc, "Detailed chemical kinetic models for the low-temperature combustion of hydrocarbons with application to gasoline and diesel fuel surrogates," *Progress in Energy and Combustion Science*, vol. 34, no. 4, pp. 440–498, 2008.

[3] C. K. Westbrook and F. L. Dryer, "Chemical kinetic modeling of hydrocarbon combustion," *Progress in Energy and Combustion Science*, vol. 10, no. 1, pp. 1–57, 1984.

[4] H. Machrafi, S. Cavadias, and J. Amouroux, "Influence of fuel type, dilution and equivalence ratio on the emission reduction from the auto-ignition in an Homogeneous Charge Compression Ignition engine," *Energy*, vol. 35, no. 4, pp. 1829–1838, 2010.

[5] D. Ganesh and G. Nagarajan, "Homogeneous charge compression ignition (HCCI) combustion of diesel fuel with external mixture formation," *Energy*, vol. 35, no. 1, pp. 148–157, 2010.

[6] S. Cong, G. P. McTaggart-Cowan, C. P. Garner, E. Wahab, and M. Peckham, "Experimental investigation of low temperature diesel combustion processes," *Combustion Science and Technology*, vol. 183, no. 12, pp. 1376–1400, 2011.

[7] Y. Ohkubo, Y. Idota, and Y. Nomura, "Evaporation characteristics of fuel spray and low emissions in a lean premixed-prevaporization combustor for a 100 kW automotive ceramic gas turbine," *Energy Conversion and Management*, vol. 38, pp. 1297–1309, 1997.

[8] T. B. Gradinger, A. Inauen, R. Bombach, B. Käppeli, W. Hubschmid, and K. Boulouchos, "Liquid-fuel/air premixing in gas turbine combustors: experiment and numerical simulation," *Combustion and Flame*, vol. 124, no. 3, pp. 422–443, 2001.

[9] L. Hartmann, K. Lucka, and H. Köhne, "Mixture preparation by cool flames for diesel-reforming technologies," *Journal of Power Sources*, vol. 118, no. 1-2, pp. 286–297, 2003.

[10] A. Naidja, C. R. Krishna, T. Butcher, and D. Mahajan, "Cool flame partial oxidation and its role in combustion and reforming of fuels for fuel cell systems," *Progress in Energy and Combustion Science*, vol. 29, no. 2, pp. 155–191, 2003.

[11] J. Matos da Silva, I. Hermann, C. Mengel, K. Lucka, and H. Köhne, "Autothermal reforming of gasoline using a cool flame vaporizer," *AIChE Journal*, vol. 50, no. 5, pp. 1042–1050, 2004.

[12] E. J. D'Onofrio, "Cool flame and autoignition in glycols," *Journal of Loss Prevention in the Process Industries*, vol. 13, pp. 89–97, 1979.

[13] A. A. Pekalski and H. J. Pasman, "Distinction between the upper explosion limit and the lower cool flame limit in determination of flammability limit at elevated conditions," *Process Safety and Environmental Protection*, vol. 87, no. 1, pp. 47–52, 2009.

[14] P. G. Lignola and E. Reverchon, "Cool flames," *Progress in Energy and Combustion Science*, vol. 13, no. 1, pp. 75–96, 1987.

[15] J. F. Griffiths, "Reduced kinetic models and their application to practical combustion systems," *Progress in Energy and Combustion Science*, vol. 21, no. 1, pp. 25–107, 1995.

[16] A. J. Harrison and L. R. Cairnie, "The development and experimental validation of a mathematical model for predicting hot-surface autoignition hazards using complex chemistry," *Combustion and Flame*, vol. 71, no. 1, pp. 1–21, 1988.

[17] P. Dagaut, M. Reuillon, and M. Cathonnet, "Experimental study of the oxidation of n-heptane in a jet stirred reactor from low to high temperature and pressures up to 40 Atm," *Combustion and Flame*, vol. 101, no. 1-2, pp. 132–140, 1995.

[18] R. S. Sheinson and F. W. Williams, "Chemiluminescence spectra from cool and blue flames: electronically excited formaldehyde," *Combustion and Flame*, vol. 21, no. 2, pp. 221–230, 1973.

[19] M. J. Pilling, Ed., *Low-Temperature Combustion and Autoignition*, vol. 35 of *Comprehensive Chemical Kinetics*, Elsevier, Amsterdam, The Netherlands, 1997.

[20] K. Lucka and H. Koehne, "Usage of cold flames for the evaporation of liquid fuels," in *Proceedings of the 5th International Conference on Clean Air Technology*, pp. 207–213, Lisbon, Portugal, 1999.

[21] H. J. Curran, P. Gaffuri, W. J. Pitz, and C. K. Westbrook, "A comprehensive modeling study of n-heptane oxidation," *Combustion and Flame*, vol. 114, no. 1-2, pp. 149–177, 1998.

[22] H. J. Curran, P. Gaffuri, W. J. Pitz, and C. K. Westbrook, "A comprehensive modeling study of iso-octane oxidation," *Combustion and Flame*, vol. 129, no. 3, pp. 253–280, 2002.

[23] E. Ranzi, P. Gaffuri, T. Faravelli, and P. Dagaut, "A wide-range modeling study of n-heptane oxidation," *Combustion and Flame*, vol. 103, no. 1-2, pp. 91–106, 1995.

[24] D. I. Kolaitis and M. A. Founti, "On the assumption of using n-heptane as a "surrogate fuel" for the description of the cool flame oxidation of diesel oil," *Proceedings of the Combustion Institute*, vol. 32, pp. 3197–3205, 2009.

[25] P. Zhao and C. K. Law, "The role of global and detailed kinetics in the first-stage ignition delay in NTC-affected phenomena," *Combustion and Flame*, vol. 160, no. 11, pp. 2352–2358, 2013.

[26] C. K. Law and P. Zhao, "NTC-affected ignition in nonpremixed counterflow," *Combustion and Flame*, vol. 159, no. 3, pp. 1044–1054, 2012.

[27] D. M. A. Karwat, S. W. Wagnon, M. S. Wooldridge, and C. K. Westbrook, "Low-temperature speciation and chemical kinetic studies of n-heptane," *Combustion and Flame*, vol. 160, pp. 2693–2706, 2013.

[28] O. Herbinet, B. Husson, Z. Serinyel et al., "Experimental and modeling investigation of the low-temperature oxidation of n-heptane," *Combustion and Flame*, vol. 159, no. 12, pp. 3455–3471, 2012.

[29] S. K. Aggarwal, "A review of spray ignition phenomena: present status and future research," *Progress in Energy and Combustion Science*, vol. 24, no. 6, pp. 565–600, 1998.

[30] A. Cuoci, M. Mehl, G. Buzzi-Ferraris, T. Faravelli, D. Manca, and E. Ranzi, "Autoignition and burning rates of fuel droplets under microgravity," *Combustion and Flame*, vol. 143, no. 3, pp. 211–226, 2005.

[31] O. Colin, A. P. da Cruz, and S. Jay, "Detailed chemistry-based auto-ignition model including low temperature phenomena applied to 3-D engine calculations," *Proceedings of the Combustion Institute*, vol. 30, pp. 2649–2656, 2005.

[32] M. A. Founti and D. I. Kolaitis, "Numerical simulation of diesel spray evaporation exploiting the "stabilized cool flame" phenomenon," *Atomization and Sprays*, vol. 15, no. 1, pp. 1–18, 2005.

[33] D. I. Kolaitis and M. A. Founti, "Numerical modelling of transport phenomena in a diesel spray "stabilized cool flame" reactor," *Combustion Science and Technology*, vol. 178, no. 6, pp. 1087–1115, 2006.

[34] D. I. Kolaitis and M. A. Founti, "A tabulated chemistry approach for numerical modeling of diesel spray evaporation in a "stabilized cool flame" environment," *Combustion and Flame*, vol. 145, no. 1-2, pp. 259–271, 2006.

[35] D. I. Katsourinis and M. A. Founti, "CFD modelling of a "stabilized cool flame" reactor with reduced mechanisms and a direct integration approach," *Chemical Engineering Science*, vol. 63, no. 2, pp. 424–433, 2008.

[36] J. J. Hernandez, J. Sanz-Argent, J. Benajes, and S. Molina, "Selection of a diesel fuel surrogate for the prediction of auto-ignition under HCCI engine conditions," *Fuel*, vol. 87, no. 6, pp. 655–665, 2008.

[37] W. J. Pitz and C. J. Mueller, "Recent progress in the development of diesel surrogate fuels," *Progress in Energy and Combustion Science*, vol. 37, no. 3, pp. 330–350, 2011.

[38] F. Tao, V. I. Golovitchev, and J. Chomiak, "Self-ignition and early combustion process of n-heptane sprays under diluted air conditions: numerical studies based on detailed chemistry," Society of Automotive Engineers Paper 2000-01-2931, SAE Transactions, 2000.

[39] H. K. Ciezki and G. Adomeit, "Shock-tube investigation of self-ignition of n-heptane-air mixtures under engine relevant conditions," *Combustion and Flame*, vol. 93, no. 4, pp. 421–433, 1993.

[40] A. J. Donkerbroek, A. P. van Vliet, L. M. T. Somers et al., "Time- and space-resolved quantitative LIF measurements of formaldehyde in a heavy-duty diesel engine," *Combustion and Flame*, vol. 157, no. 1, pp. 155–166, 2010.

[41] A. J. Donkerbroek, A. P. van Vliet, L. M. T. Somers, N. J. Dam, and J. J. ter Meulen, "Relation between hydroxyl and formaldehyde in a direct-injection heavy-duty diesel engine," *Combustion and Flame*, vol. 158, no. 3, pp. 564–572, 2011.

[42] N. A. Slavinskaya and O. J. Haidn, "Modeling of n-heptane and iso-octane oxidation in air," *Journal of Propulsion and Power*, vol. 19, no. 6, pp. 1200–1216, 2003.

[43] Y. Muharam and J. Warnatz, "Kinetic modelling of the oxidation of large aliphatic hydrocarbons using an automatic mechanism generation," *Physical Chemistry Chemical Physics*, vol. 9, no. 31, pp. 4218–4229, 2007.

[44] M. Chaos, A. Kazakov, Z. Zhao, and F. L. Dryer, "A high-temperature chemical kinetic model for primary reference fuels," *International Journal of Chemical Kinetics*, vol. 39, no. 7, pp. 399–414, 2007.

[45] T. Zeuch, G. Moréac, S. S. Ahmed, and F. Mauss, "A comprehensive skeletal mechanism for the oxidation of n-heptane generated by chemistry-guided reduction," *Combustion and Flame*, vol. 155, no. 4, pp. 651–674, 2008.

[46] A. M. Kanury, *Introduction to Combustion Phenomena*, Gordon and Breach Science Publishers, 1994.

[47] N. Peters, *Turbulent Combustion*, Cambridge University Press, 2000.

[48] S. C. Kong, C. D. Marriott, R. D. Reitz, and M. Christensen, "Modeling and experiments of HCCI engine combustion using detailed chemical kinetics with multidimensional CFD," Tech. Rep. 2001-01-1026, SAE Transactions, 2001.

[49] R. J. Kee, F. M. Rupley, J. A. Miller et al., *CHEMKIN Collection, Release 4. 1*, Reaction Design, San Diego, Calif, USA, 2007.

[50] N. Steinbach, *Untersuchungen zum Zuendverhalten von Heizoel EL-Luft-Gemischen unter atmosphaerischem Druck [Ph.D. thesis]*, RWTH , Aachen, Germany, 2002.

[51] R. Edenhofer, K. Lucka, and H. Kohne, "Low temperature oxidation of diesel-air mixtures at atmospheric pressure," *Proceedings of the Combustion Institute*, vol. 31, pp. 2947–2954, 2007.

[52] S. W. Benson, "The kinetics and thermochemistry of chemical oxidation with application to combustion and flames," *Progress in Energy and Combustion Science*, vol. 7, no. 2, pp. 125–134, 1981.

[53] A. Kazakov, M. Chaos, Z. Zhao, and F. L. Dryer, "Computational singular perturbation analysis of two-stage ignition of large hydrocarbons," *Journal of Physical Chemistry A*, vol. 110, no. 21, pp. 7003–7009, 2006.

Combustion Characteristics for Turbulent Prevaporized Premixed Flame Using Commercial Light Diesel and Kerosene Fuels

Mohamed S. Shehata,[1] Mohamed M. ElKotb,[2] and Hindawi Salem[2]

[1] Mechanical Engineering Department, Faculty of Engineering, King Khalid University, Abha 61413, Saudi Arabia
[2] Mechanical Power Engineering Department, Faculty of Engineering, Cairo University, University Street, Giza 12316, Egypt

Correspondence should be addressed to Mohamed S. Shehata; m3ohamed4@yahoo.com

Academic Editor: Yiguang Ju

Experimental study has been carried out for investigating fuel type, fuel blends, equivalence ratio, Reynolds number, inlet mixture temperature, and holes diameter of perforated plate affecting combustion process for turbulent prevaporized premixed air flames for different operating conditions. CO_2, CO, H_2, N_2, C_3H_8, C_2H_6, C_2H_4, flame temperature, and gas flow velocity are measured along flame axis for different operating conditions. Gas chromatographic (GC) and CO/CO_2 infrared gas analyzer are used for measuring different species. Temperature is measured using thermocouple technique. Gas flow velocity is measured using pitot tube technique. The effect of kerosene percentage on concentration, flame temperature, and gas flow velocity is not linearly dependent. Correlations for adiabatic flame temperature for diesel and kerosene-air flames are obtained as function of mixture strength, fuel type, and inlet mixture temperature. Effect of equivalence ratio on combustion process for light diesel-air flame is greater than for kerosene-air flame. Flame temperature increases with increased Reynolds number for different operating conditions. Effect of Reynolds number on combustion process for light diesel flame is greater than for kerosene flame and also for rich flame is greater than for lean flame. The present work contributes to design and development of lean prevaporized premixed (LPP) gas turbine combustors.

1. Introduction

Turbulent premixed flames exhibit phenomena not found in other turbulent flows. It is natural to suppose that a premixed flame is strongly influenced by the turbulence into which it is propagating. In some circumstances, a thin flame sheet forms a connected but highly wrinkled surface that separates the reactants from the products. The environmental aspect becomes always more important for the development of energetic formation of the pollutants in the combustion chamber and consequently their emissions. Many researches are focused on new concepts for ultralow emissions combustors for gas turbine, with developments in fuel preparation and wall cooling techniques. A possible technological solution for the reduction of pollution is the use of lean prevaporized premixed (LPP) technology. Nevertheless, this new emergent technology is affected by many problems that must be solved in order to make it reliable for commercial engines. El Bakali et al. [1] studied premixed natural gas flames at stoichiometric and pressure of 10.6 kPa. Mole fraction profiles of stable species and temperature were measured using molecular beam/mass spectrometry gas chromatography and thermocouple technique. They also studied the experimental results obtained from a jet-stirred reactor at atmospheric pressure and variable equivalence ratios ($\Phi = 0.75, 1.0, 1.5$) for oxidation of methane/ethane mixture using natural gas. Combustion of natural gas generates low level of unwanted pollutants. In addition, economic aspect of natural gas combustion is attractive. Therefore, natural gas constitutes a serious alternative to conventional liquid fuel because of its high octane number. The recent studies show that the addition of natural gas to diesel fuel reduces pollutants in the exhaust gas. The details of these studies can be found in Papagiannakis and Hountals [2]. El-Sherif [3] studied the effect of minor alkanes on the combustion of premixed flames using natural gas as a fuel. The authors reported a significant effect of initial ethane concentration on the peaks of CO and NO. In jet stirred reactor conditions, Tan et al. [4] measured mole fraction profiles for oxidation of methane/ethane and methane/ethane/propane mixtures under a wide range of operating conditions (101.3–1013 kPa, 800–1400 K). The results showed that methane/ethane/propane is the best model fuel to represent natural gas under these conditions. Mole fraction profiles of stable, atomic, and radical species are obtained by coupling molecular beam/mass spectrometry with gas chromatography analyses. Analysis of premixed turbulent combustion includes both effects of combustion on turbulence and effects of turbulence on chemical reaction rate which is still under investigation. Chen and Ihme [5] developed and applied a combustion model for prediction of partially premixed combustion process in turbulent reacting flows. Large-eddy simulation (LES) is employed to describe the turbulent reacting flow field. The combustion process in the three-stream burner configurations is modeled using a flamelet combustion model in which details of reaction chemistry are represented in terms of set scalar quantities. Simulation results for temperature and major products species show good agreement with the experimental data. Author noticed that the scalar boundary conditions significantly affect the mixture composition in the burner. Sadanandan et al. [6] studied combustion behavior of natural gas flames with H_2 as an admixture. Author used an optically accessible combustor which operates under gas turbine relevant conditions. Author investigated the influence of various parameters like mixture composition, degree of premixing, and velocity on the pollutant emissions. The results showed that the degree of premixing and the recirculation rate of burned gases play an important role on NO_X and CO emissions. The degree of premixing and jet velocity for optimum combustion should lead to a recirculation and mixing rates such that ignition delay is long enough to promote mixing of the fresh fuel/air with the burned gases before the flame but short enough to ensure the flame stability. At the same time, combustor residence time should be short in order to reduce the thermal NO_X formation but long enough to enable complete combustion. Obodeh and Isaac [7] investigated performance characteristics of diesel engine fuelled with diesel/kerosene blends. Pressure data for 30% kerosene blend were higher than that for 40% kerosene blend from about 80 degrees after top dead center. The exhaust gas temperature at 100% rated load was 16.7% higher at 30% kerosene blend as compared with that obtained with diesel fuel. Brake power increased with rated load for all the fuel blends. The brake power at 100% rated load was 19.8% higher at 30% kerosene blend than that obtained when the engine was run on diesel fuel. Specific fuel consumption at 100% rated load was 7.5% lower at 30% kerosene blend than that obtained when compared with diesel fuel. It was deduced that the use of 30% kerosene along with diesel fuel will result in 10% saving on fuel cost. The several studies by Ghormade and Deshpande [8] and Kumar Reddy [9] have been done by using different vegetable oil blends with kerosene to improve the performance of a small type high speed diesel engine under high load conditions. They worked with four blends (20%, 40%, 60%, and 80% by volume) of soybean oil with kerosene as well as rapeseed oil with kerosene and compared the results with those of pure diesel fuel. They also studied the spray distribution of each blend in atmosphere using four whole nozzle injectors. The result showed that a blend of 20% vegetable oil with 80% kerosene by volume fairly improves the thermal efficiency of the test engine under high load. They were recommended to use 20% to 40% vegetable oil blends as a successful alternative fuel. Spray characteristics were studied both under high and low pressure injection in atmosphere where the low pressure injection showed better performance, Kumar Reddy [9]. Azad et al. [10] carried out experimental comprehensive study of DI diesel engine performance using biodiesel from mustard oil blends with kerosene. The vegetable oil without transesterification reaction has been blended with kerosene fuel by volume in some percentage like M20, M30, M40, and M50. Several engine parameters like brake specific fuel consumption, brake power, break mean effective pressure, exhaust gas temperature, lube oil temperature, sound level, and so forth have been determined. A comparison has been made for diesel engine performance of different biodiesel blends with kerosene. Authors recommended that it can be possible to run diesel engine with mustard and kerosene fuels blends without any modification of the engine. Poovannan and Kalivarathan [11] studied combustion process through designing two combustion chambers for lean premixed (LP) for gaseous and lean premixed prevaporized (LPP) for liquid fuels. Authors measured nitric oxides (NO_X), unburned hydrocarbons (UHC), carbon monoxide (CO), particulate matter (PM), and smoke. The design is verified using numerical analysis tools. Reasonable agreement between predictions from the preliminary design and numerical analysis is achieved which indicated that the design procedures are developed successfully. Gollahalli et al. [12] studied effects of turbulence on combustion characteristics of blends canola methyl ester (CME) and diesel fuel in a partially premixed flame environment. The experiments are conducted with mixtures of prevaporized fuel and air at an initial equivalence ratio of 7 and three Reynolds numbers of 2700, 3600, and 4500. Three blends with 25, 50, and 75% volume concentration of CME are studied. The soot volume fraction is the highest for the pure diesel flame and did not change significantly with Reynolds number due to the mutually compensating effect of increasing carbon input rate and increasing interned air with increased Reynolds number. The global NO_X emission index is the highest and the CO emission index is the lowest for pure CME flame and varied nonmonotonically with biofuel percentage in the blends. The mean temperature and NO_X concentrations at three-quarter flame height are generally correlated, indicating that the thermal mechanism of NO_X formation is dominant in the turbulent biofuel flames. The measurements indicate that the combustion characteristics of the turbulent flames of CME/diesel blends cannot predict accurately based on the blend ratio and properties of pure CME and diesel flames. Details of radical concentrations are required to understand the formation of CO and NO_X in the flames of blended fuels.

The pollutants coming out from premixed flames as a result of combustion have negative impact on the global environment and consequently human health. Therefore, controlling the combustion process is essential to reduce pollutants emissions. In the present study, the combustion phenomena and flow field for prevaporized premixed flame are analyzed experimentally. The basic principle of these phenomena is that the fuel is prepared and ready for combustion due to fuel vaporization and complete mixing with air. This research is basically applied to study the combustion phenomena of commercial fuels mainly light diesel, kerosene fuels, and their blends to acquire a fundamental understanding of the combustion process and flow field for gas turbine combustor.

2. Experimental Setup

Experimental work is carried out at Mechanical Engineering Department, Faculty of engineering, Cairo University, Egypt. The analysis of data is carried out at Mechanical Engineering Department, Faculty of Engineering, King Khalid University, Saudi Arabia. The experimental setup is shown in Figure 1. It consists of a flat flame grid burner, fuel injector, mixing chamber, and two 7 kW electrical heaters. Preheated air is used to create hot flat flame without entrained air. Water cooled stainless steel isokinetic sampling probe with outer diameter 9 mm is used to suck samples of gases at different locations through flame axis and radius to define species concentrations for different operating conditions. Sample probe is connected to gas chromatograph (GC) detector through heated line provided by sampling valve and heater controller. Burner is designed to improve mixing of fuel and air to improve flame stability by (1) using three perforated plates located at bottom of the burner at burner throttle area and at burner exit. The three perforated plates are used as flame holder to prevent flame flash back and also to generate turbulence, (2) inserting stainless steel balls between upper and medial perforated plates to improve mixing of fuel and air. The burner converges to the middle plate and then diverges to plate of flame holder for many reasons: (1) it improves mixing of fuel and air and (2) improves flame stability by increasing the stability of the flame lift-off height and increases the size of the recirculation zone. A recirculation zone behind the perforated plate provides conditions favorable for flame holding, for example, lower velocities, heat recycling to the flame stabilizing region, and enhanced mixing of fuel/air and hot product of combustion. The recirculation zone has a temperature gradient because of the cold air flow surrounding the burner tube. Liquid fuel is sprayed into mixing tube where preheated air is required for combustion. Fuel is vaporized and mixed with air before entering into combustion space through perforated plate which is used as a flame holder. Complete vaporization of mixture at burner exit is tested to ensure that all fuel is presented in gaseous phase. Also, concentrations are measured at burner exit without combustion using CO/CO_2 analyzer to check low temperature reactions. The results indicated that no low temperature reactions occurred inside mixing tube or burner. The residence time for reacting mixture from point of fuel injection in mixing tube to exit from burner tip is nearly 1-2 ms for different operating conditions. Two base fuels are

light diesel and kerosene fuels and three blends are studied. Blend A_1 (75% light diesel + 25% kerosene), blend A_2 (50% light diesel + 50% kerosene), and blend A_3 (25% light diesel + 75% kerosene) are also studied. According to Egyptian International Research Center, the physical and chemical properties of Egyptian diesel and kerosene fuels are listed in Table 1. From chemical formula of diesel and kerosene fuels stoichiometric fuel/air ratios are determined. For blending fuels, stoichiometric fuel/air ratios are determined according to the percentage of diesel and kerosene presented in the blend. From measuring fuel and air flow rates with calculating stoichiometric fuel/air ratios, the equivalence ratios (Φ) are determined for different operating conditions. Gas chromatograph Parker Elmer model Sigma 300 with data acquisition system model LCI-100 are used for gas analysis using two-column Poropak Q 1/8 in × 6 ft. and Poropak R 1/8 in × 8 ft. connected in series. Gas chromatograph is equipped with constant volume gas sampling valve to control volume of the sample. An infrared CO/CO_2 analyzer is also used for measuring CO and CO_2. A Pt/Pt-13% Rh thermocouple and water cooled pitot tube are used for flame temperature and gas velocity measurements. Concentrations, flame temperature, and gas velocity are measured at different locations along flame axis. Optimum condition for gas chromatography analysis, no entrained air along the flame, and flame flatness are determined and checked before measurements, Elkotb et al. [13]. Holes diameter of the perforated plate have been changed during experiments keeping air/fuel ratio and mean inlet flow velocity constant. Three hole diameters of 3.7, 2.5, and 2 mm have been chosen for changing the solidity ratios (blocked area/total area) to S_1 = 0.4786, S_2 = 0.719, and S_3 = 0.876. The solidity ratio is calculated as follows:

$$
\begin{aligned}
\text{The solidity ratio } (S) &= \frac{\text{Blockae area}}{\text{Total area}} \\
&= \frac{\text{Total Area} - \sum \text{Holes area}}{\text{Total area}} \\
&= 1 - n\left(\frac{d}{D}\right)^2.
\end{aligned}
\tag{1}
$$

Each perforated plate has diameter of 38 mm and 5 mm thickness. It is provided with 55 holes of 5 mm center-to-center distance. Pressure loss across the flame stabilizer is a function of the created turbulent energy and the injected fuel into the flow. Turbulence also influences the thickness of the premixed turbulent reaction zone. The recirculation zone size is nearly close to the perforated plate diameter. The central hole will give an annular recirculation zone rotating outward and the baffle will give a central recirculation zone rotating inward.

Equivalence ratio, inlet Reynolds number, and inlet mixture temperature are also studied. The inlet mixture temperature is kept constant at 673 K for light diesel and blends A_1 and A_2 while for kerosene and blend A_3 it is kept at 603 K during study effect of fuel type, fuel blends, equivalence ratio, and Reynolds number. During study effect of Reynolds number and inlet mixture temperature, the solidity ratio and equivalence ratio are kept constant. Although the used

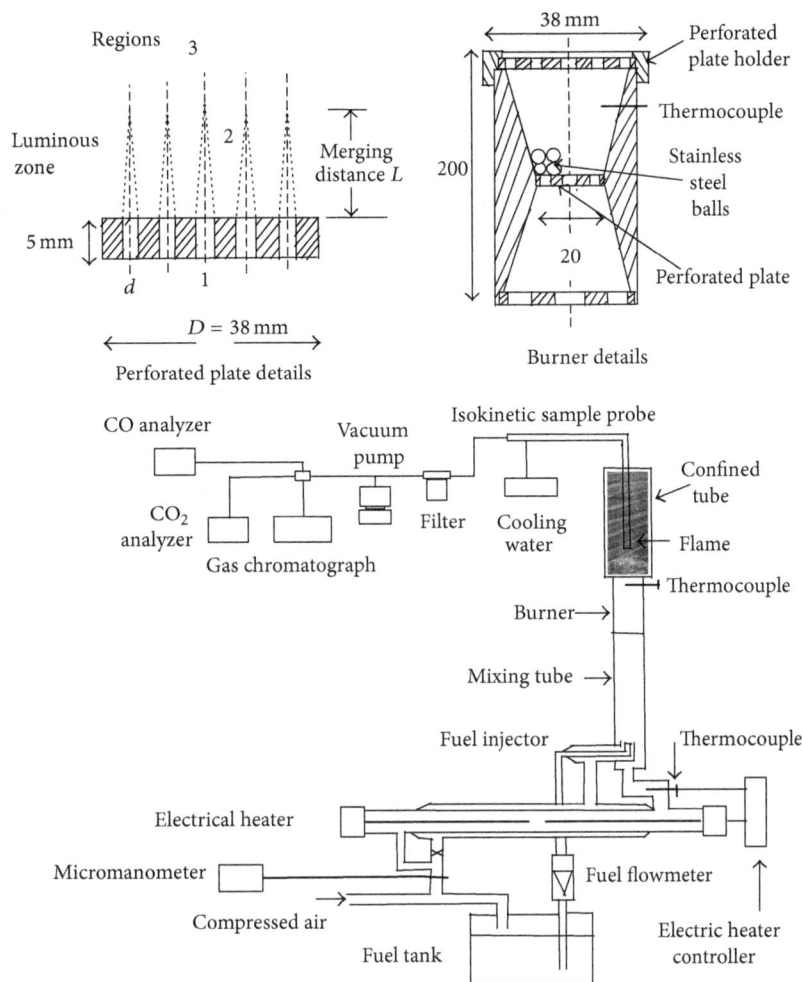

FIGURE 1: Experimental setup.

technique in this study is useful for turbulent premixed air flames, it can be generalized for other systems (gas turbines, rockets and spark ignition engines, etc.), where coupling between chemical reaction resulting from combustion and turbulent flow process exists. The aim of present work is to study effect of different operating parameters as fuel type (light diesel, kerosene fuels, and their blends), equivalence ratio, inlet Reynolds number, inlet mixture temperature, and perforated plate holes diameter on combustion process for turbulent prevaporized premixed air flames.

3. Results and Discussions

3.1. Concentrations, Flame Temperature, and Flow Velocity.
Concentration, flame temperature, and gas flow velocity for light diesel-air flames are shown in Figure 2. In the main reaction zone, CO_2 formation rate increases with increased flame temperature. In postflame zone, CO_2 formation rate increases to the flame end due to decomposition of heavy fuel into light hydrocarbon that rapidly oxidizes to CO and CO_2. Maximum CO_2 concentration appeared at the flame end. In the post-flame zone the high formation rate of CO_2 relative to the main

reaction zone attribute to heat dissipation to surrounding is not significant due to confined tube around flame. Near flame end, flame temperature drops due to complete combustion, and equilibrium condition lead to increase CO_2 concentration. The formation of CO_2 from CO and O_2 as well as decomposition is kinetically limited and sensitive to high temperature due to high activation energy. Near burner tip, high CO_2 concentration attributes to the hot probe tip that influences rate of reaction. In the main reaction zone CO formation rate is significantly higher than in the postflame zone. The maximum formation rate of CO appears at the early stage of combustion where hydrogen and free carbons atoms are found. CO concentration expects to increase at the expense of CO_2 due to reaction $C + CO_2 \longleftrightarrow 2CO$ in the region of low CO_2 formation rate with high flame temperature and consumption of all oxygen atoms. CO concentration exceeded equilibrium concentration for many reasons Howe et al. [14] (1) CO which burns out slowly in the flame itself, (2) separation by molecular diffusion, and (3) the maximum emission of C, CH, and OH, localizing in the middle of the oxidation zone. In the postflame zone CO concentration begins to decrease due to (1) consumption of all oxygen and fuel, (2) equilibrium

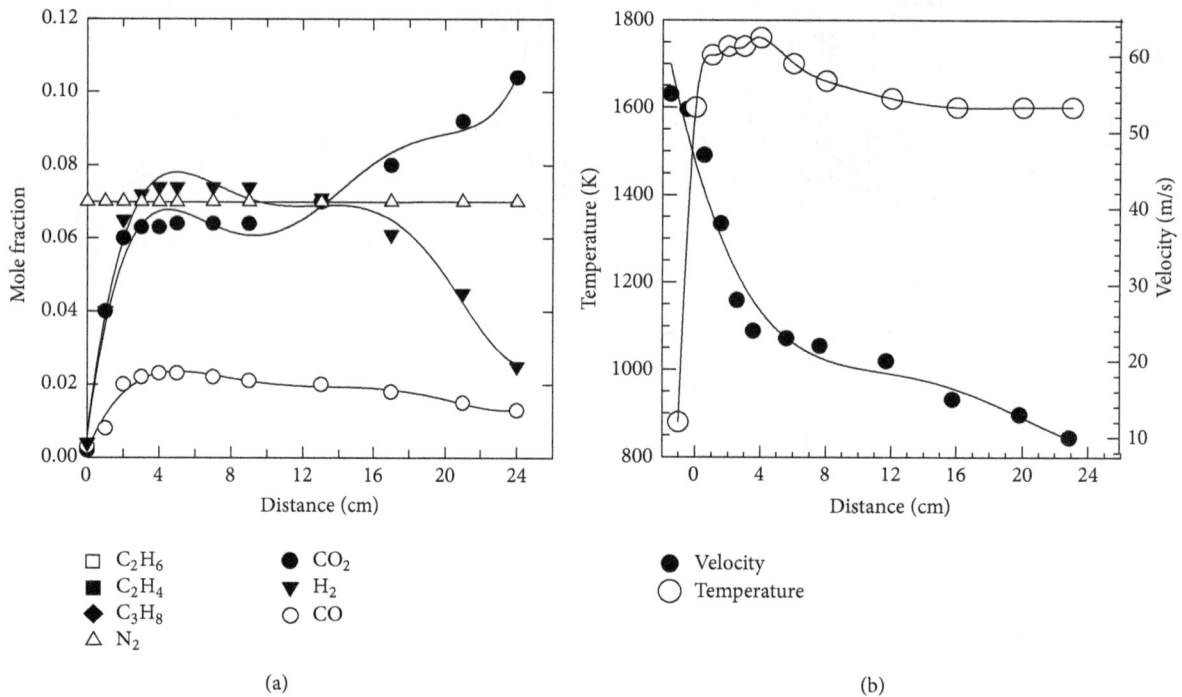

FIGURE 2: Concentrations, gas flow velocity, and flame temperature for turbulent premixed light diesel-air flame at Reynolds number 12.139×10^3, inlet temperature 673 K, solidity ratio 0.68, equivalence ratio 1.25, and pressure 1 bar.

TABLE 1: Properties of diesel and kerosene fuels.

Test properties	Test method	Diesel fuel	Kerosene
Chemical formula		$C_{12}H_{26}$	$C_{10}H_{22}$
Cetane number	ASTM D613	44.5	45.4
Flash point (°C)	ASTM D93	52	45.6–46.4
Pour point (°C)	ASTM D97	−32	−49
Boiling point (°C)	Oil weathering system	369	200–260
Self-ignition temperature (°C)	ASMT D97	725	640
Distillation temp. (°C) 90%	STM D86	228–338	153–245
Lower heating value (MJ/kg)	STM D240	45.9	45.6
Density at 40°C, (kg/m^3)	1298 TMDA	830	760–810
viscosity at 40°C, (mm^2/S)	ASTMD445	3.1	1–1.9
Sulfur content, wt%	ASTMD545	0.22	0.04
Carbon content, wt%	ASTM D5291	84.7	84.5
Hydrogen content, (wt%)	ASTM D5291	15.3	15.5
Carbon/hydrogen ratio		5.53	5.45
Sulfur content, %	ASTM D5453-39	0.16	0.04
Paraffin content, Vol.%	ASTM D 2007	41.3–52.4	47–55
Aromatic content, Vol.%	ASTM method D 2007	23.6–24.7	15.5–19.6
Olefin content, Vol.%	ASTM method D 2007	18–30	1.3–2.5
Naphthalene, content Vol.%	ASTM method D 2007	2.8–8.2	2.8
Saturates content, Vol.%	ASTM method D 2007	79.3–75.3	79–82

condition for all stable species, and (3) quenching effect due to heat loss to the surrounding. Many elementary reactions explain the conversion of CO to CO_2 as follows:

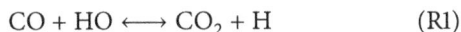

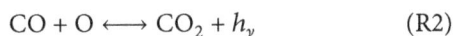

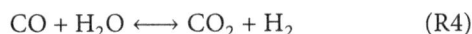

$$CO + HO \longleftrightarrow CO_2 + H \qquad (R1)$$

$$CO + O \longleftrightarrow CO_2 + h_\nu \qquad (R2)$$

$$CO + O + M \longleftrightarrow CO_2 + M \qquad (R3)$$

$$CO + H_2O \longleftrightarrow CO_2 + H_2 \qquad (R4)$$

Reaction (R2) contributes by 3–5% to CO conversion, El Kotb et al. [13]. The contribution of reactions (R1)–(R4) to the conversion rate is in order of 1×10^{-5} mole cm^{-3} s^{-1}. CO conversion is ascribed to reactions (R1)–(R4) even in the early stage of flame. In the main reaction zone, H_2 formation rate is significantly higher due to heavy fuel which decomposes into light hydrocarbons and H_2. This is true where the peak value of H_2 attained after disappeared or decreased light hydrocarbons.

The principal peak of H_2 formation rate which locates close to the burner tip is due to high rate of primary reaction as follows:

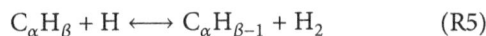

$$C_\alpha H_\beta + H \longleftrightarrow C_\alpha H_{\beta-1} + H_2 \qquad (R5)$$

In the main reaction zone, the high rate of hydrogen formation is explained by molecular diffusion. H_2 and CO formation rates increase while H_2O formation rate decreases due to the occurrence of the following reaction:

$$C + H_2O \longleftrightarrow H_2 + CO \qquad (R6)$$

Also, H_2 formation rate increases with increased CO formation rate and vice versa. H_2 formation starts nearly after one cm (0.5 ms) from burner tip due to low flame temperature and is explained by probe quenching effect. Formation rate of unburned hydrocarbons (UHC) decreases rapidly after 5 cm (6 ms) from burner tip. The reaction rates of (R1)–(R6) decrease rapidly due to decreasing flame temperature. In the postflame zone, the primary reaction is as follows:

$$H_2O + H \longleftrightarrow H_2 + OH \qquad (R7)$$

This reaction has relatively high activation energy of 8484 kJ/mole. In the postflame zone the rate of reaction decreases rapidly due to decreasing flame temperature. The relative decrease in H_2 formation rate is a result of combined effect of reactions (R5) and (R7). Near the flame end (after long residence time), all heavy fuel, light hydrocarbons, and intermediate species are nearly consumed. Consequently, all H_2 molecules approximately burned and disappeared. Near burner tip, unburned hydrocarbon (ethane, ethylene, and propane) formation rates are high due to decomposition of the existing heavy fuel, maximum O_2 concentration, and high gradient of flame temperature. Once the unburned mixture issues from the burner tip and ignition starts, many reactions in the pyrolysis zone produce unstable intermediate species which disintegrates to produce products. The chemical structure of these species depends upon the rate of reaction of

each species which mainly depends on the flame temperature, pressure, fuel structure, and time lag after start of the reaction. In rich flame, the concentrations of unstable hydrocarbon components are higher than in lean flame. The amount of unstable intermediate hydrocarbons having an even number of carbon atoms has more frequency to react than those with an odd number regardless of the fuel used, Akrich et al. [15]. The hydrocarbon profiles are steeper in formation and dissipation due to high gradient of flame temperate which increases the activity and tendency of hydrocarbons to react with the oxidizer. Propane attains its maximum value before ethane because propane is heavy than ethane and is considered as a source for ethane formation. Because propane has carbon/hydrogen ratio higher than ethane, so propane decomposes to ethane component and other intermediate components according to the following reaction:

$$C_3H_8 \longleftrightarrow C_2H_6 + CH_2 \qquad (R8)$$

Also, ethylene attains its maximum value before propane because ethylene has high carbon/hydrogen ratio and also double bond between carbon atoms. Consequently, propane decomposes into ethylene and other intermediate components according to the following reaction:

$$C_3H_8 \longleftrightarrow C_2H_4 + CH_4 \qquad (R9)$$

Primarily, ethane forms in the flame by reaction involving methyl radicals. The concentrations of these species and their primary intermediate products C_2H_5, C_2H_4, C_2H_3, C_2H_2, CH_2, and CH are small. These low levels underscore the dominance of the mechanistic paths which proceed by way of CH_2OH and CH_2O, rather than through CH_3, Westbrook and Dryer [16]. The relative rates of individual elementary reactions vary across the flame and induction zones.

The flame temperature at exit of the burner is higher than temperature of the exit mixture from the perforated plate. The increase in temperature attributes to preheating of unburned mixture within holes of the perforated plate and radiative heat to thermocouple by the plate of the flame holder. In the main reaction zone, the burning rate is high due to high gradient of flame temperature. But in the postflame zone, gradient of flame temperature is low due to complete combustion and equilibrium condition. In the main reaction zone, based on heat and mass transfer from the recirculation zone, gradient of flame temperature is high. In the recirculation zone (regions between each two adjacent potential cores), the hot gases mix with cool gases at a point of high O-atoms concentrations, and, consequently, high gradient of the flame temperature. Radhakrishnan et al. [17] observed that flame temperature increased by 200 K corresponding to less than 25% of recirculation gases; consequently radical species are consumed early. In the main reaction zone, the over temperature attributes to the actual finite rates of reactions and the diffusion of the heat release due to chemical reactions. In the neighborhood of the main reaction zone, where most of the unburned gases react rapidly and not appear as cold eddies, the flame temperature remains high. Furthermore, in the main reaction zone, the over temperature attributes to high temperature fluctuation resulting from (1)

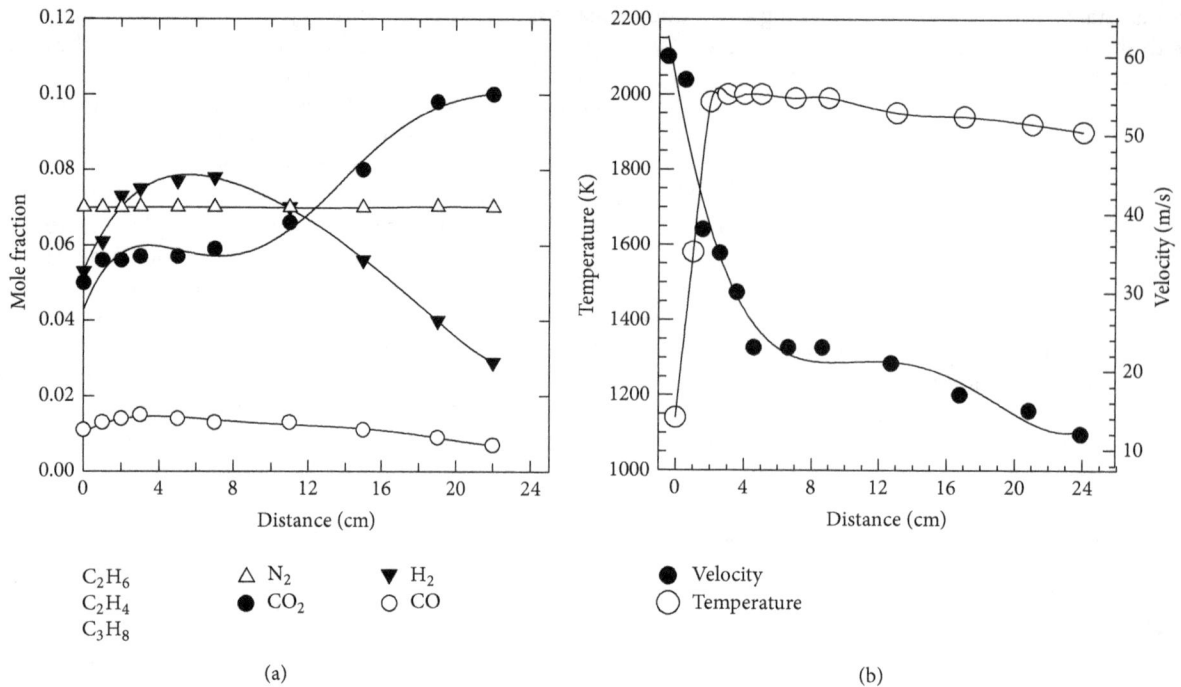

C$_2$H$_6$
C$_2$H$_4$
C$_3$H$_8$

△ N$_2$ ▼ H$_2$
● CO$_2$ ○ CO

● Velocity
○ Temperature

(a) (b)

FIGURE 3: Concentrations, gas flow velocity, and flame temperature for turbulent premixed kerosene-air flame at Reynolds number 12.139×10^3, inlet temperature 573 K, solidity ratio 0.68, equivalence ratio 1.25, and pressure 1 bar.

the variations of the local gases concentrations, (2) increasing rate of reaction due to presence of cold and hot eddies in turbulent flame, and (3) high temperature gradient with a high degree of turbulence. Outside the main reaction zone, a rapid decrease in temperature is observed as the energy convection and conduction away from the hot flame region. Also, near end of flame, flame temperature decreases due to heat transfer and cooling effect of the surrounding.

The axial velocity remains nearly constant for 2 cm (0.5 ms) from burner tip due to potential core effect. Actually, mixture passing through the perforated plate with hole diameter 2.9 mm gives a potential core length equals 20 times the hole diameter extending downstream the hot air flow. Furthermore, as unburned mixture exits from burner tip, flow velocity increases to high value due to increase flame temperature and decrease gas density. Thereafter, flame cross section area increases due to expand and spread of the flow and, consequently, flow velocity decreases steeply. The drop in flow velocity occurs early and rate of decreasing length of the potential core for flame is higher than that for cold flow. Moreover, at the end of the potential core and the beginning of the main reaction zone, flow velocity decreases due to effect of turbulence, where small change in axial and radial flow velocities and existing of recirculating gases. Beyond the main reaction zone, gas flow velocity profile decreases continuously until (at distance from burner tip that is nearly 31 times hole diameter) its shape becomes uniform and nearly no change occurs any longer with increased distance from burner tip.

3.2. Effect of Fuel Type on Combustion Process. The effect of fuel type on combustion process for turbulent prevaporized premixed light diesel-air flame is shown in Figure 2 and

for kerosene-air flame is shown in Figure 3. Light diesel-air flame has reaction zone length longer than kerosene-air flame due to high C/H ratio and double bonds between carbon atoms which need longer residence time and high flame temperature for complete combustion. Also, maximum burning velocity decreases with increased carbon atoms in the fuels. For lean flame, kerosene-air flame has reaction zone length longer than light diesel-air flame due to cooling effect of excess air which decreases rate of reaction and burning velocity. In the main reaction zone, light diesel-air flame emits UHC and H$_2$ higher than kerosene-air flame due to high percentage of olefins and aromatics. The rate of reaction decreases with increased carbon bonds, so UHC profiles extend to long distance along the flame. Also, for light diesel-air flame, rate of decreasing CO is lesser than for kerosene-air flame. This is due to existence of UHC in flame which is considered as a source for CO formation. For kerosene-air flame after 3 cm (one millisecond) all hydrocarbons are nearly burned and CO decreases due to absence sources of CO formation. Regarding light diesel-air flame, rate of formation of CO$_2$ increases continuously in the postflame zone, but for kerosene-air flame rate of formation of CO$_2$ decreases to low value where all fuel is consumed in the main reaction zone and equilibrium occurred early. Furthermore, since kerosene fuel contains paraffin higher than light diesel fuel, it is easy to crack atomic bonds and need less residence time for complete combustion. In contrast, kerosene-air flame emits high H$_2$ concentration especially in the main reaction zone where kerosene fuel has (H/C) ratio higher than light diesel fuel. For kerosene-air flame, the decreasing rate of H$_2$ is lower than light diesel-air flame where it attains equilibrium condition (complete combustion) early. Also, near end of flame, kerosene-air flame has flame temperature higher than light diesel-air flame. Also, in the main reaction zone, light diesel-air flame has flame temperature lower than kerosene-air flame due to pyrolysis of fuel to free radicals and intermediate species is lesser and the reaction zone length is longer. For light diesel-air flame, flame temperature reaches its maximum value after that its decreases with higher rate than kerosene-air flame. This is due to early burning of paraffin content near burner tip with the same burning rate as for kerosene fuel. Thereafter, olefins and aromatics begin to burn but with lower rates than paraffin. So, for light diesel-air flame, rate of heat release and rate of increasing flame temperature are lesser than kerosene-air flame. High efforts are done to correlate adiabatic flame temperature for light diesel and kerosene-air flames. From plotting experimental data of flame temperature along flame axis for different operating conditions (different equivalence ratio, different Reynolds number, different inlet temperatures, and different holes diameter of perforated plate) the adiabatic flame temperatures are correlated as follows.

For light diesel fuel: consider the following.

(1) For $\phi > 1$

$$T_{\text{adiabatic}} = 2100\phi^{-0.219}\zeta\theta^Z\Psi^Y, \qquad (2)$$

where $\zeta = \text{Exp}(0.4038\phi - 1.502)^2$; $Y = 0.01 - 0.13\phi + 0.08\phi$; $Z = 0.571 - 0.552\phi + 0.131\phi^2$ and

(2) for $\phi < 1$

$$T_{\text{adiabatic}} = 2310\phi^{-0.052}\zeta\theta^Z\Psi^Y, \qquad (3)$$

where $\Psi = $ H/C, and $\theta = (T_i/T_0)$; $\zeta = \text{Exp}(-(1.05\phi - 1.253)^2)$; $Y = 0.573 - 0.552\phi + 0.132\phi^2$; $Z = 0.011 - 0.133\phi + 0.081\phi^2$.

For kerosene fuel: consider the following.

(1) For $\phi > 1$

$$T_{\text{adiabatic}} = 2310\phi^{0.3029}\zeta\theta^Z\Psi^Y, \qquad (4)$$

where $\zeta = \text{Exp}(0.40038\phi - 1.422)^2$; $Y = 0.22 - 0.442\phi + 0.132\phi^2$; $Z = 0.181 - 0.312\phi + 0.0918\phi^2$ and

(2) for $\phi < 1$

$$T_{\text{adiabatic}} = 2310\phi^{0.12}\zeta\theta^Z\Psi^Y, \qquad (5)$$

where $\zeta = \text{Exp}(-(0.9097\phi - 1.07)^2)$; $Y = 0.011 - 0.131\phi + 0.0832\phi^2$; $Z = 0.3961 - 0.441\phi + 0.141\phi^2$,

where T_i and T_o are inlet and atmospheric temperatures, respectively, K.

3.3. Effect of Fuels Blends on Combustion Process. The purpose of fuel blending is to improve quality of combustion process to increase flame temperature. Diesel and kerosene fuels have different physical and chemical properties such as chemical structures, viscosity, heating value, self-ignition temperature, boiling temperature, adiabatic flame temperature, difference contents of paraffin, aromatic, olefin, naphthalene, and saturates hydrocarbon as listed in Table 1. There are many engineering applications for blending diesel/kerosene/gasoline fuels for internal combustion engines, Obodeh and Isaac [7], Ghormade and Deshpande [8], and Kumar Reddy [9]. So, blending kerosene with light diesel fuel is one of the major parameters for the present work. Effects of fuels blends on combustion process for rich and lean turbulent prevaporized premixed hydrocarbon-air flames at different residence time are shown in Figures 4, 5, and 6. Actually concentrations, flow velocity, and flame temperature are measured for different operating conditions along flame axis. After that from flow velocity and distance above burner tip for different operating conditions, the residence time at each location is calculated to obtain Figure 4 to Figure 10 for different residence time with different operating conditions.

For rich mixture, reaction zone length increases with increased light diesel percentage in the blend. So, residence time for complete combustion increases with increased light diesel percentage. Consequently, CO and UHC concentrations increase with increased light diesel percentage. Also, CO_2 profiles for different blends are nonsmooth due to different percentages of paraffin, olefin, and aromatic in blends where all these components have different rates of reaction and burned at different stages in flame (paraffin, olefin, and aromatic, resp.). So, paraffin burns near burner tip while olefin begins to burn after 4 ms from burner tip. Furthermore, flame temperature and flow velocity profiles increase with increased kerosene percentage. In the postflame zone (16 ms), rate of decreasing flame temperature increases with decreased light diesel percentage. For rich mixture, length of reaction zone increases with increased light diesel percentage. So, residence time for complete combustion increases with increased light diesel percentage. So, percentage of CO and UHC increases with increased light diesel percentage. At residence time of 3 ms, CO_2 profile is steeper with increased kerosene percentage due to increase flame temperature.

Also, H_2 concentration increases with increased kerosene percentage due to increasing (H/C) ratio. Maximum flame temperature increases with increased kerosene percentage in the blend. In the main reaction zone (3 ms), effect of fuel blending on flow velocity is less than in the postflame zone (16 ms) and for lean flame is higher than for rich flame. Effect of blending on flame temperature and flow velocity is higher than effect of blending on concentrations except for H_2. Effect of blending on combustion process in main reaction zone (3 ms) is smaller than in the postflame zone. The effects of kerosene percentage for lean and rich flames are not linear dependent. As light diesel percentage increases, CO and CO_2 concentrations increase. As light diesel percentage in the flame increases, rate of increasing CO_2 for rich flame is lesser than for lean flame due to insufficient oxygen, which decreases radical concentrations and flame temperature. In the main reaction zone, rate of increasing CO for lean and rich flames is steeper with light diesel percentage greater than 50%. As light diesel percentage increases, rate of decreasing H_2 for rich flame is higher than for lean flame especially in the

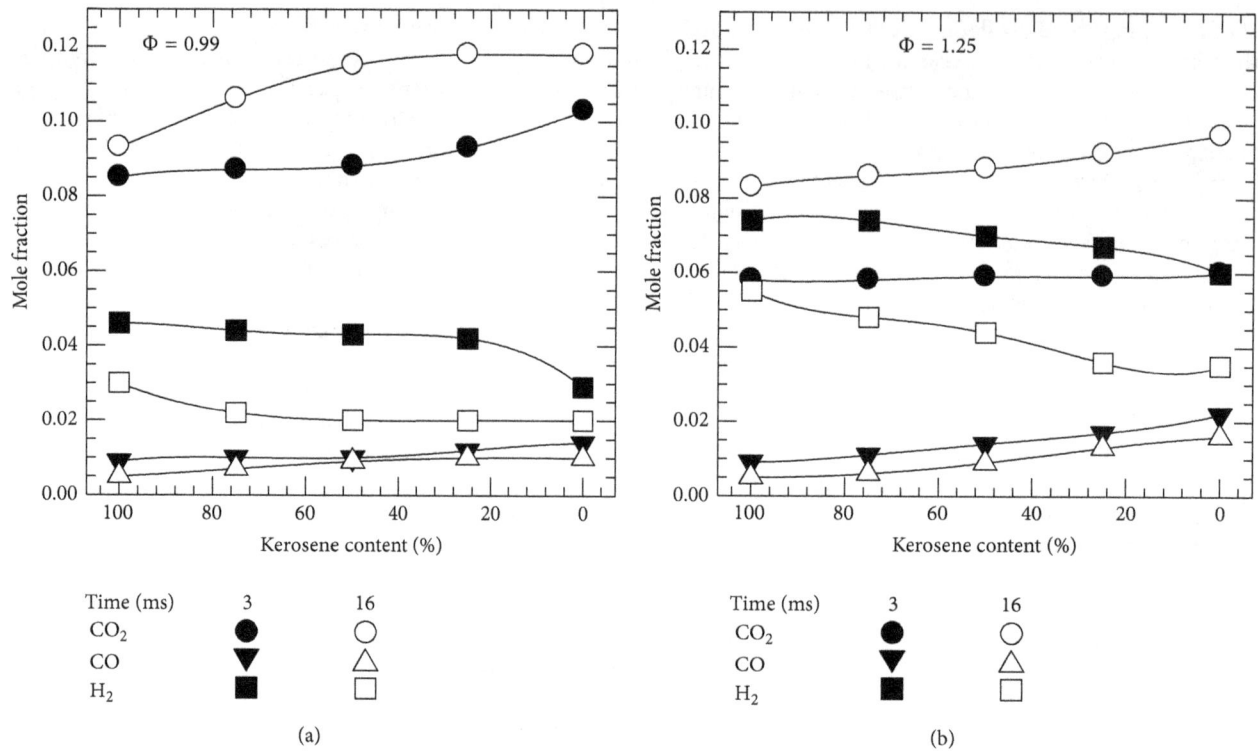

FIGURE 4: Effect of kerosene content on concentrations for turbulent premixed kerosene light diesel blends-air flame for different equivalence ratios and residence time with Re = $11.08 \times 10^3 00$ and solidity ratio = 0.68.

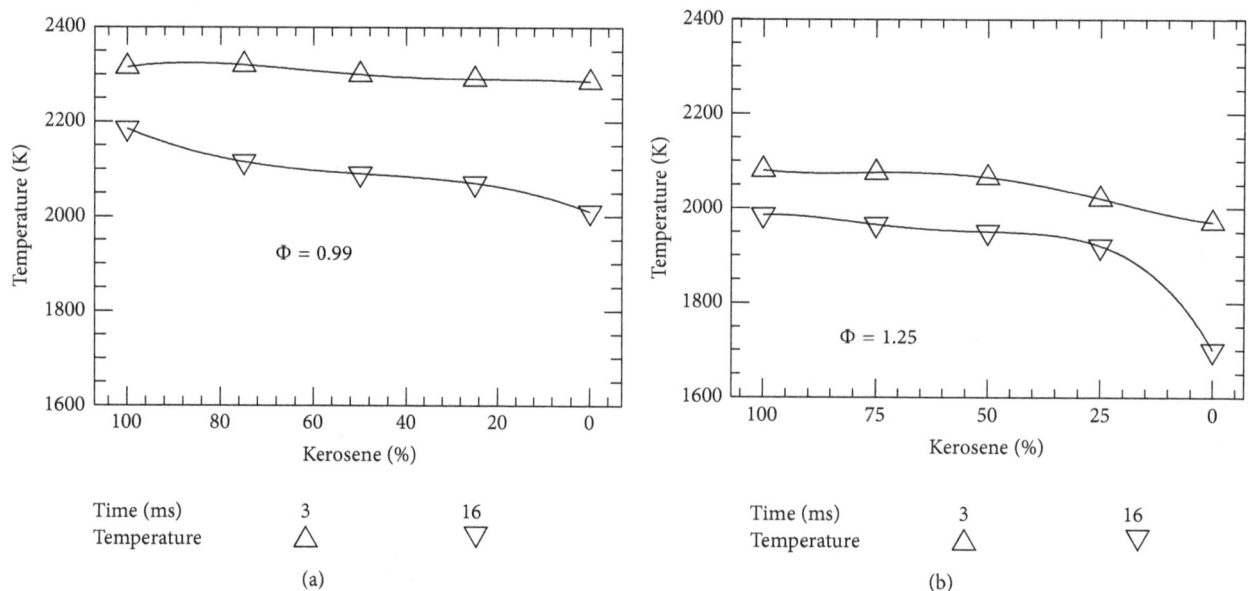

FIGURE 5: Effect of kerosene content on flame temperature for turbulent premixed kerosene light diesel blends-air flame for different equivalence ratios and residence time with Re = 11.08×10^3 and solidity ratio = 0.68.

postflame zone. As light diesel percentage increases, rate of decreasing flame temperature and flow velocity for lean flame are higher than for rich flame especially in the postflame zone.

3.4. Effect of Equivalence Ratio on Combustion Process.
Effects of equivalence ratio on combustion process for light diesel

and kerosene-air flames at different residence times are shown in Figure 7. For kerosene-air flame rate of increasing H_2 is higher than for light diesel-air flame due to high (H/C) ratio. The maximum CO_2 concentration and flame temperature for light diesel-air flame are more compatible than for kerosene-air flame. For rich flame, rate of decreasing CO_2 for

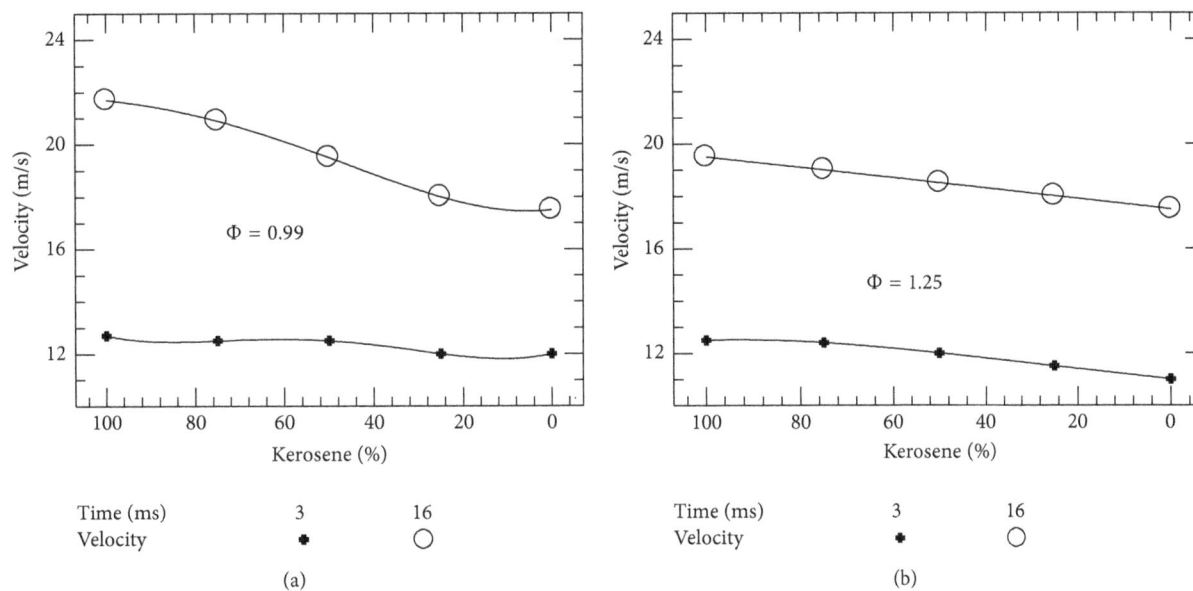

Time (ms)	3	16
Velocity	✦	○

(a)

Time (ms)	3	16
Velocity	✦	○

(b)

FIGURE 6: Effect of kerosene content on flow velocity for turbulent premixed kerosene light diesel blends-air flame for different equivalence ratios and residence time with Re = 11.08×10^3 and solidity ratio = 0.68.

kerosene-air flame is steeper than light diesel-air flame due to low (C/H) ratio. In contrast, for rich flame, in the postflame zone (at 20 ms) (not present here), rate of decreasing CO_2 for light diesel-air flame is steeper than for kerosene-air flame due to equilibrium which occurs earlier than light diesel-air flame. So, flame temperature profile for kerosene-air flame is flatter than light diesel-air flame. Concentration profiles for lean flame are steeper than for rich flame due to excess O_2 and O concentrations. For rich flame, the high organic radicals' concentrations lead to high formation rate of CO. On the other hand, the low O-atoms concentrations explain the low formation rate of CO_2. For lean flame, as equivalence ratio decreases, length of reaction zone increases while residence time for the postflame zone decreases. For slightly rich mixture (adjacent to stoichiometric) CO_2 has maximum value. For lean flame (at equivalence ratio less than 0.9) the drop in flame temperature is high due to cooling effect of excess air which increases length of reaction zone and decreases CO_2 concentration. Also, for rich flame, CO_2 gradually decreases due to decreasing radicals concentrations and flame temperature. The rate of conversion of CO to CO_2 decreases due to poor oxygen and low flame temperature. Also, rate of formation of CO_2 is low and its influences by the drop in flame temperature. In the main reaction zone, rate of formation of CO is fast due to fuel dissociation and excess O_2 concentration. For rich flame, unsaturated hydrocarbons contain more carbon atoms as a source for CO formation. The maximum concentration of CO increases with increased equivalence ratio. In the main reaction zone, rate of formation of CO decreases due to oxidization of CO to CO_2 according to (R1). In the postflame zone, CO concentration is minimum (equilibrium condition) due to complete combustion. At stoichiometric condition, CO concentration is high due to dissociation of CO_2 at high temperature. For rich flame, rate of formation CO increases due to lack of oxygen

concentration. For lean flame, CO concentration is high due to decrease rate of reaction of (R1) (thermal quenching). For lean flame, CO concentration is considered as an indication of thermal quenching in postflame zone. According to reaction (R1) rate of oxidation CO decreases due to decreasing flame temperature leading to high CO concentration. For rich flame, rate of formation of CO increases due to decreasing flame temperature and insufficient O_2 concentration for complete combustion leading to decreased rate of reaction. As equivalence ratio decreases, rate of decreasing H_2 is steeper due to increasing flame temperature and rate of reaction. But, for rich flame as equivalence ratio increases, flame length increases while flame temperature decreases. Consequently, high UHC concentration exists as a source for H_2 formation. In the main reaction zone (at 4.5 ms), H_2 profile is steeper than in the postflame zone (20 ms) due to decreasing rate of reaction of different species and equilibrium condition. For rich flame, there is low tendency to form free radical species HCO, O, OH, CH, H, HO, and HCN which forms many intermediate reactions for oxidation UHC and H_2. The rapid decays of HCN do not take place until UHC is almost disappeared. So, stable intermediate species CO, H_2, C_2H_4, C_2H_6, and C_3H_8 increase monotonically with increased equivalence ratio. For rich flame, as equivalence ratio increases, H_2O decreases while CO and H_2 concentrations increase according to (R6). For lean flame, flame temperature decreases with decreased equivalence ratio due to cooling effect of excess air. As equivalence ratio increases by 12%, maximum flame temperature increases by 16%. For rich flame, as equivalence ratio increases by 16%, maximum flame temperature decreases by 35%.

Therefore, for rich flame, effect of equivalence ratio on the maximum flame temperature is greater than for lean flame. In the main reaction zone (at 5.4 ms), flame temperature is steeper than in the postflame zone (at 20 ms) due to

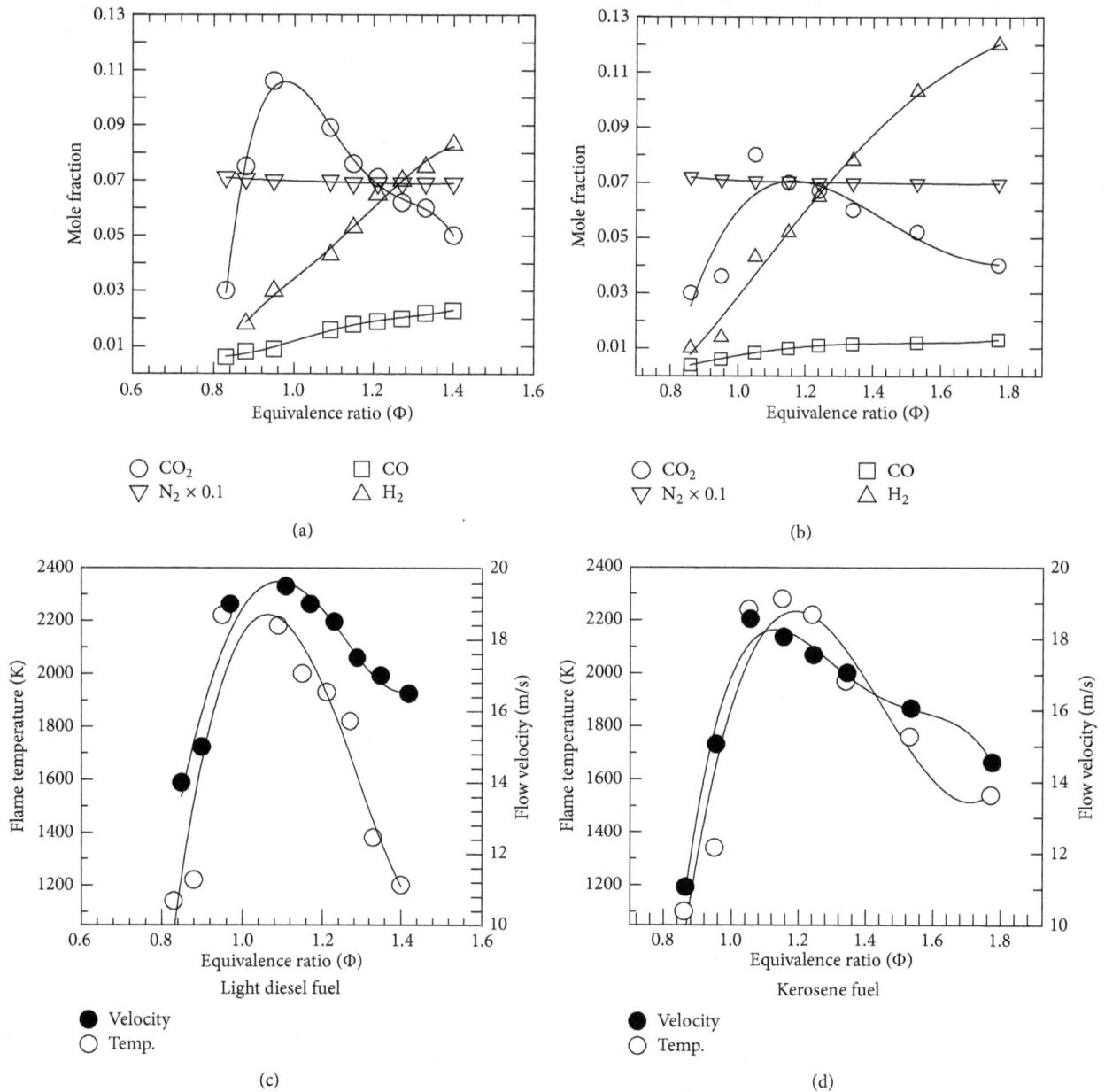

FIGURE 7: Effect of equivalence ratio on flame structure for turbulent premixed-air flame at residence time 5.4 ms, Reynolds number 11.08×10^3, solidity ratio 0.68, and pressure 1 bar.

equilibrium condition. For lean flame at 5.4 ms, flow velocity profile is steep and increases with increased equivalence ratio due to increasing flame temperature. For rich flames, as equivalence ratio increases flow velocity decreases by rate lower than for lean flame. In the postflame zone (at 20 ms), flow velocity profile is nearly straight line and decreases with increased equivalence ratio due to decreasing flame temperature overcoming the increase in fuel mass flow rate. The main effect of equivalence ratio on flow velocity appears in the main reaction zone with maximum flame temperature. CO concentration is high at low equivalence ratio due to low oxidation rate which associates with decrease in flame temperature. As equivalence ratio increases, flame temperature

increases which accelerates oxidation rate and CO declines. At temperature higher than 1800 K, formation of CO due to dissociation of CO_2 is significant. At high equivalence ratios CO concentration is high due to equilibrium condition. Only in the fairly narrow range of equivalence ratios of 0.7 to 0.9 CO concentration is low.

3.5. *Effect of Inlet Reynolds Number on Combustion Process.* In the present work Reynolds (Re) number is calculated as follows:

$$Re = \frac{4\left(\dot{m}_{air} + \dot{m}_{fuel}\right)}{\pi \mu n d}, \qquad (6)$$

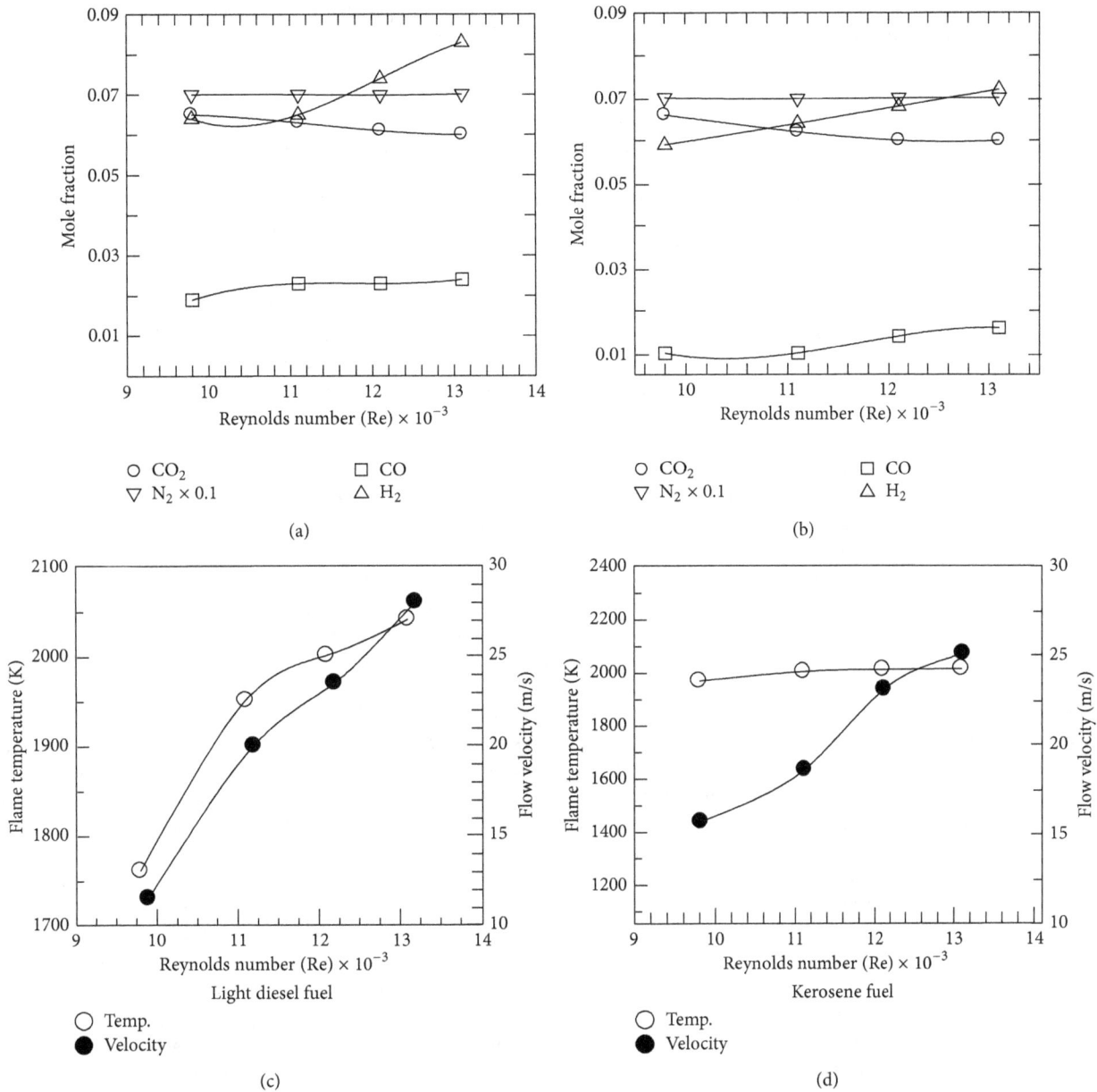

FIGURE 8: Effect of Reynolds number on flame structure for turbulent premixed air flame at residence time 5.4 ms, solidity ratio 0.68, and pressure 1 bar.

where

$\dot{m}_{air}$ is air mass flow rate, kg/s,

$\dot{m}_{fuel}$ is fuel mass flow rate, kg/s,

d is hole diameter in flame holder perforated plate, m,

n is number of holes in flame holder perforated plate,

μ is kinematic viscosity of air at exit temperature from burner tip, mm^2/s.

Effect of inlet Reynolds number on combustion process for light diesel and kerosene-air flames is shown in Figure 8. As Reynolds number increases, the increasing rate of CO and H$_2$ is higher than for CO$_2$. As Reynolds number increases eddies inside the flame increase which increases flame temperature and rate of reaction of different species. As Reynolds number increases heat transfer between reacting and recirculating hot gases increases causing rate of flame propagation to increase. Consequently, flame temperature and rate of reaction of stable species increase. Also, high Reynolds number (high turbulence intensities) causes overshooting of OH, H, and O radicals concentrations leading to increasing rate of reaction and flame temperature. On the other hand, for high Reynolds number, UHC, H$_2$, and CO concentrations increase and they are not easily oxidized in the downstream due to decay flame temperature rapidly. This is due to high mixing of burned and unburned gases in recirculation zone and flame temperature is sufficiently high to oxidize UHC, CO, and H_2. In contrast, in the postflame zone (20 ms not present here) H_2 and CO concentrations decrease with increased Reynolds number due to increasing flame temperature which increases oxidization rate of H_2 and CO. Effect of Reynolds number on concentration for light diesel-air flame is greater than for kerosene-air flames. Light diesel-air flame is more stable under high flame temperature than kerosene-air flames due to high C/H ratio. As Reynolds number increases, the decreasing rate of CO_2 for light diesel-air flame is lesser than for kerosene-air flame and this is due to high (C/H) ratio in light diesel fuel. In contrast, as Reynolds number increases, increasing rate of H_2 for kerosene-air flames is less than for light diesel-air flame.

Gas flow velocity increases with increased Reynolds number due to increasing air mass flow rate. Consequently, more expansion occurs inside the flame leading to early mixing of burned and unburned gases which improves combustion efficiency. Near burner tip, as inlet mixture velocity increases, recirculating gas increases which improves combustion efficiency due to increasing eddies and turbulence intensity. Indeed, the reaction rate is strongly dependent on relation between time (t) of chemical conversion and frequency of velocity fluctuation. The rate of chemical reaction depends on the inhomogeneity of temperature and composition caused by turbulent fluctuation. The turbulent burning velocity increases linearly with increased inlet flow velocity and turbulence intensity. The gases along the flame axis accelerate by the prevailing buoyant which produces a large recirculation cell. The actual velocity measurements verify the existence of these cells. The presence of recirculating cells indicates the importance of buoyant effect on determining the overall flow pattern in the system. Due to these cells, the area of flow decreases and consequently flow velocity increases.

As Reynolds number increases, flame temperature increases gradually to Re = 11×10^3. After that, rate of increasing flame temperature decreases due to dissociation effect. Furthermore, there are contradictions between CO_2 profile on one hand and CO and H_2 profiles on the other hand. In main reaction zone (4 ms), as Reynolds number increases, CO_2 decreases but CO and H_2 increase due to decrease residence time. In the postflame zone (20 ms), CO_2 concentration increases but CO and H_2 decrease due to complete combustion and equilibrium condition. In the main reaction and the postflame zones, Reynolds number effects on H_2 are higher than other species. For light diesel-air flame, Reynolds number effects on flame temperature are greater than for kerosene-air flame especially in the main reaction zone. So, effect of kerosene fuel on flow velocity is smaller than light diesel fuel. In the main reaction zone, as Reynolds number increases by 24% (from 9.912×10^3 to 13.1×10^3) flame temperatures of light diesel, kerosene-air flames increase by 200, 50 K (18%, 10%), respectively. In the postflame zone, flames temperatures of light diesel and kerosene-air flames increase by 220, 180 K, respectively, for the same increase of Reynolds number.

3.6. Effect of Perforated Plate Holes Diameter on Combustion Process. Holes diameter is a main parameter for turbulence generation with premixed flame. For constant air and fuel mass flow rate, as holes diameter decreases inlet mixture velocity increases. The turbulent intensity varies according to plate geometry. At the burner exit, turbulence produces mainly by the perforated plate so the turbulent intensity is lager for a plate of small holes diameter. Turbulence decays in the core of the jet although its intensity increases in the core of the jet with increased distance from burner tip, because the turbulence is generated by shear mixing in the boundary of the jet. Many runs are carried out to determine the optimum holes diameter. If the flow velocity increases above a certain limit, the flames blow off from the burner rim like the conventional flame. On the other hand, if the flow velocity decreases to certain value, flame length decreases and finally flame becomes fuzzy at the burner tip. For small holes, diameter flow velocity is very high especially in main reaction zone, so there is an instantaneous boundary between burned and unburned gases, which is the flame front. The turbulent flame velocity increases with increased turbulence intensity due to the fact that turbulence intensity is proportional to the inlet flow velocity. Flow spreading angle determines the length of negative pressure zone (recirculation zone). Flow spreading angle decreases with decreased holes diameter so an increase in the spreading angle leads to decrease in recirculation zone length. High feed velocity gives small spreading angle and increases the negative pressure zone length, which increases recirculation of unburned gases at burner tip. Increasing recirculation zone length increases turbulence intensity, improves mixing of burned and unburned gases near burner tip, and also improves combustion efficiency. Moreover, as holes diameter decreases, flame length decreases causing an increase in the rate of heat release. On the other hand, for large holes diameter, reaction zone length is considerably longer than those observed for small holes diameter because of no rapidity of mixing burnt and unburnt gases as a result of absent recirculation effect. So, due to increasing reaction zone length, the residence time in the postflame zone reduces and finial CO_2 level is found to be low. Studies of circulation zone sizes for grid plates are reasonable to assume as a function of holes diameter. The holes diameter is a more convenient parameter on base study of the recirculation zone size and the alternative is the web distance between holes which is a variable around the holes circumference. The central holes gave an annular recirculation zone rotating outwards and the web gives a central recirculation zone rotating inwards. Consequently mixing of burned and unburned gases improves. Effect of holes diameter on concentration, flame temperature, and gas flow velocity for turbulent premixed air flame of light diesel and kerosene-air flame are shown in Figure 9. From this figure, it can be noticed that the effect of holes diameter are mainly on gas flow velocity and flame temperature. Where gas flow velocity and flame temperature are compatible with each other and nearly decrease with the same rate. At the same time H_2, CO increase as holes diameter increases but CO increases up to 3 mm as hole diameter, after that CO is kept constant. Furthermore, CO_2 decreases with increased holes diameter up to 3 mm after that it is kept constant due to decreasing flame temperature, burning velocity, rate of

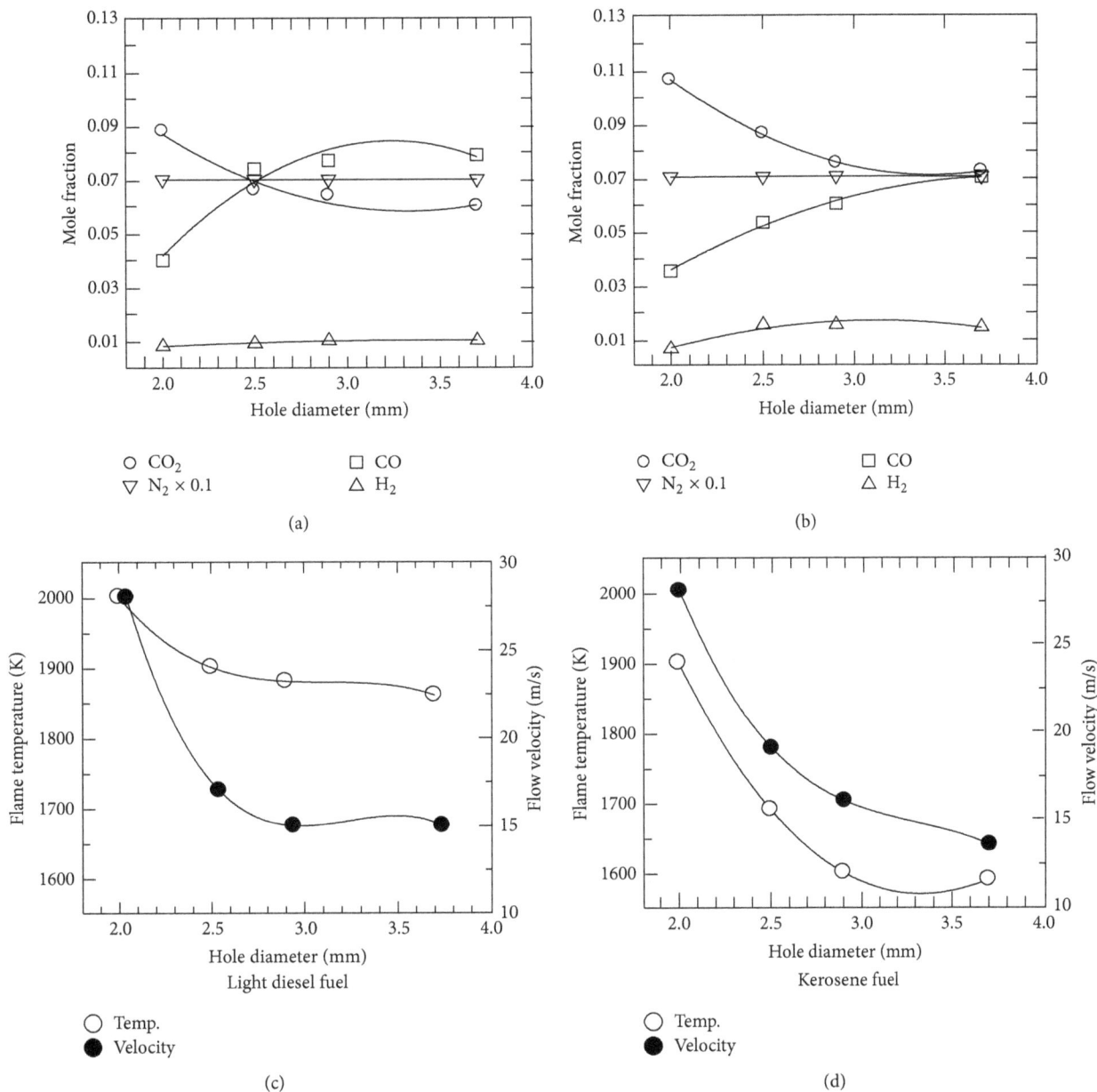

FIGURE 9: Effect of perforated plate holes diameter on flame structure for turbulent premixed air flame at residence time 5 ms, mean Reynolds number 11.08×10^3, equivalence ratio 1.27, and pressure 1 bar.

reaction, and radical concentrations. In contrast, at holes diameter 2 mm effect of holes diameter on gas flow velocity is greater than effects of flame temperature and this is due to flame blow-off effect. Also the changes of holes diameter have no effect on N_2 because equivalence ratio is kept constant. From our discussion, the optimum holes diameter for present work is 2.9 mm which corresponds to 0.68 as solidity ratio.

3.7. Effect of Inlet Mixture Temperature on Combustion Process.
Effects of inlet mixture temperature on combustion process for light diesel and kerosene-air flames at residence time 7 ms are shown in Figure 10. Concentration, gas flow velocity, and flame temperature are measured for many runs to determine the range of inlet mixture temperature from point of complete

evaporation to avoid condensation of fuel in mixing tube to maximum allowable inlet mixture temperature which affects flame stability where overheating of mixture and preflame reactions will reduce burning velocity. The main effect of inlet mixture temperature is for flame temperature and gas flow velocity. Due to the fact that all operating parameters are kept constant at burner tip except inlet mixture temperature so gas flow velocity increases due to increasing inlet temperature. Also effect of inlet mixture temperature on gas flow velocity for lean mixture is greater than for rich mixture and this effect decreases gradually with increased inlet mixture temperature. CO and H_2 increase with increased inlet mixture temperature to 570 and 670 K for kerosene and light diesel-air flames, respectively. After that, CO and H_2 are kept constant due

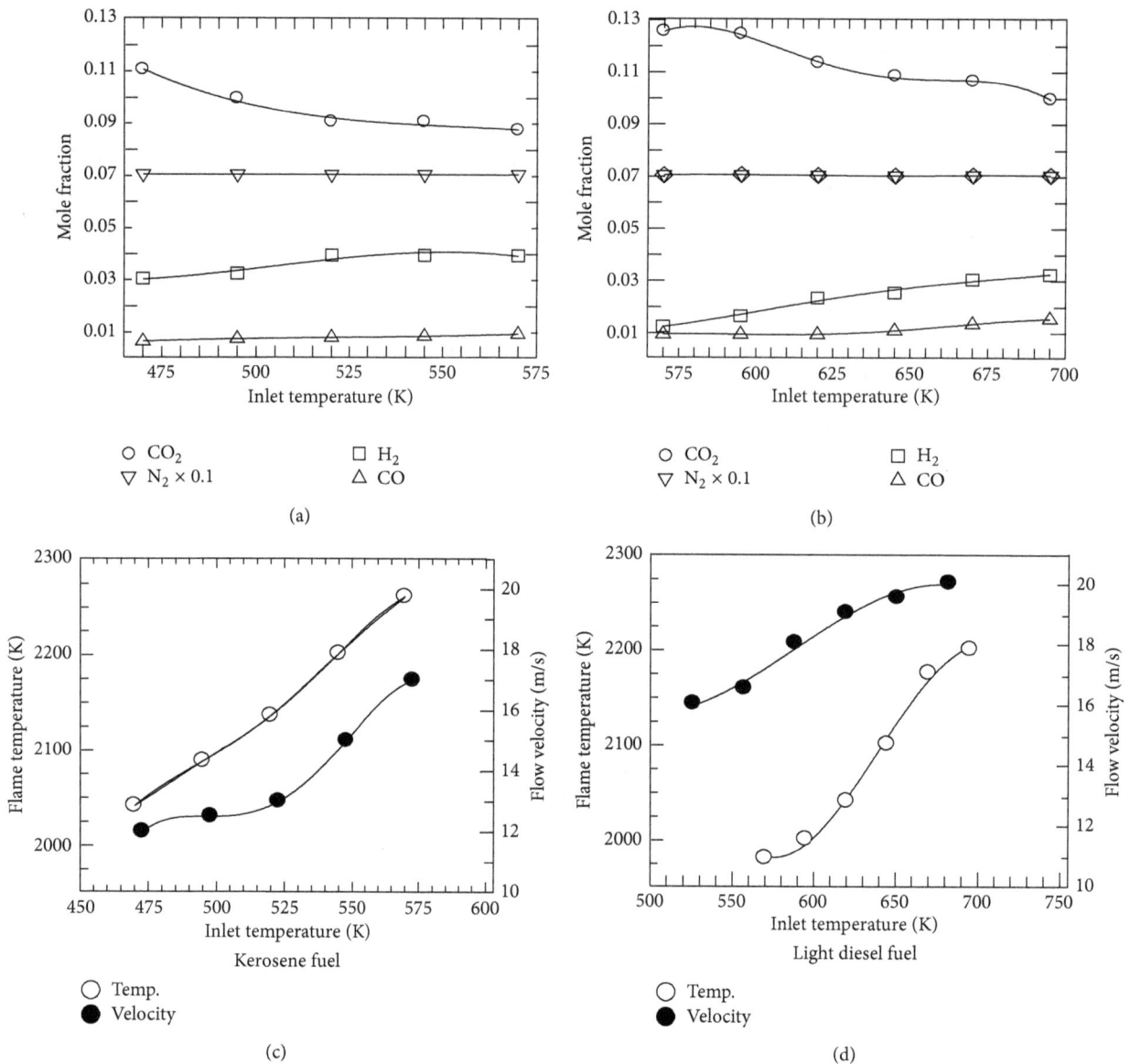

FIGURE 10: Effect of inlet temperature on flame structure for turbulent premixed air flame at residence time 7 ms, Reynolds number 11.08×10^3, solidity ratio 0.68, equivalence ratio 1.13, and pressure 1 bar.

to dissociation effect and decrease in burning velocity. So, formation rate of CO_2 decreases with high rate after 570 and 670 K for both kerosene and light diesel-air flames. Effect of inlet mixture temperature on N_2 is not appeared. For rich flames, effect of inlet mixture temperature on combustion process is high in main reaction zone and decreases gradually to end of flame. CO_2 concentration decreases with increased inlet mixture temperature due to increasing dissociation where air mass flow rate holds constant. Inlet flow velocity increases with increased inlet temperature due to increase in specific volume of mixture. As inlet mixture temperature increases residence time of the min reaction zone decreases and complete combustion occurs in postflame zone. Flame temperature increases as inlet mixture temperature increases especially in pyrolysis zone due to incomplete combustion. So, rate of reaction of intermediate species, CO and H_2,

increases due to fuel pyrolysis. CO concentration decreases with increased inlet mixture temperature due to increase in flame temperature which accelerates conversion rate of CO to CO_2. At stoichiometric and rich mixtures, high flame temperature promotes CO formation by dissociation. CO concentration increases as inlet mixture temperature increases.

4. Conclusions

(1) The main operating parameters as fuel type, fuel blends, equivalence ratio, Reynolds number, inlet mixture temperature, and design parameters as perforated plate holes diameter are major factors affecting combustion process for turbulent prevaporized premixed air flames.

(2) C_2H_4, C_2H_6, and C_3H_8 concentrations for light diesel-air flame are higher than for kerosene-air flame while formation and dissipation rates of C_2H_4, C_2H_6, and C_3H_8 for kerosene-air flame are higher than for light diesel-air flame. Kerosene-air flames have flame temperature higher than light diesel-air flames.

(3) Light diesel-air flame is more stable under high flame temperature than kerosene-air flame due to high C/H ratio.

(4) Effects of kerosene percentage for lean and rich flames are not linearly dependent for concentration, flame temperature, and gas flow velocity. Flame temperature and flow velocity increase with increased kerosene percentage in the blends. As light diesel percentage increases in the blends, decreasing rate of flame temperature and flow velocity for lean flame are higher than for rich flame especially in the postflame zone.

(5) Effect of equivalence ratio on the maximum flame temperature for rich flame is greater than for lean flame. For lean flame, as equivalence ratio increases by 12%, maximum flame temperature increases by 16%. For rich flame, as equivalence ratio increases by 16%, maximum flame temperature decreases by 35%.

(6) As Reynolds number increases, mixing rate of burned and unburned gases increases which increase flame temperature and rate of reaction of different species. In main reaction zone, CO_2 concentration decreases with increased Reynolds number, while H_2 and CO concentrations increase with increased Reynolds number. Effect of Reynolds number on concentration for light diesel-air flame is greater than for kerosene-air flames

(7) For light diesel-air flame, effect of Reynolds number on flame temperature is greater than for kerosene-air flame especially in the main reaction zone.

(8) In the main reaction zone, as Reynolds number increases by 24% (from 9.912×10^3 to 13.1×10^3) flame temperatures of light diesel, kerosene-air flames increase by 200 and 50 K (18%, 10%), respectively. In the postflame zone, flames temperatures of light diesel and kerosene-air flames increase by 220 and 180 K, respectively, for the same increase of Reynolds number.

(9) Gas flow velocity and flame temperature are more influenced by holes diameter, where gas flow velocity and flame temperature are compatible with each other and nearly decrease with the same rate. The optimum holes diameter is 2.9 mm which corresponds to 0.68 as solidity ratio.

(10) Inlet mixture temperature mainly affects maximum flame temperature and elementary rate of reaction leads to increase in fuel burning rate. Effect of inlet mixture temperature on gas flow velocity for lean mixture is greater than for rich mixture and this effect decreases gradually with increased inlet mixture temperature.

Conflict of Interests

The authors declare that there is no conflict of interests regarding the publication of this paper.

References

[1] A. El Bakali, P. Dagaut, L. Pillier et al., "Experimental and modeling study of the oxidation of natural gas in a premixed flame, shock tube, and jet-stirred reactor," *Combustion and Flame*, vol. 137, no. 1-2, pp. 109–127, 2004.

[2] G. R. Papagiannakis and T. D. Hountals, "Experimental investigation concerning the effect of natural gas percentage on performance and emissions of a DI dual fuel diesel engine," *Applied Thermal Engineering*, vol. 23, no. 3, pp. 353–365, 2003.

[3] A. S. El-Sherif, "Effects of natural gas composition on the nitrogen oxide, flame structure and burning velocity under laminar premixed flame conditions," *Fuel*, vol. 77, no. 14, pp. 1539–1547, 1998.

[4] Y. Tan, P. Dagaut, M. Cathonnet, and C. J. Boettner, "Natural gas and blends oxidation and ignition: experiments and modeling," *Proceedings of the Combustion Institute*, vol. 25, pp. 1563–1569, 1963.

[5] Y. Chen and M. Ihme, "Flame characterization of a piloted premixed jet burner," in *Proceedings of the Spring Technical Meeting of the Central States Section of the Combustion Institute*, pp. 22–24, April 2012.

[6] R. Sadanandan, R. Lückerath, W. Meier, and C. Wahl, "Flame characteristics and emissions in flameless combustion under gas turbine relevant conditions," *Journal of Propulsion and Power*, vol. 27, no. 5, pp. 970–980, 2011.

[7] O. Obodeh and O. F. Isaac, "Investigation of performance characteristics of Diesel Engine Fueled with Diesel-Kerosene blends," *Journal of Emerging Trends in Engineering and Applied Sciences*, vol. 2, no. 2, pp. 318–322, 2011.

[8] T. K. Ghormade and N. V. Deshpande, "Soybean oil as an alternative fuels for I. C. engines," in *Proceedings of the Recent Trends in Automotive Fuels*, Nagpur, India, 2002.

[9] K. V. Kumar Reddy, *Experimental investigations on the use of vegetable oil fuels in a 4-stroke single cylinder diesel engine [Ph.D. thesis]*, JNTU, Anantapur, India, 2000.

[10] K. A. Azad, M. S. Ameer Uddin, and M. M. Alam, "Experimental study of DI diesel engine performance using biodiesel blends with kerosene," *International Journal of Energy and Environment*, vol. 4, no. 2, pp. 265–278, 2013.

[11] S. Poovannan and G. Kalivarathan, "Performance evaluation of lean premixed prevapourised combustion chamber," *International Journal of Mechanical Engineering and Technology*, vol. 4, no. 2, pp. 127–133, 2013.

[12] S. R. Gollahalli, N. Dhamale, and R. N. Parthasarathy, "Effects of turbulence on the combustion properties of partially premixed flames of canola methyl ester and diesel blends," *Journal of Combustion*, vol. 2011, Article ID 697805, 13 pages, 2011.

[13] M. M. El Kotb, H. Salem, and S. M. Shehata, "Effect of flame stabilizer geometry on emissions of turbulent premixed blended flames," *Experimental Thermal and Fluid Science*, vol. 27, no. 4, pp. 343–353, 2003.

[14] N. M. Howe Jr., C. W. Shipman, and A. Vranos, "Turbulent mass transfer and rates of combustion in confined turbulent flames," *Symposium (International) on Combustion*, vol. 9, no. 1, pp. 36–47, 1963.

[15] R. Akrich, C. Vovelle, and R. Delbourgo, "Flame profiles and combustion mechanisms of methanol-air flames under reduced pressure," *Combustion and Flame*, vol. 32, pp. 171–179, 1978.

[16] C. K. Westbrook and F. L. Dryer, "Prediction of laminar flame properties of methanol-air mixtures," *Combustion and Flame*, vol. 37, pp. 171–192, 1980.

[17] K. Radhakrishnan, J. B. Heywood, and R. J. Tabaczynski, "Premixed turbulent flame blowoff velocity correlation based on coherent structures in turbulent flows," *Combustion and Flame*, vol. 42, pp. 19–33, 1981.

3

Numerical Simulation of the Deflagration-to-Detonation Transition in Inhomogeneous Mixtures

Florian Ettner, Klaus G. Vollmer, and Thomas Sattelmayer

Lehrstuhl für Thermodynamik, Technische Universität München, 85748 Garching, Germany

Correspondence should be addressed to Florian Ettner; florian.ettner@gmail.com

Academic Editor: Yiguang Ju

In this study the hazardous potential of flammable hydrogen-air mixtures with vertical concentration gradients is investigated numerically. The computational model is based on the formulation of a reaction progress variable and accounts for both deflagrative flame propagation and autoignition. The model is able to simulate the deflagration-to-detonation transition (DDT) without resolving all microscopic details of the flow. It works on relatively coarse grids and shows good agreement with experiments. It is found that a mixture with a vertical concentration gradient can have a much higher tendency to undergo DDT than a homogeneous mixture of the same hydrogen content. In addition, the pressure loads occurring can be much higher. However, the opposite effect can also be observed, with the decisive factor being the geometric boundary conditions. The model gives insight into different modes of DDT. Detonations occurring soon after ignition do not necessarily cause the highest pressure loads. In mixtures with concentration gradient, the highest loads can occur in regions of very low hydrogen content. These new findings should be considered in future safety studies.

1. Introduction

Due to its potentially catastrophic consequences, the accidental release of hydrogen is a major concern in process engineering [1, 2], power generation [3–5], and future automotive concepts [6, 7]. A small heat source can be sufficient to initiate a deflagration that causes a sudden temperature and pressure rise. The severity of the consequences depends on the propagation speed of the deflagration. In general it can be expected that the flame accelerates due to a positive feedback loop: the expansion of the combustion products induces motion in the unburned gas which generates turbulence and thereby increases the burning speed of the flame. This process is further complicated through instabilities [8–10], formation of shocks, and interaction with the surrounding structure. Under certain circumstances the deflagration can undergo a transition to a detonation. The hazardous potential of a detonation (and especially that of the unsteady transition process) is considerably higher than that of a deflagration.

Therefore, the probability of the occurrence of a deflagration-to-detonation transition (DDT) needs to be considered in safety studies involving scenarios with accidental hydrogen release. In the past, many researchers have contributed impressive DDT simulation studies by applying a very high grid resolution to observe all microscopic flow details [11–15]. This approach has helped a lot to understand the formation of DDT but is only applicable to very small domains. When using simple reaction models like one-step Arrhenius kinetics, many publications can be found using a grid resolution of 8 or less computational cells per half-reaction length. However, it is generally agreed that when employing one-step Arrhenius kinetics, a resolution of at least 20 computational cells per half-reaction length is required in inviscid flow to sufficiently resolve the flow structure and obtain correct results for heat release profile, flame-shock interaction, detonation cell size, and so forth [16–18]. This is already challenging to achieve in large computational domains. Even though it is often argued that diffusive effects are negligible when simulating established detonations, they are of importance in fast deflagrations and the onset of detonations. Mazaheri et al. [19] showed that the inclusion of diffusive terms increases the required spatial resolution to approximately 25 to 300 computational cells (depending on the activation energy of the mixture) per half-reaction length. Powers and Paolucci [20] demonstrated that when using a detailed chemical mechanism instead of simplified kinetics, the demand for spatial resolution increases to an order of magnitude of approximately 10^3 cells per half-reaction length. Due to the enormous computational costs, such "direct" simulations of DDT (even when using only simplified kinetics at 20 cells per half-reaction length) cannot be applied to large scale engineering simulations and are expected to remain impossible for years to come. Thus, the current state of the art of large scale accident simulations (e.g., in nuclear safety studies) is as follows [3, 4].

(1) Determination of spatial hydrogen distribution.

(2) Usage of empirical criteria to determine whether DDT can occur or not.

(3) Simulation of the combustion with *either* a deflagration *or* a detonation code.

Step 1 can be performed with a lumped parameter code [5] or a CFD code [21, 22]. Step 2 is performed by applying empirical criteria [23, 24] which have been obtained from experiments in explosion tubes [25–29] to the geometry enclosing the hydrogen cloud. This step can only be performed with considerable uncertainty regarding not only the different scale and geometry (explosion tube with regular, periodic obstacles versus intricate three-dimensional building structure), but also the fact that virtually all criteria have been gained from experiments with perfectly homogeneous mixtures. In accident scenarios, however, mixtures are likely to include strong concentration gradients [30, 31]. Step 3 can be performed with a variety of CFD codes. While slower deflagrations can be simulated satisfactorily with commercial codes, simulations of fast deflagrations (where gas dynamics have a major influence) and especially simulations of detonations are usually performed more accurately with in-house codes that are generally not made available to the public.

In an OECD report prepared by a group of experts this "significant lack of numerical tools available to safety analysts" [32] has been criticized and the following requirements have been identified.

(1) Development of approximate but reliable methods for simulating both flame acceleration and detonation in such a fashion that the simulation can be run within a single software framework.

(2) Development of reliable combustion models that can be used to model DDT without the judgment and intervention of the simulator.

As the current situation is very unsatisfying, the present study was aimed at developing a CFD solver capable of simulating both deflagrations and detonations and especially the transition between both regimes. While the application of 3D computations at full reactor scale remains a long-term objective, this project goes a first step into this direction: the development of a solver to show the technical feasibility of simulating DDT experiments in explosion tubes in 2D

without resolving the microstructure of the flow in the CFD grid. This forms an important prerequisite for the future simulation of flame acceleration and DDT in large, three-dimensional domains which will necessarily be performed on underresolved grids.

In a second step, this new solver is used to investigate the influence of concentration gradients on DDT formation. While many earlier studies considered a homogeneous mixture a worst-case scenario and consequently neglected the role of inhomogeneities, the influence of concentration gradients has only recently come into the focus of research [29, 33, 34].

The results of this project are shown in this paper.

2. Model Description

As the direct initiation [35] of a detonation is very unlikely, a detonation usually occurs after a turbulent flame (deflagration) has accelerated to a sufficiently high velocity [32]. This deflagration-to-detonation transition is a complex phenomenon including flame instabilities, interaction with turbulence, and gas dynamic phenomena. With all these effects occurring at very high Reynolds numbers, it is virtually impossible to resolve them completely in numerical computations. However, not all phenomena occurring on microscopic scale are necessarily relevant for an accurate simulation of macroscopic DDT events. Thomas [36] concluded from a comparison of experimental and numerical results that a reliable DDT model does not have to resolve all details of the flow. Instead, only correct turbulent burning rates, local density increase due to shocks and the capability of giving rise to detonations as a result of blast waves have been identified as necessary features of an accurate DDT model.

For the present work, the CFD code *OpenFOAM* [37] has been chosen as a basis for model development. One reason for this choice was that (contrary to many proprietary detonation codes) *OpenFOAM* has the built-in capability of dealing with unstructured grids—a clear advantage in view of the intended future application to sophisticated geometries. Another reason for the usage of *OpenFOAM* was that the code is free and open source and thus also the model developed in this project can be made available to the scientific community at no cost.

The density-based code developed under *OpenFOAM* solves the unsteady, compressible Navier-Stokes equations. All convective fluxes are determined using the HLLC scheme [38] with multidimensional slope limiters ("cellMDLimited" [37, 39]). This scheme is very suitable for the simulation of high Mach number compressible flow as it leads to much better shock capturing than the standard schemes used in most pressure-based codes like the PISO scheme [40]. As an example, the results of a 1D shock-tube calculation are compared in Figures 1 and 2. The initial data for this shock-tube problem is an ideal gas with molecular weight $M = 28.85$ kg/kmol and specific heat ratio $\gamma = 1.4$ at $p = 10$ bar, $T = 800$ K, $u = 0$ m/s for $x \leq 0$ and $p = 1$ bar, $T = 300$ K, $u = 0$ m/s for $x > 0$. The results shown have been obtained on an equidistant grid with 1.0 mm spacing. For comparison, the analytical solution is displayed as a light gray line. It can be seen from Figure 1 that the PISO scheme not only predicts a wrong shock location (i.e., wrong propagation speed), but also is in general very dissipative and displays overshoots at discontinuities. This can be attributed to the nonconservative formulation of the Navier-Stokes equations which is inherent to the pressure-based scheme. With regard to the intended application to transonic reactive flow (including autoignition caused by shocks) this has to be regarded as a very critical issue. It can be concluded that standard pressure-based solvers are not suitable for the simulation of fast deflagrations and detonations.

Density-based solvers, especially Riemann solvers like the HLLC scheme employed in Figure 2, show a much better performance, producing accurate shock propagation speeds and far less dissipation at discontinuities. Thus it was decided to use the HLLC scheme as a basis for the DDT solver developed in this work. The only disadvantage of the scheme is that it does not work in very low Mach number flow. Therefore, the PISO scheme is also implemented and can be used to start computations in stagnant flow and then switch to the HLLC scheme once a combustion-induced flow has developed (see Section 3).

Realistic material properties for the reacting hydrogen-air mixture are obtained from the *Chemkin* database [41] and molecular transport coefficients are determined using the Sutherland correlation [42].

Combustion is described via a reaction progress variable c [43]. $c = 0$ corresponds to an unburned mixture, $c = 1$ to a completely burned mixture. Within the context of Favre-averaging [44, 45], $\tilde{c}(\vec{x}, t)$ can be interpreted as the density-weighted probability of encountering burned gas at a particular instance of space and time. The transport equation of the reaction progress variable reads

$$\frac{\partial}{\partial t}\left(\overline{\rho}\tilde{c}\right) + \frac{\partial}{\partial x_j}\left(\overline{\rho}\tilde{c}\tilde{u}_j\right) = \frac{\partial}{\partial x_j}\left(\overline{\rho}D_{\text{eff}}\frac{\partial\tilde{c}}{\partial x_j}\right) + \overline{\omega}_{c,\text{def}} + \overline{\omega}_{c,\text{ign}},$$

(1)

where the overbar denotes Reynolds-averaging and the tilde denotes Favre-averaging. The equation contains two source terms that account for deflagrative and detonative combustion, respectively. This concept is related to the simulation of autoignition in gas turbines or internal combustion engines as proposed by [46–48].

The deflagrative source term $\overline{\omega}_{c,\text{def}}$ is modelled using the RANS version of the Weller combustion model [49], extended by a factor $0 \le G \le 1$ [50, 51] which takes quenching of turbulent flames into account [52]:

$$\overline{\omega}_{c,\text{def}} = \overline{\rho}_u s_T |\nabla\tilde{c}| G,$$

(2)

$$s_T = \xi s_L.$$

(3)

ρ_u is the density of the unburned mixture and s_T the turbulent burning speed which is modelled as the product of the

laminar burning speed s_L and a flame wrinkling factor ξ. The latter is obtained from a transport equation:

$$\frac{\partial}{\partial t}\left(\overline{\rho}\xi\right) + \frac{\partial}{\partial x_j}\left(\overline{\rho}\xi\tilde{u}_j\right) = \frac{\partial}{\partial x_j}\left(\overline{\rho}D_{\text{eff}}\frac{\partial\xi}{\partial x_j}\right) + \overline{\rho}P_\xi\xi - \overline{\rho}R_\xi\xi^2.$$

(4)

Details about the expressions P_ξ and R_ξ describing the generation and destruction of flame surface can be found in [37, 49]. Due to the gradient ansatz, (2) leads to correct consumption rates even on underresolved grids. This means, even if the flame thickness is smeared out over several computational cells and thus thicker than the physical flame brush, the overall consumption rate is not affected [49, 52]. An adjustment of model constants to grid size or domain geometry is not required.

Experimental values for the laminar burning speed of hydrogen-air flames at standard conditions (temperature $T_0 = 298$ K and pressure $p_0 = 1.013$ bar) have been published by Konnov [53]. The dependence of flame speed on molar hydrogen fraction x_{H2} can be approximated as a polynomial [52]:

$$s_{L,0} = \begin{cases} \begin{aligned} &\left(-488.9x_{\text{H2}}^4 + 285.0x_{\text{H2}}^3\right. \\ &\left.-21.92x_{\text{H2}}^2 + 1.351x_{\text{H2}}\right. \\ &\left.-0.040\right)\text{ m/s}, & x_{\text{H2}} \le 0.35 \\ &\left(-160.2x_{\text{H2}}^4 + 377.7x_{\text{H2}}^3\right. \\ &\left.-348.7x_{\text{H2}}^2 + 140.0x_{\text{H2}}\right. \\ &\left.-17.45\right)\text{ m/s}, & x_{\text{H2}} > 0.35. \end{aligned} \end{cases}$$

(5)

In inhomogeneous mixtures it is essential to compute a correct flame speed in partially burned cells. Therefore, in the computation of the laminar flame speed, x_{H2} is not to be based on the actual hydrogen content, but on the mixture fraction f_{H}, that is, the amount of hydrogen that would be present if the cell was completely unburned [54]. This is achieved by evaluating the hydrogen content x_{H2} (molar fraction of the hydrogen molecule) based on the mixture fraction f_{H} (mass fraction of the hydrogen atom). The spatial distribution of the mixture fraction is described by a transport equation [54]:

$$\frac{\partial}{\partial t}\left(\overline{\rho}\tilde{f}_{\text{H}}\right) + \frac{\partial}{\partial x_j}\left(\overline{\rho}\tilde{f}_{\text{H}}\tilde{u}_j\right) = \frac{\partial}{\partial x_j}\left(\overline{\rho}D_{\text{eff}}\frac{\partial\tilde{f}_{\text{H}}}{\partial x_j}\right).$$

(6)

The dependence of laminar flame speed on pressure and temperature can be approximated as [55]:

$$s_L = s_{L,0}\left(\frac{T}{T_0}\right)^\alpha\left(\frac{p}{p_0}\right)^\beta.$$

(7)

Constant values of $\alpha = 1.75$ and $\beta = -0.2$ have been used in this study.

The detonative source term $\overline{\omega}_{c,\text{ign}}$ in (1) accounts for autoignition effects. Autoignition occurs after the expiry of the local autoignition delay time t_{ign}. The autoignition delay time of a gaseous mixture is a function of local temperature T, pressure p, and mixture composition. In pure hydrogen-air mixtures, the mixture composition can be described using

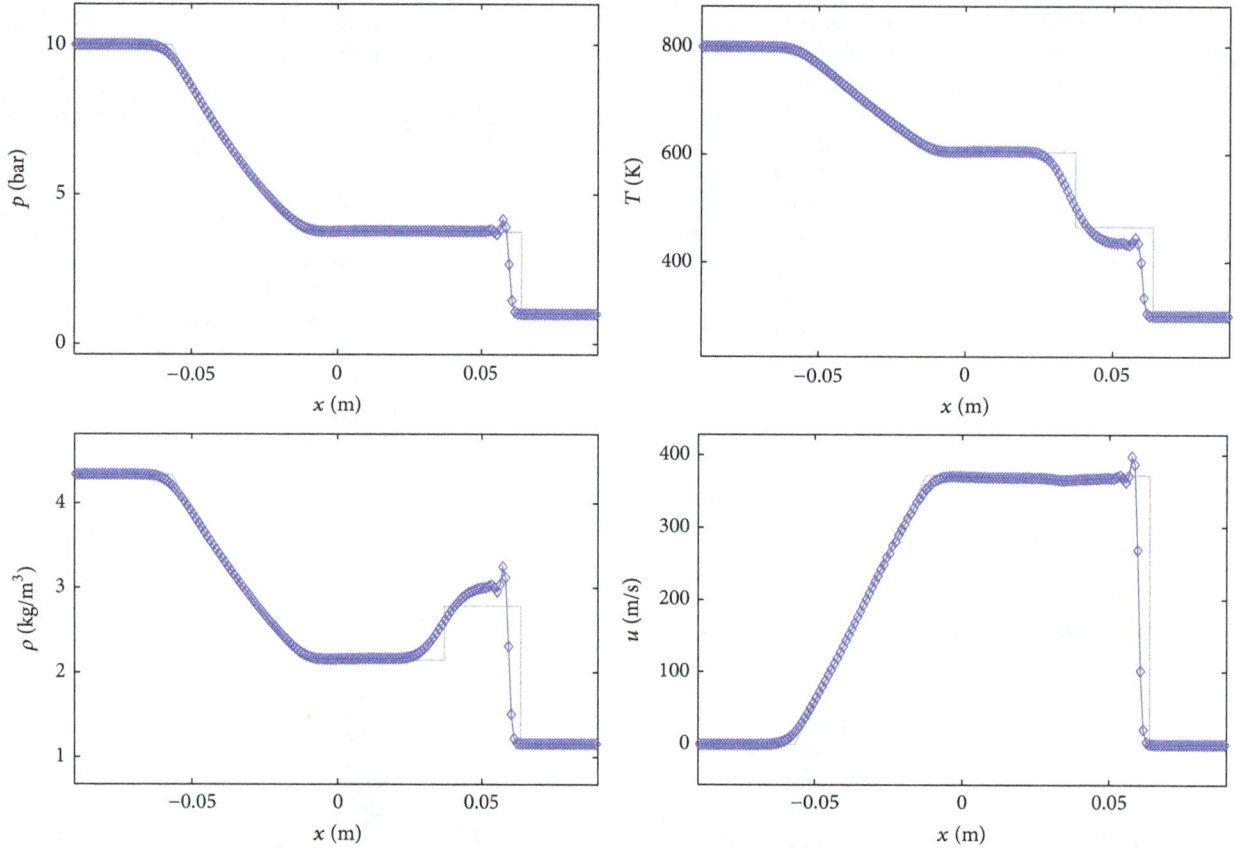

FIGURE 1: Results of shock-tube calculation with PISO scheme.

the mixture fraction f_H (6). In order to avoid frequent recomputation of the local ignition delay time, a table of t_{ign} as a function of T, p, and f_H is generated using *Cantera* [56] and the Ó Conaire mechanism [57]. This mechanism is valid for pressures up to 87 atm and temperatures up to 2700 K. The flow solver can access the table and query the local autoignition delay time $t_{ign}(T, p, f_H)$ in each computational cell.

An alternative means of modelling autoignition is the concept of a virtual radical species R preferred by some authors [46, 47]. While in this study the concept of a tabulated ignition delay time is preferred, it can be shown that both models equivalently lead to the definition of a dimensionless variable describing the autoignition process [52]:

$$\tau = \frac{t}{t_{ign}} = \frac{y_R}{y_{R,critical}}. \tag{8}$$

The ignition variable τ reaches unity when the ignition delay time has passed (equivalently, it can be thought of the virtual ignition radical reaching a critical concentration). As long as the ignition delay time has not been reached yet ($t < t_{ign}$), there is no impact on the flow properties. Only if $\tau = 1$ is reached, the mixture is ignited.

Due to the fact that autoignition in a DDT context can be triggered by shock-induced heating, a submodel is introduced that increases the accuracy of autoignition modelling on coarse grids. Tosatto and Vigevano [58] demonstrated that, while the average temperature in a computational cell can be high enough to trigger autoignition, it is important for detonation simulations to account for the fact that the shock causing the temperature rise might not have traversed the entire computational cell yet. Consequently a model that predicts autoignition of a computational cell based on average temperature and average pressure leads to incorrect results. Due to the large disparity of scales, that is, the shock being far too thin to be resolved on the computational grid, each computational cell is divided into two parts: in one part (volume fraction α) temperature and pressure are elevated to T_{high} and p_{high}; in the other part (volume fraction $1 - \alpha$) the values remain at T_{low} and p_{low}. p_{high} is defined by the highest value and p_{low} by the lowest value that can be found in the surrounding computational cells. From consistency with the average pressure $\bar{p}$ in the computational cell, the volume fraction α can be determined:

$$\alpha = \frac{\bar{p} - p_{low}}{p_{high} - p_{low}}. \tag{9}$$

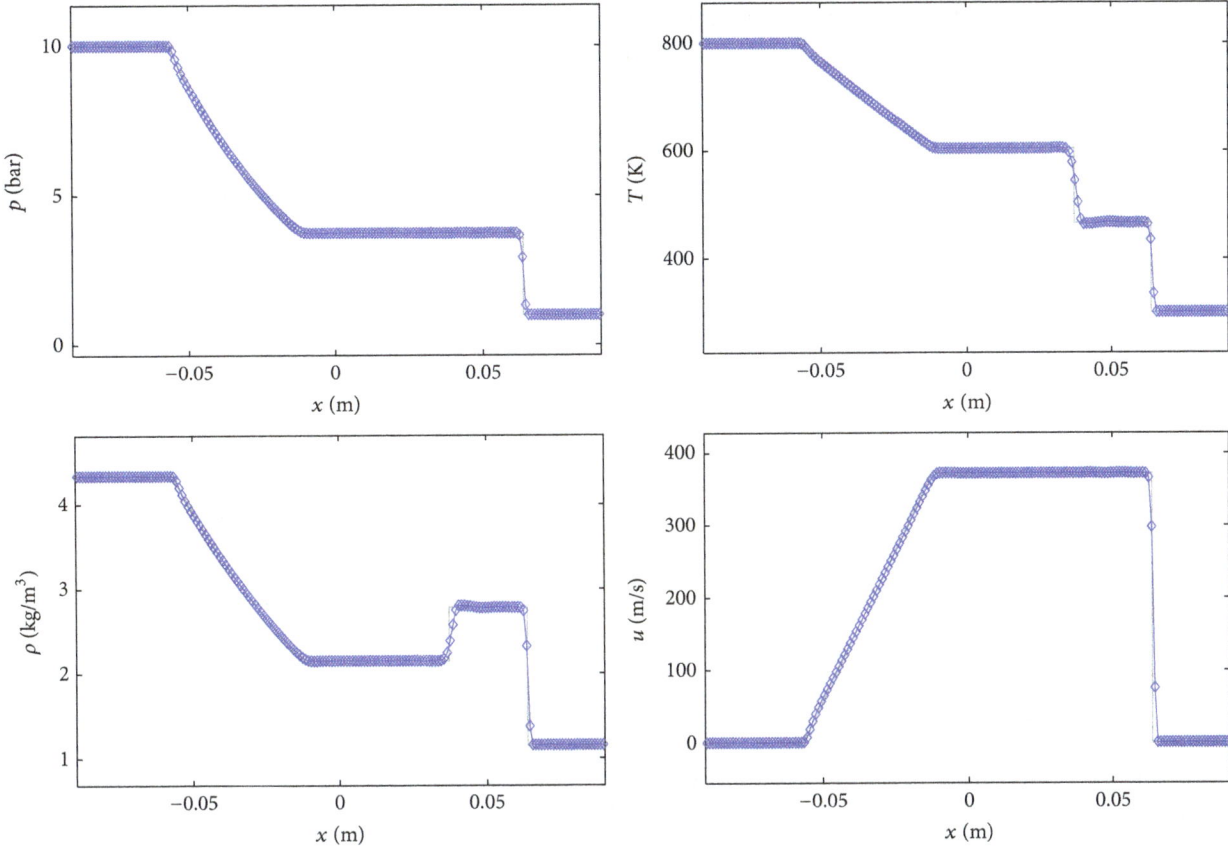

FIGURE 2: Results of shock-tube calculation with HLLC scheme.

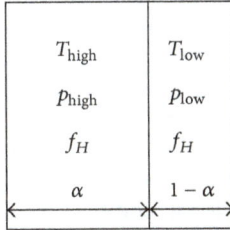

FIGURE 3: Model illustration: a shock divides a computational cell into a volume fraction α of high temperature and pressure and a volume fraction $1 - \alpha$ of low temperature and pressure.

T_{high} and T_{low} are computed subsequently from gas dynamic shock relations for an ideal gas with heat capacity ratio γ [59]:

$$\frac{p_{\text{high}}}{p_{\text{low}}} = 1 + \frac{2\gamma \left(\text{Ma}^2 - 1 \right)}{\gamma + 1}, \tag{10}$$

$$\frac{T_{\text{high}}}{T_{\text{low}}} = \left(1 + \frac{2\gamma \left(\text{Ma}^2 - 1 \right)}{\gamma + 1} \right) \left(1 - \frac{2 \left(1 - \text{Ma}^{-2} \right)}{\gamma + 1} \right). \tag{11}$$

When p_{high} and p_{low} are known, (10) can be solved for the shock Mach number Ma and (11) yields the according temperature ratio $T_{\text{high}}/T_{\text{low}}$. The temperatures can be determined by

requiring consistency of T_{high} and T_{low} with the cell-averaged temperature $\widetilde{T}$:

$$\widetilde{T} = \alpha T_{\text{high}} + (1 - \alpha) T_{\text{low}}. \tag{12}$$

The resulting representation of a computational cell is given in Figure 3. While the presence of a shock alters temperature and pressure, the mixture fraction f_{H} is not affected. The ignition delay time is evaluated separately on each side of the shock:

$$t_{\text{ign,high}} = t_{\text{ign}} \left(T_{\text{high}}, p_{\text{high}}, f_{\text{H}} \right),$$
$$t_{\text{ign,low}} = t_{\text{ign}} \left(T_{\text{low}}, p_{\text{low}}, f_{\text{H}} \right). \tag{13}$$

It is important to note that this model does not only work in computational cells where a shock is present but can also be applied to the entire computational domain. The reason is that in the case of low pressure differences, $p_{\text{high}} \gtrsim \overline{p} \gtrsim p_{\text{low}}$, the temperature rise computed from (10) and (11) is quasi-identical to the temperature rise gained from an isentropic compression:

$$\frac{T_{\text{high}}}{T_{\text{low}}} = \left(\frac{p_{\text{high}}}{p_{\text{low}}} \right)^{(\gamma-1)/\gamma}. \tag{14}$$

In each grid cell, the process of autoignition is evaluated separately on both sides of the discontinuity. Transport and mixing effects are accounted for by solving transport equations for τ_{high} and τ_{low}:

$$\frac{\partial}{\partial t}\left(\overline{\rho}\widetilde{\tau}_{\text{high}}\right)+\frac{\partial}{\partial x_j}\left(\overline{\rho}\widetilde{\tau}_{\text{high}}\widetilde{u}_j\right) = \frac{\partial}{\partial x_j}\left(\overline{\rho}D_{\text{eff}}\frac{\partial\widetilde{\tau}_{\text{high}}}{\partial x_j}\right)+\frac{\overline{\rho}}{t_{\text{ign,high}}},$$

$$\frac{\partial}{\partial t}\left(\overline{\rho}\widetilde{\tau}_{\text{low}}\right)+\frac{\partial}{\partial x_j}\left(\overline{\rho}\widetilde{\tau}_{\text{low}}\widetilde{u}_j\right) = \frac{\partial}{\partial x_j}\left(\overline{\rho}D_{\text{eff}}\frac{\partial\widetilde{\tau}_{\text{low}}}{\partial x_j}\right)+\frac{\overline{\rho}}{t_{\text{ign,low}}}.$$

$$(15)$$

If the critical value of $\tau = 1$ is reached on one side of the discontinuity (i.e., $\tau_{\text{high}} = 1$ or $\tau_{\text{low}} = 1$), only the corresponding volume fraction of a computational cell is ignited. Consequently the autoignition source term in (1) can be formulated as

$$\overline{\omega}_{c,\text{ign}} = \alpha\frac{1-\widetilde{c}}{\Delta t}\text{H}\left(\widetilde{\tau}_{\text{high}}-1\right) + (1-\alpha)\frac{1-\widetilde{c}}{\Delta t}\text{H}\left(\widetilde{\tau}_{\text{low}}-1\right).$$

$$(16)$$

Here, Δt represents the current time step and $\text{H}(x)$ represents the Heaviside function:

$$\text{H}\left(x\right) = \begin{cases} 0, & x < 0 \\ 1, & x \geq 0. \end{cases} \qquad (17)$$

The Heaviside function activates only the part of the computational cell in which the local ignition delay time has expired. This is achieved by weighting the volumetric source term by the volume fraction of either α or $1 - \alpha$.

3. Experimental and Numerical Setup

The model has been tested against experimental results gained in a closed rectangular channel of length $L = 5.4\,\text{m}$, height $H = 60\,\text{mm}$, and width $W = 300\,\text{mm}$. The channel is equipped with flat plate obstacles (thickness 12 mm) of height h spaced at a distance of $S = 300\,\text{mm}$ from each other. The first obstacle is placed at $x = 0.25\,\text{m}$ from the front plate where a spark plug ignites the mixture. The last obstacle is placed at $x = 2.05\,\text{m}$ and the remaining part of the channel is unobstructed (see Figure 4). The obstacle blockage ratio BR is determined by the obstacle height h:

$$\text{BR} = \frac{2h}{H}. \qquad (18)$$

The top wall of the channel is equipped with 42 UV sensitive photodiodes and 6 pressure transducers operated at a sampling rate of 250 k Samples/s. Another pressure transducer is mounted head-on in the center of the end wall ($x = 5.4\,\text{m}$). A flame position versus time (x versus t) correlation is obtained from the photodiode measurements. The flame velocity between two subsequent photodiodes is calculated by applying a first order derivative:

$$v\left(x = \frac{x_i + x_{i+1}}{2}\right) = \frac{x_{i+1}-x_i}{t_{i+1}-t_i}. \qquad (19)$$

Here, t_i and t_{i+1} represent the time at which the flame passes the photodiodes located at x_i and x_{i+1}, respectively. The same

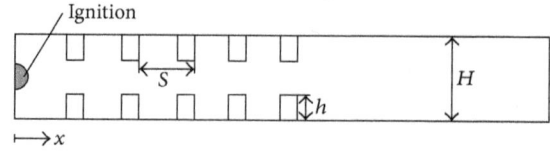

FIGURE 4: Schematic sketch of the channel geometry (side view).

procedure is applied for evaluating the flame velocity in the numerical simulations.

Within the channel defined vertical concentration gradients can be generated. The overall amount of hydrogen is controlled via the partial pressure method. First, the air-filled channel is partially evacuated. Then, hydrogen is injected through several nozzles located at the top wall. The injection velocity is constant due to a choked nozzle upstream of the point of injection. The injection time defines the amount of hydrogen injected. Subsequently there is a defined time interval (waiting time t_w) during which diffusion takes place. Due to the strong density difference between hydrogen and air, a defined vertical concentration gradient is achieved while horizontal concentration gradients remain negligible. Finally the mixture is ignited by a spark plug. For a more detailed description of the experimental setup, the procedure of hydrogen injection, and mixture generation it is referred to the publications of Vollmer et al. [29, 60].

In order to determine the hydrogen distribution before ignition for many different hydrogen/air ratios and different waiting times, numerical simulations of the injection process have been conducted. Exemplary results for local hydrogen mole fraction over channel height at a waiting time $t_w = 3\,\text{s}$ (the strongest gradient under investigation) are shown in Figure 5. For waiting times $t_w > 30\,\text{s}$ the mixture can be considered as homogeneous. The results of the injection simulations are stored as polynomials (hydrogen content versus channel height) which are used as initial conditions for the combustion simulations.

In the two-dimensional combustion simulations presented in the following section the channel is discretized with a uniform, rectangular grid of 2 mm grid spacing. Test runs showed that this resolution is the minimum resolution required to achieve grid independence with respect to the location of DDT. On coarser grids, DDT occurred mostly later or not at all. On finer grids, the location of DDT did not vary any more. However, at higher grid resolution, pressure peaks still got a little sharper. This should be kept in mind for the interpretation of the pressure plots shown in this paper.

Initially the fluid is at rest at a temperature of 293 K and a pressure of 1.01 bar. The boundary conditions are defined as adiabatic no-slip walls. Turbulence is modelled using the k-ω-SST model which is known for its good performance for both free-stream jets and wall-bounded flow [61, 62]. The initial hydrogen distribution either is homogeneous or corresponds to a concentration gradient of waiting time $t_w = 3\,\text{s}$ (see Figure 5).

Ignition is modelled by patching the site of ignition at $x = 0$ with a burned mixture ($c = 1$, see Figure 4). The initial turbulence is vanishingly small and consequently ξ

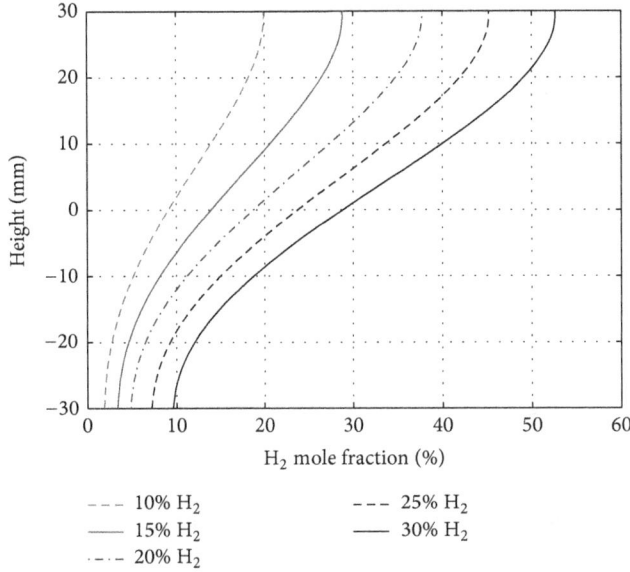

FIGURE 5: Local hydrogen content versus channel height after a waiting time of 3 s. The legend displays the overall hydrogen content of each mixture.

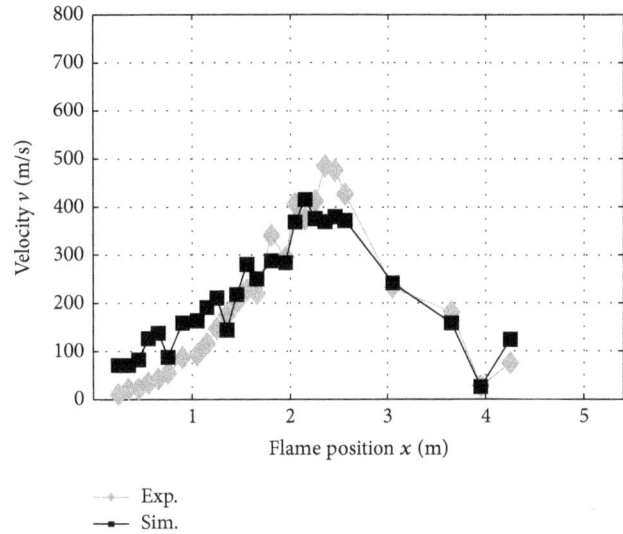

FIGURE 6: Flame propagation in a mixture with 15% H_2 (homogeneous).

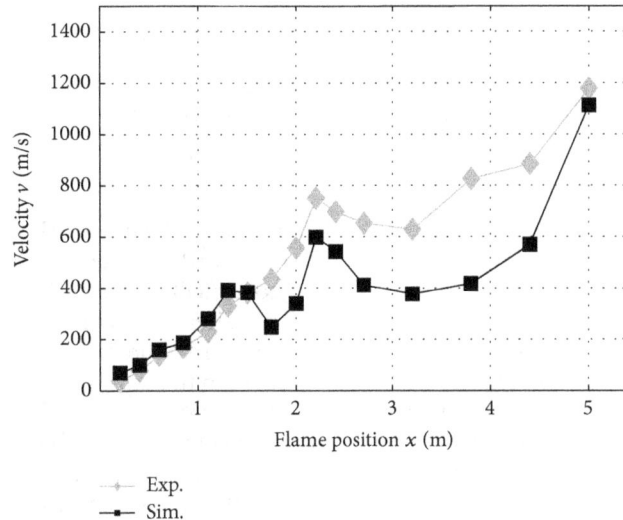

FIGURE 7: Flame propagation in a mixture with 15% H_2 (max. concentration gradient).

equals unity so that $s_T = s_L$ follows from (3). This means, the flame starts to propagate at laminar flame speed. However, turbulence is quickly generated by the flow itself so that the flame starts to accelerate. Test runs showed that the actual choice of initial turbulence variables is insignificant as long as $\xi = 1$ is ensured. As the HLLC scheme gets unstable in the incompressible limit where no coupling between pressure and density exists, the first few time steps are calculated with a pressure-based solver [37] using the PISO scheme [40]. Before the flame reaches the first obstacle, the combustion-driven flow is usually strong enough to switch to the HLLC scheme that enables better shock capturing. Test runs showed that the transition between both schemes is smooth if it occurs while the maximum Mach number in the flow is in the range of 0.05 < Ma < 0.10.

4. Results and Discussion

Experimental and numerical results for a homogeneous case with 15% hydrogen (volumetric) and blockage ratio BR = 30% are shown in Figure 6. It can be seen that the agreement between experiment and simulation is very good. The flame velocity rises continually in the obstructed part of the channel ($x \leq 2.05$ m). This can be attributed to the mutual amplification of combustion-induced expansion and turbulence generation due to interaction with obstacles. Shortly after passing the final obstacle the flame speed reaches a maximum and then decreases slowly. At $x \approx 4$ m the flame comes to a nearly complete rest before it accelerates again. This can be explained as follows: after passing the final obstacle, turbulence generation is diminished so that decelerating effects like friction outweigh the accelerating ones. The flame continuously gets slower. Simultaneously, while the flame has been consuming fresh gas, it generated

pressure waves and displaced the unburned gas into the positive x direction. Shocks were generated that propagated towards the end wall from where they are being reflected. These reflected shocks now generate fluid flow in negative x direction. When the leading, backwards-running shock reaches the flame (this happens at $x \approx 4$ m), negative flow velocity and positive burning speed nearly cancel out so that the resulting net propagation velocity approaches zero. However, as there is still unburned gas in front of the flame, it recovers and accelerates again. The maximum flame propagation velocity of approximately 500 m/s indicates that no DDT occurred and the combustion process remained entirely deflagrative.

Figure 7 shows the results for a mixture that contains an average hydrogen content of 15% as well, but with a

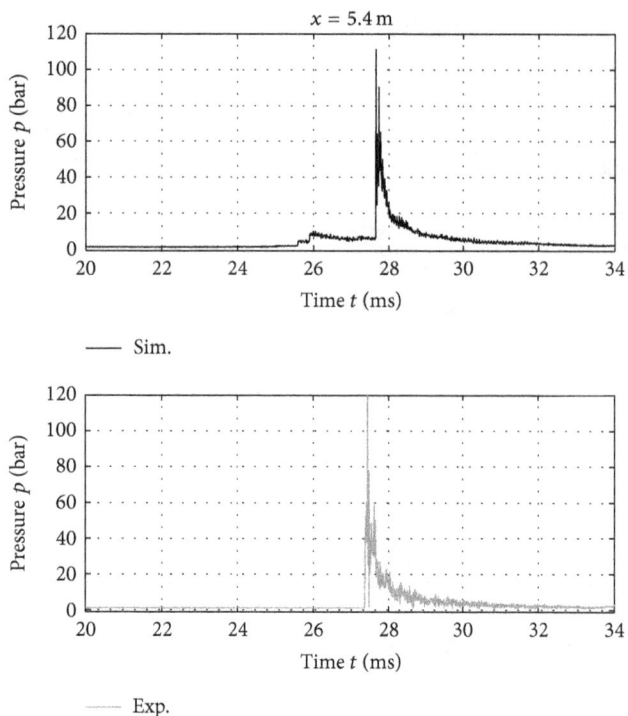

FIGURE 8: Pressure records from the sensor mounted on the end wall. Mixture with 15% H_2 (max. concentration gradient).

vertical concentration gradient as shown in Figure 5. All other parameters are kept identical. In the early acceleration phase the flame velocity increases continually. This is in good agreement with the experiment. Then the flame is decelerated for the first time, due to a first shock front reflected from the end wall. The difference in the experiment can be attributed to the different ignition process: the spark generated by the spark plug in the experiment is considerably smaller than the ignition patch used in the simulation which is limited by the grid resolution. Thus the initial pressure wave generation in the experiment might be a little different from the initial pressure rise caused by the ignition in the numerical simulation. At the end of the obstacle region the flame speed peaks and then loses some driving force but eventually recovers. Although there is a considerable velocity difference between experiment and simulation in the unobstructed part, the final velocity is nearly the same. The pressures recorded by the sensor in the end wall reach extremely high values close to 120 bar (see Figure 8). In the homogeneous case, for comparison, the maximum pressure is in the range of 10 bar. The reason for the extreme pressure rise in the inhomogeneous case is revealed in Figure 9 where the temperature and pressure distribution in the rear part of the explosion channel (4.9 m < x < 5.4 m) is displayed.

At t = 27.15 ms the flame approaches the end wall. Due to the inhomogeneous fuel distribution the flame is highly asymmetric and propagates mainly in the upper part of the channel. A leading shock has already been reflected from the end wall and moves towards the propagating flame. At t = 27.25 ms it reaches the flame. From this point onwards

the flame burns into a precompressed mixture where the heat release rate is increased due to the increased density and increased laminar burning velocity (see (2) and (7)). The increased reaction rate leads to a strong pressure rise and causes an explosion at t = 27.40 ms. A radial detonation wave emanates from the explosion center and ignites the gas over the whole channel height. The newly formed detonation front runs towards the end wall where it causes an enormous pressure rise. This DDT mechanism has been suggested as one possible explanation for the high pressure loads observed in the experimental work of Eder [63]. In Eder's work, high pressure loads on the end wall of an explosion channel have been observed, but the flame velocity measurements indicated only a fast deflagration, not a detonation. As in the present simulation, the DDT in Eder's experiments obviously occurred so late (behind the final photo diode) that the DDT was not identified as one; only the high pressures on the end wall gave rise to speculation. Recent experimental investigations of Boeck et al. [64] support the conclusion that a DDT mechanism as identified in Figure 9 is responsible for the high pressure peak.

Due to the limited spatial resolution the present simulation does not resolve the interaction with the boundary layer. Moreover, it does not capture the shock-flame interaction in such a detailed manner as previous numerical studies on highly resolved grids (e.g., [12, 13]). Nevertheless the model is able to correctly predict the consequence of the backwards-running shock hitting the flame: an increased reaction rate due to precompression and intensified mixing which consequently triggers DDT.

From the pressure records in Figure 8 it can be concluded that there is a slight difference between experiment and simulation: the initial pressure rise in the simulation at t ≈ 26 ms (caused by the reflection of the leading shock) does not appear in the experimental record. Due to the highly nonlinear dependence of ignition delay time on temperature and pressure, the higher propagation velocity in the experiment (Figure 7) is obviously sufficient to cause a strong autoignition quasi-instantaneously when the leading shock reaches the end wall. The resulting pressure load on the end wall, however, is nearly the same in experiment and simulation.

Increasing the hydrogen content leads to an earlier occurrence of DDT. At a hydrogen content of 25% (again with a concentration gradient as described in Figure 5) it can be seen from Figure 10 that the flame velocity rises continually to approximately 1000 m/s in the obstructed part of the channel and then suddenly jumps to 2500 m/s and finally relaxes to approximately 2000 m/s.

This is a clear indication for the occurrence of a DDT with an initially overdriven detonation decaying to a Chapman-Jouguet detonation. The large fluctuations in the experimental velocity after the onset of DDT can be explained by small measurement errors in flame arrival time having a relatively large effect when the derivative (19) is applied to the data. Using only the x-t diagram (Figure 11) as it is common in most publications does not reveal this difference.

The DDT process occurring in this case is visualized in Figure 12. At t = 12.44 ms, the flame approaches the final

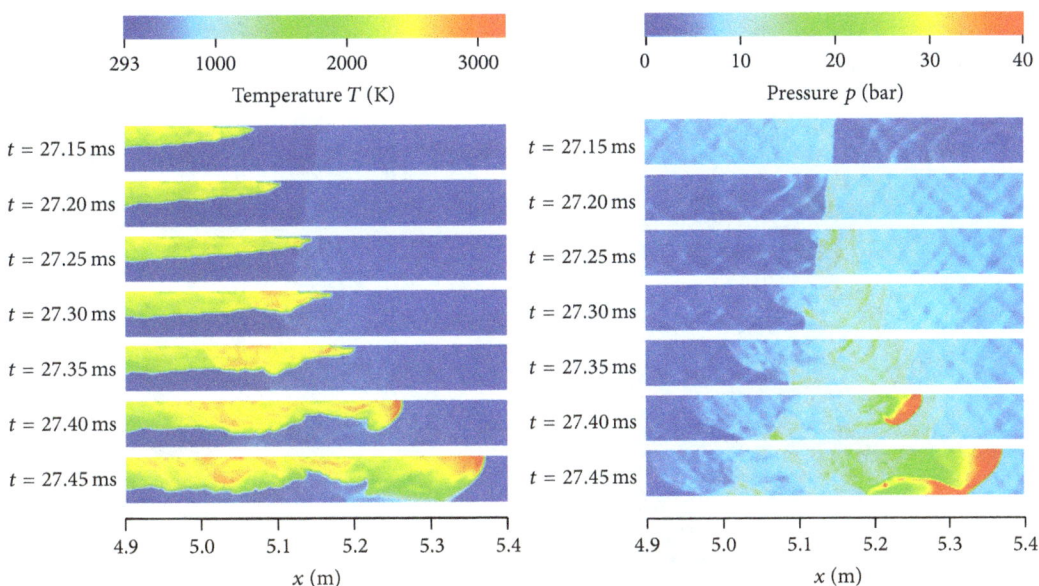

FIGURE 9: Visualization of a DDT caused by interaction of the flame with a reflected shock.

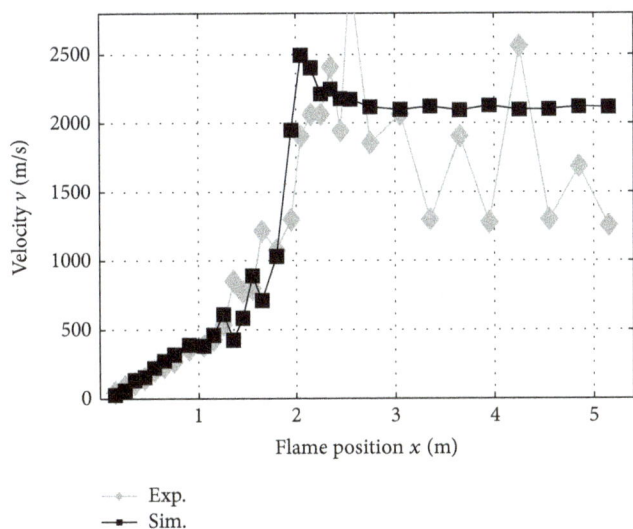

FIGURE 10: Flame velocity versus channel length for a mixture with 25% H_2 (max. concentration gradient).

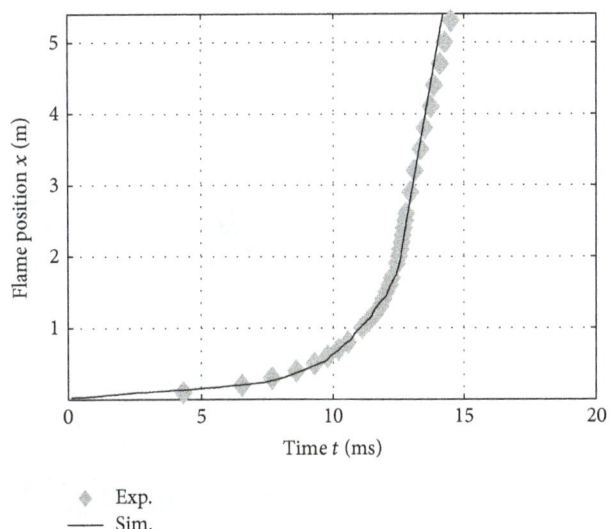

FIGURE 11: Flame position versus time for a mixture with 25% H_2 (max. concentration gradient).

obstacle. The curved shock in front of the flame is reflected from the bottom wall by forming a Mach stem. At $t = 12.45$ ms, autoignition occurs behind the Mach stem. At $t = 12.47$ ms, the oblique shock hits the upper obstacle which initiates a second autoignition event. From there a circular detonation emanates and unites with the autoignition front from the lower part of the channel. While the detonation front moves through the gap between the obstacles into the unburned gas ($t > 12.48$ ms), the opposite front of the reaction wave ("retonation wave" [65]) runs backwards and consumes the remaining fresh gas in the lower part of the channel. It is important to note that the two autoignition kernels in Figure 12 are both well ahead of the flame but occur

due to different reasons: the one on the bottom wall is due to shock compression ahead of the flame while the one on the upper wall occurs only due to reflection of the shock from the upper obstacle.

Another simulation with only six obstacles showed that the final obstacle was not necessary to achieve DDT. Instead, the autoignition occurring behind the Mach stem at $t = 12.45$ ms is sufficient to trigger DDT and is only amplified by the second autoignition event occurring on the upper obstacle. At lower fuel content (20% H_2) however, the seventh obstacle is required to obtain a DDT.

It is interesting to note that the first autoignition in Figure 12 occurs at the bottom wall where the mixture is

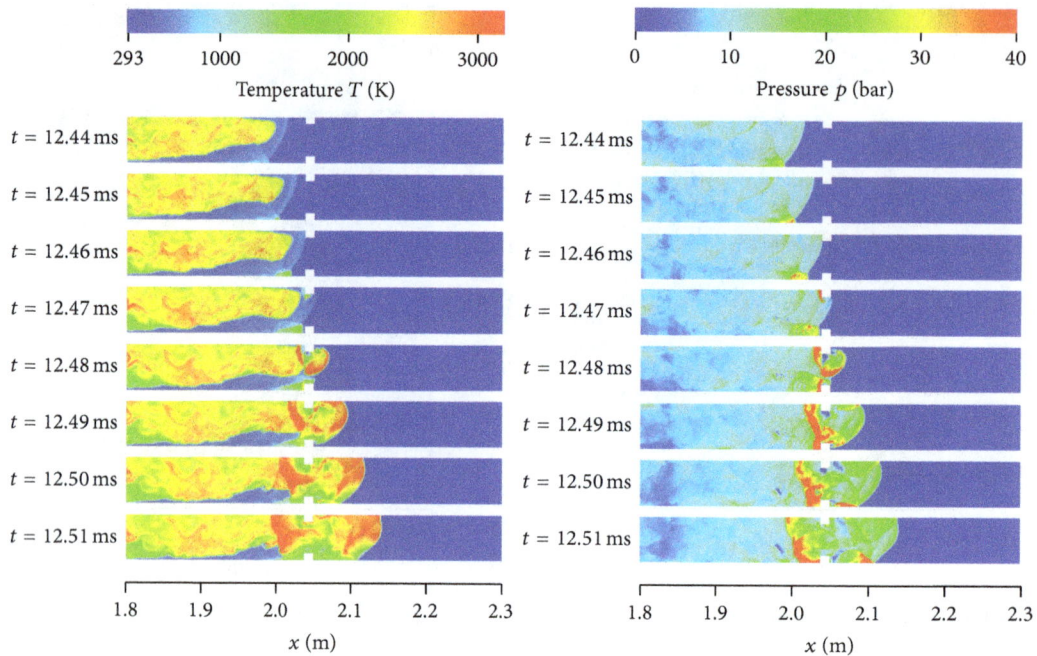

FIGURE 12: Visualization of a DDT in the vicinity of the final obstacle.

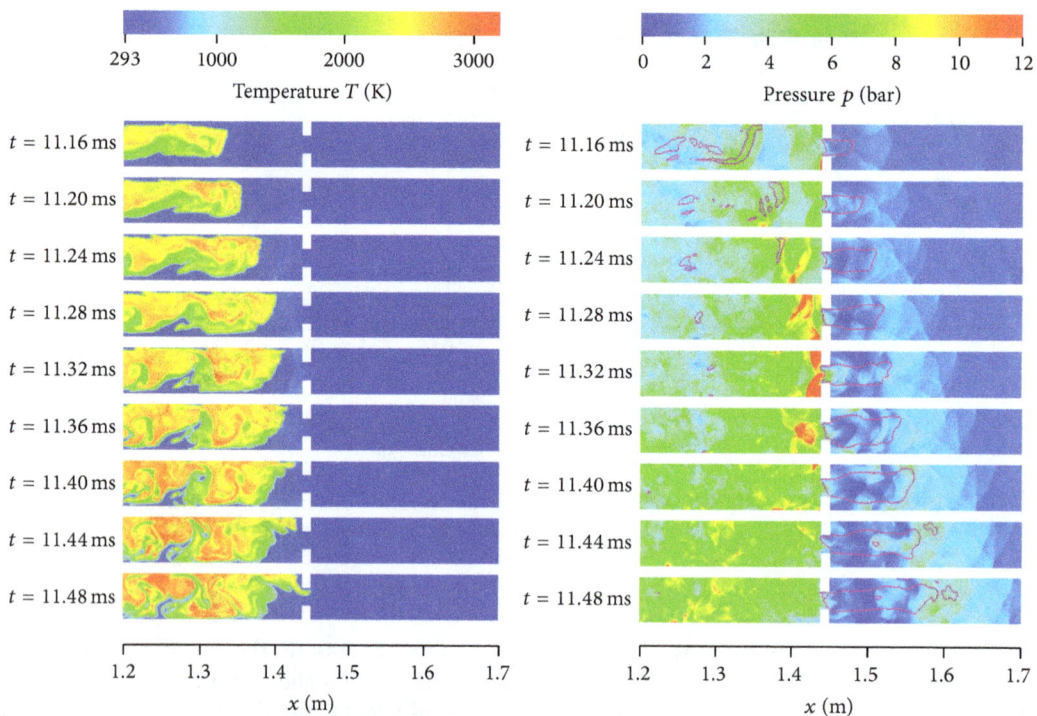

FIGURE 13: Shock-flame interaction at an obstacle of blockage ratio 60%. The pink contour shows the Ma = 1 line.

leanest. This phenomenon can be explained by taking a closer look at the shock propagation: the leading shock approaches the final obstacle at a constant speed of $v \approx 1450$ m/s. Near the bottom wall the hydrogen content is 7% (see Figure 5) which results in a local speed of sound of $a = 356$ m/s.

Thus, in the near vicinity of the bottom wall, the Mach stem can be seen as a normal shock propagating at Mach number Ma = v/a = 4.07. For fresh gas properties p_0 = 1.01 bar and T_0 = 293 K the normal shock relations (10) and (11) yield the postshock state p_1 = 19.4 bar and T_1 = 1220 K. Near the top

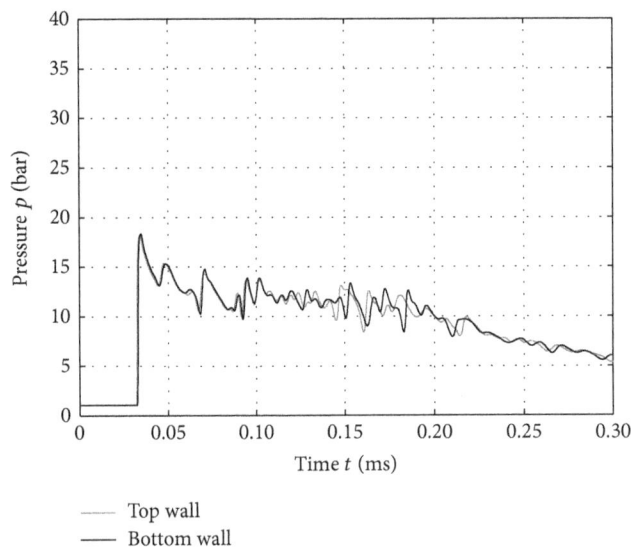

FIGURE 14: Pressure records from a steadily propagating detonation in a mixture with 25% H_2 (homogeneous).

wall where the mixture is rich (45%, see Figure 5) the local sound speed has a value of $a = 452$ m/s. This corresponds to a Mach number of Ma = 3.21 and yields a postshock state of $p_1 = 12.0$ bar and $T = 860$ K. The corresponding ignition delay time is by orders of magnitude higher than on the bottom wall. Thus, autoignition near the top wall can only be achieved via shock reflection from the upper obstacle.

Another important finding is that although DDT occurs much earlier in the case of 25% hydrogen, the pressure loads on the wall are higher in the case of 15% hydrogen. This is due to the different mode of DDT. At 25% hydrogen (Figure 12), the detonation has sufficient space to expand. At 15% hydrogen (Figure 9), shock and flame interact very close to the end wall and the overdriven detonation emerges from a preshocked mixture. As there is no space to expand, the overdriven detonation hits the end wall without losing much of its strength.

While both simulations and experiments in this configuration confirm the trend that (at equal average hydrogen content) a concentration gradient increases the DDT tendency, this must not be understood as a general conclusion. In other configurations the opposite effect can occur. For example, if the blockage ratio is increased from 30% to 60% it has been observed that the probability of DDT decreases. The reason for this phenomenon can be seen in Figure 13. At $t = 11.16$ ms, the flame approaches the obstacle. Some pressure waves have already been generated and passed the obstacle. The constriction formed by the obstacle acts like a Laval nozzle behind which an area of supersonic flow (visualized by the pink line representing the Ma = 1 contour) is generated. The obstacles cause the pressure waves to be reflected on 60% and to pass only on 40% of the cross-sectional area. Due to the gas dynamic constraint formed by the Ma = 1 line, the total mass flow through the orifice between the obstacles is limited. At $t = 11.16$ ms, a shock front is reflected from the obstacle which leads to a shock propagation into negative

x direction. This process is repeated with an even stronger shock between $t = 11.24$ ms and 11.32 ms. As the parts of the shock that are reflected from the upper and lower obstacle unite, a backwards-running shock over the whole channel height is formed. In the trailing flow behind this shock an area of negative flow velocity causes the approaching flame to decelerate, while the part of the shock that passed in between the obstacles to the right continues to propagate at nearly unaffected speed. Thus a high blockage ratio destroys the coupling between shock and flame which existed before approaching the obstacle. At $t = 11.48$ ms, the flame finally manages to enter the region of high flow speed that prevails between the obstacles. It is convected into the next cavity where strong shocks are generated and the process described repeats.

These numerical results explain the experimental observation by Vollmer et al. [29] that a concentration gradient can either increase or decrease the tendency towards DDT with the decisive factor being the obstacle geometry. If the obstacles are too large, they can lessen the DDT tendency as the majority of the strong pressure waves causing DDT are blocked. This is especially valid for an inhomogeneous mixture as shown in Figure 13, where the flame mainly burns in the upper part of the channel. In this case the obstacles are more obstructive than in a case with homogeneous mixture where the flame can be expected to propagate through the center of the channel where no obstacles are present.

Another question that has been addressed with the newly developed solver concerns the pressure loads that are caused by a steadily propagating detonation front, that is, after the occurrence of DDT. Therefore a look is taken at a detonation propagating in an unobstructed channel. First, this is demonstrated for a homogeneous mixture with 25% hydrogen. Pressure records are taken from the top and the bottom wall of the channel while a detonation passes. Test runs showed that the axial location of the pressure sensors did not influence the result any more as soon as a steadily propagating detonation was achieved. The result is shown in Figure 14. As the detonation front is nearly planar, the pressure records from the bottom and the top wall are virtually simultaneous. Upon arrival of the detonation front the pressure jumps to approximately 18 bar. The following expansion lets the pressure decrease slowly.

A completely different picture is found for a case with the same average hydrogen content, but a strong concentration gradient (Figure 15). As the leading shock is curved, it reaches the pressure sensors on the top wall earlier. They show maximum values of approximately 15 bar. On the bottom wall, however, a pressure of nearly 38 bar is reached. This is especially striking as the hydrogen content on the lower wall is only 7% and a homogeneous mixture with 7% hydrogen is basically nondetonable. Here, however, the lack of fuel does not lead to lower, but to higher, pressure loads. Again, the reason for this seeming paradox can be found in the particular structure of the leading shock front: on the bottom wall it is reflected via a Mach stem. Due to the lower speed of sound this causes a higher pressure rise on the bottom wall than on the top wall. After a short decline of the pressure, a second pressure rise is observed on both walls. This is due to secondary reflections of the leading shock that can be seen in the pressure field in Figure 16. Behind the secondary reflections the pressure equalizes so far that it drops simultaneously on the bottom and the top wall.

The results demonstrate that the pressure loads caused by a detonation in an inhomogeneous mixture can be considerably higher than in a homogeneous mixture of the same hydrogen content. Moreover, the location of the highest impact can be in fuel-lean regions. Further calculations showed that even if the hydrogen content on the bottom wall is reduced to zero, the maximum pressures observed there can still exceed those of the homogeneous mixture: the concentration gradient only needs to be strong enough to form a Mach stem. A simple method for predicting whether a detonation front in an inhomogeneous mixture develops a Mach stem can be found in [66].

5. Summary, Critical Analysis, and Outlook

Motivated by the current lack of suitable tools for DDT-related safety studies [32], this paper presented a newly developed solver able to simulate flame acceleration, deflagration-to-detonation transition, and detonation propagation within a single run. The target was not to obtain detailed insight and maximum accuracy of the complex interaction between flow and reaction on microscopic scale, but to obtain a tool for engineering purposes that works on comparatively coarse grids and enables numerical safety studies at acceptable computational costs. The applicability to coarse grids is achieved by the inclusion of subgrid models. The agreement with experimental results is very good and the simulation gives additional insight into phenomena which cannot be easily observed in experiments. Although the simulations presented do not resolve all details of the flow, they are able to capture fundamental phenomena known from highly resolved simulations and experiments (e.g., [12, 13, 36]): DDT due to shock compression/Mach stem formation ahead of the flame, DDT due to shock reflection from obstacles, and DDT due to shock-flame interaction.

It has been found that concentration gradients, which are likely to occur in accident scenarios, can have a considerable effect on the nature of flame propagation. Depending on the enclosing geometry, the presence of a concentration gradient can decrease or increase flame propagation velocities, the probability of DDT, and the pressure loads associated with it. Thus, existing safety criteria developed for homogeneous mixtures can be inaccurate and nonconservative. Neither does a homogeneous mixture pose the highest threat regarding the probability of DDT nor does it cause the highest pressure loads. Due to gas dynamic phenomena within inhomogeneous mixtures, fuel-lean regions can be more DDT-prone than stoichiometric or rich regions. This should be taken into account in future safety studies.

However, although the general agreement with experiments is good, it has to be kept in mind that all results have been gained on relatively coarse grids, without resolving the induction distance between shock and reaction in a

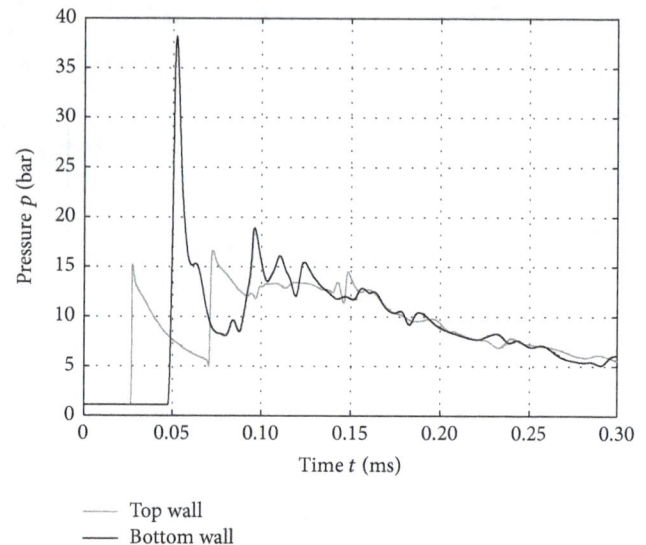

FIGURE 15: Pressure records from a steadily propagating detonation in a mixture with 25% H_2 (max. concentration gradient).

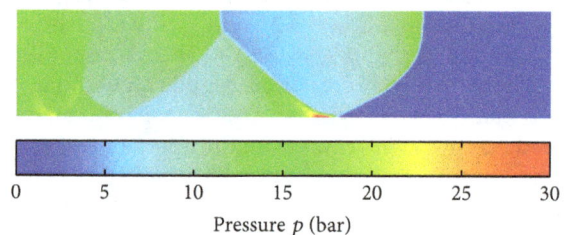

FIGURE 16: Pressure distribution in a steadily propagating detonation in a mixture with 25% H_2 (max. concentration gradient).

detonation front. Boundary layers are not resolved either and they are known to have an effect on the onset of detonations. Moreover, all the results presented in this study have been obtained on 2D grids. Therefore, the authors have deliberately chosen a geometry which is relatively wide (300 mm) compared to its height (60 mm). Nevertheless transversal waves are expected to play a role in flame acceleration and the onset of DDT. While this project has already finished, a follow-up project has started where 3D simulations are conducted [67] and also simulations in large, complex geometries with the aim of reproducing realistic accident scenarios. Approaches are being developed to use even coarser grids by applying subgrid models not only to shock propagation as shown in this paper, but also to deflagrative flame propagation.

The advantage of the solver developed and its implementation in *OpenFOAM* is that it is not limited to structured grids and thus can be applied to intricate geometries using unstructured grids as well. The solver and its source code are made freely available to the public [68].

Conflict of Interests

The authors declare that there is no conflict of interests regarding the publication of this paper.

Acknowledgments

This project is funded by the German Federal Ministry of Economics and Technology on the basis of a decision of the German Bundestag (Project no. 1501338) which is gratefully acknowledged. The authors would like to thank Oliver Borm for sharing a Riemann solver code within the framework of OpenFOAM which provided a valuable basis for this study.

References

[1] Y.-L. Liu, J.-Y. Zheng, P. Xu et al., "Numerical simulation on the diffusion of hydrogen due to high pressured storage tanks failure," *Journal of Loss Prevention in the Process Industries*, vol. 22, no. 3, pp. 265–270, 2009.

[2] B. P. Xu, J. X. Wen, S. Dembele, V. H. Y. Tam, and S. J. Hawksworth, "The effect of pressure boundary rupture rate on spontaneous ignition of pressurized hydrogen release," *Journal of Loss Prevention in the Process Industries*, vol. 22, no. 3, pp. 279–287, 2009.

[3] W. Breitung and P. Royl, "Procedure and tools for deterministic analysis and control of hydrogen behavior in severe accidents," *Nuclear Engineering and Design*, vol. 202, no. 2-3, pp. 249–268, 2000.

[4] M. Manninen, A. Silde, I. Lindholm, R. Huhtanen, and H. Sjövall, "Simulation of hydrogen deflagration and detonation in a BWR reactor building," *Nuclear Engineering and Design*, vol. 211, no. 1, pp. 27–50, 2002.

[5] H. Dimmelmeier, J. Eyink, and M.-A. Movahed, "Computational validation of the EPR combustible gas control system," *Nuclear Engineering and Design*, vol. 249, pp. 118–124, 2012.

[6] A. G. Venetsanos, D. Baraldi, P. Adams, P. S. Heggem, and H. Wilkening, "CFD modelling of hydrogen release, dispersion and combustion for automotive scenarios," *Journal of Loss Prevention in the Process Industries*, vol. 21, no. 2, pp. 162–184, 2008.

[7] W. G. Houf, G. H. Evans, E. Merilo, M. Groethe, and S. C. James, "Releases from hydrogen fuel-cell vehicles in tunnels," *International Journal of Hydrogen Energy*, vol. 37, no. 1, pp. 715–719, 2012.

[8] S. B. Dorofeev, "Flame acceleration and explosion safety applications," *Proceedings of the Combustion Institute*, vol. 33, no. 2, pp. 2161–2175, 2011.

[9] S. B. Margolis and B. J. Matkowsky, "Nonlinear stability and bifurcation in the transition from laminar to turbulent flame propagation," *Combustion Science and Technology*, vol. 34, no. 1-6, pp. 45–77, 1983.

[10] G. I. Sivashinsky, "Instabilities, pattern formation and turbulence in flames," *Annual Review of Fluid Mechanics*, vol. 15, pp. 179–199, 1983.

[11] E. S. Oran and A. M. Khokhlov, "Deflagrations, hot spots, and the transition to detonation," *Philosophical Transactions of the Royal Society A: Mathematical, Physical and Engineering Sciences*, vol. 358, no. 1764, pp. 3539–3551, 2000.

[12] E. S. Oran and V. N. Gamezo, "Origins of the deflagration-to-detonation transition in gas-phase combustion," *Combustion and Flame*, vol. 148, no. 1-2, pp. 4–47, 2007.

[13] V. N. Gamezo, T. Ogawa, and E. S. Oran, "Flame acceleration and DDT in channels with obstacles: effect of obstacle spacing," *Combustion and Flame*, vol. 155, no. 1-2, pp. 302–315, 2008.

[14] M. A. Liberman, A. D. Kiverin, and M. F. Ivanov, "On detonation initiation by a temperature gradient for a detailed chemical reaction models," *Physics Letters A: General, Atomic and Solid State Physics*, vol. 375, no. 17, pp. 1803–1808, 2011.

[15] M. C. Gwak and J. J. Yoh, "Effect of multi-bend geometry on deflagration to detonation transition of a hydrocarbon-air mixture in tubes," *International Journal of Hydrogen Energy*, vol. 38, no. 26, pp. 11446–11457, 2013.

[16] P. Hwang, R. P. Fedkiw, B. Merriman, T. D. Aslam, A. R. Karagozian, and S. J. Osher, "Numerical resolution of pulsating detonation waves," *Combustion Theory and Modelling*, vol. 4, no. 3, pp. 217–240, 2000.

[17] G. J. Sharpe, "Transverse waves in numerical simulations of cellular detonations," *Journal of Fluid Mechanics*, vol. 447, pp. 31–51, 2001.

[18] G. Cael, H. D. Ng, K. R. Bates, N. Nikiforakis, and M. Short, "Numerical simulation of detonation structures using a thermodynamically consistent and fully conservative reactive flow model for multi-component computations," *Proceedings of the Royal Society A: Mathematical, Physical and Engineering Sciences*, vol. 465, no. 2107, pp. 2135–2153, 2009.

[19] K. Mazaheri, Y. Mahmoudi, and M. I. Radulescu, "Diffusion and hydrodynamic instabilities in gaseous detonations," *Combustion and Flame*, vol. 159, no. 6, pp. 2138–2154, 2012.

[20] J. M. Powers and S. Paolucci, "Accurate spatial resolution estimates for reactive supersonic flow with detailed chemistry," *AIAA Journal*, vol. 43, no. 5, pp. 1088–1099, 2005.

[21] M. Manninen, R. Huhtanen, I. Lindholm, and H. Sjövall, "Hydrogen in BWR reactor building," in *Proceedings of the 8th International Conference on Nuclear Engineering (ICONE '00)*, Baltimore, Md, USA, 2000.

[22] D. M. Prabhudharwadkar, K. N. Iyer, N. Mohan, S. S. Bajaj, and S. G. Markandeya, "Simulation of hydrogen distribution in an Indian Nuclear Reactor Containment," *Nuclear Engineering and Design*, vol. 241, no. 3, pp. 832–842, 2011.

[23] S. B. Dorofeev, A. S. Kochurko, A. A. Efimenko, and B. B. Chaivanov, "Evaluation of the hydrogen explosion hazard," *Nuclear Engineering and Design*, vol. 148, no. 2-3, pp. 305–316, 1994.

[24] S. B. Dorofeev, M. S. Kuznetsov, V. I. Alekseev, A. A. Efimenko, and W. Breitung, "Evaluation of limits for effective flame acceleration in hydrogen mixtures," *Journal of Loss Prevention in the Process Industries*, vol. 14, no. 6, pp. 583–589, 2001.

[25] A. Eder and N. Brehm, "Analytical and experimental insights into fast deflagrations, detonations, and the deflagration-to-detonation transition process," *Heat and Mass Transfer*, vol. 37, no. 6, pp. 543–548, 2001.

[26] J. Chao and J. H. S. Lee, "The propagation mechanism of high speed turbulent deflagrations," *Shock Waves*, vol. 12, no. 4, pp. 277–289, 2003.

[27] M. Kuznetsov, V. Alekseev, I. Matsukov, and S. Dorofeev, "DDT in a smooth tube filled with a hydrogen-oxygen mixture," *Shock Waves*, vol. 14, no. 3, pp. 205–215, 2005.

[28] C. R. L. Bauwens, L. Bauwens, and I. Wierzba, "Accelerating flames in tubes—an analysis," *Proceedings of the Combustion Institute*, vol. 31, pp. 2381–2388, 2007.

[29] K. G. Vollmer, F. Ettner, and T. Sattelmayer, "Deflagration-to-detonation transition in hydrogen/air mixtures with a concentration gradient," *Combustion Science and Technology*, vol. 184, pp. 1903–1915, 2012.

[30] O. Auban, R. Zboray, and D. Paladino, "Investigation of large-scale gas mixing and stratification phenomena related to LWR containment studies in the PANDA facility," *Nuclear Engineering and Design*, vol. 237, no. 4, pp. 409–419, 2007.

[31] D. C. Visser, M. Houkema, N. B. Siccama, and E. M. J. Komen, "Validation of a FLUENT CFD model for hydrogen distribution in a containment," *Nuclear Engineering and Design*, vol. 245, pp. 161–171, 2012.

[32] W. Breitung, C. K. Chan, S. Dorofeev et al., "Flame acceleration and deflagration-to-detonation transition in nuclear safety," Tech. Rep. NEA/CSNI/R(2000)7, OECD State-of-the-Art Report by a Group of Experts, 2000.

[33] F. Ettner, K. G. Vollmer, and T. Sattelmayer, "Numerical investigation of DDT in inhomogeneous hydrogen-air mixtures," in *Proceedings of the 8th International Symposium on Hazards, Prevention and Mitigation of Industrial Explosions*, Yokohama, Japan, 2010.

[34] D. A. Kessler, V. N. Gamezo, and E. S. Oran, "Gas-phase detonation propagation in mixture omposition gradients," *Philosophical Transactions of the Royal Society A: Mathematical, Physical and Engineering Sciences*, vol. 370, no. 1960, pp. 567–596, 2012.

[35] H. Soury and K. Mazaheri, "Utilizing unsteady curved detonation analysis and detailed kinetics to study the direct initiation of detonation in H_2-O_2 and H_2-Air mixtures," *International Journal of Hydrogen Energy*, vol. 34, no. 24, pp. 9847–9856, 2009.

[36] G. Thomas, "Some observations on the initiation and onset of detonation," *Philosophical Transactions of the Royal Society A: Mathematical, Physical and Engineering Sciences*, vol. 370, no. 1960, pp. 715–739, 2012.

[37] OpenFOAM, "User guide, version 2.1.1," 2012, http://www.openfoam.org.

[38] E. F. Toro, M. Spruce, and W. Speares, "Restoration of the contact surface in the HLL-Riemann solver," *Shock Waves*, vol. 4, no. 1, pp. 25–34, 1994.

[39] OpenFOAM Wiki, "Limiters," 2010, http://openfoamwiki.net/index.php/OpenFOAM_guide/Limiters.

[40] R. I. Issa, "Solution of the implicitly discretised fluid flow equations by operator-splitting," *Journal of Computational Physics*, vol. 62, no. 1, pp. 40–65, 1986.

[41] R. J. Kee, F. M. Rupley, J. A. Miller et al., "The chemkin thermodynamic database, chemkin Collection, release 3.6," 2000, http://www.sandia.gov/chemkin.

[42] R. B. Bird, W. E. Stewart, and E. N. Lightfoot, *Transport Phenomena*, Wiley, New York, NY, USA, 2001.

[43] K. N. C. Bray and J. B. Moss, "A unified statistical model of the premixed turbulent flame," *Acta Astronautica*, vol. 4, no. 3-4, pp. 291–319, 1977.

[44] A. Favre, "Equations des gaz turbulents compressibles," *Journal de Mécanique*, vol. 4, pp. 361–390.

[45] C. Chen, J. J. Riley, and P. A. McMurtry, "A study of Favre averaging in turbulent flows with chemical reaction," *Combustion and Flame*, vol. 87, no. 3-4, pp. 257–277, 1991.

[46] M. Brandt, W. Polifke, B. Ivancic, P. Flohr, and B. Paikert, "Auto-ignition in a gas turbine burner at elevated temperature," in *Proceedings of the ASME Turbo Expo*, pp. 195–205, Atlanta, Ga, USA,, June 2003.

[47] O. Colin, A. Pires da Cruz, and S. Jay, "Detailed chemistry-based auto-ignition model including low temperature phenomena applied to 3-D engine calculations," *Proceedings of the Combustion Institute*, vol. 30, no. 2, pp. 2649–2656, 2005.

[48] J.-B. Michel, O. Colin, and C. Angelberger, "On the formulation of species reaction rates in the context of multi-species CFD codes using complex chemistry tabulation techniques," *Combustion and Flame*, vol. 157, no. 4, pp. 701–714, 2010.

[49] H. G. Weller, G. Tabor, A. D. Gosman, and C. Fureby, "Application of a flame-wrinkling LES combustion model to a turbulent mixing layer," *Symposium (International) on Combustion*, vol. 27, pp. 899–907, 1998.

[50] K. N. C. Bray, "Complex chemical reaction systems," in *Chapter Methods of Including Realistic Chemical Reaction Mechanisms in Turbulent Combustion Models*, vol. 47 of *Springer Series in Chemical Physics*, pp. 356–375, Springer, 1986.

[51] V. L. Zimont and A. N. Lipatnikov, "A numerical model of premixed turbulent combustion of premixed gases," *Chemical Physics Reports*, vol. 14, pp. 993–1025, 1995.

[52] F. Ettner, *Effiziente numerische simulation des deflagrations-detonations-Übergangs [Ph.D. thesis]*, TU München, 2013.

[53] A. A. Konnov, "Remaining uncertainties in the kinetic mechanism of hydrogen combustion," *Combustion and Flame*, vol. 152, no. 4, pp. 507–528, 2008.

[54] W. Polifke, P. Flohr, and M. Brandt, "Modeling of inhomogeneously premixed combustion with an extended TFC model," *Journal of Engineering for Gas Turbines and Power*, vol. 124, no. 1, pp. 58–65, 2002.

[55] S. R. Turns, *An Introduction to Combustion*, McGraw-Hill, 2000.

[56] D. Goodwin, "Cantera: an object-oriented software toolkit for chemical kinetics, thermodynamics and transport processes," 2009, http://code.google.com/p/cantera.

[57] M. Ó Conaire, H. J. Curran, J. M. Simmie, W. J. Pitz, and C. K. Westbrook, "A comprehensive modeling study of hydrogen oxidation," *International Journal of Chemical Kinetics*, vol. 36, no. 11, pp. 603–622, 2004.

[58] L. Tosatto and L. Vigevano, "Numerical solution of under-resolved detonations," *Journal of Computational Physics*, vol. 227, no. 4, pp. 2317–2343, 2008.

[59] J. D. Anderson, *Modern Compressible Flow*, McGraw-Hill, 2004.

[60] K. G. Vollmer, F. Ettner, and T. Sattelmayer, "Influence of concentration gradients on flame acceleration in tubes," in *Proceedings of the 8th International Symposium on Hazards, Prevention and Mitigation of Industrial Explosions*, Yokohama, Japan, 2010.

[61] F. R. Menter, "Two-equation eddy-viscosity turbulence models for engineering applications," *AIAA Journal*, vol. 32, no. 8, pp. 1598–1605, 1994.

[62] F. R. Menter, "Review of the shear-stress transport turbulence model experience from an industrial perspective," *International Journal of Computational Fluid Dynamics*, vol. 23, no. 4, pp. 305–316, 2009.

[63] A. Eder, *Brennverhalten schallnaher und überschall-schneller Wasserstoff-Luft Flammen [Ph.D. thesis]*, TU München, 2001.

[64] L. R. Boeck, J. Hasslberger, F. Ettner, and T. Sattelmayer, "Investigation of peak pressures during explosive combustion of inhomogeneous hydrogen-air mixtures," in *Proceedings of the 7th International Fire and Explosion Hazards Seminar*, Providence, RI, USA, 2013.

[65] A. K. Oppenheim, A. J. Laderman, and P. A. Urtiew, "The onset of retonation," *Combustion and Flame*, vol. 6, pp. 193–197, 1962.

[66] F. Ettner, K. G. Vollmer, and T. Sattelmayer, "Mach reflection in detonations propagating through a gas with a concentration gradient," *Shock Waves*, vol. 23, pp. 201–206, 2013.

[67] J. Hasslberger, F. Ettner, L. R. Boeck, and T. Sattelmayer, "2D and 3D flame surface analysis of flame acceleration and deflagration-to-detonation transition in hydrogenair mixtures with concentration gradients," in *Proceedings of the 24th International Conference on the Dynamics of Explosions and Reactive Systems (ICDERS '13)*, Taipei, Taiwan, 2013.

[68] F. Ettner and T. Sattelmayer, ddtFoam, 2013, http://sourceforge.net/projects/ddtfoam.

4

Local Strain Rate and Curvature Dependences of Scalar Dissipation Rate Transport in Turbulent Premixed Flames: A Direct Numerical Simulation Analysis

Y. Gao,[1] N. Chakraborty,[1] and N. Swaminathan[2]

[1] *School of Mechanical and Systems Engineering, Newcastle University, Claremont Road, Newcastle-Upon-Tyne NE1 7RU, UK*
[2] *Cambridge University Engineering Department, Trumpington Street, Cambridge CB2 1PZ, UK*

Correspondence should be addressed to N. Chakraborty; nilanjan.chakraborty@newcastle.ac.uk

Academic Editor: Michael Fairweather

The statistical behaviours of the instantaneous scalar dissipation rate N_c of reaction progress variable c in turbulent premixed flames have been analysed based on three-dimensional direct numerical simulation data of freely propagating statistically planar flame and V-flame configurations with different turbulent Reynolds number Re_t. The statistical behaviours of N_c and different terms of its transport equation for planar and V-flames are found to be qualitatively similar. The mean contribution of the density-variation term T_1 is positive, whereas the molecular dissipation term $(-D_2)$ acts as a leading order sink. The mean contribution of the strain rate term T_2 is predominantly negative for the cases considered here. The mean reaction rate contribution T_3 is positive (negative) towards the unburned (burned) gas side of the flame, whereas the mean contribution of the diffusivity gradient term (D) assumes negative (positive) values towards the unburned (burned) gas side. The local statistical behaviours of N_c, T_1, T_2, T_3, $(-D_2)$, and $f(D)$ have been analysed in terms of their marginal probability density functions (pdfs) and their joint pdfs with local tangential strain rate a_T and curvature k_m. Detailed physical explanations have been provided for the observed behaviour.

1. Introduction

Scalar dissipation rate (SDR) plays a pivotal role in turbulent reacting flows [1, 2] and thus its statistical behaviour is of fundamental importance to the modelling of turbulent premixed combustion. In turbulent premixed combustion the mean/filtered reaction rate $\overline{\dot{w}}$ of a reaction progress variable c is directly related to the Favre mean/filtered value of SDR $\widetilde{N}_c = \overline{\rho D \nabla c \cdot \nabla c}/\overline{\rho}$ [1–4], where ρ is the fluid density and D is the progress variable diffusivity with the overbar indicating a Reynolds averaging/large eddy simulation (LES) filtering process as applicable. It is well known that strain rate and curvature can significantly affect the local flame propagation behaviour and $|\nabla c|$ statistics in turbulent premixed flames [5–16]. Thus, strain rate and curvature are expected to have appreciable influences on local statistics of SDR N_c and its

transport. The transport equation of the instantaneous SDR of reaction progress variable N_c is given as [3, 17]

$$
\frac{\partial (\rho N_c)}{\partial t} + \frac{\partial (\rho u_j N_c)}{\partial x_j}
$$

$$
= \frac{\partial}{\partial x_j}\left(\rho D \frac{\partial N_c}{\partial x_j} \right) - \frac{2D}{\rho}\frac{\partial \rho}{\partial x_j}\frac{\partial c}{\partial x_j}\left[\dot{w} + \nabla \cdot (\rho D \nabla c) \right]
$$

$$
- 2\rho D \frac{\partial c}{\partial x_i}\frac{\partial u_i}{\partial x_j}\frac{\partial c}{\partial x_j} + 2D \frac{\partial \dot{w}}{\partial x_j}\frac{\partial c}{\partial x_j}
$$

$$
- 2\rho D^2 \frac{\partial^2 c}{\partial x_i \partial x_j}\frac{\partial^2 c}{\partial x_i \partial x_j} + f(D),
$$

(1a)

where

$$f(D) = \underbrace{2D\frac{\partial c}{\partial x_k}\frac{\partial(\rho D)}{\partial x_k}\frac{\partial^2 c}{\partial x_j \partial x_j}}_{T_{D1}}$$

$$+ \underbrace{2D\frac{\partial c}{\partial x_k}\frac{\partial^2(\rho D)}{\partial x_j \partial x_k}\frac{\partial c}{\partial x_j}}_{T_{D2}} \underbrace{-\frac{\partial}{\partial x_j}\left(\rho N_c \frac{\partial D}{\partial x_j}\right)}_{T_{D3}}$$

$$\underbrace{-2\rho D\frac{\partial D}{\partial x_j}\frac{\partial(\nabla c \cdot \nabla c)}{\partial x_j}}_{T_{D4}} + \underbrace{\rho \nabla c \cdot \nabla c \left[\frac{\partial D}{\partial t} + u_j\frac{\partial D}{\partial x_j}\right]}_{T_{D5}}.$$

(1b)

The first two terms on the left hand side of (1a) represent the transient and advection effects, whereas the first term on the right hand side (i.e., $D_1 = \nabla \cdot (\rho D \nabla N_c)$) denotes molecular diffusion of SDR. The second term on the right hand side of (1a) (i.e., $T_1 = -2D\nabla\rho \cdot \nabla c[\dot{w} + \nabla \cdot (\rho D \nabla c)]/\rho$) originates due to density variation and will henceforth be referred to as the density variation term. The third term on the right hand side of (1a) (i.e., $T_2 = -2\rho D(\partial c/\partial x_i)(\partial u_i/\partial x_j)(\partial c/\partial x_j)$) represents the effects of fluid-dynamic straining, whereas the fourth term (i.e., $T_3 = 2D(\partial\dot{w}/\partial x_i)(\partial c/\partial x_i)$) denotes the reaction rate contribution to the SDR transport. The penultimate term on the right hand side of (1a) (i.e., $-D_2 = -2\rho D^2(\partial^2 c/\partial x_i \partial x_j)(\partial^2 c/\partial x_i \partial x_j)$) denotes molecular dissipation of N_c, and the terms involving temporal and spatial gradients of diffusivity are collectively referred to as $f(D)$ (see (1b)).

Although the statistical behaviours of $|\nabla c|$ and the terms of its transport equation were analysed earlier, the terms of N_c transport equation are fundamentally different from the terms of the $|\nabla c|$ transport equation, which can be written for a given c isosurface in the following manner [11, 13–16]:

$$\frac{\partial|\nabla c|}{\partial t} + \frac{\partial\left(u_j|\nabla c|\right)}{\partial x_j}$$

(2)

$$= \left(\delta_{ij} - n_i n_j\right)\frac{\partial u_i}{\partial x_j}|\nabla c| + S_d\frac{\partial n_i}{\partial x_i} - \frac{\partial\left(S_d n_i |\nabla c|\right)}{\partial x_i},$$

where $n_i = -(\partial c/\partial x_i)/|\nabla c|$ is the ith component of flame normal vector and $S_d = (\dot{w} + \nabla \cdot (\rho D \nabla c))/\rho|\nabla c|$ is the local flame displacement speed. It is evident from (1a) and (1b) and (2) that the statistical behaviour of N_c transport is likely to be different from $|\nabla c|$ transport although the quantities N_c and $|\nabla c|$ are closely related to each other (i.e., $N_c = D|\nabla c|^2$).

It is often necessary to solve a transport equation for $\widetilde{N}_c$ in the context of Reynolds averaged Navier-Stokes (RANS) simulations and LES [17–30]. The transport equation for $\widetilde{N}_c$

can be obtained by Reynolds averaging or LES filtering of (1a) and (1b) as

$$\frac{\partial\left(\bar{\rho}\widetilde{N}_c\right)}{\partial t} + \frac{\partial\left(\bar{\rho}\tilde{u}_j\widetilde{N}_c\right)}{\partial x_j}$$

$$= \frac{\partial}{\partial x_j}\left(\overline{\rho D\frac{\partial N_c}{\partial x_j}}\right) - \frac{\partial\left[\overline{\rho u_i N_c} - \bar{\rho}\tilde{u}_i\widetilde{N}_c\right]}{\partial x_i} \quad (3)$$

$$+ \overline{T}_1 + \overline{T}_2 + \overline{T}_3 - \overline{D}_2 + \overline{f(D)}.$$

The terms $\overline{T}_1$, $\overline{T}_2$, $\overline{T}_3$, $(-\overline{D}_2)$, and $\overline{f(D)}$ are unclosed and therefore it is important to understand the statistical behaviours of N_c, T_1, T_2, T_3, $(-D_2)$, and $f(D)$ (since $\lim_{\Delta\to 0}\widetilde{N}_c = N_c$, $\lim_{\Delta\to 0}\overline{T}_1 = T_1$, $\lim_{\Delta\to 0}\overline{T}_2 = T_2$, $\lim_{\Delta\to 0}\overline{T}_3 = T_3$, $\lim_{\Delta\to 0}(-\overline{D}_2) = (-D_2)$, and $\lim_{\Delta\to 0}\overline{f(D)} = f(D)$, where Δ is the LES filter width) and their local strain rate and curvature dependences in order to model these quantities in the context of LES, where the local strain rate and curvature dependences of these terms need to be adequately captured. The local strain rate and curvature dependences of N_c and the terms of its transport equation (i.e., T_1, T_2, T_3, and $(-D_2)$) are yet to be analysed in detail in the existing literature. This paper aims to address this gap by analysing local tangential strain rate $a_T = (\delta_{ij} - n_i n_j)\partial u_i/\partial x_j$ and curvature $\kappa_m = 0.5(\partial n_i/\partial x_i)$ (for the above definition of κ_m, the flame elements convex towards the reactants has a positive curvature) dependences of N_c, T_1, T_2, T_3, $(-D_2)$, and $f(D)$ at different locations within the flame using direct numerical simulations (DNS) data of turbulent premixed freely propagating statistically planar flame and turbulent V-flame configurations. In this respect, the main objectives of this study are as follows:

(1) to analyse local statistical behaviours of instantaneous SDR (i.e., N_c) and the terms of its transport equation T_1, T_2, T_3, $(-D_2)$, and $f(D)$;

(2) to explain the observed strain rate and curvature dependences of N_c, T_1, T_2, T_3, $(-D_2)$, and $f(D)$;

(3) to compare the statistical behaviours of instantaneous SDR and the terms of its transport equation obtained from DNS in a canonical configuration with constant thermophysical properties with DNS of a laboratory configuration (e.g., turbulent V-flame configuration) with temperature-dependent thermophysical properties.

The rest of the paper will be organised as follows. The necessary mathematical modelling and the information related to the numerical implementation of DNS simulations will be presented in the next section. This will be followed by the presentation of the results and the subsequent discussion. The main findings will be summarised and conclusions will be drawn in the final section of this paper.

2. Mathematical Background and Numerical Implementation

DNS simulations of turbulent reacting flows should address both the three-dimensionality of turbulence and detailed chemical structure of the flames. However, limitation of computer hardware until recently restricted DNS of turbulent reacting flows either to two dimensions with detailed chemistry or to three dimensions with simplified chemistry. Although it is now possible to carry out three-dimensional DNS simulations with detailed chemistry, they remain extremely expensive [31] and are often not suitable for a detailed parametric analysis especially for simulations in relatively complex configurations (e.g., V-flame). Here, three-dimensional simulations with single step Arrhenius type chemistry have been considered for an extensive parametric analysis. The parametric analysis based on freely propagating statistically planar flames in a canonical configuration has been carried out using a well-proven compressible DNS code SENGA [32]. In the context of simple chemistry, the species field is uniquely represented by a reaction progress variable c, which can be defined in terms of a suitable reactant (product) mass fraction $Y_R(Y_P)$ as $c = (Y_{R0} - Y_R)/(Y_{R0} - Y_{R\infty})(c = (Y_P - Y_{P0})/(Y_{P\infty} - Y_{P0})$, where the subscripts 0 and ∞ are used to denote the values in unburned reactants and fully burned products, respectively. For the simulations of freely propagating statistically planar flames (i.e., cases P1–P5, where "P" denotes the statistically planar flames), a rectangular domain of size $36.1\delta_{th} \times 24.1\delta_{th} \times 24\delta_{th}$ is considered, where $\delta_{th} = (T_{ad} - T_0)/\text{Max}|\nabla\widehat{T}|_L$ is the thermal flame thickness with T_{ad}, T_0, and $\widehat{T}$ being the adiabatic flame, unburned reactant, and instantaneous dimensional temperatures, respectively, and the subscript "L" refers to the unstrained laminar flame quantities. For the thermochemistry used in cases P1–P5, the thermal flame thickness δ_{th} is found to be $1.785D_0/S_L$ (i.e., $\delta_{th} = 1.785D_0/S_L$), where D_0 is the mass diffusivity in the unburned gas.

The simulation domain for cases P1–P5 is discretised using a uniform Cartesian grid of $345 \times 230 \times 230$. The largest side of the domain is taken to align with the mean direction of flame propagation and the boundaries in that direction are taken to be partially nonreflecting. The partially nonreflecting boundary conditions are specified using the Navier-Stokes characteristic boundary conditions (NSCBC) technique [33]. The transverse directions are taken to be periodic and thus do not need any separate boundary conditions. A 10th order central-difference scheme is used to evaluate spatial derivatives at the internal grid points but the order of differentiation gradually drops to a one-sided 4th order scheme near nonperiodic boundaries. The time-advancement is carried out using a 3rd order low storage Runge-Kutta scheme [34]. One does not obtain any spurious fluctuations due to the 10th order central difference scheme and its transition to the lower-order finite difference scheme for sufficiently small grid spacing (e.g., $\Delta x \leq \eta$, where Δx and η are the grid spacing and the Kolmogorov length scale, respectively). Thus it was not necessary to use numerical filter to eliminate spurious oscillations. The flames in cases P1–P5 remain sufficiently away from the domain boundaries whereas the major part of the reactive region in cases V1–V3 does not interact with the nonperiodic boundaries except for flame crossing the outlet boundary. For the present analysis, the regions of flame crossing nonperiodic boundary are not considered for extracting SDR statistics in cases V1–V3. Thus, the evaluation of SDR and the terms of its transport equation at a given point of time is nominally 10th order accurate in this analysis. It is worth noting that similar numerical schemes for spatial discretisation and time advancement were used in several previous studies [4–17, 22–32].

The initial values of root-mean-square turbulent velocity fluctuation normalised by unstrained laminar burning velocity u'/S_L, integral length scale to flame thickness ratio l/δ_{th}, turbulent Reynolds number $Re_t = \rho_0 u'l/\mu_0$, Damköhler number Da $= lS_L/u'\delta_{th}$ and Karlovitz number Ka $= (u'/S_L)^{3/2}(l/\delta_{th})^{-1/2}$, heat release parameter $\tau = (T_{ad} - T_0)/T_0$, and Zel'dovich number $\beta = T_{ac}(T_{ad} - T_0)/T_{ad}^2$ for cases P1–P5 are provided in Table 1, where ρ_0 and μ_0 are the unburned gas density and viscosity, respectively, and T_{ac} is the activation temperature. As Re_t scales as $Re_t \sim \text{Da}^2\text{Ka}^2$ [35], the change in turbulent Reynolds number in cases P1–P5 is brought about by modifying Da and Ka independently of each other (e.g., Da (Ka) is kept unaltered in cases P1, P3, and P5 (P2, P3, and P4)). In cases P1–P5, the flame-turbulence interaction takes place under decaying turbulence, which necessitates a simulation time $t_{sim} \geq \text{Max}(t_f, t_c)$, where $t_f = l/u'$ is the initial eddy turn over time and $t_c = \delta_{th}/S_L$ is the chemical time scale. In all cases, statistics were extracted after one chemical time scale t_c, which corresponds to a time equal to $2.0t_f$ in case P4, $3.0t_f$ in cases P1, P3, and P5, and $4.34t_f$ for case P2. It is worth noting that the chemical time scale t_c remains the same for all cases due to identical thermochemistry. The present simulation time is comparable to the simulation times used for several previous DNS studies [5–9, 12, 36–39]. The global level of turbulent velocity fluctuation had decayed by 52.66%, 61.11%, 45%, 24%, and 34% in comparison to the initial values for cases P1–P5, respectively. By contrast, the integral length scale increased by factors between 1.5 and 2.25, ensuring that sufficient numbers of turbulent eddies were retained in each direction to obtain useful statistics. The values for u'/S_L, l/δ_{th}, and δ_{th}/η at the time when statistics were extracted have been presented elsewhere [39] and thus are not repeated here. For cases P1–P5, the thermal flame thickness δ_{th} is greater than the Kolmogorov length scale η at the time of the analysis, and this suggests that combustion in these cases takes place in the thin reaction zones regime [35]. The temporal evolutions of turbulent kinetic energy evaluated over the whole domain and the global burning rate were shown in [39], which demonstrate that these quantities were not varying rapidly with time when the statistics were extracted. It was also shown in [39] that the flame propagation statistics remain unchanged halfway through the simulation.

The V-flame cases (i.e., cases V1, V2, and V3, where "V" denotes V-shape flames here) are simulated using an updated version of SENGA and SENGA2 [40, 41] with ability to handle complex chemistry. However, the V-flames were simulated

TABLE 1: Initial values of simulation parameters and nondimensional numbers relevant to the DNS database considered here.

Case	Domain size/δ_{th}^3	Grid size	u'/S_L	l/δ_{th}	τ	Re_t	Da	Ka	β
P1	$36.1 \times 24.1 \times 24.1$	$345 \times 230 \times 230$	5.0	1.67	4.5	22.0	0.33	8.67	6.0
P2	$36.1 \times 24.1 \times 24.1$	$345 \times 230 \times 230$	6.25	1.44	4.5	23.5	0.23	13.0	6.0
P3	$36.1 \times 24.1 \times 24.1$	$345 \times 230 \times 230$	7.5	2.50	4.5	48.0	0.33	13.0	6.0
P4	$36.1 \times 24.1 \times 24.1$	$345 \times 230 \times 230$	9.0	4.31	4.5	100	0.48	13.0	6.0
P5	$36.1 \times 24.1 \times 24.1$	$345 \times 230 \times 230$	11.25	3.75	4.5	110	0.33	19.5	6.0
V1	$29.7 \times 29.7 \times 29.7$	$512 \times 512 \times 512$	1.0	3.57	2.52	18	3.59	0.529	7.1
V2	$29.7 \times 29.7 \times 29.7$	$512 \times 512 \times 512$	2.0	3.62	2.52	37	1.81	1.487	7.1
V3	$29.7 \times 29.7 \times 29.7$	$512 \times 512 \times 512$	6.0	3.43	2.52	92	0.57	7.936	7.1

using a single step chemistry to keep the comparison with statistically planar flames consistent. All the nonperiodic boundaries are specified using the NSCBC technique [33]. Nonreflecting outflows, modified to accommodate the presence of flame on the boundary, were applied to the transverse and downstream faces [40, 41]. Inlet turbulence was taken from a precomputed simulation of fully developed homogeneous isotropic turbulence, and the velocity components were interpolated onto the inlet using a high-order scheme to ensure that the structure of the turbulence was preserved. The computational domain in cases V1-V3 is taken to be cubic with sides equal to $L = 29.7\delta_{th}$, where $\delta_{th} = 3.563D_0/S_L$ for the thermochemistry used in these cases. A Cartesian grid of $512 \times 512 \times 512$ with uniform grid spacing is used. The numerical schemes used for spatial discretisation and time-integration in cases V1-V3 are similar to those used for cases P1-P5. The flame holder centre is located at $x_1 = 3.48\delta_{th}$ and has an approximate radius $R = 1.2\delta_{th}$. At the flame holder, the reaction progress variable and mean velocity distributions were imposed using a Gaussian function. It is worth noting that formation of boundary layer around the flame holder and its effect on the flow and flame dynamics are not represented in the simulation due to prohibitive computational cost. However, the possible influence of these effects on the results reported in this study is minimised by carefully selecting the region for the analysis. In the selected regions, the statistical distributions of strain and curvature experienced by flame elements are similar to those for freely propagating statistically planar flames under comparable local conditions [40, 41]. The values of turbulent Reynolds number $\mathrm{Re}_{t,\mathrm{inlet}} = \rho_0 u'_{\mathrm{inlet}} l/\mu_0$, Karlovitz number $\mathrm{Ka} = (u'_{\mathrm{inlet}}/S_L)^{3/2}(l/\delta_{th})^{-1/2}$, and Damköhler number $\mathrm{Da} = lS_L/u'_{\mathrm{inlet}}\delta_{th}$ based on the root-mean-square turbulent velocity fluctuation u'_{inlet} at the inlet are provided in Table 1 along with the values of τ and β.

To ensure that initial transients had decayed and a stationary state had been reached, the simulation was carried out for a period of one flow-through time $\tau_{FT} = L/\overline{U}_{\mathrm{in}}$ before data were collected for analysis, where $\overline{U}_{\mathrm{in}}$ is the mean inlet velocity. In the V-flame configuration, the flame is continuously developing downstream from the flame holder, and so the present analysis is restricted to a region spanning $14.9\delta_{th} \leq x_1 \leq 29.1\delta_{th}$ in the streamwise direction, thus ensuring sufficient time for the flame to develop following ignition. For the purpose of ensuring adequate convergence of the statistics, four snapshots from the simulation were used

to obtain SDR statistics presented in the next section, which are taken at an interval of $0.2\tau_{FT}$ after the initial flow-through time. Standard values have been taken for Prandtl number (Pr = 0.7) and ratio of specific heats, $\gamma = 1.4$. The global Lewis number is taken to be unity for all cases considered in this analysis.

The grid spacing Δx for all cases ensures 10 grid points within δ_{th}. As Karlovitz number can be scaled as $\mathrm{Ka} \sim \delta_{th}^2/\eta^2$, the grid spacing Δx can be taken to be $\Delta x \leq \delta_{th}/10 \sim \eta\sqrt{\mathrm{Ka}}/10$. This indicates that $\eta/\Delta x \sim 10/\sqrt{\mathrm{Ka}}$ assumes the smallest value in case P5 amongst the cases considered here as the value Ka is the highest in case P5. For case P5, Δx remained $\Delta x \sim \eta/2$ throughout the duration of the simulation. For other cases, the Kolmogorov scale is resolved by more than two grid points due to smaller value of Ka than in case P5. The above discussion suggests that the grid size chosen for the cases considered here is sufficient to resolve turbulence structures.

The thermophysical properties such as thermal conductivity (λ), dynamic viscosity (μ), and density-weighted mass diffusivity (ρD) are taken to be constant and independent of temperature in cases P1-P5, whereas these quantities in cases V1-V3 are taken to be temperature dependent and the temperature dependence approximated by 5th order polynomials following the CHEMKIN formats [40, 41]. It is worth noting that the cases P1-P5 and cases V1-V3 were originally developed independently (see [39] for cases P1-P5 and [40, 41] for cases V1-V3), but here these cases are considered together to assess if the SDR statistics obtained from DNS data with constant thermophysical properties in a canonical configuration remain qualitatively valid in a laboratory-scale configuration (e.g., V-flame configuration) with temperature-dependent thermophysical properties.

3. Results and Discussion

3.1. Flame-Turbulence Interaction. The contours of c in the central $x_1 - x_2$ plane for cases P1-P5 and V1-V3 are shown in Figures 1(a)-1(h), respectively. It is evident from Figures 1(a)-1(e) that the level of wrinkling increases with increasing $u'/S_L \sim \mathrm{Re}_t^{1/4}\mathrm{Ka}^{1/2} \sim \mathrm{Re}_t^{1/2}/\mathrm{Da}^{1/2}$. Turbulent eddies penetrate into the preheat zone in the thin reaction zones regime combustion (Ka $\sim \delta_{th}^2/\eta^2 \sim \mathrm{Re}_t^{1/2}/\mathrm{Da} > 1$), but the reaction zone remains unperturbed because the Kolmogorov length scale is larger than the reaction zone

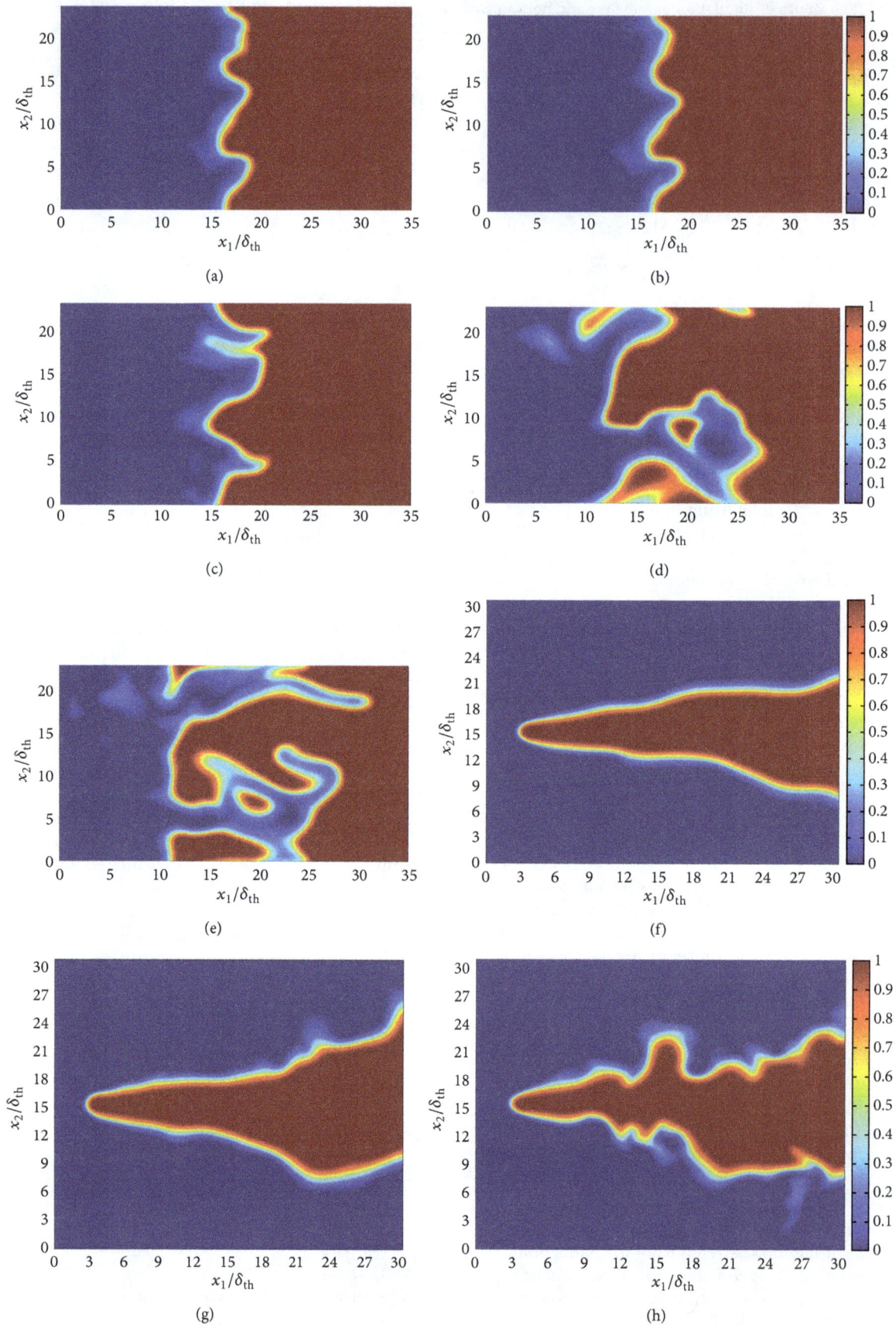

FIGURE 1: Distributions of c in the central $x_1 - x_2$ plane for cases P1–P5 (a–e) at $t_c = \delta_{th}/S_L$ and for cases V1–V3 (f–h).

thickness. The isosurfaces of c representing the preheat zone (i.e., $c \leq 0.5$) show more distortion than the isosurfaces representing the reaction zones (i.e., $0.7 \leq c \leq 0.9$) due to penetration of turbulent eddies within the preheat zone. However, this tendency is more prevalent for high values of u'/S_L and Ka (e.g., cases P3, P4, P5, and V3) but the isosurfaces of c remain mostly parallel to each other for small values of u'/S_L and Ka (e.g., cases P1, P2, V1, and V2) indicating that the internal flame structure is weakly affected by turbulence in these cases.

3.2. Statistical Behaviour of the Mean Values of N_c and the Unclosed Terms of Its Transport Equation. The variations of $\langle N_c \rangle \times \delta_{th}/S_L$ with c for statistically planar flames and V-flames are shown in Figures 2(a) and 2(b), where $\langle Q \rangle$ indicates the mean value of Q, which is obtained by ensemble averaging the quantity in question on a given c isosurface in the manner previously used by Boger et al. [42], Chakraborty and Cant [10, 11], and Chakraborty and Klein [15, 16]. It is worth noting that $\langle Q \rangle$ should not be confused with either Reynolds averaging or conventional conditional averaging operation in the context of RANS simulations because $\langle Q \rangle$ is evaluated using all the samples for a given c value over the whole domain. Figures 2(a) and 2(b) show that the variations of $\langle N_c \rangle \times \delta_{th}/S_L$ for statistically planar and V-flames are qualitatively similar to each other. For both statistically planar and V-flame configurations, the location of the maximum value of $\langle N_c \rangle \times \delta_{th}/S_L$ is skewed slightly towards the burned gas side of the flame (i.e., $c \approx 0.7$). The peak magnitude of $\langle N_c \rangle \times \delta_{th}/S_L$ does not change significantly in response to u'/S_L as the standard deviation for the case in the middle of the parameter range (i.e., cases P3 and V2) is found to exceed the difference in $\langle N_c \rangle \times \delta_{th}/S_L$ values for the cases considered here for both statistically planar and V-flame configurations. In order to understand the distribution of $\langle N_c \rangle \times \delta_{th}/S_L$ across the flame front, the variations of the mean values of the terms $\langle T_1 \rangle$, $\langle T_2 \rangle$, $\langle T_3 \rangle$, $\langle (-D_2) \rangle$, and $\langle f(D) \rangle$ conditional on c for planar and V-flames are shown in Figure 3. The variations of the mean values of the terms in cases P2, P3, and P4 (cases V2) are qualitatively similar to those in cases P1 and P5 (case V1) and thus are not explicitly shown here. It is evident from Figure 3 that the qualitative behaviour of these terms remains similar for all cases considered here. In all cases, $\langle T_1 \rangle$ remains positive throughout the flame. By contrast, $\langle (-D_2) \rangle$ assumes negative values throughout the flame in all cases as dictated by (1a). Expressing $\rho = \rho_0/(1 + \tau c)$ for low Mach number, unity Lewis number flames give rise to an alternative expression for T_1 [3, 17, 25, 28, 29]:

$$T_1 = 2\rho \frac{\partial u_j}{\partial x_j} N_c. \tag{4}$$

As dilatation rate $\partial u_j/\partial x_j$ is predominantly positive in premixed flames, $\langle T_1 \rangle$ for all values of c is positive across the flame and vanishes on both ends of the flame.

The quantity $\langle T_2 \rangle$ assumes negative values throughout the flame front for cases P1 and V1. Although $\langle T_2 \rangle$ remains negative for the major portion of the flame, small positive values can be discerned in cases P5 and V3. In order to

understand this behaviour, the term T_2 can be expressed in the following manner [3, 22–24, 28, 30]:

$$T_2 = -2\rho N_c \left(e_\alpha \cos^2\alpha + e_\beta \cos^2\beta + e_\gamma \cos^2\gamma \right), \tag{5}$$

where e_α, e_β, and e_γ are the most extensive, intermediate, and most compressive principal strain rates and α, β, and γ are the angles of these principal strain rates with ∇c. Equation (5) demonstrates that the predominant alignment of e_α (e_γ) with ∇c leads to a negative (positive) contribution to T_2.

It has been discussed in the previous analyses [23, 24, 28, 30] that the alignment of ∇c with e_α and e_γ is determined by relative strengths of the strain rate induced by flame normal acceleration a_{chem} and turbulent straining a_{turb}. It has been demonstrated earlier that ∇c preferentially aligns with $e_\alpha(e_\gamma)$ when $a_{chem}(a_{turb})$ dominates over $a_{turb}(a_{chem})$. The strain rate induced by flame normal acceleration due to chemical heat release can be scaled as $a_{chem} \sim \tau f(\text{Ka})S_L/\delta_{th}$, where $f(\text{Ka})$ is expected to decrease with increasing Ka [43]. Following Meneveau and Poinsot [44], a_{turb} can be scaled as $a_{turb} \sim u'/l$, which gives rise to $a_{chem}/a_{turb} \sim \tau f(\text{Ka})S_L l/u'\delta_{th} \sim \tau f(\text{Ka})\text{Da} \sim \tau f(\text{Re}_t^{1/2}/\text{Da})\text{Da}$. Alternatively, turbulent straining can be scaled as [45] $a_{turb} \sim u'/\lambda$ (where λ is the Taylor microscale), which yields $a_{chem}/a_{turb} \sim \tau f(\text{Ka})S_L\lambda/u'\delta_{th} \sim \tau f(\text{Ka})\text{Da}/\text{Re}_t^{1/2} \sim \tau f(\text{Re}_t^{1/2}/\text{Da})\text{Da}/\text{Re}_t^{1/2} \sim \tau f(\text{Ka})/\text{Ka}$. The above scaling relations suggest that a_{chem} strengthens with respect to a_{turb} with increasing Da for a given value of Re_t. Previous analyses [22–24, 28, 30] demonstrated that ∇c predominantly aligns with e_α for $\text{Da} \gg 1$ flames, whereas ∇c aligns with e_γ in $\text{Da} < 1$ flames for comparable values of Re_t. Both $a_{chem}/a_{turb} \sim \tau f(\text{Ka})\text{Da}$ and $a_{chem}/a_{turb} \sim \tau f(\text{Re}_t^{1/2}/\text{Da})\text{Da}/\text{Re}_t^{1/2} \sim \tau f(\text{Ka})/\text{Ka}$ indicate that an increase in $\text{Ka} \sim \text{Re}_t^{1/2}/\text{Da}$ for a given value of Da (e.g., cases P1, P3, and P5) gives rise to weakening of a_{chem} in comparison to a_{turb}. This increases the extent of ∇c alignment with e_γ with increasing Ka when Da is held constant as in cases P1, P3, and P5. In cases P1 and P3, ∇c predominantly aligns with e_α; however the extent of this alignment decreases from P1 to P3. This predominant alignment of ∇c with e_α in cases P1 and P3 leads to a negative contribution of $\langle T_2 \rangle$ in these cases. In case P5, ∇c predominantly aligns with e_γ in the unburned and fully burned gases but a_{chem} overcomes a_{turb} in the regions of intense heat release close to the middle of the flame and as a result ∇c aligns with e_α in the reaction zone. Thus the mean value of $\langle T_2 \rangle$ in case P5 assumes positive values towards both the unburned and burned gas sides, whereas the mean contribution of $\langle T_2 \rangle$ remains negative close to the middle of the flame. The relation $a_{chem}/a_{turb} \sim \tau f(\text{Ka})\text{Da}/\text{Re}_t^{1/2}$ indicates that a_{chem} weakens in comparison to a_{turb} with decreasing $\tau\text{Da}/\text{Re}_t^{1/2}$. The quantity $\tau\text{Da}/\text{Re}_t^{1/2}$ assumes values equal to 0.96, 0.55, and 0.49 for cases P2, P3, and P4, respectively, when the statistics were extracted. This leads to larger extent of ∇c aligning with e_γ in case P4 (case P3) than in case P3 (case P2). This leads to predominantly negative contribution of $\langle T_2 \rangle$ in cases P2 and P3, whereas $\langle T_2 \rangle$ assumes positive values towards the unburned and burned gas sides of the flame in case P4. However, a_{chem} overcomes a_{turb} in the regions of intense heat release at the middle of

(a)

(b)

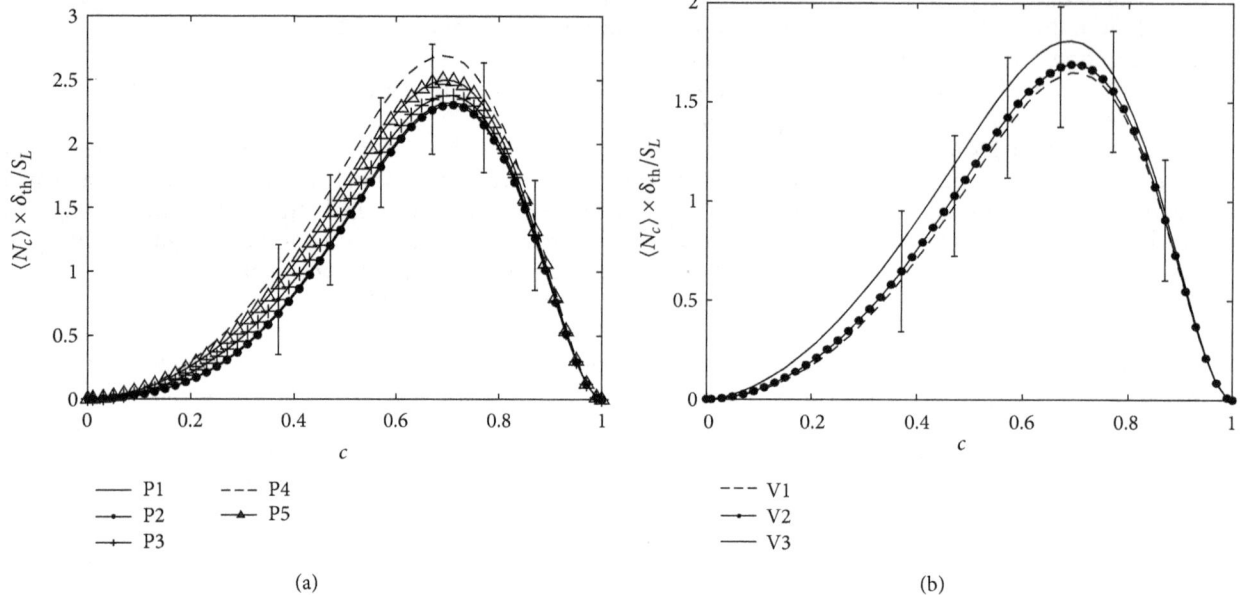

FIGURE 2: Variation of the mean value of $\langle N_c \rangle \times \delta_{\text{th}}/S_L$ conditional on c values across the flame front for (a) statistically planar cases with the bar indicating the standard deviation for case P3 and (b) V-flame cases with the bar indicating the standard deviation for case V2.

the flame and ∇c starts to align with e_α in the reaction zone giving rise to negative values of $\langle T_2 \rangle$ in case P4. In cases V1 and V2, the values of $a_{\text{chem}}/a_{\text{turb}} \sim \tau f(\text{Ka})\text{Da}/\text{Re}_t^{1/2}$ are larger than the corresponding value in case V3 (see the parameters in Table 1). Thus, the extent of ∇c alignment with $e_\alpha(e_\gamma)$ decreases (increases) from case V1 to case V3. This gives rise to positive values of $\langle T_2 \rangle$ towards both unburned and burned gas sides of the flame in case V3. This tendency is less prevalent in cases V1 and V2 due to smaller extent of ∇c alignment with e_γ than in case V3. However, the mean contribution of $\langle T_2 \rangle$ is negative in the middle of the flame for cases V1–V3 due to the alignment of ∇c with e_α in the heat releasing zone.

The contribution of $\langle T_3 \rangle$ remains positive (negative) towards the unburned (burned) gas side of the flame with the transition from positive to negative value taking place close to $c \approx 0.85$. In order to explain this behaviour, T_3 can be rewritten as

$$T_3 = -2Dn_i \frac{\partial \dot{w}}{\partial x_i}|\nabla c| = -2D\frac{\partial \dot{w}}{\partial n}|\nabla c|, \qquad (6)$$

where n is the spatial coordinate in the local flame normal direction and the flame normal vector $\vec{n}$ points towards the unburned gas side of the flame. For single step chemistry considered here, the maximum $\dot{w}$ occurs close to $c \approx 0.85$ [10, 14]. This suggests that the probability of finding negative (positive) values of $\partial \dot{w}/\partial n$ is significant for $c < 0.85$ ($c > 0.85$), which gives rise to positive (negative) value of $\langle T_3 \rangle$ towards the unburned (burned) gas side of the flame.

Figure 3 shows that $\langle f(D) \rangle$ is weakly negative towards the unburned gas side before becoming positive towards the burned gas side in all the cases. The magnitude of the mean contribution of $\langle f(D) \rangle$ remains comparable to that of $\langle T_1 \rangle$ in all cases indicating that $\langle f(D) \rangle$ cannot be neglected even

for cases P1–P5, where ρD is considered to be constant. In cases P1–P5, $\rho N_c [\partial D/\partial t + u_j \partial D/\partial x_j]$ can be expressed using $\rho = \rho_0/(1+\tau c)$ for globally adiabatic Le = 1.0 flames as $T_1/2 = \rho N_c \partial u_j/\partial x_j$ (i.e., $\rho N_c [\partial D/\partial t + u_j \partial D/\partial x_j] = \rho N_c (\partial u_j/\partial x_j)$ for constant ρD) and the first two terms on the right hand side of (1b) vanish for constant values of ρD. The contributions of $\langle (T_{D3} + T_{D4}) \rangle$ are responsible for the change in sign of $\langle f(D) \rangle$ in cases P1–P5. These terms are also principally responsible for sign change of $\langle f(D) \rangle$ in cases V1–V3.

3.3. Local Behaviour of N_c and Its Curvature and Strain Rate Dependences. The marginal probability density functions (pdfs) of normalised N_c^+ (i.e., $N_c \times \delta_{\text{th}}/S_L$) for different c iso-surfaces across the flame are shown in Figures 4(a) and 4(b) in log-log scale for cases P3 and V2, respectively. The pdfs of N_c in cases P1, P2, P4, and P5 (cases V1 and V3) are qualitatively similar to those in case P3 (case V2) and thus are not explicitly shown here. The pdfs for $c < 0.5$ are not shown in Figures 4(a) and 4(b), as N_c assumes small values in the preheat zone of the flame due to small magnitude of scalar gradient ∇c. It is evident from Figures 4(a) and 4(b) that the pdfs of N_c are qualitatively similar for statistically planar and V-flames and in both cases the probability of finding high values of N_c is most prevalent in the middle of the flame with slight skewness towards the burned gas side (i.e., $c \approx 0.7$) and the probability of finding high values of N_c decreases on both unburned and burned gas sides of the flame front. This is consistent with the observed behaviour of the mean values of N_c conditional on c shown in Figure 2. It can be seen in Figure 4 that a log-normal distribution captures the qualitative behaviour of the pdf of N_c although there are some disagreements in the pdf tails. This is consistent with several previous experimental [46–52] and numerical [53–55] studies

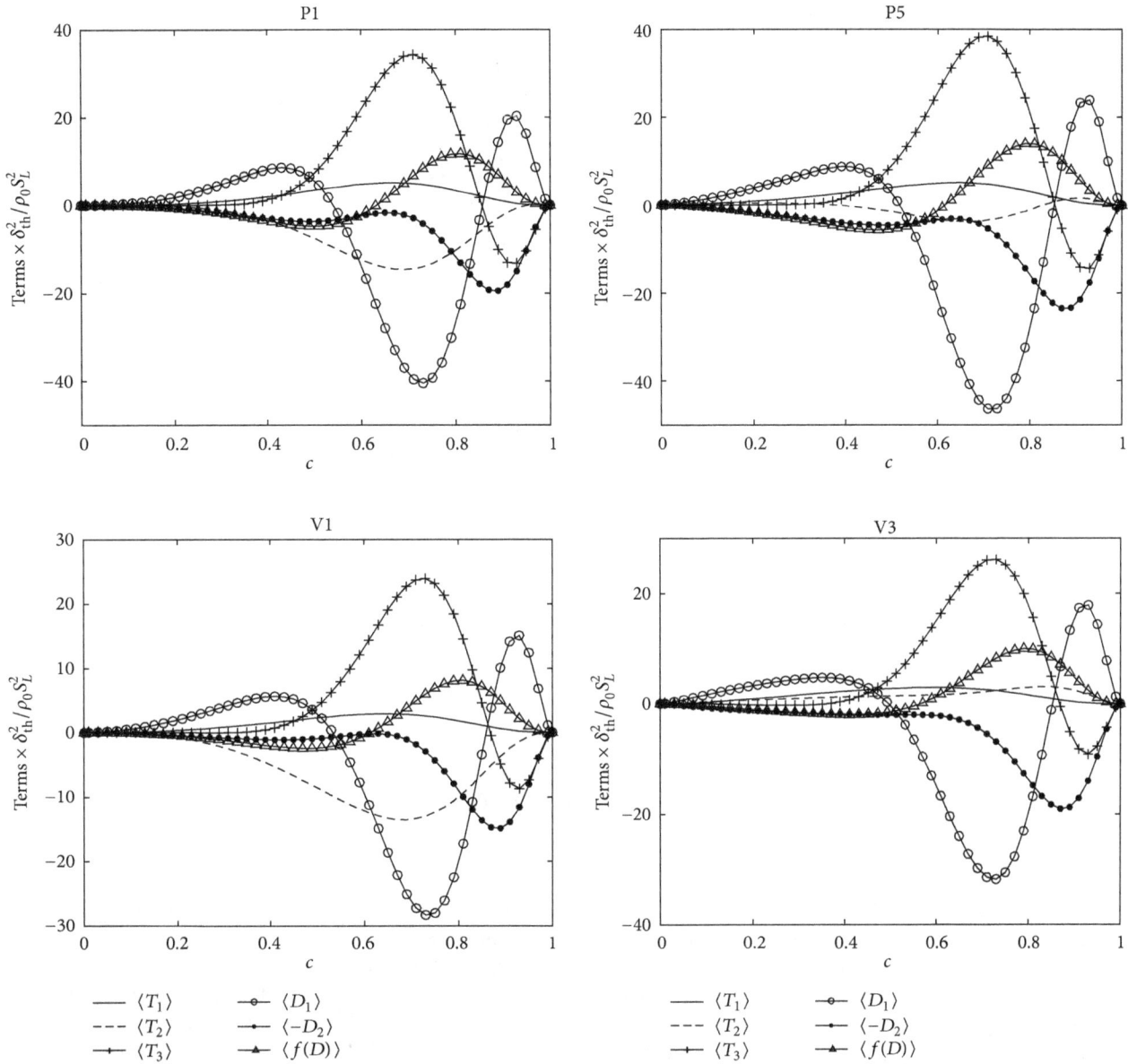

FIGURE 3: Variation of the mean values of $\langle T_1 \rangle$, $\langle T_2 \rangle$, $\langle T_3 \rangle$, $\langle -D_2 \rangle$, and $\langle f(D) \rangle$ conditional on c values across the flame for cases P1, P5, V1, and V3. All the terms of the transport equation of N_c are normalised with respect to the respective values of $\rho_0 S_L^2 / \delta_{th}^2$.

investigating the scalar dissipation rate pdf of a passive scalar. An approximate log-normal distribution of SDR in turbulent premixed flames has also been reported in a previous analysis [56].

The joint pdfs of N_c and tangential strain rate a_T for cases P1, P5, V1, and V3 are shown in Figure 5(a) for $c = 0.8$ isosurface, which is close to the most reactive region for the present thermochemistry. It can be seen from Figure 5(a) that N_c and a_T are positively correlated on $c = 0.8$ isosurface for cases P1, P5, V1, and V3 and similar qualitative behaviour has been observed also for other c isosurfaces in all cases considered here. This positive correlation between N_c and a_T can be explained in the following manner.

(i) The dilatation rate $\nabla \cdot \vec{u}$ can be expressed as $\nabla \cdot \vec{u} = a_T + a_n$, where $a_n = n_i n_j \partial u_i / \partial x_j$ is the normal strain rate. For unity Lewis number flames, $\nabla \cdot \vec{u}$ can be scaled as $\nabla \cdot \vec{u} \sim a_{chem} \sim \tau f(Ka) S_L / \delta_{th}$, whereas a_T can be taken to scale with turbulent strain rate a_{turb} (i.e., $a_T \sim a_{turb} \sim u'/l$ according to Meneveau and Poinsot [44] and $a_T \sim a_{turb} \sim u'/\lambda$ according to Tennekes and Lumley [45]).

(ii) Above scalings indicate that $\nabla \cdot \vec{u}/a_T$ scales as $\nabla \cdot \vec{u}/a_T \sim \tau f(Re_t^{1/2}/Da)Da$ and $\nabla \cdot \vec{u}/a_T \sim \tau f(Re_t^{1/2}/Da) Da/Re_t^{1/2} \sim \tau f(Ka)/Ka$ according to the scaling arguments by Meneveau and Poinsot [44] and Tennekes and Lumley [45], respectively. Both

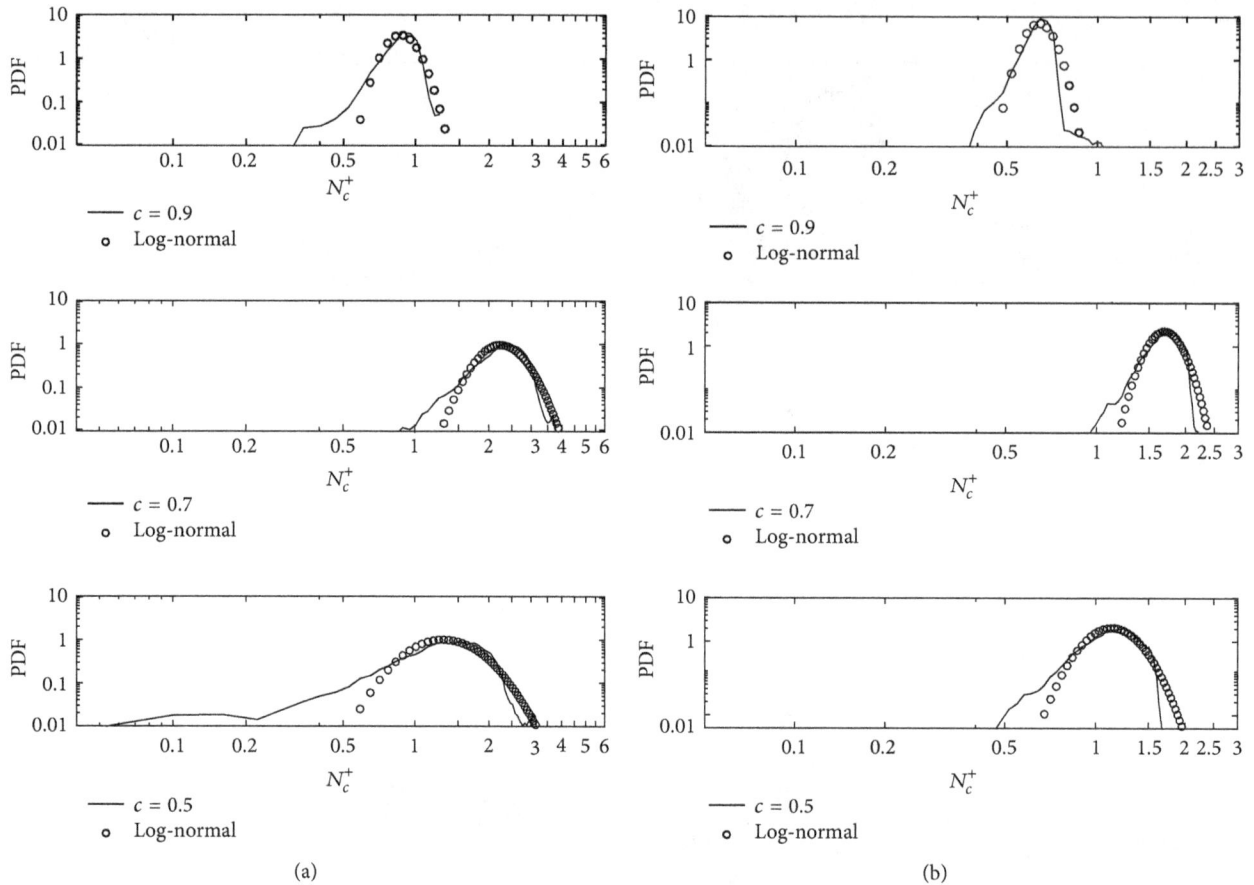

Figure 4: The marginal pdf of normalised N_c^+ (i.e., $N_c \times \delta_{th}/S_L$) and the log-normal distribution in log-log scale for $c = 0.5, 0.7,$ and 0.9 across the flame for cases (a) P3 and (b) V2.

$\nabla \cdot \vec{u}/a_T \sim \tau f(\mathrm{Re}_t^{1/2}/\mathrm{Da})\mathrm{Da}$ and $\nabla \cdot \vec{u}/a_T \sim \tau f(\mathrm{Re}_t^{1/2}/\mathrm{Da})\ \mathrm{Da}/\mathrm{Re}_t^{1/2} \sim \tau f(\mathrm{Ka})/\mathrm{Ka}$ suggest that the magnitude of a_T is likely to supersede the magnitude of $\nabla \cdot \vec{u}$ in most locations within the flame for small values of Da and high values of Ka.

(iii) It has been shown in several previous analyses [10, 30] that both $\nabla \cdot \vec{u}$ and a_T assume predominantly positive values and thus a higher magnitude of a_T than $\nabla \cdot \vec{u}$ induces a negative (i.e., compressive) normal strain rate a_n. Thus, an increase in a_T often leads to a decrease in $a_n = \nabla \cdot \vec{u} - a_T$ for small (high) values of Da (Ka). Thus, the isoscalar lines come close to each other under the action of decreasing a_n, which leads to increase in the magnitude of scalar gradient ∇c. This is reflected in the positive correlation between N_c and a_T.

The joint pdfs between N_c and curvature κ_m for cases P1, P5, V1, and V3 are shown in Figure 5(b) for $c = 0.8$ isosurface. Cases P2 and P3 (case V2) are not explicitly shown here due to their similarities to cases P1 and P5 (case V1), respectively. It can be seen from Figure 5(b) that the joint pdf between N_c and κ_m exhibits both positive and negative correlating branches on $c = 0.8$ isosurface for cases P5 and V3, and as a result of this, the net correlation between N_c and κ_m

remains weak. The positive correlation branch between N_c and κ_m remains weak for small values of u'/S_L in statistically planar flames (see Figure 5(b) for case P1) and this branch disappears completely in the V-flames with small values of u'/S_L (see Figure 5(b) for case V1). Similar behaviour is observed for other c isosurfaces in all cases considered here and the correlation between N_c and κ_m is weak throughout the flame for high values of u'/S_L (e.g., cases P3–P5 and V3). However, the disappearance of the positive correlating branch in the joint pdf of N_c and κ_m in Figure 5(b) indicates that N_c and κ_m are negatively correlated with each other throughout the flame for small values of u'/S_L (e.g., cases P1, P2, V1, and V2). The observed behaviour can be explained based on the following physical mechanisms.

(i) Previous analyses (e.g., [57]) demonstrated that both a_T and $\nabla \cdot \vec{u}$ remain negatively correlated with κ_m in turbulent premixed flames, and thus the behaviour of a_n at locations with large positive curvature is principally determined by a_T since $\nabla \cdot \vec{u}$ is small in these zones due to defocusing of heat. Small values of a_T are associated with high values of κ_m at these locations, which lead to small values of N_c at high values of positive κ_m due to positive correlation between N_c

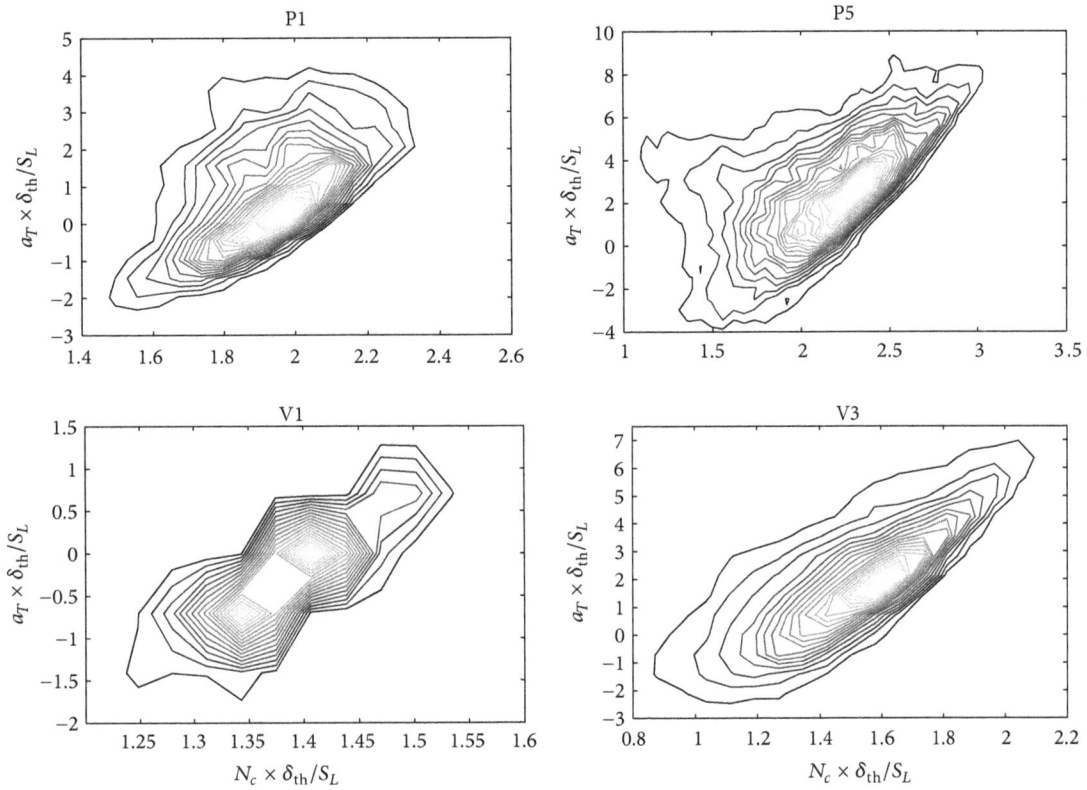

(a)

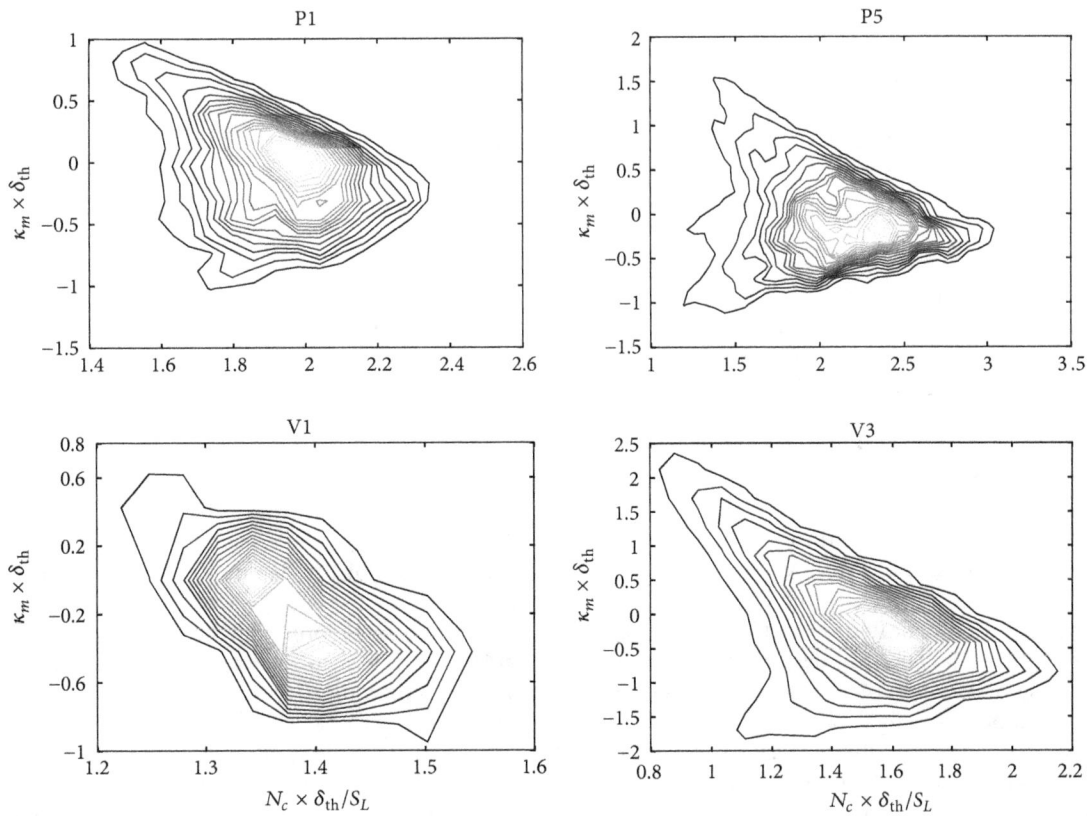

(b)

FIGURE 5: (a) Joint pdfs between $N_c \times \delta_{th}/S_L$ and normalised tangential strain rate $a_T \times \delta_{th}/S_L$ on $c = 0.8$ isosurface for cases P1, P5, V1, and V3. (b) Joint pdf between $N_c \times \delta_{th}/S_L$ and normalised curvature $\kappa_m \times \delta_{th}$ on $c = 0.8$ isosurface for cases P1, P5, V1, and V3.

and a_T. This leads to a negative correlating branch between N_c and κ_m at the positively curved zones.

(ii) The dilatation rate $\nabla \cdot \vec{u}$ is large in the negatively curved locations due to strong focussing of heat and the magnitude of $\nabla \cdot \vec{u}$ can locally be high enough to supersede the magnitude of a_T, which leads to a positive value of a_n. This tendency strengthens with decreasing κ_m, especially in the zones with large negative curvature, which gives rise to an increase in a_n with decreasing curvature. As the distance between the isoscalar lines increases with increasing a_n, the magnitude of scalar gradient ∇c decreases with decreasing κ_m in the negatively curved zones. This leads to the positive correlating branch in the joint pdf of N_c and κ_m (see Figure 5(b) for cases P5 and V3).

(iii) The relative strengths of the positive and negative correlating branches ultimately determine the net correlation between N_c and κ_m in the high u'/S_L cases. The probability of finding high negative curvature remains small for small values of u'/S_L and as a result the probability of finding high values of $\nabla \cdot \vec{u}$, which locally overcomes a_T, to induce a positive value of a_n, becomes rare (e.g., cases P1 and V1). Thus the combination of positive correlations between N_c and a_T and negative correlations between a_T and κ_m leads to a predominantly negative correlating branch between N_c and κ_m in the low u'/S_L cases (e.g., cases P1 and V1; see Figure 5(b)).

The strain rate and curvature dependences of N_c discussed above, in turn, affect the local statistical behaviours of $T_1, T_2, T_3, (-D_2)$, and $f(D)$ in response to a_T and κ_m. The curvature and strain rate dependences of $T_1, T_2, T_3, (-D_2)$, and $f(D)$ are discussed next.

3.4. Local Behaviour of T_1 and Its Curvature and Strain Rate Dependences. The marginal pdfs of T_1 for different c isoscalars across the flame are shown in Figures 6(a) and 6(b) for cases P3 and V2, respectively. The pdfs of T_1 in cases P1, P2, P4, and P5 (cases V1 and V3) are qualitatively similar to those in case P3 (case V2) and thus are not explicitly shown here. It is evident from Figures 6(a) and 6(b) that the pdfs of T_1 are qualitatively similar for statistically planar and V-flames and in both cases $T_1 = 2\rho(\nabla \cdot \vec{u})N_c$ assumes predominantly positive values throughout the flame. As dilatation rate $\nabla \cdot \vec{u}$ is principally positive due to thermal expansion in premixed flames [10, 30], the contribution of $T_1 = 2\rho(\nabla \cdot \vec{u})N_c$ is predominantly positive throughout the flame. Moreover, Figures 6(a) and 6(b) demonstrate that the probability of finding high values of T_1 is most prevalent in the middle of the flame with slight skewness towards the burned gas side (i.e., $c \approx 0.7$) and the probability of finding high values of T_1 decreases on both unburned and burned gas sides of the flame. This is consistent with the observed behaviour of the mean values of T_1 conditional on c shown in Figure 3. The probability of finding large magnitudes of $\nabla \cdot \vec{u}$ is the highest at a location which is slightly skewed towards the burned gas side of the flame [30]. As the distributions

of N_c and $\nabla \cdot \vec{u}$ are slightly skewed towards the burned gas side of the flame, the probability of finding large values of $T_1 = 2\rho(\nabla \cdot \vec{u})N_c$ becomes high around $c \approx 0.7$.

The joint pdfs between T_1 and a_T for cases P3 and V2 are shown in Figures 6(c) and 6(d), respectively, for $c = 0.8$ isosurface. It can be seen from Figures 6(c) and 6(d) that T_1 and a_T are positively correlated on $c = 0.8$ isosurface for cases P3 and V2 and similar qualitative behaviours have been observed for other c isosurfaces in all cases considered here. Both $\nabla \cdot \vec{u}$ and a_T are positively correlated for all flames considered here, which along with positive correlation between N_c and a_T (see Figure 5) gives rise to a positive correlation between $T_1 = 2\rho(\nabla \cdot \vec{u})N_c$ and a_T.

The joint pdfs between T_1 and κ_m for cases P3 and V2 are shown in Figures 6(e) and 6(f), respectively, for $c = 0.8$ isosurface. It can be seen from Figures 6(e) and 6(f) that the joint pdf between T_1 and κ_m exhibits a negative correlation on $c = 0.8$ isosurface for cases P3 and V2, and similar qualitative behaviour has been observed for other c isosurfaces in all cases considered here. In all cases, the net correlation between N_c and κ_m is weak (see Figure 5(b)), but $\nabla \cdot \vec{u}$ assumes high (small) values at negatively (positively) curved locations because of focussing (defocusing) of heat. This leads to a predominantly negative correlation between $\nabla \cdot \vec{u}$ and κ_m [57]. The negative correlation between $\nabla \cdot \vec{u}$ and κ_m is principally responsible for the negative correlation between $T_1 = 2\rho(\nabla \cdot \vec{u})N_c$ and κ_m.

3.5. Local Behaviour of T_2 and Its Curvature and Strain Rate Dependences. The marginal pdfs of T_2 for different c isosurfaces across the flame are shown in Figures 7(a) and 7(b) for cases P3 and V2, respectively. The pdfs of T_2 in cases P1, P2, P4, and P5 (cases V1 and V3) are qualitatively similar to those in case P3 (case V2) and thus are not explicitly shown here. Figures 7(a) and 7(b) show that the probability of finding negative values of T_2 supersedes the probability of finding positive values. The probability of finding negative values of T_2 increases as the heat releasing zone (see the pdfs for $c = 0.7$ isosurface) is approached. It has been discussed earlier that the effects of a_{chem} overcome the effects of a_{turb} in the heat releasing zone to give rise to a preferential alignment of ∇c with e_α even for small values of Da. This preferential alignment of ∇c with e_α in these zones gives rise to negative values of T_2 according to (5). The extent of ∇c alignment with $e_\alpha(e_\gamma)$ decreases (increases) towards both unburned and burned gas sides of the flame due to diminishing effects of a_{chem}.

The contours of joint pdfs between T_2 and a_T for $c = 0.8$ are shown in Figures 7(c) and 7(d) for cases P3 and V2 and the correlation coefficients between T_2 and a_T for different c isosurfaces across the flame for all cases are shown in Table 2. It is evident from Figures 7(c) and 7(d) and Table 2 that T_2 and a_T are positively correlated for high u'/S_L cases (e.g., cases P5 and V3) although the strength of the correlation changes through the flame. However, T_2 and a_T are weakly correlated with each other within the flame, where the effects of heat release are significant for cases with small and moderate values of u'/S_L (see Table 2). In order to explain

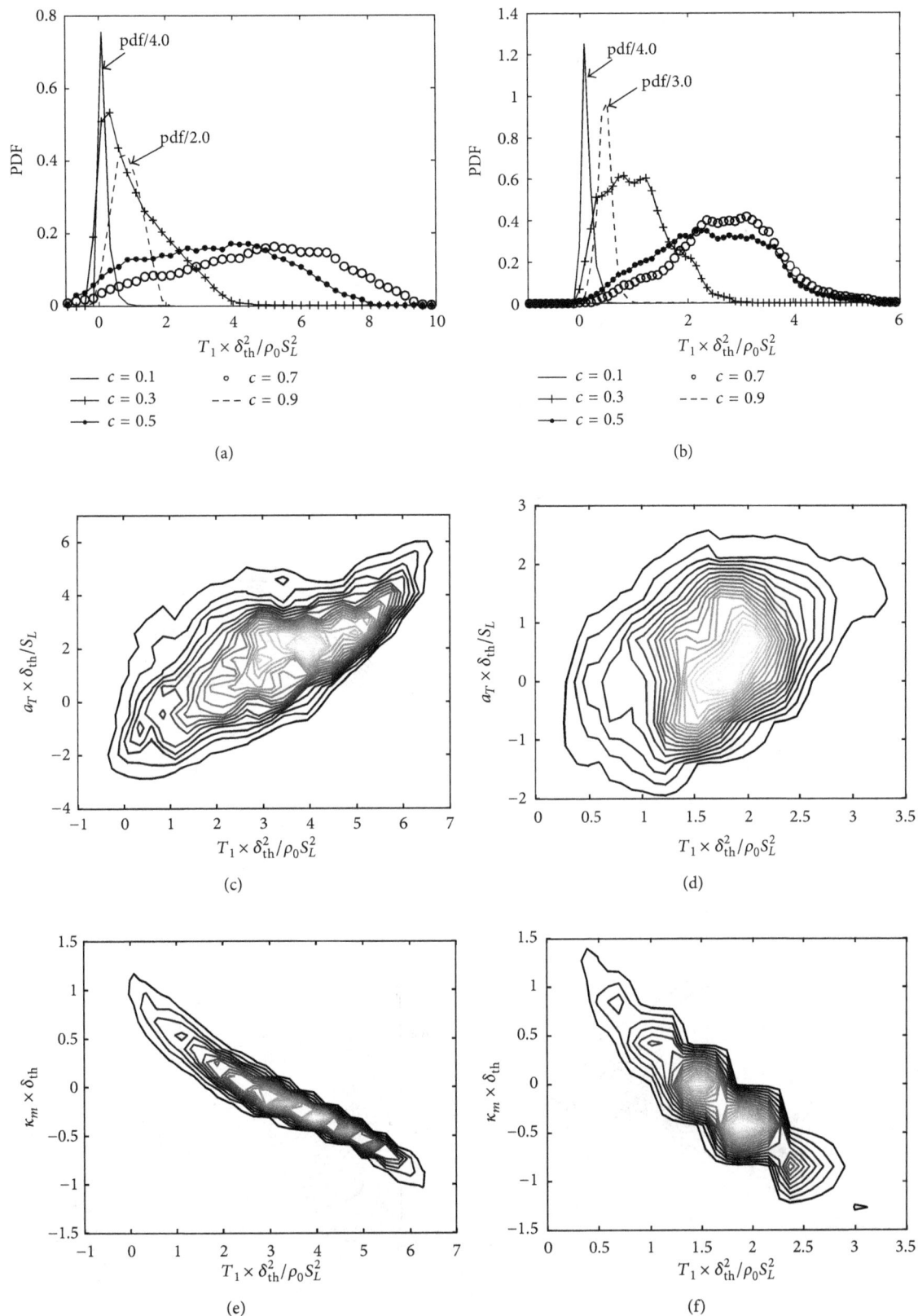

FIGURE 6: The marginal pdfs of $T_1 \times \delta_{th}^2/\rho_0 S_L^2$ for $c = 0.1, 0.3, 0.5, 0.7,$ and 0.9 for cases (a) P3 and (b) V2. Joint pdf between $T_1 \times \delta_{th}^2/\rho_0 S_L^2$ and normalised tangential strain rate $a_T \times \delta_{th}/S_L$ on $c = 0.8$ isosurface for cases (c) P3 and (d) V2. Joint pdf between $T_1 \times \delta_{th}^2/\rho_0 S_L^2$ and normalised curvature $\kappa_m \times \delta_{th}$ on $c = 0.8$ isosurface for cases (e) P3 and (f) V2.

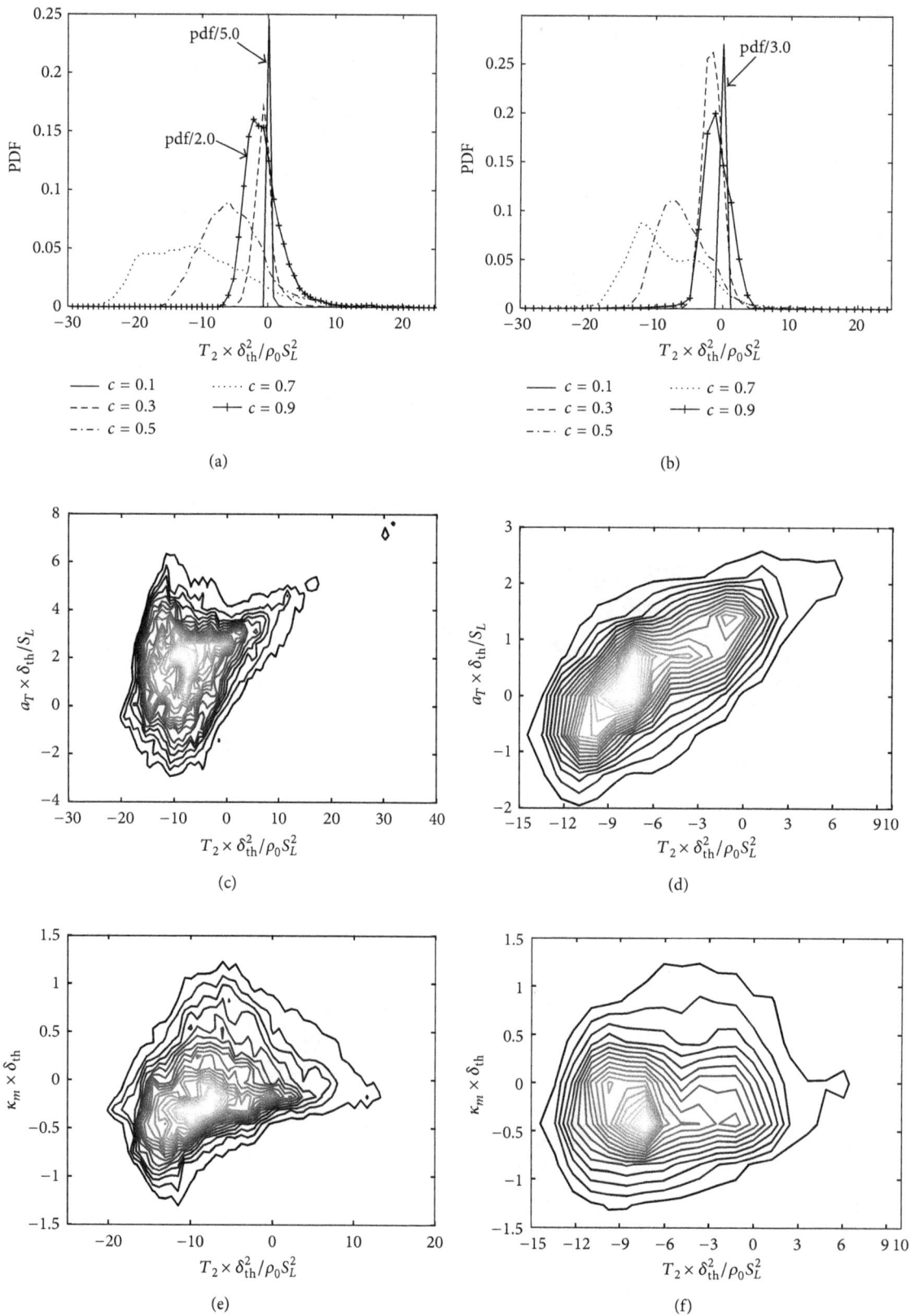

FIGURE 7: The marginal pdfs of $T_2 \times \delta_{th}^2/\rho_0 S_L^2$ for $c = 0.1, 0.3, 0.5, 0.7$, and 0.9 for cases (a) P3 and (b) V2. Joint pdf between $T_2 \times \delta_{th}^2/\rho_0 S_L^2$ and normalised tangential strain rate $a_T \times \delta_{th}/S_L$ on $c = 0.8$ isosurface for cases (c) P3 and (d) V2. Joint pdf between $T_2 \times \delta_{th}^2/\rho_0 S_L^2$ and normalised curvature $\kappa_m \times \delta_{th}$ on $c = 0.8$ isosurface for cases (e) P3 and (f) V2.

TABLE 2: Correlation coefficients between T_2 and a_T and between T_2 and κ_m on c = 0.1, 0.3, 0.5, 0.7, and 0.9 isosurfaces.

Case	$T_2 - a_T$					$T_2 - \kappa_m$				
	$c = 0.1$	$c = 0.3$	$c = 0.5$	$c = 0.7$	$c = 0.9$	$c = 0.1$	$c = 0.3$	$c = 0.5$	$c = 0.7$	$c = 0.9$
P1	0.642	−0.098	−0.217	−0.092	0.685	0.141	0.509	0.720	0.614	−0.250
P2	0.673	−0.090	−0.208	−0.084	0.676	0.116	0.506	0.719	0.616	−0.227
P3	0.751	0.376	0.263	0.263	0.648	0.544	0.252	0.412	0.423	−0.065
P4	0.802	0.593	0.616	0.689	0.827	0.052	0.196	0.235	0.198	−0.027
P5	0.783	0.662	0.614	0.616	0.787	0.028	0.137	0.223	0.242	−0.014
V1	0.245	0.367	0.749	0.816	0.971	0.612	0.623	0.354	0.135	−0.043
V2	0.591	0.335	0.495	0.699	0.920	0.345	0.565	0.497	0.328	−0.067
V3	0.767	0.721	0.738	0.768	0.888	−0.009	0.057	0.077	0.028	−0.155

this behaviour, it is useful to rewrite T_2 in the following manner:

$$T_2 = -2\rho a_n N_c. \tag{7}$$

Based on (7) the strain rate dependences of T_2 can be explained in the following manner.

(i) It has already been demonstrated that N_c and a_T are positively correlated with each other (see Figure 5(a)). The quantity $(-a_n) = a_T - \nabla \cdot \vec{u}$ tends to increase with increasing a_T in the regions where the effects of $\nabla \cdot \vec{u}$ are weak. This along with positive correlation between N_c and a_T leads to a positive correlation between T_2 and a_T for both unburned and burned gas sides of the flame for all cases.

(ii) The magnitudes of $\nabla \cdot \vec{u}$ and a_T increase with decreasing κ_m, and thus $(-a_n) = a_T - \nabla \cdot \vec{u}$ might not increase (even decrease) with increasing a_T in the heat releasing zone of the flame where the effects of $\nabla \cdot \vec{u}$ are strong. The a_T dependences of $(-a_n)$ and N_c ultimately determine the nature of the correlation between T_2 and a_T. The strain rate and curvature dependences of $\nabla \cdot \vec{u}$ weaken with increasing u'/S_L [58], so $(-a_n) = a_T - \nabla \cdot \vec{u}$ increases with increasing a_T, which leads to a positive correlation between T_2 and a_T for the major portion of the flame for cases with high values of u'/S_L (see Table 2).

The joint pdfs between T_2 and κ_m for cases P3 and V2 are shown in Figures 7(e) and 7(f), respectively, for c = 0.8 isosurface and the correlation coefficients between T_2 and κ_m for different c isosurfaces across the flame are shown in Table 2 for all cases considered here. It is evident from Figures 7(e) and 7(f) and Table 2 that T_2 and κ_m remain weakly positively correlated except the burned gas side of the flame. The observed curvature dependence of T_2 could be explained based on the following physical mechanisms.

(i) The effects of dilatation rate $\nabla \cdot \vec{u}$ and thermal expansion are particularly strong in the negatively curved regions due to focussing of heat. By the same token, the effects of heat release are weak in the positively curved zones due to defocusing of heat. Thus, the effects of a_{chem} are more likely to dominate over the effects of a_{turb} in the negatively curved zones,

which increase the extent of ∇c alignment with e_α as demonstrated earlier by Hartung et al. [58]. Weakening of the heat release effects at positively curved zones due to defocusing of heat leads to a greater (lesser) extent of ∇c alignment with $e_\gamma(e_\alpha)$ in the positively curved zones. The extent of ∇c alignment with e_α increases in the negatively curved zones, which in turn makes T_2 increasingly negative (see (5)) and the magnitude of the negative contribution of T_2 decreases for positive curvature locations. This gives rise to a positive correlation between T_2 and κ_m, as observed from Figures 7(e) and 7(f) and Table 2.

(ii) However, the effects of a_{turb} are more likely to dominate over the effects of a_{chem} towards the burned gas side and thus the extent of ∇c alignment with e_γ is determined by local turbulent flow conditions. The effects of flame-generated turbulence become stronger at the negatively curved zones due to stronger thermal expansion effects resulting from focussing of heat especially in the heat releasing zone. The straining induced by flame-generated turbulence may overcome relatively weak effects of $\nabla \cdot \vec{u}$ towards the burned gas side, which can give rise to an increasing extent of ∇c alignment with e_γ increases in the negative curved zones. This in turn gives rise to an increase in T_2 (see (5)) with decreasing κ_m towards the burned gas side and leads to a negative correlation between T_2 and κ_m (see Table 2).

3.6. Local Behaviour of T_3 and Its Curvature and Strain Rate Dependences. The marginal pdfs of normalised T_3 for different c isosurfaces across the flame are shown in Figure 8 for cases P3 and V2, respectively. The pdfs of T_3 in cases P1, P2, P4, and P5 (cases V1 and V3) are qualitatively similar to those in case P3 (case V2) and thus are not explicitly shown here. The pdfs for $c < 0.5$ are not shown in Figure 8 because T_3 assumes negligible value in the preheat zone of the flame due to negligible magnitude of $\dot{w}$. It is evident that T_3 assumes positive values for the major portion of the flame for both statistically planar and V-flames and the probability of finding high positive values increases towards the most reactive zone (e.g., c = 0.7 in Figure 8) of the flame front. However, T_3 assumes negative values only towards the burned gas side

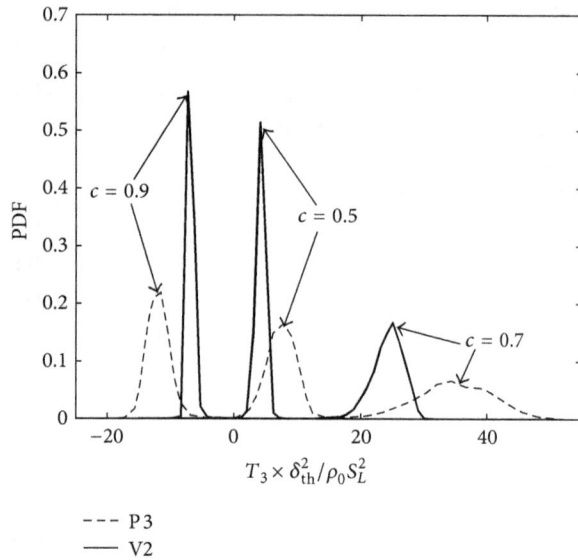

FIGURE 8: The marginal pdfs of $T_3 \times \delta_{th}^2/\rho_0 S_L^2$ for $c = 0.5, 0.7$, and 0.9 for cases P3 and V2.

(e.g., $c = 0.9$) of the flame front for both planar and V-flames. This is consistent with the behaviour of $\langle T_3 \rangle$ shown in Figure 3. The physical mechanism behind the transition from positive to negative values of the mean contribution of T_3 (see (6)) is also responsible for obtaining negative (positive) values of T_3 towards the burned (unburned) gas side of the flame.

The contours of joint pdfs between T_3 and a_T for $c = 0.5$, 0.7, and 0.9 isosurfaces are shown in Figures 9(a)–9(f) for cases P3 and V2 and similar qualitative behaviour has been observed for other cases considered here. It is evident from Figures 9(a)–9(f) that T_3 and a_T remain positively correlated for the part of the flame where finding positive values of T_3 is prevalent. On the other hand, T_3 and a_T are negatively correlated with each other towards the burned gas side of the flame where T_3 is predominantly negative. The observed a_T dependence of T_3 can be explained in the following manner.

(i) It has been demonstrated earlier that N_c and a_T are positively correlated with each other which suggests that $|\nabla c| = |\partial c/\partial n|$ increases with increasing a_T. For low Mach number, unity Lewis number flames $\dot{w}$ depend only on c and thus high values of $|\partial \dot{w}/\partial n|$ are associated with high values of $|\nabla c| = |\partial c/\partial n|$ and N_c.

(ii) As N_c and a_T are positively correlated with each other, the magnitude of reaction rate contribution $|T_3| = |2D(\partial \dot{w}/\partial n)|\nabla c||$ is positively correlated with tangential strain rate a_T. Thus, T_3 is positively (negative) correlated with a_T, where T_3 assumes positive (negative) values.

The joint pdfs between T_3 and κ_m for cases P3 and V2 are shown in Figure 10 for $c = 0.5$, 0.7, and 0.9 isosurfaces and similar qualitative behaviour has been observed for other cases considered here. It is evident from Figure 10 that the joint pdf of T_3 and κ_m exhibits both positive and negative correlating branches and the net correlation is weak

throughout the flame. The physical explanations for the observed κ_m dependence of T_3 can be summarised in the following manner.

(i) The term $|T_3| = |2D(\partial \dot{w}/\partial n)|\nabla c||$ is expected to be positively (negatively) correlated with curvature κ_m at negatively (positively) curved locations for high values of u'/S_L, as in the case of N_c (see cases P5 and V3 in Figure 5(b)), because high values of $|\partial \dot{w}/\partial n|$ are associated with high values of N_c and $|\nabla c| = |\partial c/\partial n|$.

(ii) As a result of the aforementioned physical mechanisms, the term T_3 and κ_m remain positively (negatively) correlated with curvature κ_m at negatively (positively) curved locations in the planar flames where T_3 assumes positive values. By contrast, the joint pdfs of T_3 and κ_m exhibit negative (positive) correlation with curvature κ_m at negatively (positively) curved locations within the flame where T_3 assumes negative values for the planar flames considered here (see Figure 10(c)). However, N_c remains predominantly negatively correlated with κ_m for V-flame cases (see Figure 5(b)) and thus T_3 shows positive (negative) correlation with curvature where T_3 assumes negative (positive) values (see Figures 10(d)–10(f)).

3.7. *Local Behaviour of $(-D_2)$ and Its Curvature and Strain Rate Dependences.* The marginal pdfs of $(-D_2)$ for c iso-surfaces representative of leading edge, reaction zone, and trailing edge of the flame (e.g., $c = 0.3$, 0.7 and 0.9 isosurfaces) are shown in Figure 11(a) for cases P3 and V2. The pdfs of $(-D_2)$ in cases P1, P2, P4, and P5 (cases V1 and V3) are qualitatively similar to that in case P3 (case V2) and thus are not explicitly shown here. Figure 11(a) shows that $(-D_2)$ assumes negative values throughout the flame and the probability of finding high magnitude of $(-D_2)$ increases from unburned gas side towards a region of the flame which is severely skewed towards the burned gas side (e.g., $c = 0.9$ isosurface). This behaviour is found to be consistent with the mean behaviour of $(-D_2)$ shown in Figure 3. It can further be seen from Figure 11(a) that the pdfs of $(-D_2)$ for statistically planar and V-flames are qualitatively similar to each other.

The contours of joint pdfs between $(-D_2)$ and a_T for $c = 0.8$ isosurface are shown in Figures 11(b) and 11(c) for cases P3 and V2 and the correlation coefficients between $(-D_2)$ and a_T for different c isosurfaces across the flame for all cases considered here are shown in Table 3. Figures 11(b) and 11(c) and Table 3 show that $(-D_2)$ and a_T are predominantly negatively correlated throughout the flame but the strength of this negative correlation weakens with increasing u'/S_L and the correlation becomes weakly positive at the middle of the flame for high values of u'/S_L (e.g., cases P4 and P5). This behaviour can be explained in the following manner.

(i) The instantaneous SDR N_c and the molecular dissipation term $(-D_2)$ can be taken to scale as $N_c \sim D/\delta^2$ and $(-D_2) \sim (-\rho D^2/\delta^4) \sim (-\rho N_c^2)$ (where δ is the typical local flame thickness) because in premixed flames, the gradient of progress variable is

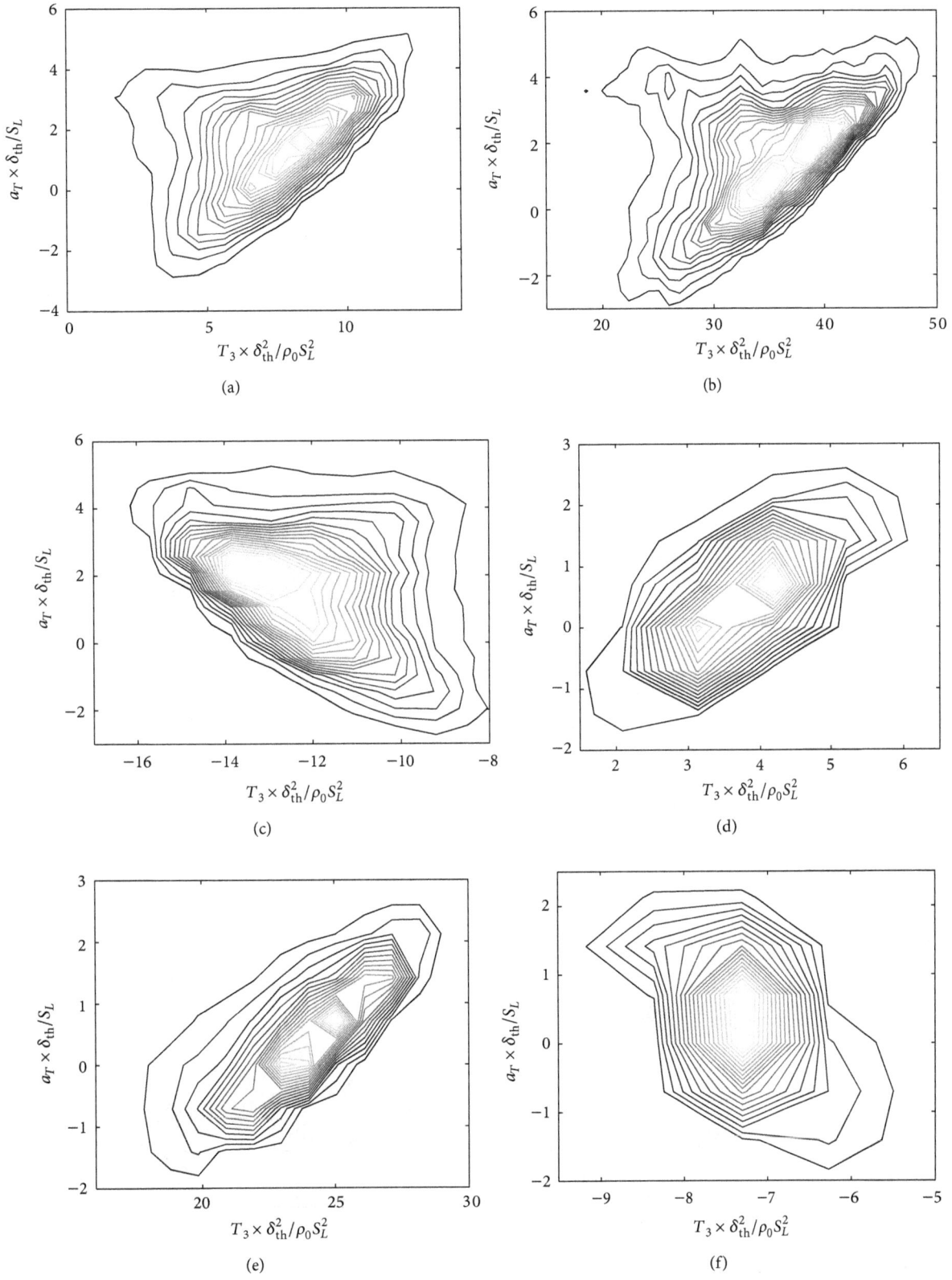

FIGURE 9: Joint pdfs between $T_3 \times \delta_{th}^2/\rho_0 S_L^2$ and normalised tangential strain rate $a_T \times \delta_{th}/S_L$ on (a) $c = 0.5$, (b) 0.7, and (c) 0.9 isosurfaces for case P3. Joint pdfs between $T_3 \times \delta_{th}^2/\rho_0 S_L^2$ and normalised tangential strain rate $a_T \times \delta_{th}/S_L$ on (d) $c = 0.5$, (e) 0.7, and (f) 0.9 isosurfaces for case V2.

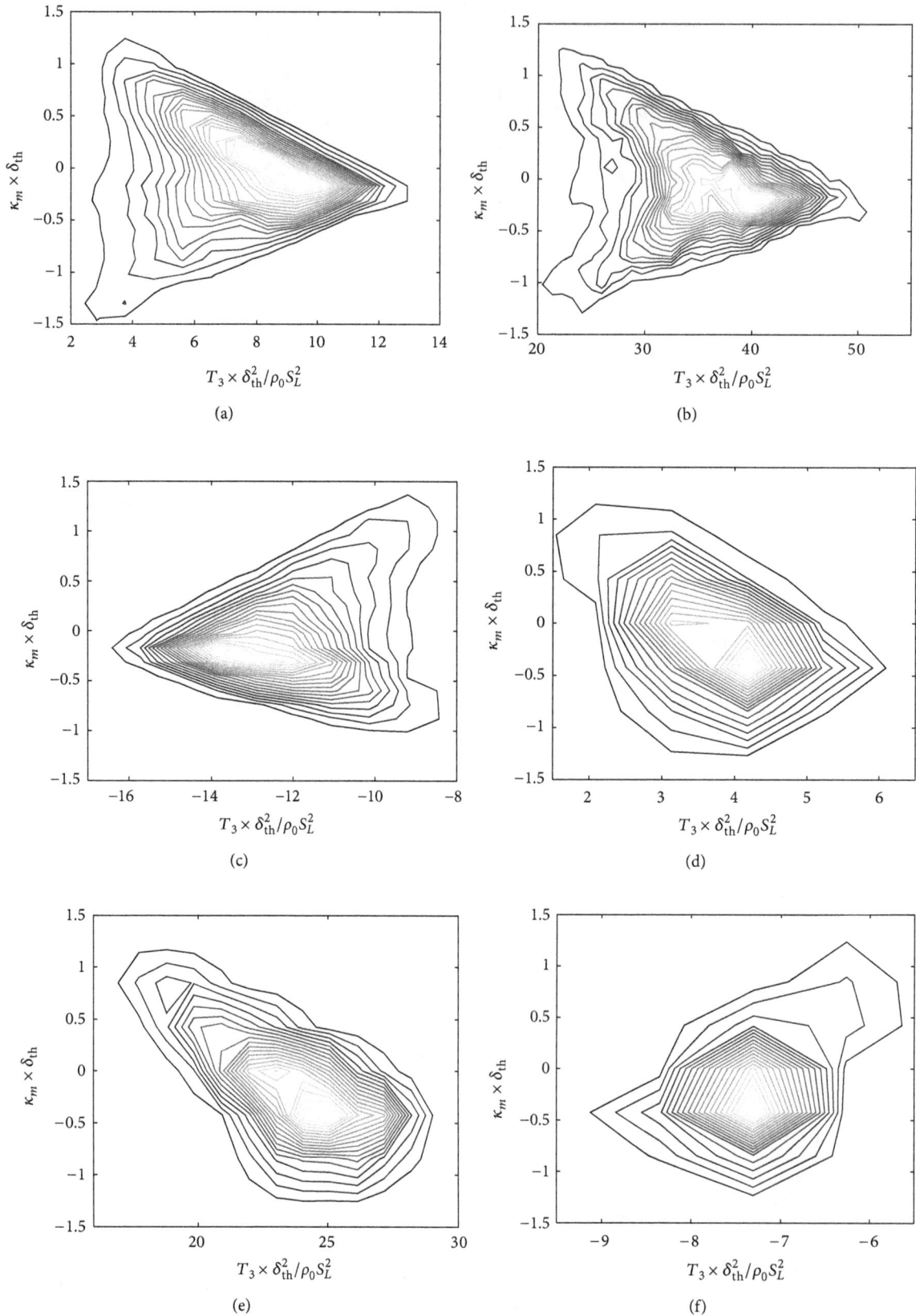

FIGURE 10: Joint pdfs between $T_3 \times \delta_{th}^2/\rho_0 S_L^2$ and normalised curvature $\kappa_m \times \delta_{th}$ on (a) $c = 0.5$, (b) 0.7, and (c) 0.9 isosurfaces for case P3. Joint pdfs between $T_3 \times \delta_{th}^2/\rho_0 S_L^2$ and normalised curvature $\kappa_m \times \delta_{th}$ on (d) $c = 0.5$, (e) 0.7, and (f) 0.9 isosurfaces for case V2.

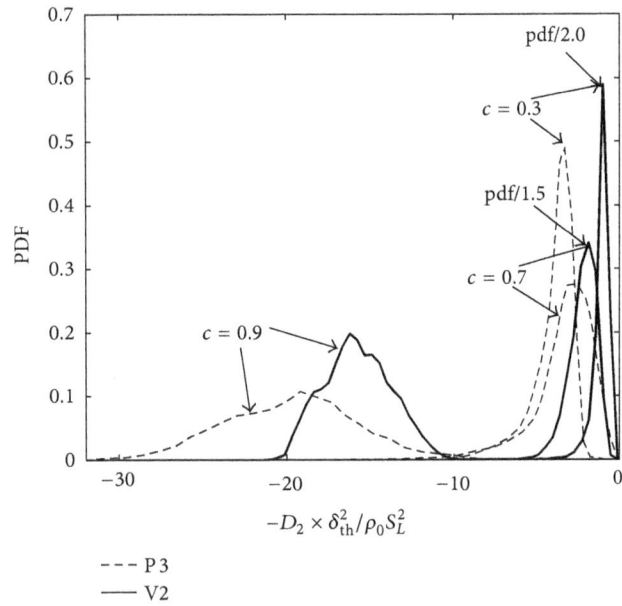

(a)

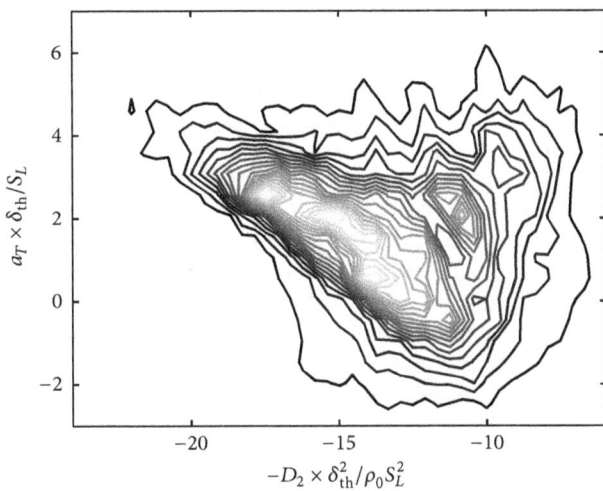

(b)

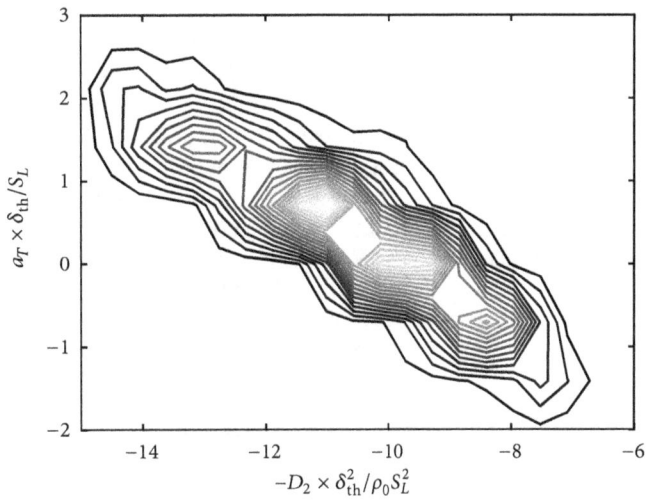

(c)

FIGURE 11: The marginal pdfs of $(-D_2) \times \delta_{th}^2/\rho_0 S_L^2$ for $c = 0.3$, 0.7, and 0.9 for cases (a) P3 and V2. Joint pdfs between $(-D_2) \times \delta_{th}^2/\rho_0 S_L^2$ and normalised tangential strain rate $a_T \times \delta_{th}/S_L$ on $c = 0.8$ isosurface for cases (b) P3 and (c) V2.

TABLE 3: Correlation coefficients between $(-D_2)$ and a_T and between $(-D_2)$ and κ_m on $c = 0.1$, 0.3, 0.5, 0.7, and 0.9 isosurfaces.

Case	$(-D_2) - a_T$					$(-D_2) - \kappa_m$				
	$c = 0.1$	$c = 0.3$	$c = 0.5$	$c = 0.7$	$c = 0.9$	$c = 0.1$	$c = 0.3$	$c = 0.5$	$c = 0.7$	$c = 0.9$
P1	−0.598	−0.809	−0.368	−0.183	−0.522	0.280	0.577	0.582	−0.020	0.261
P2	−0.581	−0.806	−0.389	−0.218	−0.488	0.272	0.554	0.600	0.023	0.225
P3	−0.546	−0.699	−0.028	0.050	−0.296	0.283	0.422	0.364	−0.234	−0.001
P4	−0.513	−0.652	0.221	−0.078	−0.483	0.231	0.338	−0.074	−0.412	−0.100
P5	−0.472	−0.581	0.111	−0.031	−0.378	0.205	0.288	0.174	−0.317	−0.107
V1	−0.712	−0.695	−0.069	−0.616	−0.679	0.700	0.825	−0.176	0.619	0.766
V2	−0.628	−0.525	0.089	−0.445	−0.703	0.475	0.696	0.033	0.159	0.633
V3	−0.452	−0.355	0.130	−0.312	−0.697	0.289	0.481	0.049	−0.104	0.260

only existent within the flame thickness. Alternatively, $(-D_2)$ can be considered to be governed by small-scale eddies and thus the characteristic length scale can be taken to be the Kolmogorov length scale η. Thus one obtains $(-D_2) \sim (-\rho D^2/\eta^4) \sim (-\rho N_c^2)\text{Ka}^2$ when dissipation processes are taken to be governed by η but $\delta/\eta \sim \text{Ka}^{1/2}$ remains of the order of unity for all cases considered here (see Table 1) and both scalings of $(-D_2)$ (i.e., $(-D_2) \sim (-\rho D^2/\delta^4) \sim (-\rho N_c^2)$ and $(-D_2) \sim (-\rho D^2/\eta^4) \sim (-\rho N_c^2)\text{Ka}^2)$ suggest that high magnitudes of the dissipation term $|-D_2|$ are associated with high values of a_T due to positive correlation between N_c and a_T (see Figure 5(a)).

(ii) As $(-D_2)$ assumes negative values, the quantities $(-D_2)$ and a_T are predominantly negatively correlated throughout the flame due to positive correlation between $|-D_2|$ and a_T. However, the negative correlation between $(-D_2)$ and a_T weakens with increasing u'/S_L due to weakening of positive correlation between N_c and a_T. Thus the correlation between $(-D_2)$ and a_T becomes weakly positive at the middle of the flame for high values of u'/S_L (e.g., cases P4 and P5).

The joint pdfs of $(-D_2)$ and curvature κ_m for cases P3 and V2 are shown in Figure 12 for $c = 0.3$, 0.7, and 0.9 isosurfaces and the correlation coefficients between $(-D_2)$ and κ_m for different c isosurfaces across the flame are shown in Table 3 for all cases considered here. The joint pdfs of $(-D_2)$ and κ_m in cases P1, P2, P4, and P5 (cases V1 and V3) are qualitatively similar to those in case P3 (case V2) and thus are not shown here. It can be seen from Figure 12 that the quantities $(-D_2)$ and κ_m are nonlinearly related to one another. The physical explanations behind the observed behaviour are provided below.

(i) The molecular dissipation term $(-D_2)$ can alternatively be expressed as

$$(-D_2)$$

$$= -2\rho D^2 \left\{ \begin{array}{l} \left[\dfrac{\partial |\nabla c|}{\partial n}\right]^2 + 4\kappa_m |\nabla c| \dfrac{\partial |\nabla c|}{\partial n} + 4\kappa_m^2 |\nabla c|^2 \\[2ex] + 2\left[\dfrac{\partial^2 c}{\partial x_1 \partial x_2}\dfrac{\partial^2 c}{\partial x_1 \partial x_2} - \dfrac{\partial^2 c}{\partial x_1 \partial x_1}\dfrac{\partial^2 c}{\partial x_2 \partial x_2}\right] \\[2ex] + 2\left[\dfrac{\partial^2 c}{\partial x_1 \partial x_3}\dfrac{\partial^2 c}{\partial x_1 \partial x_3} - \dfrac{\partial^2 c}{\partial x_1 \partial x_1}\dfrac{\partial^2 c}{\partial x_3 \partial x_3}\right] \\[2ex] + 2\left[\dfrac{\partial^2 c}{\partial x_2 \partial x_3}\dfrac{\partial^2 c}{\partial x_2 \partial x_3} - \dfrac{\partial^2 c}{\partial x_2 \partial x_2}\dfrac{\partial^2 c}{\partial x_3 \partial x_3}\right] \end{array} \right\} . \quad (8)$$

The above expression clearly indicates that the third term on the right hand side of (8) (i.e., $-8\rho D^2 \kappa_m^2 |\nabla c|^2$) induces nonlinear curvature dependence of the molecular dissipation term $(-D_2)$.

(ii) The quantity $\partial |\nabla c|/\partial n$ remains negative (positive) towards the unburned (burned) gas side of the flame

[10, 59, 60]; thus the second term on the right hand side is positively (negatively) correlated with κ_m towards the unburned (burned) gas side of the flame. The first term on the right hand side of (8) can be taken to scale with $(-\rho N_c^2)$ (i.e., $-2\rho D^2(\partial |\nabla c|/\partial n)^2 \sim -2\rho N_c^2$). It has already been shown that the joint pdfs of N_c and κ_m exhibit both positive and negative correlating branches for high values of u'/S_L (see cases P5 and V3 in Figure 5(b)) and thus the joint pdf of $-2\rho D^2(\partial |\nabla c|/\partial n)^2$ and κ_m is also expected to show branches with both positive and negative correlations in these cases. The weak negative correlation between N_c and κ_m for small values of u'/S_L (see cases P1 and V1 in Figure 5(b)) leads to weak positive correlation between $-2\rho D^2(\partial |\nabla c|/\partial n)^2 \sim -2\rho N_c^2$ and κ_m. The last three terms on the right hand side vanish in the limit of small scale isotropy and for the present cases they remain weakly correlated with curvature.

The relative strengths of the above mechanisms determine the net curvature dependence of $(-D_2)$. Thus, both positive and negative correlations between $(-D_2)$ and κ_m have been observed within the flame front in all cases considered here.

3.8. Local Behaviour of $f(D)$ and Its Curvature and Strain Rate Dependences. The marginal pdfs of $f(D)$ for $c = 0.1$, 0.3, 0.5, 0.7, and 0.9 isosurfaces across the flame front are shown in Figure 13(a) for cases P3 and V2. The pdfs of $f(D)$ in cases P1, P2, P4, and P5 (cases V1 and V3) are qualitatively similar to those in case P3 (case V2) and thus are not explicitly shown here. It is evident from Figure 13(a) that $f(D)$ predominantly assumes negative (positive) values towards the unburned (burned) gas side of the flame (see Figure 3). The density-weighted diffusivity ρD is considered to be constant in cases P1–P5 and thus T_{D1} and T_{D2} are identically zero in these cases. The marginal pdfs of T_{D3} and T_{D4} for case P3 are shown in Figures 13(b) and 13(c), which show that both T_{D3} and T_{D4} predominantly assume positive (negative) values towards burned (unburned) gas side of the flame. As $T_{D5} = T_1/2$ in cases P1–P5, the pdfs of T_{D5} are qualitatively similar to those of T_1 and thus are not shown here. This indicates that T_{D5} shows predominant probability of finding positive values throughout the flame (see Figure 6). The pdfs of $T_{D1}, T_{D2}, T_{D3}, T_{D4}$, and T_{D5} for case V2 are shown in Figures 13(d)–13(h), respectively. It is evident from Figures 13(f) and 13(g) that both T_{D3} and T_{D4} assume positive (negative) values towards burned (unburned) gas side of the flame, whereas T_{D5} assumes positive values throughout the flame, which is qualitatively similar to the behaviour of the corresponding term in case P3, where ρD is assumed to be constant. Figures 13(d) and 13(e) show that both T_{D1} and T_{D2} assume predominantly positive (negative) values towards the unburned (burned) gas side of the flame.

The contours of joint pdfs between $f(D)$ and a_T (κ_m) for $c = 0.1$, 0.5, and 0.7 isosurfaces are shown in Figure 14 (Figure 15) for cases P3 and V2, and the correlation coefficients between $f(D)$ and a_T (κ_m) for different c isosurfaces across the flame for cases P3 and V2 are shown in Figures 16(a) and 16(b) (Figures 16(c) and 16(d)), respectively.

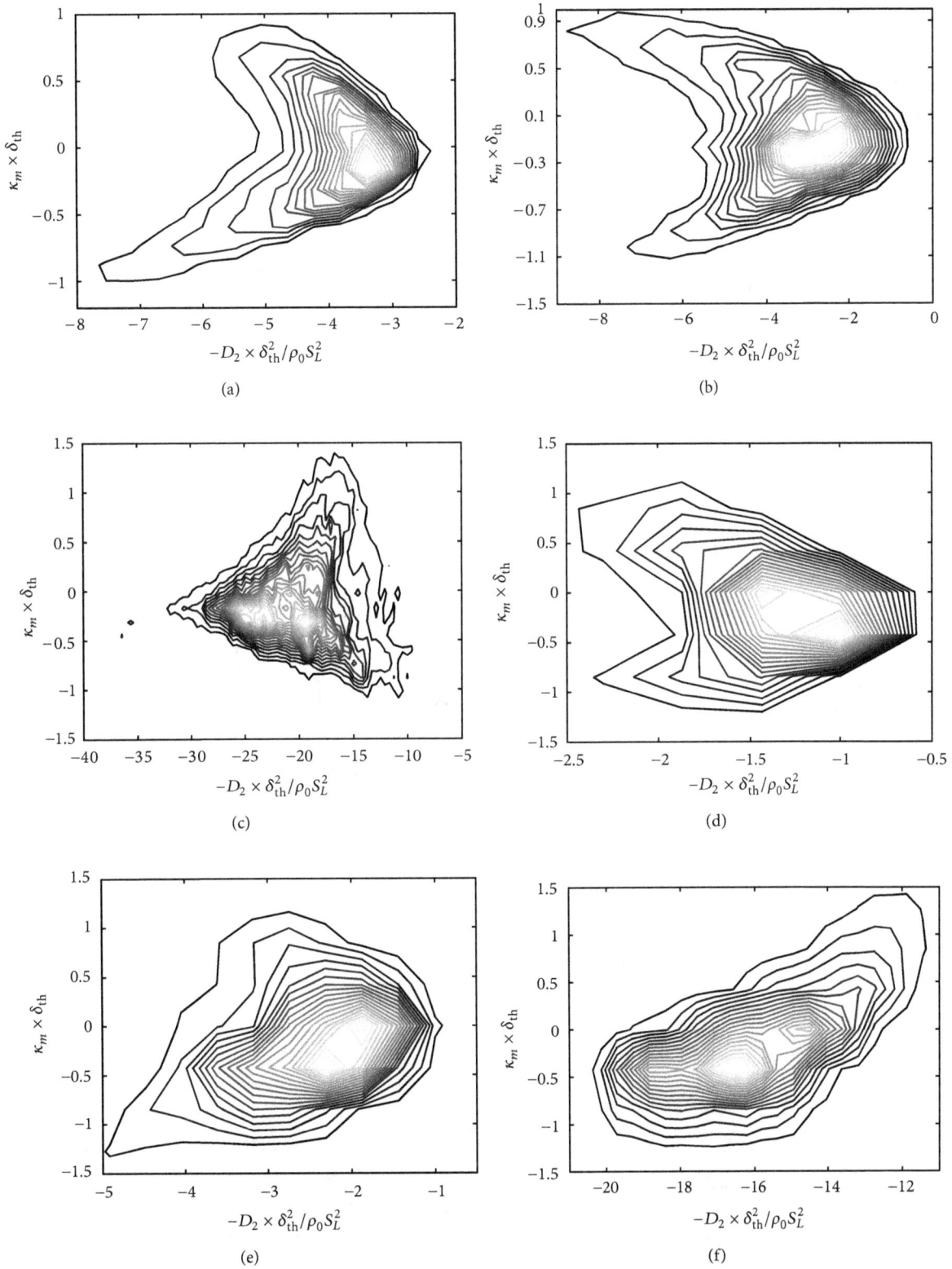

FIGURE 12: Joint pdfs between $(-D_2) \times \delta_{th}^2/\rho_0 S_L^2$ and normalised curvature $\kappa_m \times \delta_{th}$ on (a) $c = 0.3$, (b) 0.7. and (c) 0.9 isosurfaces for case P3. Joint pdfs between $(-D_2) \times \delta_{th}^2/\rho_0 S_L^2$ and normalised curvature $\kappa_m \times \delta_{th}$ on (d) $c = 0.3$, (e) 0.7, and (f) 0.9 isosurfaces for case V2.

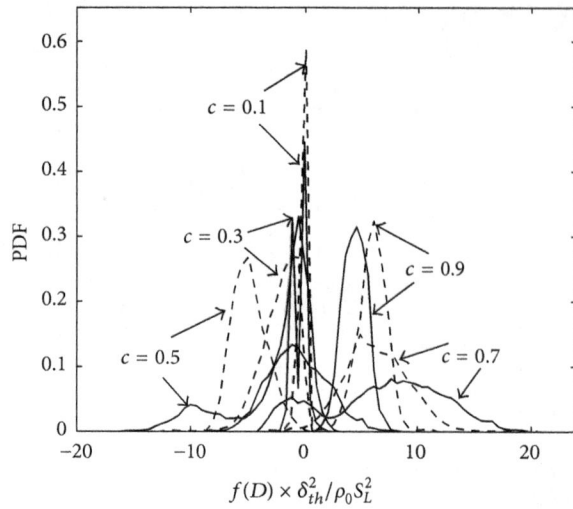

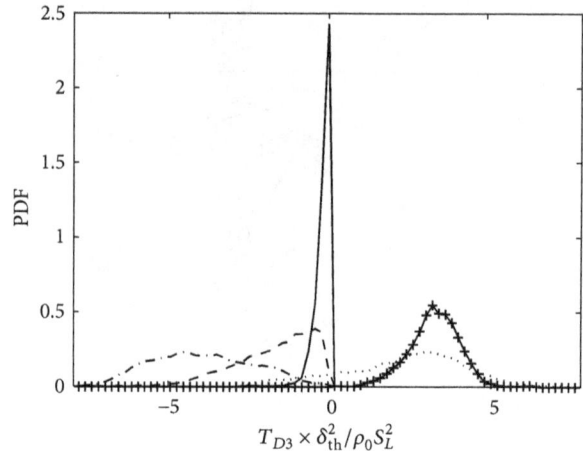

(a)

(b)

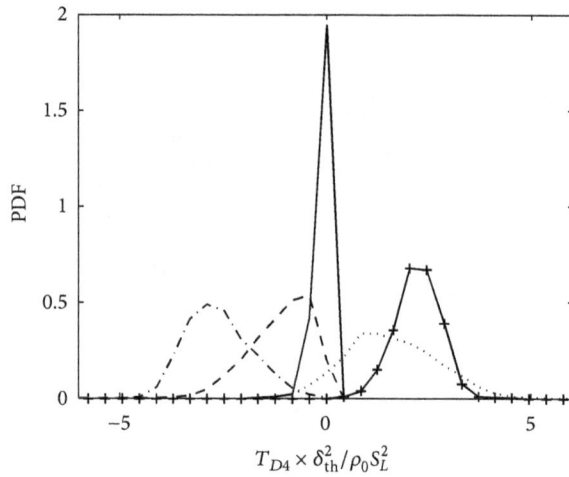

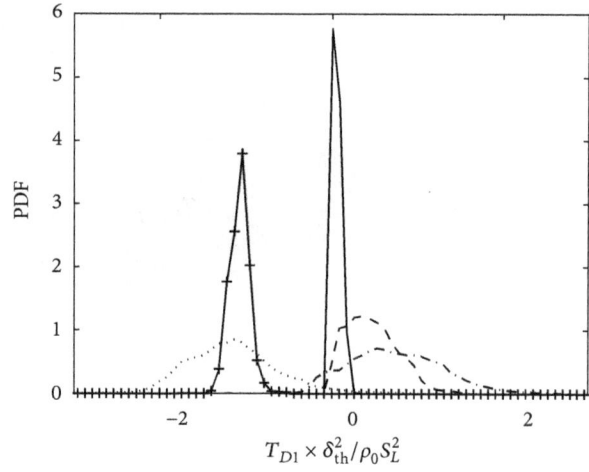

(c)

(d)

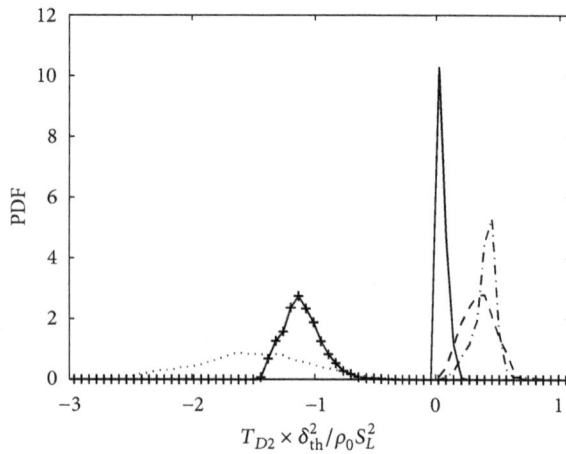

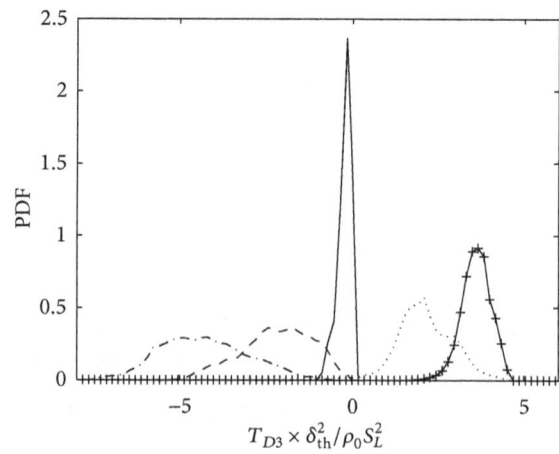

(e)

(f)

FIGURE 13: Continued.

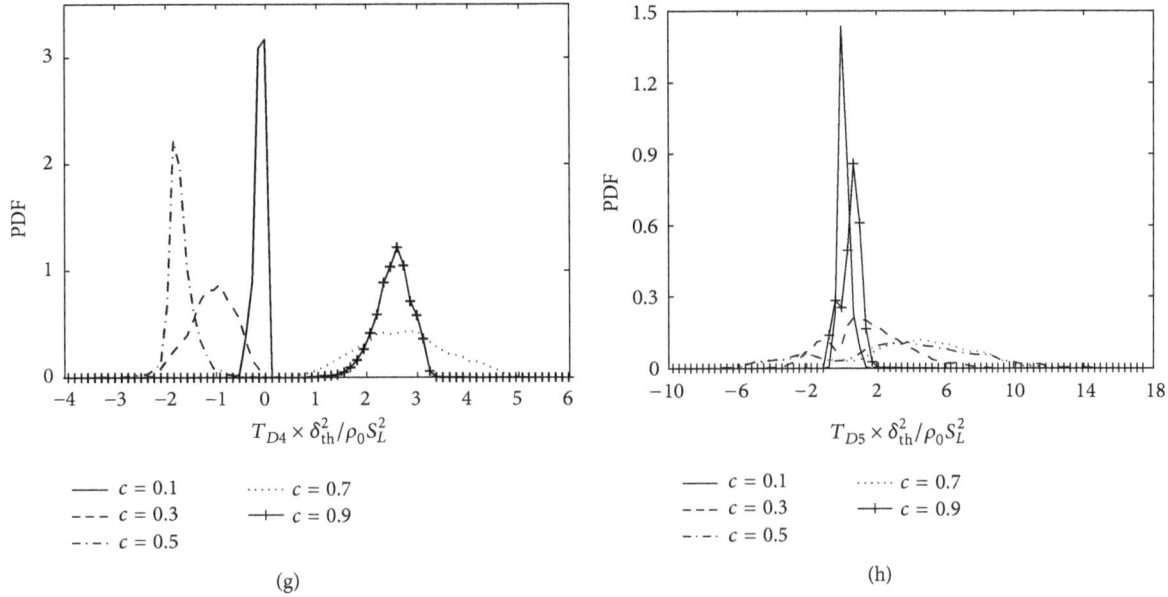

FIGURE 13: The marginal pdfs of $f(D) \times \delta_{th}^2/\rho_0 S_L^2$ for $c = 0.1, 0.3, 0.5, 0.7,$ and 0.9 for cases (a) P3 and V2. The marginal pdfs of (b) $T_{D3} \times \delta_{th}^2/\rho_0 S_L^2$ and (c) $T_{D4} \times \delta_{th}^2/\rho_0 S_L^2$ for $c = 0.1, 0.3, 0.5, 0.7,$ and 0.9 across the flame for case P3. The marginal pdfs of (d) $T_{D1} \times \delta_{th}^2/\rho_0 S_L^2$, (e) $T_{D2} \times \delta_{th}^2/\rho_0 S_L^2$, (f) $T_{D3} \times \delta_{th}^2/\rho_0 S_L^2$, (g) $T_{D4} \times \delta_{th}^2/\rho_0 S_L^2$, and (h) $T_{D5} \times \delta_{th}^2/\rho_0 S_L^2$ for $c = 0.1, 0.3, 0.5, 0.7,$ and 0.9 across the flame for case V2.

Both Figures 14 and 16 indicate that $f(D)$ and a_T are negatively (positively) correlated with each other towards the unburned (burned) gas side of the statistically planar flame (i.e., case P3), whereas $f(D)$ and a_T are positively correlated with each other from the middle to the burned gas side in the V-flame case (i.e., case V2), but this correlation remains weak towards the unburned gas side. It is evident from Figure 15 that $f(D)$ and κ_m remain weakly correlated for both statistically planar and V-flames, which is consistent with the correlation coefficient between $f(D)$ and κ_m shown in Figures 16(c) and 16(d). However, there are qualitative differences in the joints pdfs between $f(D)$ and κ_m for cases P3 and V2.

In order to explain the observed strain rate dependence of $f(D)$, the correlation coefficients between T_{D3}, T_{D4}, and T_{D5} ($T_{D1}, T_{D2}, T_{D3}, T_{D4}$, and T_{D5}) with a_T, for $c = 0.1, 0.3, 0.5, 0.7,$ and 0.9 isosurfaces, are also shown in Figure 16(a) (Figure 16(b)) for case P3 (case V2). It is evident from Figures 16(a) and 16(b) that both T_{D3} and T_{D4} remain negatively (positively) correlated with a_T towards the unburned (burned) gas side of the flame.

The strain rate dependences of $T_{D1}, T_{D2}, T_{D3}, T_{D4}$, and T_{D5} can be explained in the following manner.

(i) The magnitudes of T_{D3} and T_{D4} can be taken to scale as $|T_{D3}| \sim \rho D N_c/\delta^2 \sim \rho N_c^2$ and $|T_{D4}| \sim \rho D N_c/\delta^2 \sim \rho N_c^2$, which indicates that $|T_{D3}| \sim \rho N_c^2$ and $|T_{D4}| \sim \rho N_c^2$ remain positively correlated with a_T due to positive correlation between N_c and a_T (see Figure 5(a)). This suggests that the negative (positive) values of T_{D3} and T_{D4} (see Figures 13, 16(a), and 16(b)) lead to negative (positive) correlations of these terms with a_T due to positive correlations between N_c and

a_T (also due to positive correlation between $|T_{D3}| \sim \rho N_c^2$ ($|T_{D4}| \sim \rho N_c^2$) and a_T).

(ii) The term T_{D5} remains positively correlated with a_T throughout the flame, which is consistent with the positive correlation between T_1 and a_T shown in Figure 6(c), as $T_{D5} = 0.5T_1$ in statistically planar cases P1–P5 considered here (see Figure 16(a)). Even though ρD increases with increasing temperature in cases V1–V3, T_{D5} can still be taken to scale with T_1 (i.e., $T_{D5} \sim T_1 \sim \rho \tau N_c S_L/\delta$) and the positive correlation between T_1 and a_T leads to a positive correlation between T_{D5} and a_T in case V2 and also in cases V1 and V3 (see Figure 16(b)).

(iii) The magnitudes of the terms T_{D1} and T_{D2} can be scaled as $|T_{D1}| \sim \rho D^2/\delta^4 \sim \rho N_c^2$ and $|T_{D2}| \sim \rho D^2/\delta^4 \sim \rho N_c^2$, respectively, which suggests that $|T_{D1}|$ and $|T_{D2}|$ are expected to be positively correlated with a_T due to positive correlation between N_c and a_T (see Figure 5(a)). As T_{D1} and T_{D2} assume predominantly positive (negative) values towards the unburned (burned) gas side of the flame, these terms scale with $T_{D1} \sim T_{D2} \sim \rho N_c^2$ ($T_{D1} \sim T_{D2} \sim -\rho N_c^2$) towards the reactant (product) side of the flame. Thus the positive correlation between N_c and a_T leads to positive (negative) $T_{D1} - a_T$ and $T_{D2} - a_T$ correlations towards the unburned (burned) gas side of the flame (see Figure 16(b)).

(iv) In cases P1–P5, the terms $T_{D3}, T_{D4},$ and T_{D5} remain positively correlated with a_T towards the burned gas side of the flame (see Figure 16(a)) and these positive correlations result in a net positive correlation between $f(D)$ and a_T towards the burned gas

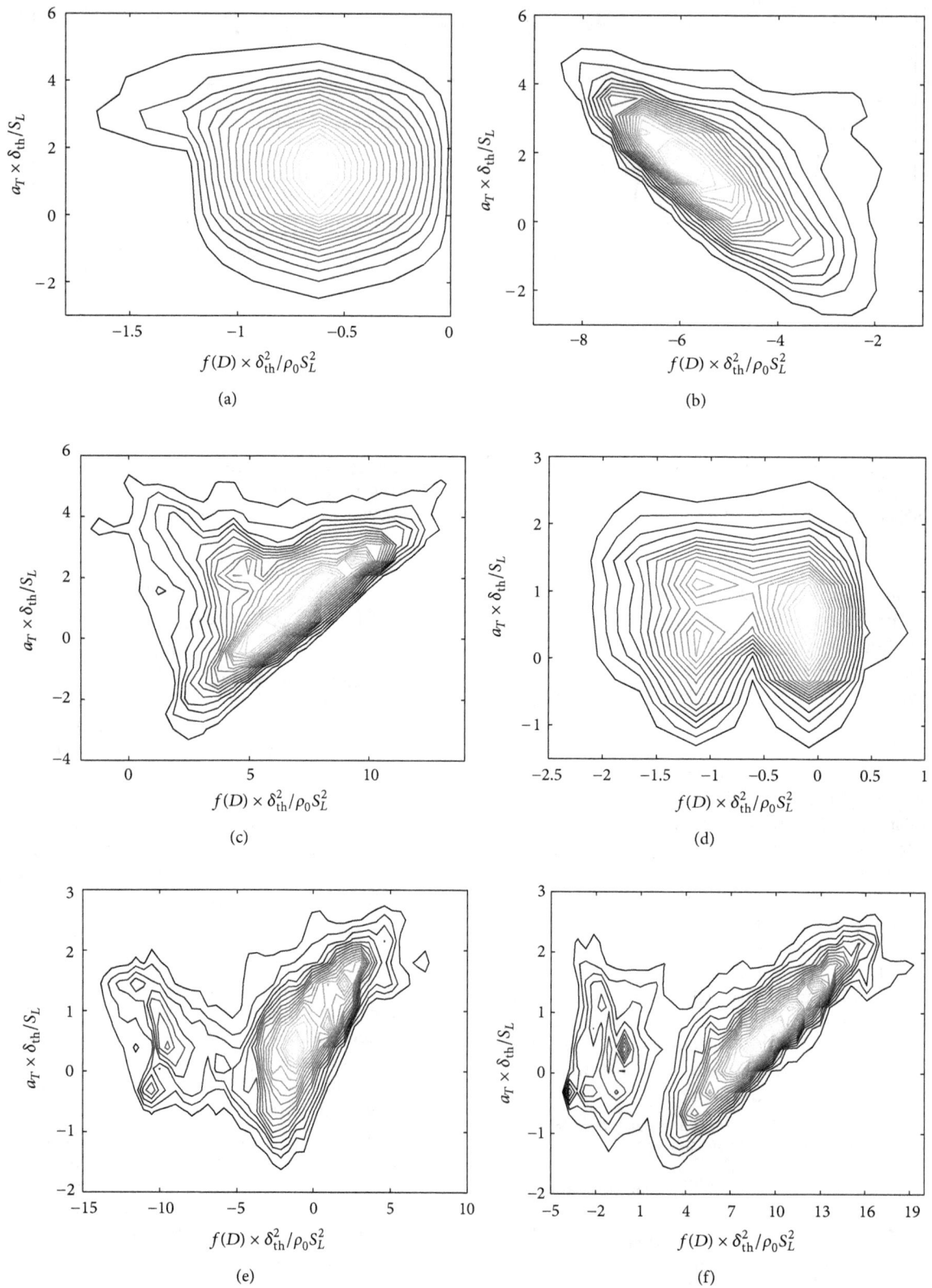

FIGURE 14: Joint pdfs between $f(D) \times \delta_{\text{th}}^2/\rho_0 S_L^2$ and normalised tangential strain rate $a_T \times \delta_{\text{th}}/S_L$ for case P3 on (a) $c = 0.1$, (b) 0.5, and (c) 0.7 isosurfaces. Joint pdfs between $f(D) \times \delta_{\text{th}}^2/\rho_0 S_L^2$ and normalised tangential strain rate $a_T \times \delta_{\text{th}}/S_L$ for case V2 on (d) $c = 0.1$, (e) 0.5, and (f) 0.7 isosurfaces.

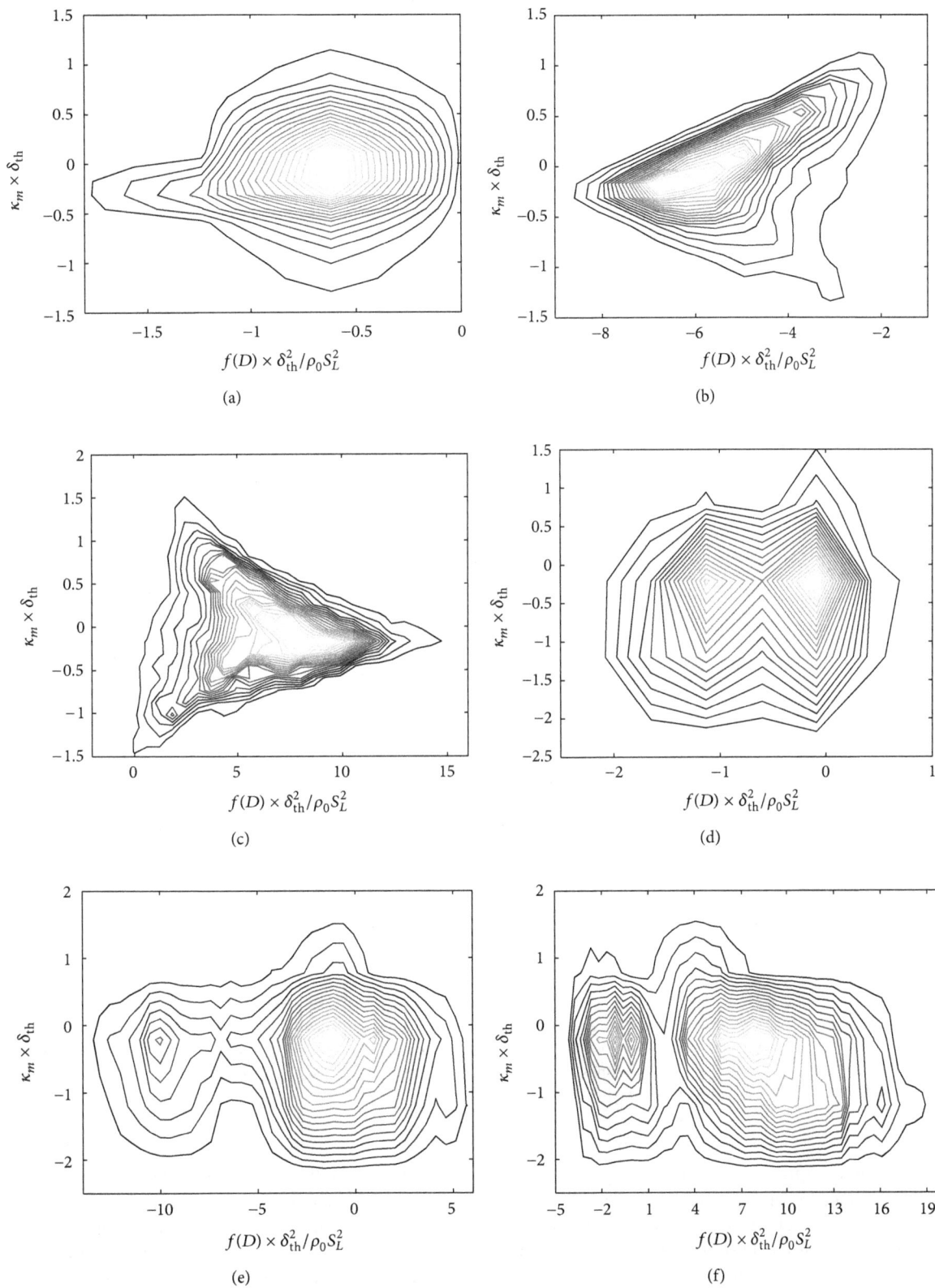

FIGURE 15: Joint pdfs between $f(D) \times \delta_{th}^2/\rho_0 S_L^2$ and normalised curvature $\kappa_m \times \delta_{th}$ on (a) $c = 0.1$, (b) 0.5, and (c) 0.7 isosurfaces for case P3. Joint pdfs between $f(D) \times \delta_{th}^2/\rho_0 S_L^2$ and normalised curvature $\kappa_m \times \delta_{th}$ on (d) $c = 0.1$, (e) 0.5, and (f) 0.7 isosurfaces for case V2.

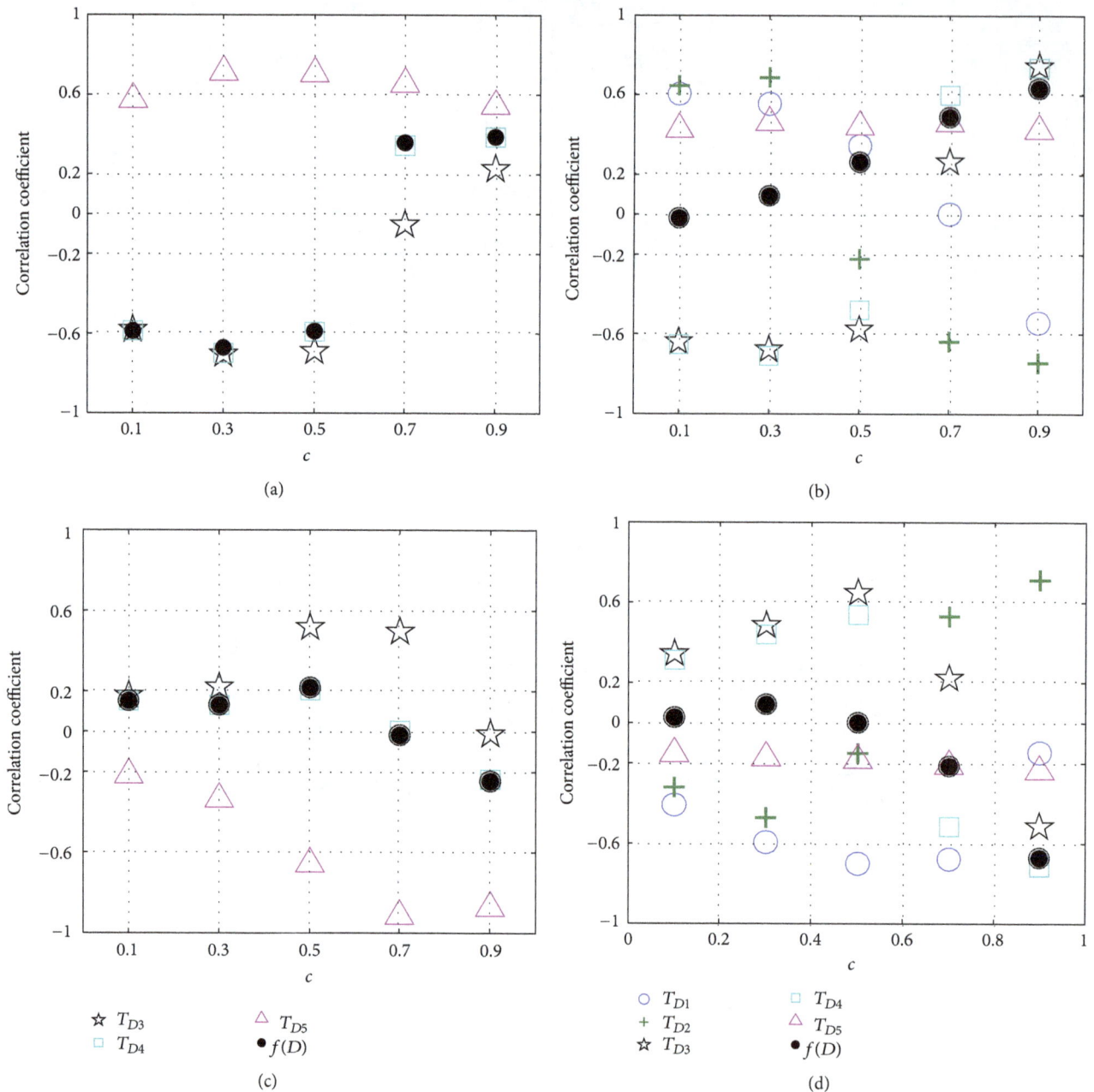

FIGURE 16: (a) Correlation coefficients for the $T_{D3}-a_T$, $T_{D4}-a_T$, $T_{D5}-a_T$, and $f(D)-a_T$ correlations on $c = 0.1, 0.3, 0.5, 0.7$, and 0.9 isosurfaces for case P3. (b) Correlation coefficients for the $T_{D1}-a_T$, $T_{D2}-a_T$, $T_{D3}-a_T$, $T_{D4}-a_T$, $T_{D5}-a_T$, and $f(D)-a_T$ correlations on $c = 0.1, 0.3, 0.5$, 0.7, and 0.9 isosurfaces for case V2. (c) Correlation coefficients for the $T_{D3}-\kappa_m$, $T_{D4}-\kappa_m$, $T_{D5}-\kappa_m$, and $f(D)-\kappa_m$ correlations on $c = 0.1$, $0.3, 0.5, 0.7$, and 0.9 isosurfaces for case P3. (d) Correlation coefficients for the $T_{D1}-\kappa_m$, $T_{D2}-\kappa_m$, $T_{D3}-\kappa_m$, $T_{D4}-\kappa_m$, $T_{D5}-\kappa_m$, and $f(D)-\kappa_m$ correlations on $c = 0.1, 0.3, 0.5, 0.7$, and 0.9 isosurfaces for case V2.

side of the flame. On the other hand, T_{D3} and T_{D4} remain negatively correlated with a_T towards the unburned gas side of the flame (see Figure 16(a)) and these correlations dominate over the positive correlation between T_{D5} and a_T to result in a net negative correlation between $f(D)$ and a_T towards the unburned gas side of the flame.

(v) In cases V1–V3, the terms T_{D3}, T_{D4}, and T_{D5} remain positively correlated with a_T towards the burned gas

side of the flame (see Figure 16(b)) and these positive correlations overcome the negative $T_{D1}-a_T$ and $T_{D2}-a_T$ correlations to give rise to a net positive correlation between $f(D)$ and a_T towards the burned gas side of the flame. By contrast, T_{D3} and T_{D4} remain negatively correlated with a_T towards the unburned gas side of the flame (see Figure 16(b)) and these correlations oppose the positive $T_{D1}-a_T$, $T_{D2}-a_T$, and $T_{D5}-a_T$ correlations to result in a weak correlation between

$f(D)$ and a_T towards the unburned gas side of the flame.

The correlation coefficients between T_{D3}, T_{D4}, and T_{D5} (T_{D1}, T_{D2}, T_{D3}, T_{D4}, and T_{D5}) with κ_m for c = 0.1, 0.3, 0.5, 0.7, and 0.9 isosurfaces are also shown in Figure 16(c) (Figure 16(d)) for case P3 (case V2). It is evident from Figures 16(c) and 16(d) that $f(D)$ and κ_m are weakly correlated throughout the flame for both planar and V-flame cases. The curvature κ_m dependences of T_{D1}, T_{D2}, T_{D3}, T_{D4}, T_{D5}, and $f(D)$ can be explained in the following manner.

(i) Both T_{D3} and T_{D4} remain negatively (positively) correlated with a_T towards the unburned (burned) gas side of the flame for both planar and V-flame cases (see Figures 16(a) and 16(b)), whereas a_T and κ_m are negatively correlated throughout the flame. Thus, high (low) values of T_{D3} and T_{D4} are associated with high positive values of κ_m towards the unburned (burned) gas side of the flame, which gives rise to positive (negative) correlations of T_{D3} and T_{D4} with κ_m towards the unburned (burned) gas side.

(ii) As $T_{D5} = 0.5T_1$ for statistically planar flame cases (i.e., case P1–P5), a strong negative correlation between T_{D5} and κ_m has been observed near $c = 0.7$ isosurface, which is consistent with the negative correlation between T_1 and κ_m shown in Figure 6(e). Even though ρD increases with increasing temperature in V-flame cases (i.e., cases V1–V3), T_{D5} can still be taken to scale with T_1 (i.e., $T_{D5} \sim T_1 \sim \rho \tau N_c S_L/\delta$) and the negative correlation between T_1 and κ_m (see Figure 6(f)) leads to a weak negative correlation between T_{D5} and κ_m in case V2 and also in cases V1 and V3.

(iii) Both T_{D1} and T_{D2} remain positively (negatively) correlated with a_T towards the unburned (burned) gas side of the flame in case V2 (see Figure 16(b)), whereas a_T and κ_m are negatively correlated throughout the flame. Thus, small (high) values of T_{D1} and T_{D2} are associated with high positive (negative) values of κ_m towards the unburned gas side of the flame, which gives rise to negative correlations of T_{D1} and T_{D2} with κ_m towards the unburned gas side in case V2. The combination of negative correlations of T_{D1} and T_{D2} with a_T towards the burned gas side, as well as the strengthening of negative correlation between a_T and κ_m towards the burned gas side of the flame, leads to the weakening of negative correlations of T_{D1} and T_{D2} with κ_m, as the burned gas side is approached and the correlation between T_{D2} with κ_m eventually becomes positive towards the burned gas side of the flame (see Figure 16(d)).

(iv) The terms T_{D3}, T_{D4}, and T_{D5} remain negatively correlated with κ_m towards the burned gas side of the flame and these negative correlations dominate over weak positive $T_{D1} - \kappa_m$ and $T_{D2} - \kappa_m$ correlations to result in a net negative correlation between $f(D)$ and κ_m towards the burned gas side of the flame in case V2. The terms T_{D3} and T_{D4} remain positively correlated with κ_m towards the unburned gas side of the flame

(see Figure 16(d)) and these correlations are opposed by the negative $T_{D1} - \kappa_m$, $T_{D2} - \kappa_m$, and $T_{D5} - \kappa_m$ correlations to result in a weak correlation between $f(D)$ and κ_m towards the unburned gas side of the flame front in case V2 and other V-flame cases.

3.9. Modelling Significances. A modelled transport equation of $\widetilde{N}_c$ may need to be solved alongside other modelled conservation equations in RANS/LES simulations, when the rate of generation of scalar gradients does not remain in equilibrium with its destruction rate. The local strain rate and curvature dependences of SDR are expected to play important roles in LES simulations, as the necessity of capturing local behaviour of $\widetilde{N}_c$ is particularly important in the context of LES. As $\widetilde{N}_c$ approaches N_c with decreasing filter width Δ (i.e., $\lim_{\Delta \to 0} \widetilde{N}_c = N_c$) in the context of LES, the local resolved-scale strain rate and curvature dependences of $\widetilde{N}_c$ and the terms of its transport equation (i.e., $\overline{T}_1, \overline{T}_2, \overline{T}_3, (-\overline{D}_2)$, and $\overline{f(D)}$) are likely to be qualitatively similar to the local strain rate and curvature dependences of N_c and the terms of its transport equation (i.e., $T_1, T_2, T_3, (-D_2)$, and $f(D)$), respectively. The above discussion suggests that the models for $\widetilde{N}_c, \overline{T}_1, \overline{T}_2, \overline{T}_3, (-\overline{D}_2)$, and $\overline{f(D)}$ in LES should be developed in such a manner so that they are capable of capturing the resolved strain rate and curvature dependences of $\widetilde{N}_c, \overline{T}_1, \overline{T}_2, \overline{T}_3, (-\overline{D}_2)$, and $\overline{f(D)}$ for a range of different filter widths and approach local strain rate and curvature dependences of $N_c, T_1, T_2, T_3, (-D_2)$, and $f(D)$ for small filter widths.

It can further be observed from Figures 5–16 that the local strain rate and curvature dependence of N_c and the terms of its transport equation for V-flames are found to be qualitatively similar to the behaviour observed for the statistically planar flame cases. As the SDR statistics are principally governed by the small-scale molecular processes, the local statistics of N_c and the terms of its transport equation are largely independent of the flow configuration. Thus, the models for $\widetilde{N}_c$ transport developed based on data extracted from a canonical configuration might broadly be applicable to different geometries.

Moreover, it is worth noting that ρD is assumed to be constant in cases P1–P5 whereas ρD is taken to be temperature dependent in cases V1–V3. However, the statistical behaviours of N_c and the terms of its transport equation are found to be broadly similar qualitatively in all cases indicating that the models, which have been developed based on DNS databases with constant ρD, should at least be able to capture the qualitative trends of $\widetilde{N}_c$ transport.

4. Conclusions

The statistical behaviours of the instantaneous SDR N_c and the terms of its transport equation have been analysed using simple chemistry DNS databases of statistically planar and V-flames for a range of different values of turbulent Reynolds number. It has been found that the mean and local behaviours of N_c and the terms of its transport equation are similar for both statistically planar and V-flames for the range of turbulent Reynolds number explored here. In all the cases, N_c is positively correlated with tangential strain rate a_T throughout the flame. By contrast, N_c and local curvature κ_m remain negatively correlated for small values of u'/S_L but the joint pdf of N_c and κ_m shows branches with both positive and negative correlation, and the net correlation becomes weak for high values of u'/S_L. It has been found that the mean contributions of the density-variation term T_1 and the molecular dissipation term $(-D_2)$ in the transport equation of N_c are the leading order source and sink, respectively. The mean value of the strain rate contribution T_2 to the SDR transport is predominantly negative for the major part of the flame due to predominant ∇c alignment with the most extensive principal strain rate e_α, although positive contributions of T_2 were observed towards the unburned gas side for high u'/S_L cases where ∇c preferentially aligns with the most compressive principal strain rate e_γ. The mean value of the reaction rate contribution to the SDR transport T_3 remains positive for the major part of the flame before assuming negative values towards the burned gas side. The mean contribution of the term originating due to the diffusivity gradient in the SDR transport $f(D)$ remains negative towards the unburned gas side before assuming positive values towards the burned gas side of the flame.

It has been found that the density variation term T_1 remains positively correlated with tangential strain rate a_T, whereas the correlation between T_1 and the local curvature κ_m is negative throughout the flame. The strain rate term T_2 is predominantly positively correlated with both a_T and κ_m for the major part of the flame front. The qualitative nature of the local strain rate and curvature dependences of T_3, $(-D_2)$, and $f(D)$ change across the flame. The reaction rate contribution T_3 and the tangential strain rate a_T remain positively (negatively) correlated towards the unburned (burned) gas side of the flame. However, T_3 is weakly correlated with curvature κ_m throughout the flame for all cases considered here. The molecular dissipation term $(-D_2)$ is negatively correlated with a_T throughout the flame, whereas the joint pdf of $(-D_2)$ and κ_m shows branches with both positive and negative correlation and their qualitative behaviours change across the flame. The diffusivity gradient term $f(D)$ and a_T are found to be negatively (positively) correlated with each other towards the unburned (burned) gas side of the flame, whereas the joint pdfs of $f(D)$ and κ_m show a weak positive (negative) correlation in the unburned (burned) gas side of the flame. Detailed physical explanations have been provided for the strain rate and curvature dependences of N_c and the terms of its transport equation. The qualitative nature of these statistics has been found to be unaltered for the range of turbulent Reynolds number Re_t considered here, but the strength of the correlations is affected by Re_t. Moreover, the local strain rate and curvature dependence of N_c and the terms of its transport equation for V-flames are found to be broadly similar qualitatively to the behaviour observed for the statistically planar flame cases. Moreover, the assumption of mass diffusivity variation with temperature has been shown not to affect the qualitative behaviour of SDR and its transport statistics.

In the context of single step chemistry, the reaction progress variable c can be uniquely defined, but c can be defined based on different species mass fractions in the presence of detailed chemistry. However, the conclusions drawn in this analysis are unlikely to change if c is defined based on the mass fraction of a major reactant/product, which is closely correlated with density change and heat release. The present single-step Arrhenius type chemistry qualitatively captures the statistics of $|\nabla c|$ transport obtained using detailed chemistry based simulations for the flames with global Lewis number close to unity. This can be substantiated by qualitative similarities between the strain rate and curvature dependence of the terms of $|\nabla c|$ transport equation obtained from detailed chemistry [14] and single-step chemistry [15] based DNS simulations. Given the close relation between $|\nabla c|$ and SDR, it can be expected that the conclusions drawn regarding SDR transport will at least be qualitatively valid for detailed chemistry based simulations.

It is worth noting that the effects of differential diffusion of heat and mass are not addressed in the present analysis, and the presence of differential diffusion may have influences on the local strain rate and curvature dependences of N_c and the different terms of its transport equation. As the SDR statistics are principally governed by the small-scale molecular processes, the local statistics of N_c and the terms of its transport equation obtained from this analysis are expected to be qualitatively similar for higher values of Re_t than the values of turbulent Reynolds number considered here. However, this analysis has been carried out for moderate values of Re_t; thus further experimental and computational studies at large values of Re_t in the presence of detailed chemistry are needed for further confirmation and deeper understanding of the statistics of SDR transport in turbulent premixed flames.

Conflict of Interests

The authors declare that there is no conflict of interests regarding the publication of this paper.

Acknowledgment

The authors are grateful to the EPSRC, UK, for the financial support.

References

[1] R. W. Bilger, "Some aspects of scalar dissipation," *Flow, Turbulence and Combustion*, vol. 72, no. 2-4, pp. 93–114, 2004.

[2] K. N. C. Bray, "Turbulent flows with premixed reactants," in *Turbulent Reacting Flows*, P. A. Libby and F. A. Williams, Eds., pp. 115–183, Springer, New York, NY, USA, 1980.

[3] N. Chakraborty, M. Champion, A. Mura, and N. Swaminathan, "Scalar dissipation rate approach to reaction rate closure," in *Turbulent Premixed Flame*, N. Swaminathan and K. N. C. Bray, Eds., pp. 76–102, Cambridge University Press, Cambridge, UK, 1st edition, 2011.

[4] T. D. Dunstan, Y. Minamoto, N. Chakraborty, and N. Swaminathan, "Scalar dissipation rate modelling for Large Eddy

Simulation of turbulent premixed flames," *Proceedings of the Combustion Institute*, vol. 34, pp. 11193–11201, 2013.

[5] N. Peters, P. Terhoeven, J. H. Chen, and T. Echekki, "Statistics of flame displacement speeds from computations of 2-D unsteady methane-air flames," *Proceedings of the Combustion Institute*, vol. 27, pp. 833–839, 1998.

[6] J. H. Chen and H. G. Im, "Correlation of flame speed with stretch in turbulent premixed methane/air flames," *Proceedings of the Combustion Institute*, vol. 27, pp. 819–826, 1998.

[7] T. Echekki and J. H. Chen, "Analysis of the contribution of curvature to premixed flame propagation," *Combustion and Flame*, vol. 118, no. 1-2, pp. 308–311, 1999.

[8] J. H. Chen and H. G. Im, "Stretch effects on the burning velocity of turbulent premixed hydrogen/air flames," *Proceedings of the Combustion Institute*, vol. 28, no. 1, pp. 211–218, 2000.

[9] H. G. Im and J. H. Chen, "Preferential diffusion effects on the burning rate of interacting turbulent premixed hydrogen-air flames," *Combustion and Flame*, vol. 131, no. 3, pp. 246–258, 2002.

[10] N. Chakraborty and S. Cant, "Unsteady effects of strain rate and curvature on turbulent premixed flames in an inflow-outflow configuration," *Combustion and Flame*, vol. 137, no. 1-2, pp. 129–147, 2004.

[11] N. Chakraborty and R. S. Cant, "Effects of strain rate and curvature on surface density function transport in turbulent premixed flames in the thin reaction zones regime," *Physics of Fluids*, vol. 17, no. 6, pp. 1–15, 2005.

[12] E. R. Hawkes, J. H. Chen, E. R. Hawkes, and J. H. Chen, "Evaluation of models for flame stretch due to curvature in the thin reaction zones regime," vol. 30, pp. 647–653.

[13] N. Chakraborty and R. S. Cant, "A priori analysis of the curvature and propagation terms of the flame surface density transport equation for large eddy simulation," *Physics of Fluids*, vol. 19, no. 10, Article ID 105101, 2007.

[14] N. Chakraborty, E. R. Hawkes, J. H. Chen, and R. S. Cant, "The effects of strain rate and curvature on surface density function transport in turbulent premixed methane-air and hydrogen-air flames: a comparative study," *Combustion and Flame*, vol. 154, no. 1-2, pp. 259–280, 2008.

[15] N. Chakraborty and M. Klein, "Influence of Lewis number on the surface density function transport in the thin reaction zone regime for turbulent premixed flames," *Physics of Fluids*, vol. 20, no. 6, Article ID 065102, 2008.

[16] N. Chakraborty and M. Klein, "Effects of global flame curvature on surface density function transport in turbulent premixed flame kernels in the thin reaction zones regime," in *Proceedings of the 32nd International Symposium on Combustion*, pp. 1435–1443, August 2008.

[17] N. Swaminathan and K. N. C. Bray, "Effect of dilatation on scalar dissipation in turbulent premixed flames," *Combustion and Flame*, vol. 143, no. 4, pp. 549–565, 2005.

[18] R. Borghi and D. Dutoya, "On the scales of the fluctuations in turbulent combustion," *Symposium (International) on Combustion*, vol. 17, no. 1, pp. 235–244, 1979.

[19] T. Mantel and R. Borghi, "A new model of premixed wrinkled flame propagation based on a scalar dissipation equation," *Combustion and Flame*, vol. 96, no. 4, pp. 443–457, 1994.

[20] R. Borghi, "Turbulent premixed combustion: further discussions on the scales of fluctuations," *Combustion and Flame*, vol. 80, no. 3-4, pp. 304–312, 1990.

[21] A. Mura and R. Borghi, "Towards an extended scalar dissipation equation for turbulent premixed combustion," *Combustion and Flame*, vol. 133, no. 1-2, pp. 193–196, 2003.

[22] N. Swaminathan and R. W. Grout, "Interaction of turbulence and scalar fields in premixed flames," *Physics of Fluids*, vol. 18, no. 4, Article ID 045102, 2006.

[23] N. Chakraborty and N. Swaminathan, "Influence of the Damköhler number on turbulence-scalar interaction in premixed flames. I: physical insight," *Physics of Fluids*, vol. 19, no. 4, Article ID 045103, 2007.

[24] N. Chakraborty and N. Swaminathan, "Influence of the Damköhler number on turbulence-scalar interaction in premixed flames. II: model development," *Physics of Fluids*, vol. 19, no. 4, Article ID 045104, 2007.

[25] N. Chakraborty, J. W. Rogerson, and N. Swaminathan, "A priori assessment of closures for scalar dissipation rate transport in turbulent premixed flames using direct numerical simulation," *Physics of Fluids*, vol. 20, no. 4, Article ID 045106, 2008.

[26] A. Mura, K. Tsuboi, and T. Hasegawa, "Modelling of the correlation between velocity and reactive scalar gradients in turbulent premixed flames based on DNS data," *Combustion Theory and Modelling*, vol. 12, no. 4, pp. 671–698, 2008.

[27] A. Mura, V. Robin, M. Champion, and T. Hasegawa, "Small scale features of velocity and scalar fields in turbulent premixed flames," *Flow, Turbulence and Combustion*, vol. 82, no. 3, pp. 339–358, 2009.

[28] N. Chakraborty, J. W. Rogerson, and N. Swaminathan, "The scalar gradient alignment statistics of flame kernels and its modelling implications for turbulent premixed combustion," *Flow, Turbulence and Combustion*, vol. 85, no. 1, pp. 25–55, 2010.

[29] N. Chakraborty and N. Swaminathan, "Effects of lewis number on scalar dissipation transport and its modeling in turbulent premixed combustion," *Combustion Science and Technology*, vol. 182, no. 9, pp. 1201–1240, 2010.

[30] N. Chakraborty, M. Klein, and N. Swaminathan, "Effects of Lewis number on reactive scalar gradient alignment with local strain rate in turbulent premixed flames," *Proceedings of the Combustion Institute*, vol. 32, no. 1, pp. 1409–1417, 2009.

[31] J. H. Chen, A. Choudhary, B. de Supinski et al., "Terascale direct numerical simulations of turbulent combustion using S3D," *Computational Science and Discovery*, vol. 2, no. 1, Article ID 015001, 2009.

[32] K. W. Jenkins and R. S. Cant, "DNS of turbulent flame kernels," in *Proceedings of the 2nd AFOSR Conference on DNS and LES*, Knight and Sakell, Eds., pp. 192–202, Rutgers University, Kluwer Academic Publishers, 1999.

[33] T. J. Poinsot, "Boundary conditions for direct simulations of compressible viscous flows," *Journal of Computational Physics*, vol. 101, no. 1, pp. 104–129, 1992.

[34] A. A. Wray, "Minimal storage time advancement schemes for spectral methods," Report MS 202 A-1, NASA Ames Research Center, 1990.

[35] N. Peters, *Turbulent Combustion*, Cambridge University Press, Cambridge, UK, 2000.

[36] F. Charlette, C. Meneveau, and D. Veynante, "A power-law flame wrinkling model for LES of premixed turbulent combustion. Part I: non-dynamic formulation and initial tests," *Combustion and Flame*, vol. 131, no. 1-2, pp. 159–180, 2002.

[37] W. R. Grout, "An age extended progress variable for conditioning reaction rates," *Physics of Fluids*, vol. 19, no. 10, Article ID 105107, 2007.

[38] I. Han and K. Y. Huh, "Roles of displacement speed on evolution of flame surface density for different turbulent intensities and Lewis numbers in turbulent premixed combustion," *Combustion and Flame*, vol. 152, no. 1-2, pp. 194–205, 2008.

[39] N. Chakraborty, G. Hartung, M. Katragadda, and C. F. Kaminski, "Comparison of 2D and 3D density-weighted displacement speed statistics and implications for laser based measurements of flame displacement speed using direct numerical simulation data," *Combustion and Flame*, vol. 158, no. 7, pp. 1372–1390, 2011.

[40] T. D. Dunstan, N. Swaminathan, K. N. C. Bray, and R. S. Cant, "Geometrical properties and turbulent flame speed measurements in stationary premixed V-flames using direct numerical simulation," *Flow, Turbulence and Combustion*, vol. 87, no. 2-3, pp. 237–259, 2011.

[41] T. D. Dunstan, N. Swaminathan, and K. N. C. Bray, "Influence of flame geometry on turbulent premixed flame propagation: a DNS investigation," *Journal of Fluid Mechanics*, vol. 709, pp. 191–222, 2012.

[42] M. Boger, D. Veynante, H. Boughanem, and A. Trouve, "Direct numerical simulation analysis of flame surface density concept for large eddy simulation of turbulent premixed combustion," *Symposium (International) on Combustion*, vol. 1, pp. 917–925, 1998.

[43] N. Chakraborty and N. Swaminathan, "Reynolds number effects on scalar dissipation rate transport and its modelling in turbulent premixed combustion," *Combustion Science and Technology*, vol. 185, article 4, pp. 676–709, 2013.

[44] C. Meneveau and T. Poinsot, "Stretching and quenching of flamelets in premixed turbulent combustion," *Combustion and Flame*, vol. 86, no. 4, pp. 311–332, 1991.

[45] H. Tennekes and J. L. Lumley, *A First Course in Turbulence*, MIT press, Cambridge, Mass, USA, 1972.

[46] K. R. Sreenivasan and R. A. Antonia, "The phenomenology of small-scale turbulence," *Annual Review of Fluid Mechanics*, vol. 29, pp. 435–472, 1997.

[47] R. A. Antonia and K. R. Sreenivasan, "Log-normality of temperature dissipation in a turbulent boundary layer," *Physics of Fluids*, vol. 20, no. 11, pp. 1800–1804, 1977.

[48] J. Mi, R. A. Antonia, and F. Anselmet, "Joint statistics between temperature and its dissipation rate components in a round jet," *Physics of Fluids*, vol. 7, no. 7, pp. 1665–1673, 1995.

[49] L. K. Su and N. T. Clemens, "The structure of fine-scale scalar mixing in gas-phase planar turbulent jets," *Journal of Fluid Mechanics*, no. 488, pp. 1–29, 2003.

[50] A. N. Karpetis and R. S. Barlow, "Measurements of scalar dissipation in a turbulent piloted methane/air jet flame," *Proceedings of the Combustion Institute*, vol. 29, no. 2, pp. 1929–1936, 2002.

[51] D. Geyer, A. Kempf, A. Dreizler, and J. Janicka, "Scalar dissipation rates in isothermal and reactive turbulent opposed-jets: 1-D Raman/Rayleigh experiments supported by LES," *Proceedings of the Combustion Institute*, vol. 30, no. 1, pp. 681–689, 2005.

[52] C. N. Markides and E. Mastorakos, "Measurements of scalar dissipation in a turbulent plume with planar laser-induced fluorescence of acetone," *Chemical Engineering Science*, vol. 61, no. 9, pp. 2835–2842, 2006.

[53] W. P. Jones and P. Musonge, "Closure of the Reynolds stress and scalar flux equations," *Physics of Fluids*, vol. 31, no. 12, pp. 3589–3604, 1988.

[54] P. K. Yeung, S. S. Girimaji, and S. B. Pope, "Straining and scalar dissipation on material surfaces in turbulence: Implications for flamelets," *Combustion and Flame*, vol. 79, no. 3-4, pp. 340–365, 1990.

[55] E. Hawkes, R. R. Sankaran, J. C. Sutherland, and J. H. Chen, "Scalar mixing in direct numerical simulations of temporally evolving plane jet flames with skeletal CO/H_2 kinetics," *Proceedings of the Combustion Institute*, vol. 31, no. 1, pp. 1633–1640, 2007.

[56] N. Swaminathan and R. W. Bilger, "Scalar dissipation, diffusion and dilatation in turbulent H2-air premixed flames with complex chemistry," *Combustion Theory and Modelling*, vol. 5, no. 3, pp. 429–446, 2001.

[57] N. Chakraborty, M. Klein, and R. S. Cant, "Effects of turbulent reynolds number on the displacement speed statistics in the thin reaction zones regime of turbulent premixed combustion," *Journal of Combustion*, vol. 2011, Article ID 473679, 19 pages, 2011.

[58] G. Hartung, J. Hult, C. F. Kaminski, J. W. Rogerson, and N. Swaminathan, "Effect of heat release on turbulence and scalar-turbulence interaction in premixed combustion," *Physics of Fluids*, vol. 20, no. 3, Article ID 035110, 2008.

[59] K. W. Jenkins, M. Klein, N. Chakraborty, and R. S. Cant, "Effects of strain rate and curvature on the propagation of a spherical flame kernel in the thin-reaction-zones regime," *Combustion and Flame*, vol. 145, no. 1-2, pp. 415–434, 2006.

[60] M. Klein, N. Chakraborty, K. W. Jenkins, and R. S. Cant, "Effects of initial radius on the propagation of premixed flame kernels in a turbulent environment," *Physics of Fluids*, vol. 18, no. 5, Article ID 055102, 2006.

5

On Laminar Rich Premixed Polydisperse Spray Flame Propagation with Heat Loss

G. Kats and J. B. Greenberg

Faculty of Aerospace Engineering, Technion-Israel Institute of Technology, 32000 Haifa, Israel

Correspondence should be addressed to J. B. Greenberg; aer9801@technion.ac.il

Academic Editor: Ashwani K. Gupta

A mathematical analysis of laminar premixed spray flame propagation with heat loss is presented. The analysis makes use of a distributed approximation of the Arrhenius exponential term in the reaction rate expression and leads to an implicit expression for the laminar burning velocity dependent on the spray-related parameters for the fuel, gas-related parameters and the intensity of the heat losses. It is shown that the initial droplet load, the value of the evaporation coefficient, and the initial size distribution are the spray-related parameters which exert an influence on the onset of extinction. The combination of these parameters governs the manner in which the spray heat loss is distributed spatially and it is this feature that is the main factor, when taken together with volumetric heat loss, which determines the spray's impact on flame propagation and extinction.

1. Introduction

Spalding [1] was the first to treat the problem of a laminar gas flame propagating through a combustible premixture in the presence of heat losses, for example, due to heat loss by conduction to the walls of the combustion chamber or radiation. In keeping with experimental evidence it was found that for a given heat loss there exist two possible burning velocities, one stable and the other not. Extinction occurs at the point of traversal from the stable to the unstable mode of propagation. Essentially, this happens when the heat loss is too great. The theory agreed well with experiments for flame propagation and extinction in tubes. Subsequent work [2, 3] also dealt with similar problems of one-dimensional flame propagation and examined different aspects of extinction of these flames. Buckmaster [4] reexamined the aforementioned problem using asymptotic tools and was able to construct the slow and fast waves as well as to predict a simple explicit quenching criterion. Joulin and Clavin [5] considered the stability of laminar premixed flames subject to linear heat loss and found a variety of instabilities for the different regimes (slow and fast waves) examined by previous researchers. Nicoli and Clavin [6] considered the effect of variable heat loss intensities on the dynamics of a premixed flame. Clavin and Nicoli [7] investigated heat loss effects on stability limits of downward propagating premixed flames.

In the context of mathematical analysis of one-dimensional premixed spray flames, some attention was directed to the influence of heat losses [8, 9], although when the stability of such flames was considered heat losses were not accounted for [10–13]. However, in [8, 9] the linear volumetric heat losses were taken as being of order ε, where ε is inversely proportional to the activation energy of the assumed global chemical reaction, and were only applied in the region between the onset of droplet evaporation and the flame front. In addition, the sprays were taken to be monodisperse.

As pointed out by Sirignano [14] radiation impacts on *individual* droplet heating and evaporation in several ways. Primarily, droplets may be heated by radiation from high temperature gases. Or, alternatively, radiation may decrease the flame temperature so that radiative (and conductive) heat transfer to droplets will be diminished. In modeling the behavior of single droplets the radiative heating effect is expressed via a modification to the latent heat of evaporation. However, Sazhin [15], in discussing single droplets, argues

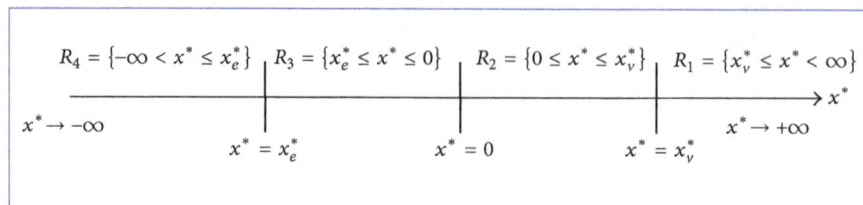

FIGURE 1: Schematic of the subregions considered for the analysis of a one-dimensional planar premixed spray flame in the presence of volumetric/radiative heat losses.

that taking them as grey opaque bodies "overlooks the fact that droplet radiative heating takes place not at their surface (as in the case of convective heating) but via the absorption of thermal radiation penetrating inside the droplets." He therefore assumes for modelling purposes that the droplet is semitransparent.

In the current paper we investigate the propagation of an off-stoichiometric rich laminar premixed polydisperse spray flame in the presence of heat loss, for the first time. In this paper we restrict our attention exclusively to the stable branch of propagation down to conditions of extinction. Our ultimate aim is to explore spray flame ignition since during the first moments of application of the igniter the role of heat losses can be rather dominant. It is a well-established fact that modern combustors in aircraft need to satisfy a large number of requirements. Of particular interest is the fact that under extreme conditions they must reignite following flame extinction without any problems and without any external help. The possibility of extinction also exists in cold and wet conditions (e.g., in a hailstorm) as well as at high altitudes due to oxygen starvation. The presence of liquid fuel in the form of a multisized spray of droplets that must first produce a sufficient amount of fuel vapor for successful ignition increases the difficulties. The current work is a prelude to such an ignition study that will be reported in the future.

In a previous publication [16] we modified a nonasymptotic mathematical approach [17, 18] to analyzing gas flame propagation and successfully applied it to examine the propagation of liquid fuel spray flames and double spray flames (i.e., both fuel and oxidizer supplied as a spray of droplets). For propagation studies this approach seems to be a viable alternative to an asymptotic approach. Here we adopt the same methodology.

The structure of the paper is as follows. We present the governing equations and the assumptions upon which they are based. We then explain how they are solved and present their solution. Finally, we examine how the combination of volumetric and spray-related heat losses influences the propagation and extinction conditions of the spray flames.

2. Governing Equations and Problem Definition

2.1. Assumptions. We consider a laminar one-dimensional premixed flame propagating into an off-stoichiometric fresh homogeneous mixture of fuel vapor, liquid fuel droplets, oxygen, and an inert gas. A schematic of the situation

considered is shown in Figure 1. The flame is taken to propagate from left to right. The droplets are viewed from a far-field vantage point; that is, their average velocity is equal to that of their host environment. For qualitative purposes this approach has been demonstrated to be quite valid [19]. The spray is taken to be polydisperse; that is, at any point in space and time there is a distinct size distribution of the spray's droplets. The temperature of the droplets is taken to be that of the surroundings; essentially the droplets heat-up time is small compared to the characteristic time associated with their motion. Droplet evaporation is assumed negligible until a prescribed reference temperature T_v (such as the boiling temperature of the liquid fuel) is attained.

The stoichiometry of the gas mixture that the flame front meets is taken to be fuel rich, so that the limiting reactant consumed by chemical reaction is oxygen. It is assumed that the various transport coefficients, such as thermal conductivity, diffusion coefficient, specific heat at constant pressure, and latent heat of vaporization of the droplets, can be satisfactorily specified by representative constant values. An overall reaction of the form $\nu_F \text{ fuel} + \nu_O \text{ oxidant} \rightarrow$ products is taken to describe the chemistry.

As the velocity of propagation of the flame is much less than the velocity of sound, dynamic compressibility effects in the mixture can be neglected. Thus, the density becomes only a function of the temperature through the gas law.

The polydisperse spray is described using the sectional method [20] in which the droplet size-distribution is divided into sections (or bins) and conservation equations are derived for the liquid fuel in each section allowing for droplet evaporation from a given section, for example, j, and addition to that section as droplets evaporate in the next section up and become eligible for membership in j.

The spatial region from $x^* \rightarrow +\infty$ to $x^* \rightarrow -\infty$ can be divided into four distinct regions:

(a) A prevaporization region where the system's temperature has not yet reached the critical temperature of the fuel at which significant evaporation begins to occur. This region is denoted by $R_1 = \{x_v^* \leq x^* < \infty\}$.

(b) A preflame region in which chemical reaction has not yet begun because the temperature is less than the temperature required to initiate it. This is region $R_2 = \{0 \leq x^* \leq x_v^*\}$.

(c) The flame region where reaction occurs, denoted by $R_3 = \{x_e^* \leq x^* \leq 0\}$.

(d) A postflame region where the reaction has essentially ceased because, due to heat loss, the temperature drops below the temperature that sustains the flame, $R_4 = \{-\infty < x^* \le x_e^*\}$.

Note that Figure 1 is simply a schematic drawing of the four aforementioned regions and does not reflect their actual scales which will be determined by the spray and gas environment parameters.

2.2. Governing Equations.
In the presence of linear heat loss the governing conservation equations can be shown to be

$$\rho c_p \frac{\partial T}{\partial t^*} = \lambda \frac{\partial^2 T}{\partial x^{*2}} + \rho q Y_O A \exp\left(-\frac{E}{RT}\right) - \widehat{\alpha}(T - T_0)$$
$$- L S_v,$$
(1)

$$\rho \frac{\partial Y_O}{\partial t^*} = \rho D_O \frac{\partial^2 Y_O}{\partial x^{*2}} - \rho Y_O A \exp\left(-\frac{E}{RT}\right),$$
(2)

$$\rho \frac{\partial Y_{d,j}}{\partial t^*} = -S_{v,j}, \quad j = 1, 2, 3, \ldots, N_S,$$
(3)

where

$$S_{vj} = \Delta_j Y_{d,j} - \Psi_j Y_{d,j+1}$$
(4)

is the source term for production of liquid fuel in droplets of section j, and the sectional vaporization coefficients are (see [20])

$$(\Delta_j, \Psi_j) = 1.5 \widehat{E}_v \left(\frac{3d_{uj} - 2d_{lj}}{d_{uj}^3 - d_{lj}^3}, \frac{d_{lj+1}}{d_{uj+1}^3 - d_{lj+1}^3}\right),$$
(5)

$$j = 1, 2, \ldots, N_S, \quad \Psi_{N_S} = 0,$$

where $\widehat{E}_v$ is the liquid fuel's evaporation coefficient and d_{lj} and d_{uj} are the lower and upper diameters, respectively, defining droplet size section j.

$$S_v = \sum_{j=1}^{N_S} S_{vj}$$
(6)

S_v is the source term for the production of fuel vapor by evaporating droplets in all sections. The sectional Damkohler numbers are based on d^2 law, which was confirmed by previous studies [21] to predict the actual vaporization history of an interacting droplet, especially in the initial period of combustion.

Note that the terms resulting from the chemical reaction are linear in the deficient reactant, which, for the rich off-stoichiometric case herein discussed, is oxidant. (The near-stoichiometric case, in which the product of the oxidant and fuel vapor concentrations is present, is not considered here since the approach to be adopted was reported [18] as yielding considerably less satisfactory results under such circumstances.)

The boundary conditions for this set of equations are

$$x^* \longrightarrow \infty: \quad Y_O = m_{Ou}, \ T = T_0, \ Y_{d,j} = \delta_j,$$
$$x^* \longrightarrow -\infty: \quad Y_O = 0, \ T = T_0, \ Y_d = 0.$$
(7)

2.3. Solution Approach.
Following [18, 22] the nonlinear nature of the chemical source terms can be alleviated by replacing the exponential temperature-dependent term $A \exp(-E/RT)$ by the step function $k(T_m)H(T - T_*)$ in which

$$k(T_m) = A \exp\left(-\frac{E}{RT_m}\right).$$
(8)

T_* is found so that there is equality of the integrals with respect to T of both the Arrhenius exponential and the step function over the entire range of temperatures (from the cold mixture temperature, T_0, to the highest temperature attained in the system, T_m):

$$\int_{T_0}^{T_m} k(T_m) H(T - T_*) dT = \int_{T_0}^{T_m} A \exp\left(\frac{-E}{RT}\right) dT.$$
(9)

It is not hard to show that this yields

$$\int_{T_0}^{T_m} k(T_m) H(T - T_*) dT = k(T_m)(T_m - T_*);$$
$$\int_{T_0}^{T_m} A \exp\left(\frac{-E}{RT}\right) dT \approx \frac{RT_m^2}{E} k(T_m)$$
(10)

and we obtain

$$T_* = T_m(1 - \varepsilon), \quad \varepsilon = \frac{RT_m}{E} \ll 1$$
(11)

so that in (1) and (2) use is made of

$$\widehat{K}_O(T) = A \exp\left(-\frac{E}{RT_m}\right) H(T - T_*)$$
$$= k(T_m) H(T - T_*).$$
(12)

For a spray flame the heat loss due to droplet evaporation also plays a role in the heat balance and it is therefore introduced via T_m, the maximum temperature attained in the system. In this way the step function specifically reflects the maximum value of the Arrhenius exponential function thereby capturing the essence of the physical meaning of that function.

The above governing equations (with their appropriate boundary conditions) describe the spray flame propagation from the perspective of laboratory coordinates. However, in order to extract a solution it proves more straightforward to rewrite the equations in coordinates attached to the flame front. Assuming the flame is moving at a constant speed u in the positive x-direction (i.e., propagation from left to right) we can define the new coordinate $\xi = x^* - u \cdot t^*$, whereby the governing set of equations reduces to the following ordinary differential equations:

$$D_T \frac{\partial^2 T}{\partial \xi^2} + u \frac{\partial T}{\partial \xi} + \frac{q}{C_p} Y_O \widehat{K}_O(T) - \frac{L}{C_p} S_v \widehat{K}_d(T)$$
$$- \alpha(T - T_0) = 0,$$
(13)

$$D_O \frac{\partial^2 Y_O}{\partial \xi^2} + u \frac{\partial Y_O}{\partial \xi} - Y_O \widehat{K}_O(T) = 0,$$
(14)

$$u \frac{\partial Y_{d,j}}{\partial \xi} - S_{vj} \widehat{K}_d(T) = 0, \quad j = 1, 2, 3, \ldots, N_S,$$
(15)

where

$$\widehat{K}_d(T) = H(T - T_v) \tag{16}$$

with H being the Heaviside function used to ensure that droplet evaporation only occurs significantly once the temperature of the fuel droplets reaches the fuel's boiling point, $D_T = \lambda/\rho c_p$ is the thermal diffusion coefficient, and

$\alpha = \widehat{\alpha}/\rho c_p$ is the heat loss coefficient. Henceforth, for convenience, we assume a unity Lewis number ($D_T = D_O$).

Note that the chemical source terms in (13) and (14) are applicable only for $T \geq T_*$ but due to our coordinate transformation we can replace $H(T - T_*)$ by $H(-\xi)$ as the system is invariant under spatial translation and we can take the location of the interface between the second and third regions as = 0.

Accordingly, the boundary and matching conditions that are applicable are

$$\xi = \xi_v: \quad [Y_O] = [Y'_O] = 0, \ [Y_{d,j}] = 0, \ T = T_v, \ [T] = [T'] = 0, \tag{17a}$$

$$\xi = 0: \quad [Y_O] = [Y'_O] = 0, \ [Y_{d,j}] = 0, \ T = T_*, \ [T] = [T'] = 0, \tag{17b}$$

$$\xi = \xi_e: \quad [Y_O] = [Y'_O] = 0, \ [Y_{d,j}] = 0, \ T = T_*, \ [T] = [T'] = 0, \tag{17c}$$

$$\xi = \xi_m: \quad [Y_O] = [Y'_O] = 0, \ [Y_{d,j}] = 0, \ T = T_m, \ [T] = [T'] = 0, \ T'(\xi_m) = 0, \tag{17d}$$

where derivatives with respect to ξ are denoted by $'$ and ξ_v, ξ_e, and ξ_m are, respectively, the interface locations where finite-rate vaporization begins and the interface between the region where the chemical reaction takes place and the region where it ceases and a spatial point where the maximum temperature T_m is attained.

Physically, ξ_m must be located in the middle of the third region $\xi_e < \xi_m < 0$, where the reaction is adding heat to the system but the volumetric/radiative heat loss is of considerable competitive importance.

3. Solution

The solution of the governing equations is found in every region separately with the matching conditions connecting the solutions. We present the results for the four relevant regions:

In $R_1 = \{\xi_v \leq \xi < \infty\}$

$$T(\xi) = T_0 + (T_v - T_0) \exp\left[\lambda_2(\xi - \xi_v)\right], \tag{18a}$$

$$Y_O(\xi) = m_{Ou}\left(1 + \frac{\mu_1}{\mu_2} \exp\left(-\frac{u}{D_O}\xi\right)\right), \tag{18b}$$

$$Y_{d,j} = \delta_j, \quad j = 1, 2, 3, \ldots, N_S. \tag{18c}$$

In $R_2 = \{0 \leq \xi \leq \xi_v\}$

$$T(\xi) = T_0 + \omega_1 \exp\left[\lambda_1(\xi - \xi_v)\right]$$
$$+ \omega_2 \exp\left[\lambda_2(\xi - \xi_v)\right] + \frac{L}{C_p}P(S_v(\xi)), \tag{19a}$$

$$Y_O(\xi) = m_{Ou}\left(1 + \frac{\mu_1}{\mu_2} \exp\left(-\frac{u}{D_O}\xi\right)\right), \tag{19b}$$

$$Y_{d,j}(\xi) = \sum_{i=j}^{N_S} \Omega_{ji} \exp\left[\frac{\Delta_i}{u}(\xi - \xi_v)\right], \tag{19c}$$

$$j = 1, 2, 3, \ldots, N_S.$$

In $R_3 = \{\xi_e \leq \xi \leq 0\}$

$$T(\xi) = T_0 + \omega_1 \exp\left[\lambda_1(\xi - \xi_v)\right] - \Gamma m_{Ou}\left(\frac{\mu_1}{\mu_2} + 1\right)$$
$$\cdot \left[\frac{\lambda_2 - \mu_1}{\lambda_1 - \lambda_2} \exp(\lambda_1 \xi) + \exp(\mu_1 \xi)\right] + \frac{L}{C_p}$$
$$\cdot P(S_v(\xi)), \tag{20a}$$

$$Y_O(\xi) = m_{Ou}\left(1 + \frac{\mu_1}{\mu_2}\right) \exp(\mu_1 \xi), \tag{20b}$$

$$Y_{d,j}(\xi) = \sum_{i=j}^{N_S} \Omega_{ji} \exp\left[\frac{\Delta_i}{u}(\xi - \xi_v)\right], \tag{20c}$$

$$j = 1, 2, 3, \ldots, N_S.$$

In $R_4 = \{-\infty < \xi \leq \xi_e\}$

$$T(\xi)$$
$$= T_0$$
$$+ \left\{T_* - T_0 - \frac{L}{C_p}P(S_v(\xi_e))\right\} \exp\left[\lambda_1(\xi - \xi_e)\right] \tag{21a}$$
$$+ \frac{L}{C_p}P(S_v(\xi)),$$

$$Y_O(\xi) = 0, \tag{21b}$$

$$Y_{d,j}(\xi) = \sum_{i=j}^{N_S} \Omega_{ji} \exp\left[\frac{\Delta_i}{u}(\xi - \xi_v)\right], \tag{21c}$$

$$j = 1, 2, 3, \ldots, N_S,$$

where

$$P(S_v(\xi)) = \sum_{j=1}^{N_S} \left\{ \Delta_j \sum_{i=j}^{N_S} \frac{\Omega_{ji} \exp\left[(\Delta_i/u)(\xi - \xi_v)\right]}{D(\Delta_i/u)^2 + u(\Delta_i/u) - \alpha} \right.$$

$$\left. - \Psi_j \sum_{i=j+1}^{N_S} \frac{\Omega_{j+1,i} \exp\left[(\Delta_i/u)(\xi - \xi_v)\right]}{D(\Delta_i/u)^2 + u(\Delta_i/u) - \alpha} \right\}, \tag{22a}$$

$$P'(S_v(\xi))$$

$$= \sum_{j=1}^{N_S} \left\{ \Delta_j \sum_{i=j}^{N_S} \frac{\Omega_{ji}(\Delta_i/u) \exp\left[(\Delta_i/u)(\xi - \xi_v)\right]}{D(\Delta_i/u)^2 + u(\Delta_i/u) - \alpha} \right. \tag{22b}$$

$$\left. - \Psi_j \sum_{i=j+1}^{N_S} \frac{\Omega_{j+1,i}(\Delta_i/u) \exp\left[(\Delta_i/u)(\xi - \xi_v)\right]}{D(\Delta_i/u)^2 + u(\Delta_i/u) - \alpha} \right\},$$

$$\Omega_{ji} = \frac{\Omega_{j+1,i}\Psi_j}{\Delta_j - \Delta_i}, \quad i \neq j, \tag{22c}$$

$$\Omega_{ji} = \delta_j - \sum_{i=j+1}^{N_S} \Omega_{ji}, \quad i = j, \tag{22d}$$

$$\omega_1 = -\frac{L}{C_p} \frac{P'(S_v(\xi_v)) - \lambda_2 P(S_v(\xi_v))}{\lambda_1 - \lambda_2}, \tag{22e}$$

$$\omega_2 = T_v - T_0 + \frac{L}{C_p} \frac{P'(S_v(\xi_v)) - \lambda_1 P(S_v(\xi_v))}{\lambda_1 - \lambda_2}, \tag{22f}$$

$$\Gamma = \frac{qk(T_m)}{C_p[k(T_m) - \alpha]}, \tag{23a}$$

$$\lambda_1 = \frac{u}{2D_T}(d_T - 1), \quad (> 0), \tag{23b}$$

$$\lambda_2 = -\frac{u}{2D_T}(d_T + 1), \quad (< 0), \tag{23c}$$

$$\phi = \frac{\alpha D_T}{u^2}, \tag{23d}$$

$$d_T = \sqrt{1 + 4\phi}, \quad (> 1), \tag{23e}$$

$$\mu_1 = \frac{u}{2D_O}(d_O - 1), \quad (> 0), \tag{23f}$$

$$\mu_2 = -\frac{u}{2D_O}(d_O + 1), \quad (< 0), \tag{23g}$$

$$d_O = \sqrt{1 + 4B}, \tag{23h}$$

$$B = \frac{k(T_m)D_O}{u^2}. \tag{23i}$$

An explicit formula for the burning velocity cannot be extracted; however, by applying matching and boundary conditions a set of coupled implicit algebraic equations are obtained through which $u, T_m, \xi_m, \xi_e, \xi_v$ can be found by using a numerical iterative method. The solution is based on the assumption that $\exp(\mu_1\xi_e) \to 0$ which can be readily verified.

In the limit of infinite vaporization coefficient ($\widehat{E}_v \to \infty$) when the spray of liquid fuel droplets evaporates in a single vaporization front, some simplification of the afore-described solutions is achieved (see Appendix for details).

4. Results and Discussion

Use was made of the analytical solution in the previous section to examine the effect of heat loss and fuel spray parameters on conditions for spray flame propagation and extinction. The data used for the calculations was as follows (unless otherwise specified):

$$q = 1.279 \times 10^7 \text{ J/kg},$$

$$L = 0.04Q,$$

$$\lambda = 0.02512 \text{ Wm/K},$$

$$A = 10^{10} \text{ s}^{-1},$$

$$E = 2 \times 10^8 \text{ J/kmol}, \tag{24}$$

$$c_p = 1255.92 \text{ J/kg K},$$

$$T_0 = 300 \text{ K},$$

$$T_v = 400 \text{ K},$$

$$\widehat{E}_v = 1.4524 \cdot 10^{-15} \text{ m}^2/\text{s}.$$

The chemical kinetic scheme employed concerns the burning of n-decane and relevant thermochemical data was taken from [23, 24]. By specifying the initial fraction of liquid fuel to the total fuel (vapor + liquid) in the fresh mixture, $\sum_{j=1}^{N_S} \delta_j/(m_{Fu} + m_{du}) = \delta$, it can be shown that the mass fractions in the fresh mixture are given by the following expressions:

$$(m_{Ou}, m_{du}, m_{Fu})$$

$$= \left(s\{1 - m_{du} - m_{Fu}\}, \frac{s\delta}{(1 + \alpha_{OF}/\varphi)}, m_{du}\frac{(1 - \delta)}{\delta} \right) \tag{25}$$

unless $\delta = 0$ for which

$$m_{Fu} = \frac{s}{(s + \alpha_{OF}/\varphi)}, \tag{26}$$

where s is the mole fraction of oxygen in the fresh mixture, α_{OF} is the stoichiometric coefficient, and φ is the equivalence ratio. Here φ is taken as 2 and $\alpha_{OF} = 3.5$.

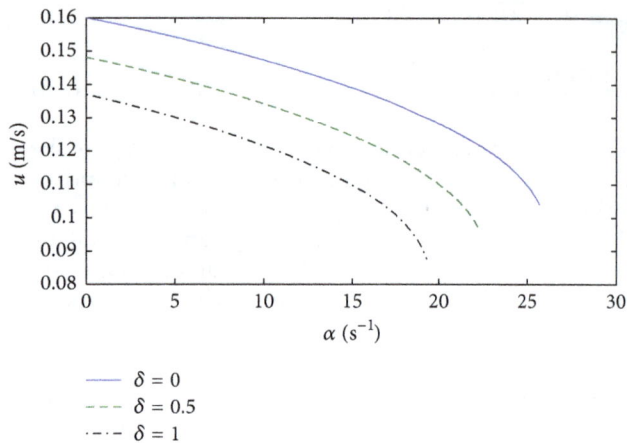

FIGURE 2: Influence of heat loss parameter on spray flame propagation velocity for different initial droplet loads—evaporation front case.

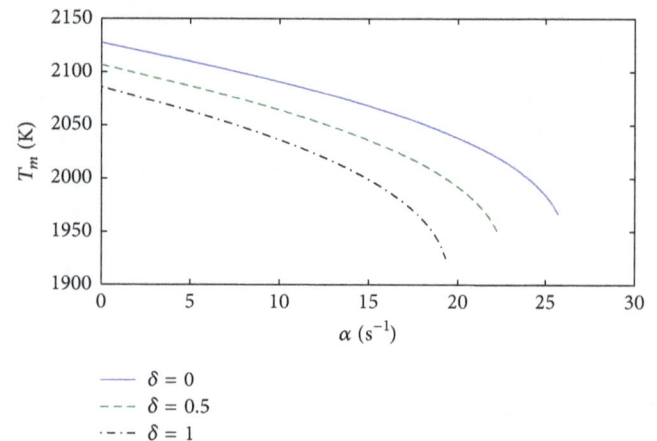

FIGURE 3: Influence of heat loss parameter on spray flame temperature for different initial droplet loads—evaporation front case.

In order to extract and highlight the various factors at work in the fuel rich spray flames under consideration here we focus on three cases: (a) evaporation of the droplets in a single vaporization front (the solution for this case is given in Appendix), (b) finite rate evaporation but with all droplets subsumed into a single section (monosectional description of the spray with solution derivable by setting $N_S = 1$ in the analysis of the previous section), and (c) the afore-detailed full polydisperse case. Case (a) is applicable when the liquid fuel is highly volatile. Case (b) applies to a less volatile fuel and is modeled to capture only the gross features of the spray impact on the combustion. Case (c) applies to a less volatile fuel but with the details of the spray size structure accounted for.

(a) Evaporation in a Front. We begin our discussion of the predictions of the theory by considering the case for which the liquid droplets evaporate in a sharp front. In Figure 2 a plot is presented of the spray flame's propagation velocity for different initial liquid fuel loads, δ, as a function of the heat loss parameter, α. For the purely gaseous case, $\delta = 0$, the behavior of the velocity follows the classical behavior (see, e.g., [4]), with a decrease in the velocity resulting from increasing the heat loss. Of course, this is to be expected as a result of the competing exothermic-endothermic mechanisms at play. Eventually, flame extinction occurs at some critical value of α. With the fuel supplied as a liquid spray only, that is, $\delta = 1$, it is readily observed that the flame propagation velocity is less than that of its gaseous counterpart for any given value of α. This is not surprising since, even without volumetric/radiative heat loss, the droplets themselves must absorb heat for evaporation, thereby automatically lowering the flame temperature and, hence, the flame velocity (see, also, [10, 16]). As the droplet evaporation takes place in a *front* this source of heat loss is quite concentrated leading to a notable influence on the flame velocity. In addition, extinction of the spray flame occurs at a lower value of the heat loss parameter α. This is understandable as both the distributed heat loss from the surroundings and

the liquid droplets heat loss combine to overcome the exothermic chemical reaction.

The case in which 50% of fuel is supplied as liquid and 50% as vapor is also illustrated in Figure 2. It can be seen that, not surprisingly, the relevant curve lies in between the limiting curves which we have discussed.

In Figure 3 the flame temperature is drawn as a function of the heat loss parameter. The curves clearly reflect the discussion of Figure 2 and show the rather drastic effect of the heat loss particularly in the proximity of extinction conditions.

It is known from the theory of gas flame propagation [4] that the extinction velocity when linear heat loss is present is $e^{-1/2}$ times the adiabatic flame velocity. For spray flames the two sources of heat loss play a role, namely, radiative/volumetric heat loss and heat loss, due to absorption of heat by the liquid droplets for evaporation. In Figure 4 we examine the relative importance of these heat losses by plotting the log of the ratio of the flame velocity at extinction to the adiabatic gas flame velocity as a function of the initial liquid spray load, δ. The green line with circles is the classical result when only radiative/volumetric heat loss is accounted for and it is readily seen to be constant at the value of $e^{-1/2}$. The blue continuous unmarked line shows the decrease in the spray flame burning velocity predicted when only droplet heat loss is accounted for. In this case the heat absorbed by the droplets for evaporation is not sufficient to extinguish the flame and the velocity ratio is always larger than $e^{-1/2}$. When both heat losses are included in the model the red line with boxes is obtained. (Note that since the flame velocity *at extinction* is used to construct this figure the value of the heat loss parameter, α, varies along the curve (cf. Figure 2).) The ratio of velocities is now evidently dependent on the initial liquid droplet load, and a factor of about $e^{-0.66}$ is found when $\delta = 1$.

In view of these findings and the underlying rationale it would seem that the latent heat of vaporization of the liquid fuel should be an important factor in determining spray flame velocity and extinction conditions. This influence is shown in

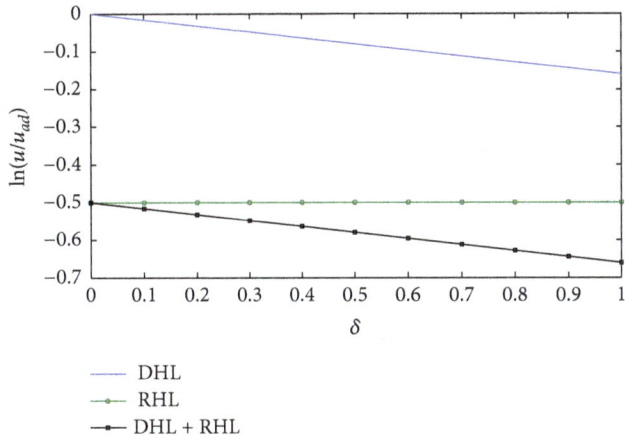

FIGURE 4: Relative importance of sources of heat loss in determining the flame velocity of laminar spray flames as function of the initial liquid fuel load, evaporation front case. DHL = droplet heat loss only, RHL = radiative/volumetric heat loss only, and DHL + RHL = combined droplet and radiative/volumetric heat loss.

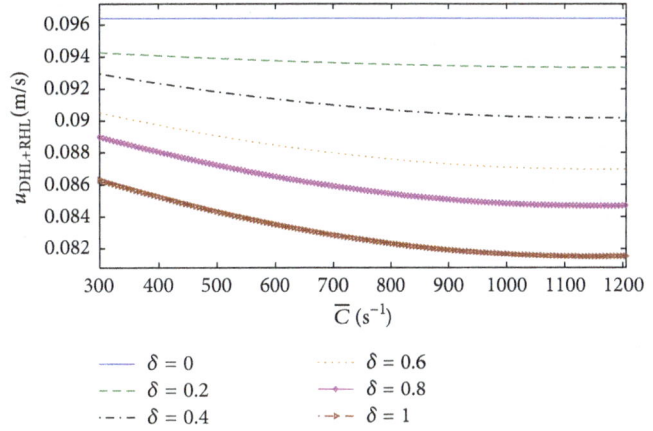

FIGURE 6: Influence of vaporization coefficient on spray flame propagation velocity at critical values of volumetric/radiative heat loss parameter for different initial droplet loads, monosectional spray.

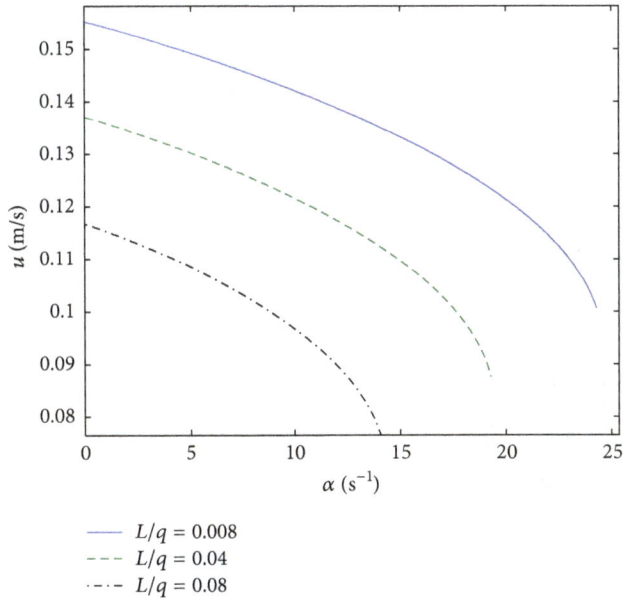

FIGURE 5: Influence of latent heat of vaporization on spray flame propagation velocity—evaporation front case.

Figure 5 in which the flame velocity is drawn as a function of the volumetric/radiative heat parameter for different values of the ratio of the latent heat to the heat of reaction. The effect is quite striking with the value of the flame velocity decreasing by a factor of 50% (for a fixed value of α) as the latent heat increases by a factor of 10. Moreover, the critical value of α for extinction decreases by a factor of 67% as the latent heat of vaporization increases tenfold.

(b) *Monosectional Spray*. The graphs presented so far related to the case in which the spray of fuel droplets evaporates in a sharp front. Attention will now be turned to the monosectional case with a finite evaporation rate. In Figure 6

the spray flame propagation velocity is drawn as a function of the evaporation coefficient, $\overline{C} = 1.5\hat{E}_v(3d_u - 2d_l)/(d_u^3 - d_l^3)$ (see (5)), at the values of the relevant critical volumetric/radiative heat loss parameter, and for different values of the initial fuel droplet load. For this case increasing the value of $\overline{C}$ is equivalent to using a more volatile fuel (i.e., increasing $\hat{E}_v$) and/or using smaller droplets in the spray. For any given load it is clear that the velocity decreases as the evaporation coefficient increases. In fact as $\overline{C} \to \infty$ the velocity levels off at the value appropriate to the evaporation front case, as anticipated. In addition, as the initial droplet load increases the critical velocity decreases, irrespective of the value of the vaporization coefficient. This, too, is in keeping with the prediction of the evaporation front case (see, e.g., Figure 2).

In Figure 7 we examine the influence of the vaporization coefficient in determining the flame velocity as a function of the initial liquid fuel load by plotting the log of the ratio of the flame velocity at extinction to the adiabatic gas flame velocity as a function of the vaporization coefficient. It is clear that as $\overline{C}$ increases, for all values of δ, the logarithm of the velocity ratio decreases below the value of -0.5 with the effect being greatest when the fuel is supplied in liquid form, that is, $\delta = 1$. In addition, it can be observed that for large values of the evaporation coefficient the relevant curve overlaps with that of Figure 4 for the case of the evaporation front. Whereas the heat loss due to volumetric/radiative heat loss is distributed throughout the entire field, the more focused the heat loss due to droplet evaporation the greater the effect on critical conditions for extinction. Once again it is evident that the influence of the spray parameter, this time in the form of the evaporation coefficient, combines with that of the volumetric/radiative heat loss so that the classical $e^{-1/2}$-factor is modified. The maximum modification corresponds to the case of an infinitely large rate of evaporation.

(c) *Polydisperse Spray*. Having established the important characteristic features of a fuel spray and their role in flame

TABLE 1: The sectional diameters d_j (μm) and initial droplet size distributions.

Section number	1	2	3	4	5	6	7	8	9
Section diameters	1–5	5–10	10–20	20–30	30–40	40–50	50–70	70–90	90–110
Distribution 1	0	0	0	0	0	1	0	0	0
Distribution 2	0	0	0.207	0	0	0	0	0	0.793
Distribution 3	0.0005	0.0005	0.0141	0.0793	0.1662	0.2464	0.2349	0.1547	0.1034

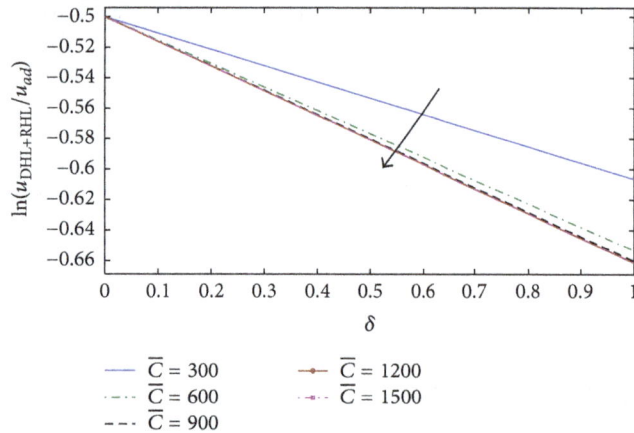

FIGURE 7: Influence of the vaporization coefficient in determining the flame velocity of laminar spray flames as function of the initial liquid fuel load, monosectional spray.

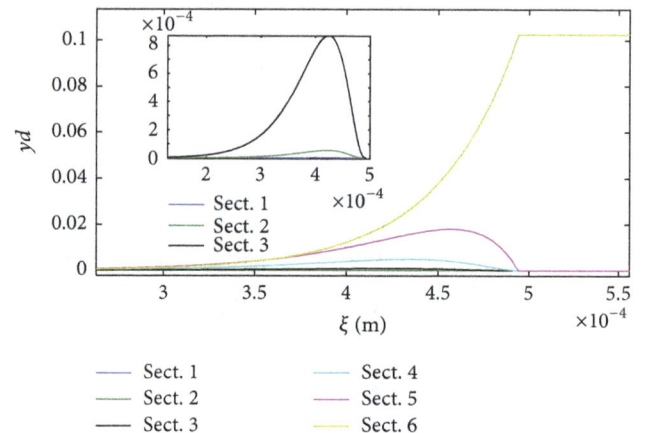

FIGURE 8: Liquid fuel sectional mass fractions profiles in one-dimensional laminar spray flame propagation, initial size distribution 1.

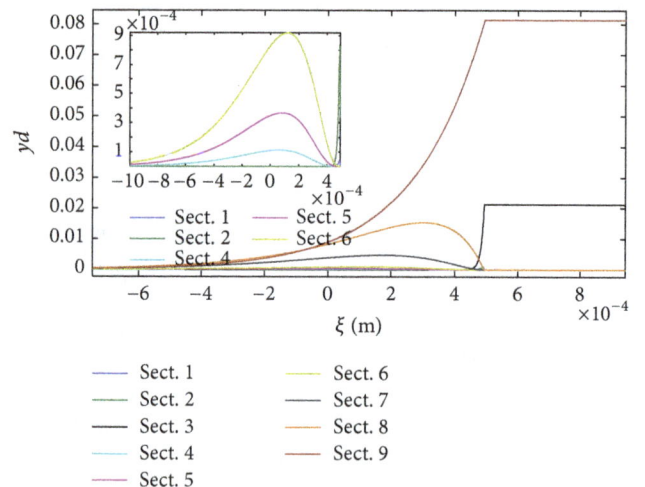

FIGURE 9: Liquid fuel sectional mass fractions profiles in one-dimensional laminar spray flame propagation, initial size distribution 2.

propagation in the presence of heat losses without recourse to the actual droplet sizes within the spray we turn our attention to the impact of the initial spray droplet size distribution. To this end we make use of three quite different initial spray distributions that are listed in Table 1, together with the definitions of the size sections utilized.

The numbers in each distribution's columns represent the fraction of liquid fuel (in the relevant size section) in the total liquid fuel supplied. Distribution 1 is initially monosectional but will become multisize once evaporation occurs and droplets become ineligible for membership in section 6. Distribution 2 is bidisperse, initially having almost 80% liquid fuel in the highest section, number 9, and the rest in much smaller droplets in section 3. Finally, the third distribution has droplets initially well spread out throughout all size sections and is probably closer to what exists in real life. These three different size distributions have a common Sauter Mean Diameter (SMD) of 44.8 μm. The SMD is the ratio of the volume of droplets in the spray to their surface area and is often used to characterize a polydisperse spray by an equivalent spray of single size droplets all of which have a diameter equal to the SMD. However, there is evidence that this characterization may sometimes be misleading [11, 25, 26].

In Figures 8, 9, and 10 the profiles of the sectional fuel mass fractions for the three initial size distributions are drawn, for comparison. m_{du} was taken as 0.1.

Recall that the flame is propagating from left to right with the close vicinity of the point $\xi = 0$ marking the intense

flame reaction zone. The redistribution of liquid droplets in all three cases as they evaporate and migrate down the size sections is apparent. For example, for distribution 1 all droplets initially occupy section 6. As they evaporate there is a transfer to the initially unoccupied section 5 in which a build of liquid fuel is readily observable. Subsequently, the relocation to lower sections from section 5 occurs as droplets evaporate and become ineligible for membership in section 5. Similarly, behavior is found for the lower sections, too.

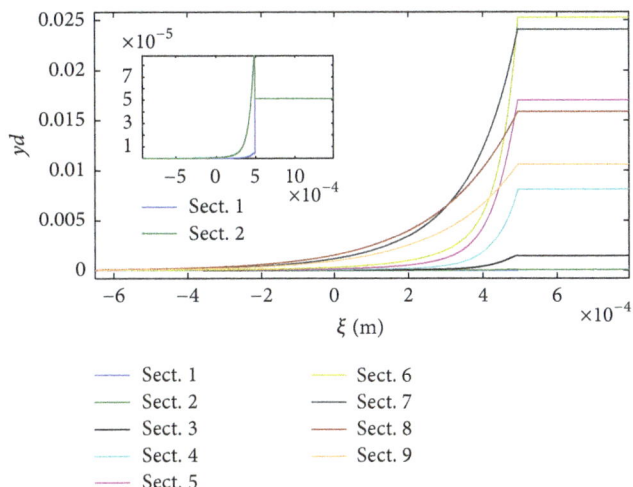

FIGURE 10: Liquid fuel sectional mass fractions profiles in one-dimensional laminar spray flame propagation; initial size distribution 3.

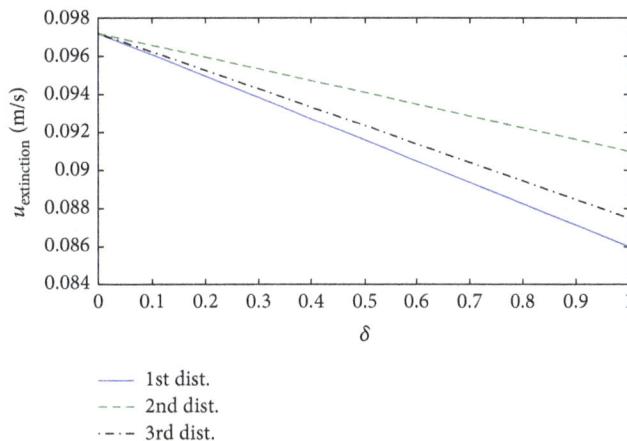

FIGURE 11: Flame velocity at extinction versus initial liquid droplet load—influence of initial droplet distribution.

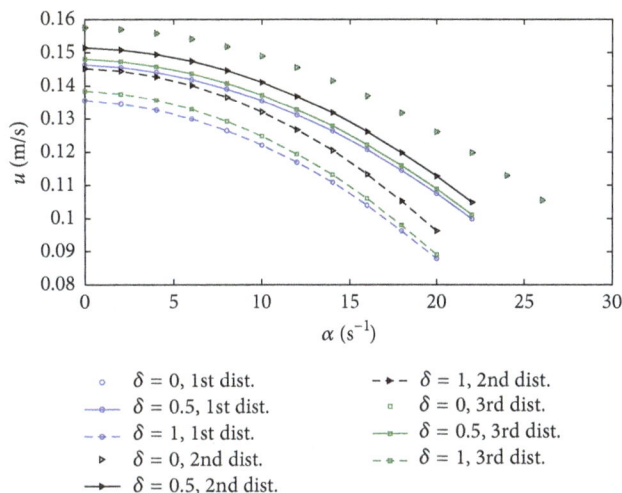

FIGURE 12: Influence of heat loss parameter on spray flame propagation velocity for different initial droplet loads—polydisperse case.

However, what should also be noted, for the data utilized here, is the spatial distribution of the liquid fuel. For distribution 1 virtually all the liquid fuel evaporates before reaching the reaction zone, whereas for distributions 2 and 3 this is not the case. Although for these latter distributions much vapor is released upstream of the flame region noticeable continued evaporation occurs downstream of the flame front as the droplets pass through the flame into the hot products region. This will lead to different spatially distributed heat release behaviors depending on the internal spray structure (i.e., size distribution) and its evolution as the droplets absorb heat for evaporation.

Consider Figure 11 in which the velocity at extinction is drawn as a function of the initial liquid load for all three initial droplet size distributions. It is clear that for $\delta \neq 0$ the initial droplet distribution has a noticeable influence over the velocity at extinction. Distribution 1 leads to the lowest velocity; distribution 2 leads to the highest velocity with distribution 3 in between. At most the discrepancy is about 9% (comparing distributions 1 and 2) and about 2% (comparing distributions 1 and 3). This ordering is also reflected in Figure 12 in which the influence of the volumetric heat loss on the flame velocity is charted for three initial liquid fuel loads and for all three initial size distributions. The underlying rationale for this behavior can be deduced from Figure 13 where the flame temperatures associated with the velocities of Figure 11 are drawn. First we isolate the effect of the initial total liquid load. For any value of α for which a flame exists and for any initial droplet size distribution the flame temperature drops as δ increases. As we have seen before this is due to the increased heat loss sustained due to the droplets heat absorption for evaporation. This, in turn, influences the speed of propagation and lowers it accordingly.

Now isolate the effect of the droplet size distribution for a fixed value of δ and α. It is clear that size distribution 1 leads to the lowest flame temperature followed by distribution 3 with the highest flame temperature supplied when distribution 2

is used. Evidently, the initially monosectional distribution produces the most focused heat loss due to the initial concentration of droplets in section 6. At the other extreme the least focused heat loss results from the bidisperse distribution 2. Although the smaller droplets in section 3 evaporate fairly rapidly thereby lowering the flame temperature somewhat, it is the large fraction of large droplets initially in section 9 which evaporate gradually that dominates the droplet heat loss mechanism thereby lessening the effect of heat loss when compared to that of distribution 1. Distribution 3, which has droplets spread throughout the size range involved, yields a situation between that generated by the other two distributions, as can be readily observed in Figures 11 and 12. Thus, it is the way in which heat loss due to droplet evaporation is spatially distributed that determines the flame temperature and velocity. This sharpens our observations for the afore-discussed cases of the evaporation front and monosectional spray for which the details of the initial size distribution were dealt with in an integral fashion.

TABLE 2: The sectional diameters d_j (μm) and initial droplet size distributions for examining the influence of droplet size on spray flame propagation.

Section number	1	2	3	4	5	6	7	8	9
Section diameters	1–5	5–10	10–20	20–30	30–40	40–50	50–70	70–90	90–110
Distribution 1	0	1	0	0	0	0	0	0	0
Distribution 2	0	0	0	1	0	0	0	0	0
Distribution 3	0	0	0	0	0	1	0	0	0
Distribution 4	0	0	0	0	0	0	0	1	0

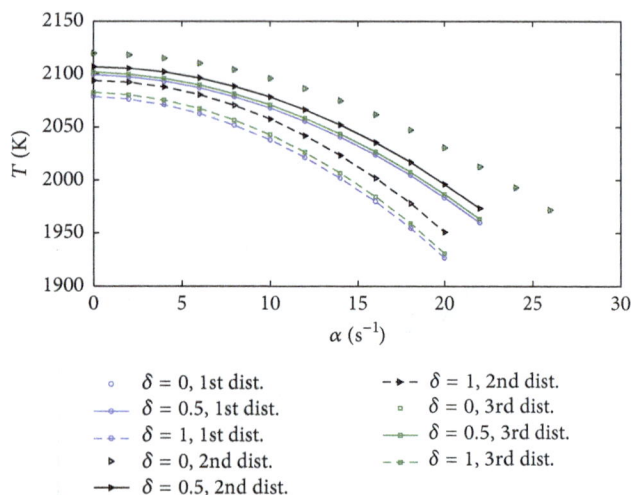

FIGURE 13: Influence of heat loss parameter on spray flame temperature for different initial droplet loads—polydisperse case.

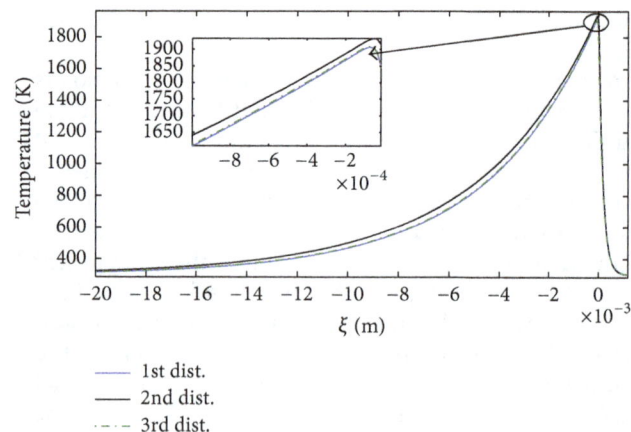

FIGURE 14: Influence of the initial droplet size distribution in determining the flame velocity of laminar spray flames as function of the initial liquid fuel load, polydisperse spray.

In Figure 14 we examine the influence of the initial droplet size distributions on the ratio of the velocity at extinction to the adiabatic flame velocity. The ordering of the curves follows the pattern we have explained and impacts on the classical factor of $e^{-1/2}$ in accordance with that ordering and the initial droplet loading.

The impact of the initial droplet size distributions on the temperature distribution is illustrated in Figure 15. Although the profiles are similar, the differences reflect the different rates of heat loss due to droplet evaporation. As mentioned previously in connection with Figures 8, 9, and 10, initial size distribution 1 provides the most concentrated heat loss so that the associated temperature profile lies below those of the other two size distributions. Distribution 2 provides the most protracted droplet heat loss behavior thereby leading to greater temperatures than associated with the other distributions. Due to this disparity in droplet-related heat loss, differences of several tens of degrees occur at the peak temperature.

We now turn to examine more explicitly the influence of droplet size on the spray flame behavior described before. For this purpose we make use of the four initial droplet size distributions listed in Table 2.

Note that these four distributions are initially monosectional. The difference between them is the size of droplets in the fresh mixture. The SMDs for these four distributions are

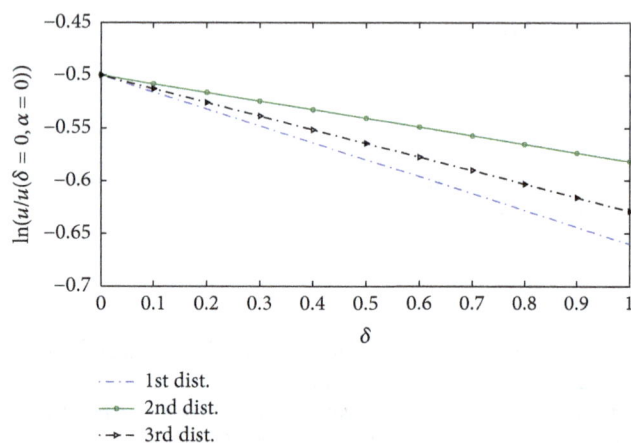

FIGURE 15: Effect of initial droplet size distribution on thermal profile in one-dimensional laminar spray flame propagation.

7.213, 24.663, 44.814, and 79.582 μm, respectively. The solution in (18a)–(21c) is, of course, applicable as the polydisperse development of the spray is independent of the initial size distribution.

In Figure 16 the spray flame velocity at extinction is plotted for all four cases of Table 2 as a function of the initial liquid fuel load, δ. As anticipated and explained previously,

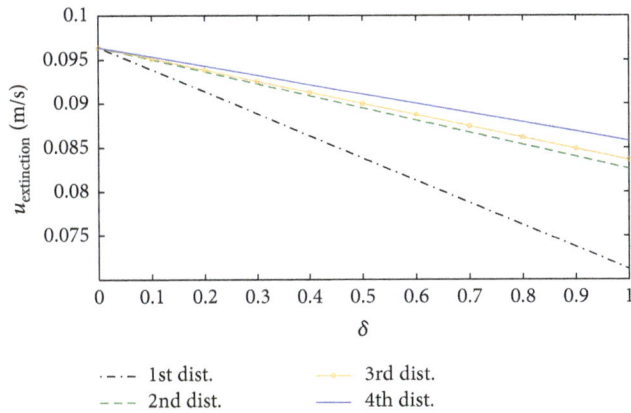

FIGURE 16: Influence of initial droplet size on extinction velocity in polydisperse spray flames.

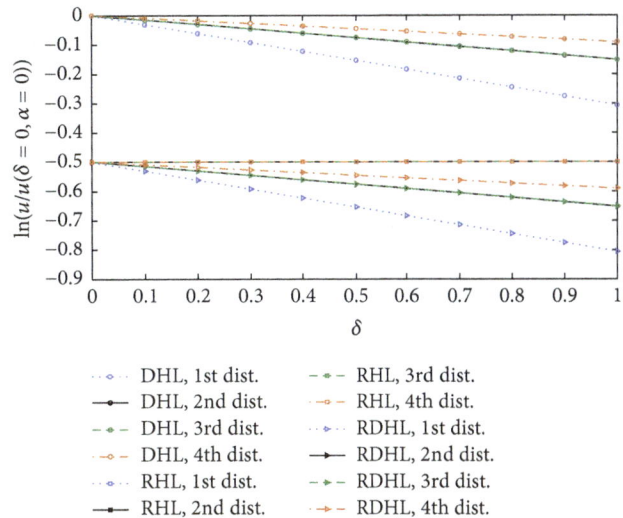

FIGURE 17: Influence of the initial droplet size in determining the flame velocity of laminar polydisperse spray flames as function of the initial liquid fuel load.

for all four initial distributions, the critical velocity decreases as the initial liquid fuel load increases. The largest effect is noted for the case of smallest droplets for which adecrease of about 26% occurs. The initially larger droplets of distributions 2–4 of Table 2 lead to more modest maximum decreases of between 14% and 11%, with the largest droplets supplying the smallest increase. This is entirely in keeping with the previous assertion that it is the way in which heat loss due to droplet evaporation is spatially distributed that determines the flame velocity. This is further confirmed by Figure 17 in which the relative importance of the volumetric and spray-related heat losses is plotted, for the initial droplet sizes of Table 2. The compound effect of both sources of heat loss is apparent. Once again it is clearly evident that the smaller the initial droplet size the more focused the droplet related heat loss, and, hence, its impact on the flame propagation velocity. The previously mentioned classical factor of $e^{-1/2}$ at extinction for a gas flame can become as small as $e^{-0.8}$ for size distribution 1 (of Table 2) and an initial liquid fuel load $\delta = 1$.

5. Conclusions

The role of both volumetric/radiative heat loss and that of heat absorbed by a spray of evaporating droplets in determining premixed spray flame propagation and extinction was investigated analytically using a nonasymptotic solution of the governing equations. This is the first analytical treatment of laminar premixed *polydisperse* spray flame propagation with volumetric/radiation heat loss that we are aware of. Such analysis is important in terms of the physical insights it is able to supply and its potential use as a benchmark for computational studies.

Calculated results indicate that the presence of the fuel droplets in the premixture reduces both the critical value of both the flame velocity prior to extinction and the heat loss, primarily due to the aforementioned heat absorption by the droplets. In addition to the initial droplet load, the value of the evaporation coefficient and the initial size distribution are the

other spray-related parameters which exert an influence on the onset of extinction. The combination of these parameters governs the manner in which the spray heat loss is distributed *spatially* and it is this which is the main factor, when taken together with the volumetric heat loss, which determines the spray's impact on flame propagation and extinction.

In addition, the analysis demonstrates that use of a Sauter Mean Diameter to characterize the behavior of a polydisperse spray flame may lead to erroneous conclusions. In fact, the actual polydispersity of the flame must be considered.

Finally, the results clearly show how the presence of the spray and volumetric heat losses lead to a reduction of the burning velocity in comparison with a single phase gas flame, due to the total heat loss, which leads to a lowering of the burned gas temperature. It should be mentioned that, interestingly, this result does not match some experimental data for laminar premixed spray flames for which flame velocity *increases* were found for rich spray flames [27–29]. However, this discrepancy is due to the fact that in the current steady state, one-dimensional theory the inherently two- or three dimensional phenomenon of flame cellularization clearly cannot be captured. This necessitates a stability analysis for the two- or three dimensional perturbed flame front, which is beyond the scope of the current work.

Appendix

Solution of Governing Equations When the Spray of Droplets Evaporates in a Sharp Front

We apply a matching condition at $\xi = \xi_v$ that reflects the presence of vaporization front:

$\xi = \xi_v$:

$$[Y_O] = [Y'_O] = 0, \quad T = T_v, \quad [T'] = \frac{L m_{du} u}{D_T c_p}. \quad \text{(A.1)}$$

The solution can then be readily found:

In $R_1 = \{\xi_v \leq \xi < \infty\}$,

$$T(\xi) = T_0 + (T_v - T_0) \exp[\lambda_2(\xi - \xi_v)],$$

$$Y_O(\xi) = m_{Ou}\left[1 + \frac{\mu_1}{\mu_2}\exp\left(-\frac{u}{D_O}\xi\right)\right]. \quad \text{(A.2)}$$

In $R_2 = \{0 \leq \xi \leq \xi_v\}$

$$T(\xi) = T_0 - \Lambda \exp[\lambda_1(\xi - \xi_v)]$$

$$+ [T_v - T_0 + \Lambda]\exp[\lambda_2(\xi - \xi_v)],$$

$$Y_O(\xi) = m_{Ou}\left[1 + \frac{\mu_1}{\mu_2}\exp\left(-\frac{u}{D_O}\xi\right)\right]. \quad \text{(A.3)}$$

In $R_3 = \{\xi_e \leq \xi \leq 0\}$

$$T(\xi) = T_0$$

$$+ \left[\frac{\Gamma m_{Ou}[\lambda_2 - (\mu_1/\mu_2)\lambda_1]}{\lambda_2 - \lambda_1} - \Lambda\exp(-\lambda_1\xi_v)\right]$$

$$\cdot \exp(\lambda_1\xi) - m_{Ou}\left(1 + \frac{\mu_1}{\mu_2}\right)\Gamma\exp(\mu_1\xi), \quad \text{(A.4)}$$

$$Y_O(\xi) = m_{Ou}\left(1 + \frac{\mu_1}{\mu_2}\right)\exp(\mu_1\xi).$$

In $R_4 = \{-\infty < \xi \leq \xi_e\}$

$$T(\xi) = T_0 + (T_* - T_0)\exp[\lambda_1(\xi - \xi_e)],$$

$$Y_O(\xi) = 0, \quad \text{(A.5)}$$

where

$$\Lambda \equiv \frac{L m_{du}}{d_T C_P}. \quad \text{(A.6)}$$

In this vaporization front case, it can be shown that *explicit* expressions are obtained for ξ_m, ξ_e, ξ_v:

$$\xi_v = -\frac{\ln[\Gamma m_{Ou}[d_O - d_T]/d_T(1 + d_O)(T_v - T_0 + \Lambda)]}{\lambda_2},$$

$$\xi_e = \frac{\ln\{(T_* - T_0)/(-\Delta_2\exp(-\lambda_1\xi_v) + \Gamma m_{Ou}(d_T + d_O)/d_T(1 + d_O))\}}{\lambda_1}, \quad \text{(A.7)}$$

$$\xi_m = \frac{\ln(\lambda_1[-\Delta_2\exp(-\lambda_1\xi_v) + \Gamma m_{Ou}(d_T + d_O)/d_T(1 + d_O)]/(2\Gamma m_{Ou}\mu_1/(1 + d_O)))}{\mu_1 - \lambda_1}.$$

Nomenclature

A:	Preexponential constant
B:	Parameter in solution (Equation (23i))
c_p:	Specific heat
d_{lj}, d_{uj}:	Lower and upper diameters of size section j
d_O, d_T:	Parameters in solution ((23e) and (23h))
D_T:	Thermal diffusion coefficient
D_O:	Oxygen mass diffusion coefficient
E:	Activation energy
$\widehat{E}_v$:	Evaporation coefficient
H:	Heaviside function
$k, K_d, \widehat{K}_O$:	Functions defined in (8), (16), and (12), respectively
L:	Latent heat of vaporization
m_{Ou}:	Initial mass fraction of gaseous oxygen
m_{du}:	Initial mass fraction of liquid fuel
N_S:	Number of size sections
P:	Spray-related function ((22a) and (22b))
q:	Heat of reaction

R:	Universal gas constant
R_1, R_2, R_3, R_4:	Solution subdomains
s:	Mole fraction of oxygen in the fresh mixture
S_v:	Total rate of droplet evaporation
$S_{v,j}$:	Rate of evaporation of droplets in section j
t^*:	Time
T:	Temperature
T_0:	Ambient temperature
u:	Velocity
x^*:	Spatial coordinate
Y:	Mass fraction
$\widehat{\alpha}, \alpha$:	Radiative heat loss coefficients
α_{OF}:	Stoichiometric coefficient
Γ:	Parameter in solution (Equation (23a))
δ_j:	Mass fraction of liquid fuel in section j
δ:	Initial ratio of mass fraction of liquid fuel to total fuel
Δ_j, Ψ_j:	Vaporization Damkohler numbers for section j
ε:	Small parameter (Equation (11))

ϕ: Parameter in solution (Equation (23d))
φ: Equivalence ratio
λ: Thermal conductivity
λ_1, λ_2: Parameters in solution (Equation (23b) and (23c))
μ_1, μ_2: Parameters in solution (Equations (23f) and (23g))
ρ: Density
ω_1, ω_2: Parameters in solution ((22e) and (22f))
Ω_{ji}: Coefficients defined in (22c) and (22d)
ξ: Flame front coordinate.

Subscripts

d, j: Relating to droplets in size section j
e: Reaction extinction point value
j: Relating to size section j
F: Fuel
m: Maximum temperature point value
O: Oxygen
u: Unburnt value
v: Vaporization front value
$*$: Value at $\xi = 0$.

Conflict of Interests

The authors declare that there is no conflict of interests regarding the publication of this paper.

Acknowledgment

J. B. Greenberg wishes to acknowledge the partial support of the Lady Davis Chair in Aerospace Engineering.

References

[1] D. B. Spalding, "A theory of inflammability limits and flame-quenching," *Proceedings of the Royal Society. Series A*, vol. 240, no. 1220, pp. 83–100, 1957.

[2] J. Adler and D. B. Spalding, "One-dimensional laminar flame propagation with an enthalpy gradient," *Proceedings of the Royal Society, Series A: Mathematical, Physical and Engineering Sciences*, vol. 261, no. 1304, pp. 53–78, 1961.

[3] J. Adler, "One-dimensional laminar flame propagation with distributed heat losses: thin flame theory," *Combustion and Flame*, vol. 7, no. 1, pp. 39–49, 1963.

[4] J. Buckmaster, "The quenching of deflagration waves," *Combustion and Flame*, vol. 26, no. C, pp. 151–162, 1976.

[5] G. Joulin and P. Clavin, "Linear stability analysis of nonadiabatic flames: diffusional-thermal model," *Combustion and Flame*, vol. 35, pp. 139–153, 1979.

[6] C. Nicoli and P. Clavin, "Effect of variable heat loss intensities in the dynamics of a premixed flame front," *Combustion and Flame*, vol. 68, no. 1, pp. 69–71, 1987.

[7] P. Clavin and C. Nicoli, "Effect of heat losses on the limits of stability of premixed flames propagating downwards," *Combustion and Flame*, vol. 60, no. 1, pp. 1–14, 1985.

[8] C.-C. Liu and T.-H. Lin, "The interaction between external and internal heat losses on the flame extinction of dilute sprays," *Combustion and Flame*, vol. 85, no. 3-4, pp. 468–478, 1991.

[9] S.-S. Hou, C.-C. Liu, and T.-H. Lin, "The influence of external heat transfer on flame extinction of dilute sprays," *International Journal of Heat and Mass Transfer*, vol. 36, no. 7, pp. 1867–1874, 1993.

[10] J. B. Greenberg, A. C. McIntosh, and J. Brindley, "Linear stability analysis of laminar premixed spray flames," *Proceedings of the Royal Society of London A*, vol. 457, pp. 1–31, 2001.

[11] J. B. Greenberg, "Stability boundaries of laminar premixed polydisperse spray flames," *Atomization and Sprays*, vol. 12, no. 1-3, pp. 123–143, 2002.

[12] C. Nicoli, P. Haldenwang, and S. Suard, "Analysis of pulsating spray flames propagating in lean two-phase mixtures with unity Lewis number," *Combustion and Flame*, vol. 143, no. 3, pp. 299–312, 2005.

[13] C. Nicoli, P. Haldenwang, and S. Suard, "Effects of substituting fuel spray for fuel gas on flame stability in lean premixtures," *Combustion and Flame*, vol. 149, no. 3, pp. 295–313, 2007.

[14] W. A. Sirignano, *Fluid Dynamics and Transport of Droplets and Sprays*, Cambridge University Press, Cambridge, UK, 1999.

[15] S. S. Sazhin, "Advanced models of fuel droplet heating and evaporation," *Progress in Energy and Combustion Science*, vol. 32, no. 2, pp. 162–214, 2006.

[16] G. Kats and J. B. Greenberg, "Application of a non-asymptotic approach to prediction of the propagation of a flame through a fuel and/or oxidant droplet cloud," *Applied Mathematical Modelling*, vol. 37, no. 12-13, pp. 7427–7441, 2013.

[17] V. A. Volpert, "Dynamics of thermal polymerization waves," in *Self-Assembly, Pattern Formation and Growth Phenomena in Nano-Sytems*, A. A. Golovin and A. A. Nepomnyashchy, Eds., pp. 195–245, Springer, Amsterdam, The Netherlands, 2006.

[18] S. Balasuriya and V. A. Volpert, "Wavespeed analysis: approximating Arrhenius kinetics with step-function kinetics," *Combustion Theory and Modelling*, vol. 12, no. 4, pp. 643–670, 2008.

[19] J. B. Greenberg and A. Kalma, "Computational aspects of the sectional modeling method for predicting spray combustion," *HTD*, vol. 361, no. 2, pp. 3–10, 1998.

[20] J. B. Greenberg, I. Silverman, and Y. Tambour, "On the origins of spray sectional conservation equations," *Combustion and Flame*, vol. 93, no. 1-2, pp. 90–96, 1993.

[21] M. Labowsky, "Calculation of the burning rates of interacting fuel droplets," *Combustion Science and Technology*, vol. 22, no. 5-6, pp. 217–226, 1980.

[22] A. Bayliss, E. M. Lennon, M. C. Tanzy, and V. A. Volpert, "Solution of adiabatic and nonadiabatic combustion problems using step-function reaction models," *Journal of Engineering Mathematics*, vol. 79, pp. 101–124, 2013.

[23] D. R. Stull and H. Prophet, *JANAF Thermochemcial Tables*, NSRDS-NBS 37, National Bureau of Standards, Washington, DC, USA, 2nd edition, 1971.

[24] N. B. Vargaftik, *Tables on the Thermophysical Properties of Liquids and Gases*, John Wiley & Sons, New York, NY, USA, 2nd edition, 1975.

[25] M. Zhu and B. Rogg, "Modelling and simulation of sprays in laminar flames," *Meccanica*, vol. 31, no. 2, pp. 177–193, 1996.

[26] V. Bykov, I. Goldfarb, V. Gol'dshtein, and J. B. Greenberg, "Auto-ignition of a polydisperse fuel spray," *Proceedings of the Combustion Institute*, vol. 31, no. 2, pp. 2257–2264, 2007.

[27] Y. Mizutani and A. Nakajima, "Combustion of fuel vapor-drop-air systems: part I-open burner flames," *Combustion and Flame*, vol. 20, no. 3, pp. 343–350, 1973.

[28] Y. Mizutani and A. Nakajima, "Combustion of fuel vapor-drop-air systems: part II-spherical flames in a vessel," *Combustion and Flame*, vol. 20, no. 3, pp. 351–357, 1973.

[29] S. Hayashi and S. Kumagai, "Flame propagation in droplet-vapor-air mixtures," *Proceedings of the Combustion Institute*, vol. 15, no. 1, pp. 445–452, 1974.

6

Experimental Study of Gas Explosions in Hydrogen Sulfide-Natural Gas-Air Mixtures

André Vagner Gaathaug,[1] **Dag Bjerketvedt,**[1] **Knut Vaagsaether,**[1] **and Sandra Hennie Nilsen**[2]

[1] *Telemark University College, Faculty of Technology, 3918 Porsgrunn, Norway*
[2] *Research, Development and Innovation, Section for Health, Safety and Water Management, Statoil ASA, 3905 Porsgrunn, Norway*

Correspondence should be addressed to André Vagner Gaathaug; andre.v.gaathaug@hit.no

Academic Editor: Constantine D. Rakopoulos

An experimental study of turbulent combustion of hydrogen sulfide (H_2S) and natural gas was performed to provide reference data for verification of CFD codes and direct comparison. Hydrogen sulfide is present in most crude oil sources, and the explosion behaviour of pure H_2S and mixtures with natural gas is important to address. The explosion behaviour was studied in a four-meter-long square pipe. The first two meters of the pipe had obstacles while the rest was smooth. Pressure transducers were used to measure the combustion in the pipe. The pure H_2S gave slightly lower explosion pressure than pure natural gas for lean-to-stoichiometric mixtures. The rich H_2S gave higher pressure than natural gas. Mixtures of H_2S and natural gas were also studied and pressure spikes were observed when 5% and 10% H_2S were added to natural gas and also when 5% and 10% natural gas were added to H_2S. The addition of 5% H_2S to natural gas resulted in higher pressure than pure H_2S and pure natural gas. The 5% mixture gave much faster combustion than pure natural gas under fuel rich conditions.

1. Introduction

Hydrogen sulfide (H_2S) may be present in various concentrations in crude oil, natural gas, and biogas; an understanding of its effects is necessary since hydrogen sulfide is a toxic, flammable, and corrosive substance. The industrial process of sulfur removal will produce a lot of sulfuric biproducts. These biproducts could be a potential hazard to factory and workers. The mixture of natural gas and hydrogen sulfide has been a safety issue in development of new oil fields recently.

Jianwen et al. [1] described three major releases of hydrogen sulfide and natural gas that caused severe accidents. To reliably calculate the hazardous consequences of a hydrogen sulphide release, knowledge of its properties is critical. Earlier work investigated detonations in hydrogen sulfide, and its laminar properties have also been studied. However, experimental data from H_2S explosions are limited. This work focuses on the turbulent combustion of hydrogen sulfide and summarizes a series of experimental investigations of explosions with H_2S mixtures. These mixtures are composed of pure H_2S, artificial natural gas (NG) (10% propane and 90% methane), and NG mixed with H_2S. All tests are mixed with air and are conducted at 1 atm initial pressure and ambient temperature. A square pipe with repeated obstacles is used to generate turbulence and increase the flame speed in the study. The experimental results provide a reference data set for verification of CFD codes and also enable a direct comparison with natural gas for the maximum pressure. As more unconventional oil sources are developed, there will be an increasing need to accurately model the combustion of natural gas and hydrogen sulfide mixtures for risk assessment.

2. Gas Explosions in Hydrogen Sulfide

Glassman and Yetter [2] provide a general discussion on sulfur combustion which describes the inhibition of oxidation of hydrogen by H_2S. The stoichiometric combustion of H_2S in oxygen can be written as the overall reaction

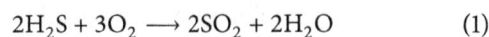

$$2H_2S + 3O_2 \longrightarrow 2SO_2 + 2H_2O \tag{1}$$

In a stoichiometric and rich mixture some of the SO_2 products may also react with H_2S to form solid S by the Claus reaction [3]

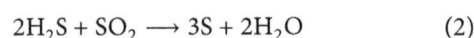

$$2H_2S + SO_2 \longrightarrow 3S + 2H_2O \tag{2}$$

Alzueta et al. [4] showed that SO_2 could either promote or inhibit the burning of CO depending on the amount of SO_2 and the stoichiometry. Selim et al. [3] investigated premixed methane-air with added H_2S, and they showed that combustion begins with the thermal and chemical decomposition of H_2S. SO_2 was also found to enhance the dimerization of CH_3 radicals to form longer hydrocarbons. A chemical reaction mechanism of sulfur and hydrocarbons has been proposed by Wendt et al. [5] and Frenklach et al. [6].

Chamberlin and Clarke [7] were early investigators of the laminar flame speed of hydrogen sulfide. Their setup was typical of the period and consisted of a tube that was 1 m long and 2.5 cm in internal diameter. The tube had a burner tip. The maximum flame speed was observed at 10% ($\phi = 0.8$) and had a value of 0.5 m/s. Also a relatively wide flammable region in H_2S-air mixtures was observed. Kurz [8] used a Bunsen burner method to investigate the effect of a hydrogen sulfide additive on the flame speed of propane, and he also included the flame speed measurements for pure H_2S-air. The flame speed decreased as H_2S was added to the propane, up to the maximum investigated concentration of 6%. However, pure H_2S resulted in a higher flame speed than the mix. A Bunsen flame was also used by Gibbs and Calcote [9] to investigate the effect of the molecular structure on the burning velocity for different equivalence ratios. These three experimental studies of H_2S flame speeds are summarized in Figure 2. As seen, there are relatively large discrepancies between the results, and it is also worth noting that none of the results consider the flame stretch effects. This work does not involve any determination of the laminar flame properties but states that the current knowledge of hydrogen sulfide flames is inconsistent. As such it does not provide a good basis for evaluation of potential hazards as compared to other gases.

There is need for further experimental investigations into the laminar burning velocities and chemical kinetics for pure H_2S gas and H_2S mixed with hydrocarbons. These studies could provide more consistent information regarding the laminar flame properties of the fuel and chemical induction delay times. Such data would be valuable as input to modelling tools and validation of chemical reaction mechanisms. Until new knowledge has been found, one must use the methods available but beware of its limitations.

Cantera software was used to calculate the constant volume combustion pressure and the constant pressure expansion ratio by the reaction mechanism of Wendt et al. [5]. These results are given in Figure 1 and are calculated for stoichiometric fuels, with the H_2S content in NG ranging from 0 (pure natural gas) to pure H_2S, using increasing additions of H_2S. It is shown that the equilibrium pressure and expansion ratio are inversely proportional to the hydrogen sulfide content in the fuel. The calculations suggest that there should be lower flame speed and pressure build-up in propagating hydrogen sulphide deflagration than natural gas mixtures.

Bozek and Rowe [10] compared fuel properties from the International Electrotechnical Commision (IEC) and the National Fire Protection Association (NFPA). Both datasets show that the flammability region of hydrogen sulfide is wider

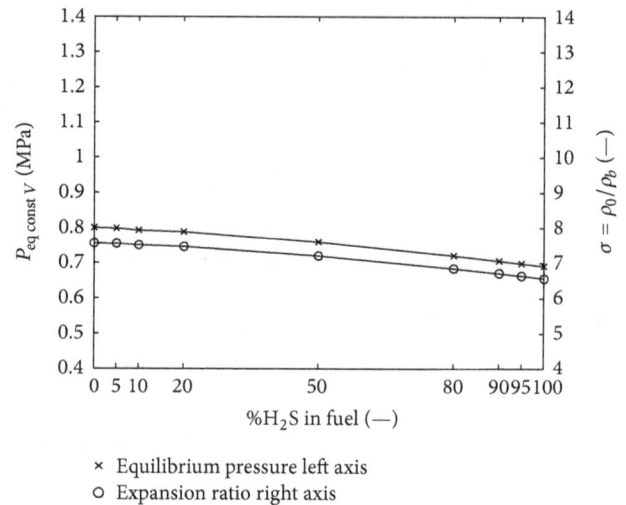

FIGURE 1: Cantera calculation. Constant volume combustion equilibrium pressure for stoichiometric fuel ranging from pure NG (left) to pure H_2S (right).

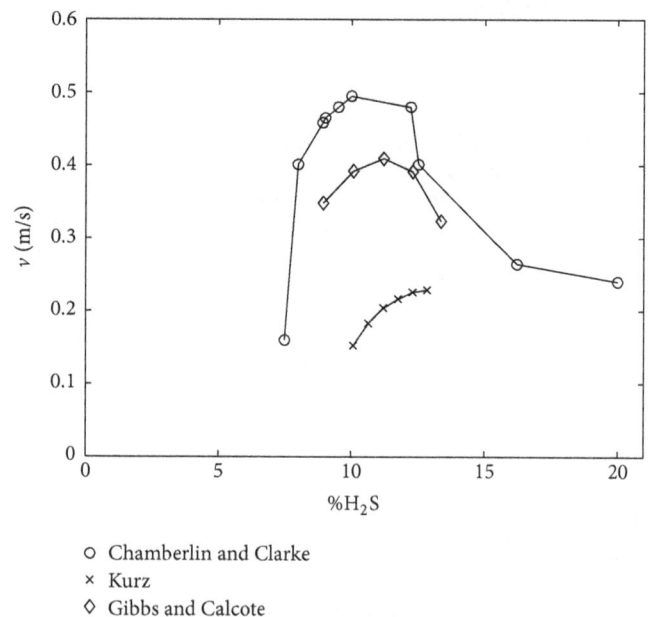

FIGURE 2: Flame speed of H_2S-air mixtures at different concentrations.

than that of methane and pentane. Pahl and Holtappels [11] at the BAM Federal Institute for Materials Research and Testing investigated the explosion limits of H_2S and air in mixtures with N_2 and CO_2. They found the upper and lower explosion limits to be 49.8% and 3.9%, respectively. When CO_2 or N_2 was added to the mixture, the measured explosion limits were higher than those found in an earlier work by Coward and Jones [12].

The minimum experimental safe gap MESG for hydrogen sulfide is lower than that for methane and pentane, which indicates the reactivity of the fuel. NFPA 68 Guide for Venting of Deflagrations (2002) provides data for the deflagration index and shows that it is higher for methane than for H_2S.

Moen and coworkers [13–17] investigated flame acceleration and detonations in H_2S mixtures. The detonation cell size of hydrogen sulfide detonations was 100 mm, while those of methane and propane were 280 mm and 69 mm, respectively. This indicates that H_2S mixtures detonate easier than methane. The deflagration to detonation transition (DDT) of H_2S mixtures has not been widely investigated. Moen et al. [16] investigated the flame acceleration of H_2S-air mixtures in a 1.8 m by 1.8 m cross-section and 15.5 m long square pipe, with obstacles made of steel pipes with diameters 500 mm and 220 mm. They compared the results to those using acetylene-air mixtures. For the hydrogen sulfide experiments they recorded overpressures of only 20 to 50 mbar and flame speeds from 36 to 81 m/s. In a comparison to acetylene they suggested that the H_2S-air mixtures could detonate if the scale was large enough, the ignition was strong enough, or sufficient confinement was present.

Shepherd et al. [17] and Vervisch et al. [18] studied the activation energy of hydrogen sulfide and compared it to other fuels. The resultant value was 109.67 kJ/mol in the Shepherd study and 92 kJ/mol in the Vervisch study. Turns [19] gave 125 kJ/mol activation energy for propane and 125 kJ/mol or 202 kJ/mol for methane.

3. Experimental Setup

The experimental setup used in this work was made from a stainless steel square pipe with inner dimensions of 84 mm. Four parts were bolted together and sealed to make an airtight compartment. Figure 4 shows a schematic of the four parts with their dimensions, obstacle spacing, and pressure transducer positions. Figure 3 shows a picture and Figure 5 shows a sketch of the assembled setup. The experimental setup was chosen to facilitate strong flame acceleration in the beginning and enough spacing in section 2 to possibly get local volume explosions or DDT. The experimental setup was tested also for propane, methane.

The pressures were recorded with two Kistler 7001 (Ch 1 and Ch 2) and four Kistler 603b (Ch 3 to Ch 6) piezoelectric transducers (Figure 2) and an oscilloscope recording at 1 MHz. The ignition system was a center-mounted 10 kV spark at the end flange of section 1. At 10 cm from the end of section 4, one obstacle was installed not only to add strength but also to reflect any shock waves and achieve DDT (if possible) at the end obstacle. DDT located at the end flange is undesirable since it would cause strain on the bolts and filling system.

The fuel-air mixture was made by evacuating air from the square pipe and filling it with fuel. All tests were done with 1 atm initial pressure and ambient temperature. A circulation pump was used to circulate and mix the gas through the system. The setup was placed with the obstacles in vertical alignment. This prevented the fuel from being "trapped" in the pockets between the obstacles at the top and bottom of the pipe. The pump and piping was isolated from the setup before ignition.

FIGURE 3: Picture of the experimental setup.

Special consideration was made regarding the toxicity of hydrogen sulfide and the sulphur dioxide combustion product. A coal filter with special coated coal was installed at the purge of the square pipe to remove sulfuric components from the gas. No H_2S was measured at the outlet of the ventilation system.

This work was part of a larger study to compare H_2S and natural gas mixtures to other more determined fuels. The fuels were acetylene, hydrogen, propane, methane, synthetic natural gas, and H_2S. All fuels were mixed with air. Four different combustion regimes were observed in the study.

To illustrate these explosion regimes, the pressure records are plotted in a diagram showing time along the x-axis and pressure plus the positions of the pressure transducers along the y-axis. This type of diagram gives a good display of the trajectory of the pressure waves, shock waves, and detonation waves in a gas explosion. Figure 6 shows these four different explosion regimes in these types of diagrams:

(i) slow flame propagation and no shock waves formed in front of the flame, which is well known as a slow flame regime;

(ii) fast flame propagation (regime) and shock wave formed but no strong local explosion due to reflection of the shock at the end of the pipe;

(iii) fast flame propagation and shock wave with local explosion and transition to detonation due to reflection of the shock wave at the end of the pipe;

(iv) fast flame propagation and transition to detonation in obstructed area or close to the exit of the obstructed part of the pipe.

Only slow and fast flames were observed in the experiments reported in this paper, but the other regimes are given to provide a qualitative justification of the assumed flame propagation.

Since there is no visual recording of the flame fronts, it is only assumed that the deflagration was similar to other reported works in a very similar setup. Details of this can be found in Lee [20]. The flame fronts become stretched and

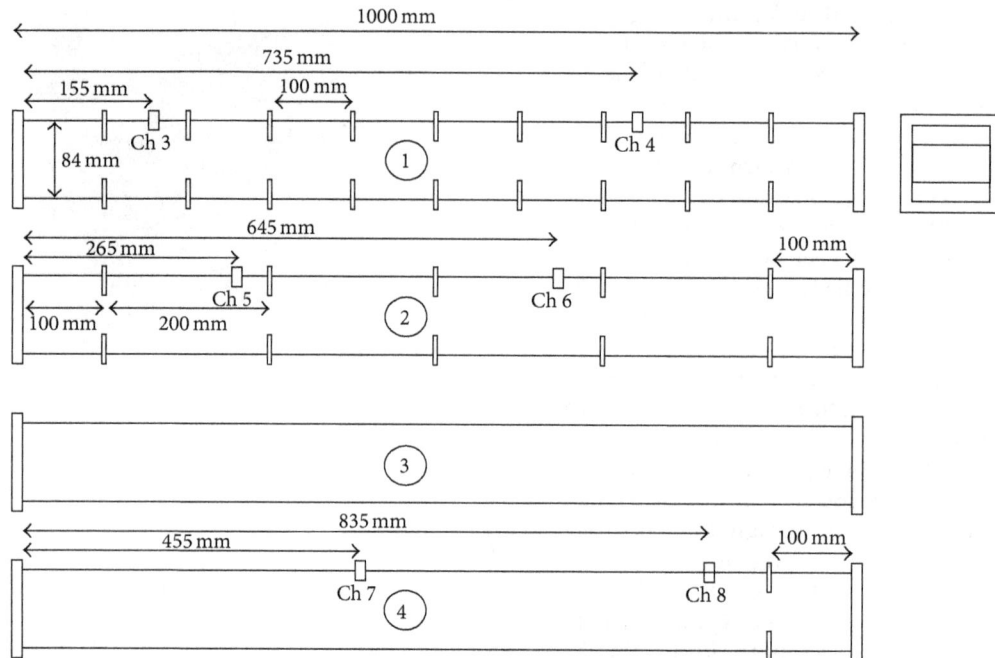

FIGURE 4: Schematic drawing of the experimental setup. Ch# = pressure recording channel number.

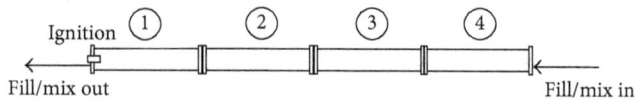

FIGURE 5: The experimental setup consisting of four stainless steel tube sections.

unstable as they propagate through the obstacles, and the flow through the obstacle openings can enhance the mixing at the flame front. Shock reflections at the solid obstacles are also well known to cause local explosions or DDT in sensitive gas mixtures.

4. Results

The fuel mixtures used in this work were pure hydrogen sulfide and fuel mixtures with artificial natural gas (premade 10% propane and 90% methane). The experiments with pure natural gas (NG) and pure H_2S in air are presented first to provide a basis for comparison. Next, results from pure H_2S are presented, and last the mixtures of H_2S and NG in air are presented.

The experimental matrix in Table 1 shows the gases, concentrations, and equivalence ratios.

4.1. Natural Gas. As reference experiments, tests were conducted using artificial natural gas. The concentrations were 6.2%, 8.3%, 9.2%, and 10.4% corresponding to equivalence ratios of $\phi = 0.72$, 0.99, 1.11, and 1.27.

Pressure records from the stoichiometric experiment are given in Figure 7. The pressure curves are offset along the vertical axis, an amount equal to the distance of the

TABLE 1: Experimental matrix.

Test #	Gas 1	Vol. %	Gas 2	Vol. %	ϕ
23	NG	8.30			0.99
24	NG	6.20			0.72
25	NG	9.20			1.11
26	NG	10.40			1.27
27	H_2S	10.00			0.79
28	H_2S	12.40			1.01
29	H_2S	9.00			0.71
30	H_2S	15.10			1.27
49	H_2S	25.00			2.38
31	H_2S	0.43	NG	8.08	1.00
32	H_2S	0.32	NG	5.99	0.72
33	H_2S	0.53	NG	9.98	1.26
34	H_2S	0.86	NG	7.74	0.99
35	H_2S	1.80	NG	7.20	1.01
36	H_2S	5.00	NG	5.00	1.00
37	H_2S	8.96	NG	2.24	1.00
39	H_2S	10.53	NG	1.17	1.00
40	H_2S	11.40	NG	0.60	1.00

transducer from the ignition end. After ignition the flame first propagated through the obstructed part of the pipe. This caused the flame to increase in surface area, and the flow ahead of the flame became turbulent. The turbulent flow caused the flame to accelerate and increase its reaction rate. This is seen in the pressure plots as the rate of pressure gradient increases. At early times a slow pressure increase was observed on channels 1 and 2, with a faster pressure rise seen on channels 3 and 4. In the smooth section a propagating shock wave was recorded on channel 5 at 5.5 ms.

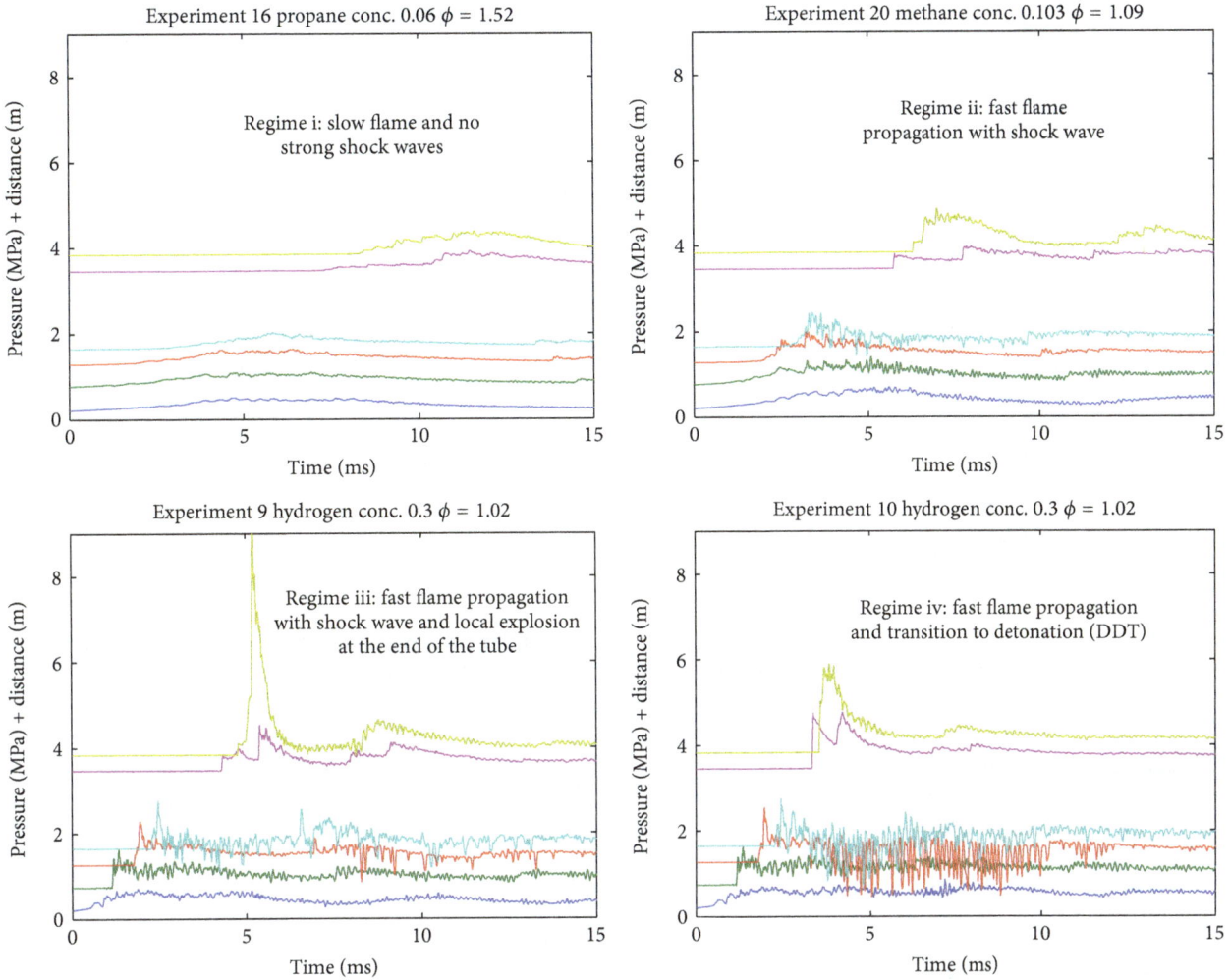

Experiment 16 propane conc. 0.06 $\phi = 1.52$

Regime i: slow flame and no strong shock waves

Experiment 20 methane conc. 0.103 $\phi = 1.09$

Regime ii: fast flame propagation with shock wave

Experiment 9 hydrogen conc. 0.3 $\phi = 1.02$

Regime iii: fast flame propagation with shock wave and local explosion at the end of the tube

Experiment 10 hydrogen conc. 0.3 $\phi = 1.02$

Regime iv: fast flame propagation and transition to detonation (DDT)

FIGURE 6: The four different explosion regimes.

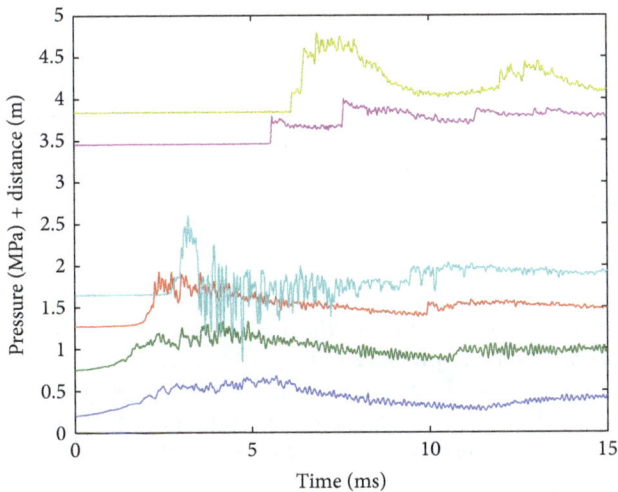

FIGURE 7: Pressure records from the stoichiometric NG-air mixture (test 23). Channels 1–6 are shown from bottom to top. The pressure levels are offset, an amount equal to the distance (m) from the ignition end.

This was generated as the flame accelerated and the displaced flow ahead was fast enough. The shock wave was recorded on channel 6 at 6.2 ms and a reflection at the end obstacle was recorded at 6.5 ms. The reflected shock wave was also recorded on channel 5 at 7.5 ms as it propagated backwards toward the ignition end. Further details on flame acceleration in obstructed pipes can be found in Ciccarelli and Dorofeev [21].

A comparison plot from the natural gas experiments with different fuel concentrations is given in Figure 8. The pressure is read on the left vertical axis and the equivalence ratio is shown on the right vertical axis. The horizontal axis shows the time. The leanest experiment ($\phi = 0.72$), with 6.2% fuel in air, showed a pressure rise of almost 0.5 MPa in the obstructed part of the experimental setup (channel 4) and a primary pressure wave of about 0.25 MPa in the smooth section. The stoichiometric experiment with 8.3% fuel in air showed the fastest pressure rise and the highest pressure (1 MPa). A 0.3 MPa shock wave was recorded in the smooth section. For 9.2% fuel in air ($\phi = 1.11$), the pressure rise in the

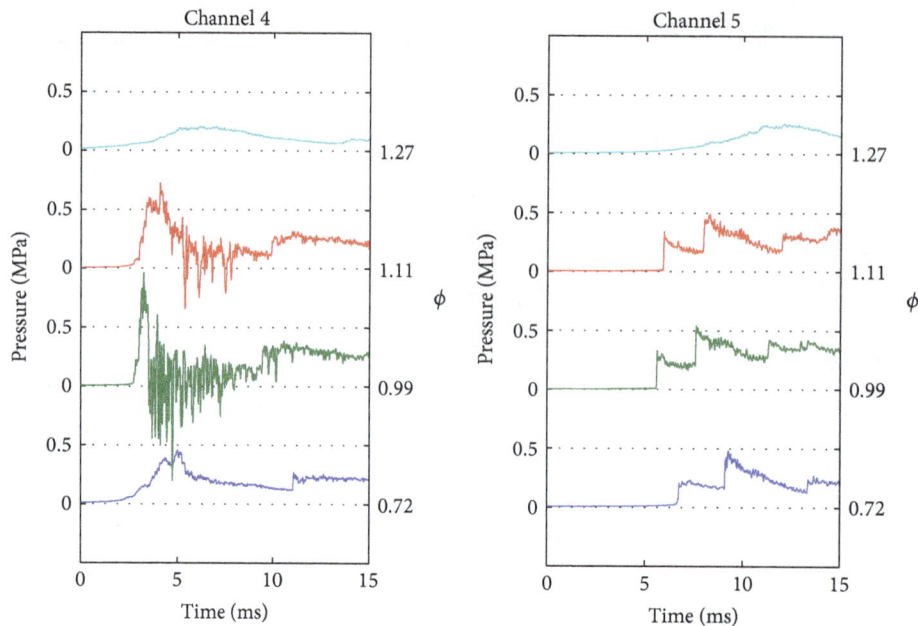

FIGURE 8: From bottom: tests 24, 23, 25, and 26. Comparison of pressure records from channels 4 and 5 for lean, stoichiometric, and rich NG-air mixtures. Pressure is shown on left vertical axis, while the equivalence ratio is given on the right vertical axis.

obstructed section was lower than that in the stoichiometric experiment, while the shock wave in the smooth section was almost equal. The richest experiment (10.4% fuel in air, ϕ = 1.27) resulted in a slow flame and a very slow pressure rise recorded on all pressure transducers.

4.2. Hydrogen Sulfide and Air Mixtures. Results from five tests with the pure H_2S-air mixture are presented. The H_2S concentration ranged from 9% to 25% (see Table 1), where 12.4% is the stoichiometric concentration. The pressure records from the stoichiometric experiment are shown in Figure 9. The overall phenomenon is similar to the stoichiometric natural gas experiment. The initial slow burning and the subsequent development to a faster turbulent flame are seen in the pressure plot. The pressure levels on channels 1 to 4 are lower than in the NG experiment, indicating that this experiment burned slower. The shock wave in the smooth section was roughly the same as in the NG experiment.

Figure 10 shows a comparison plot of the hydrogen sulfide experiments, with the pressure shown on the left vertical axis and the equivalence ratio shown on the right vertical axis. The horizontal axis shows the time. The leanest mixture was 9% H_2S in air (ϕ = 0.71) and showed a pressure rise of about 0.3 MPa. It did not result in a shock in the smooth section of the setup. The recorded pressure wave was about 0.2 MPa, and it reflected at the end wall and obstacle. The slightly richer mixture of 10% (ϕ = 0.79) showed a 0.3 MPa shock wave propagating in the smooth section of the experimental setup. In the obstructed part, 0.5 MPa was recorded at channel 4.

The stoichiometric mixture resulted in a 0.35 MPa shock in the smooth section, while 0.75 MPa was recorded in the obstructed section.

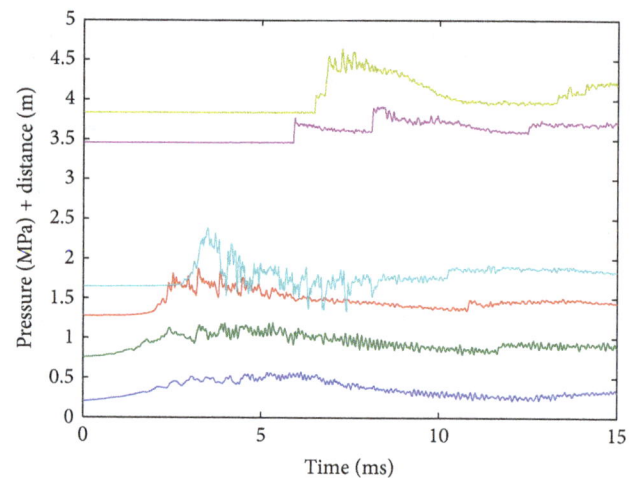

FIGURE 9: Pressure records from the stoichiometric H_2S-air mixture (test 28). Channels 1–6 are shown from bottom to top. The pressure levels are offset, an amount equal to the distance (m) from the ignition end.

The experiment with ϕ = 1.27 corresponding to 15.1% H_2S in air was very similar to the stoichiometric case, with only 0.05 MPa lower pressure in the smooth section and the obstructed section. Due to the wide flammability region of H_2S, ϕ = 2.38 was also investigated; it resulted in a very slow flame and a low pressure increase of about 0.1 MPa.

4.3. H_2S-Natural Gas-Air Experiments, Results, and Discussion. Experiments were performed on a set of nine tests, with the first three containing 5% H_2S and 95% natural gas. The equivalence ratios were ϕ = 0.72, ϕ = 1.00, and ϕ = 1.26.

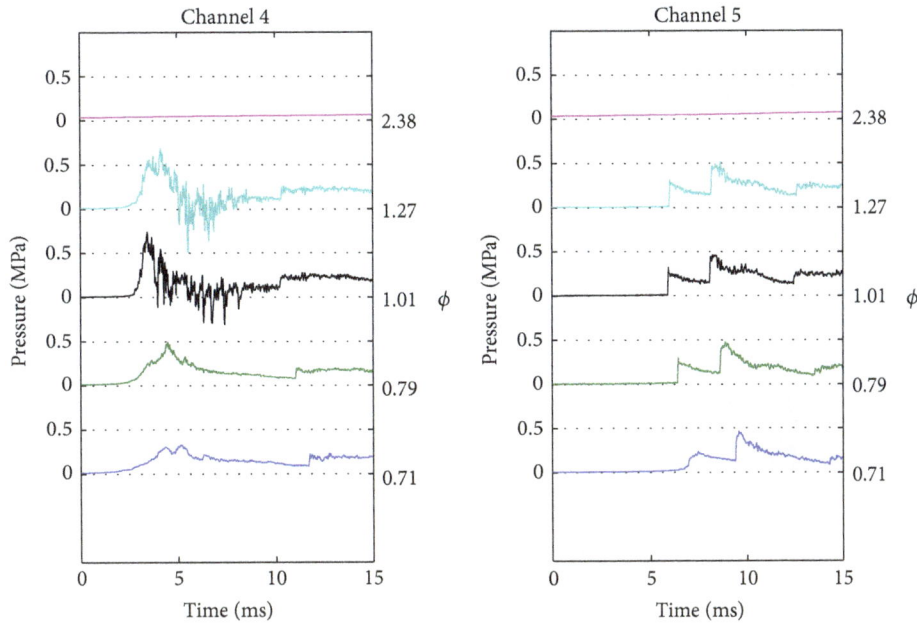

FIGURE 10: From bottom: tests 29, 27, 28, 30, and 49. Comparison of pressure records from channels 4 and 5 for lean, stoichiometric, and rich H_2S-air mixtures. Pressure is shown on left vertical axis, while the equivalence ratio is given on the right vertical axis.

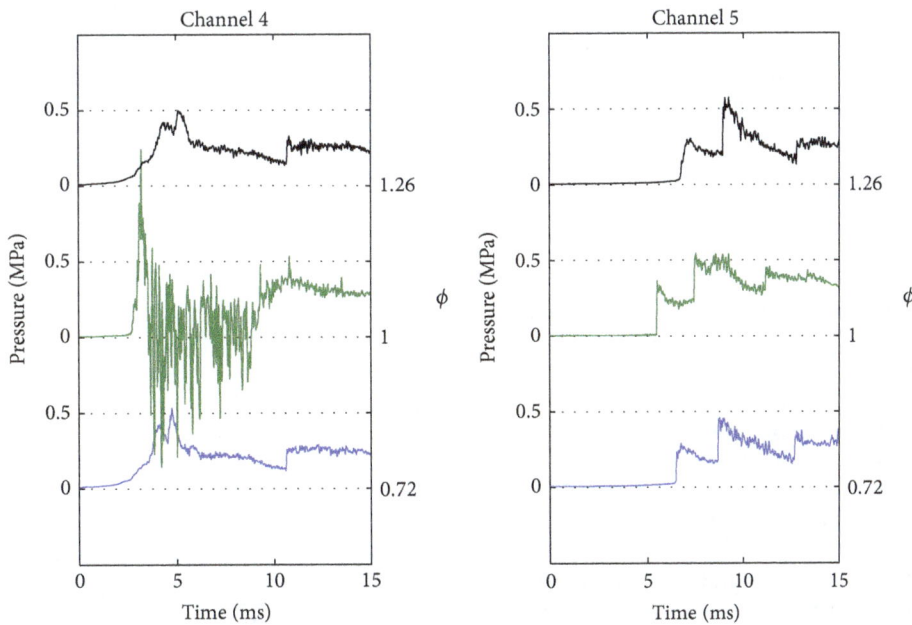

FIGURE 11: From bottom: tests 32, 31, and 33. Comparison of pressure records from channels 4 and 5. Lean, stoichiometric, and rich 5% H_2S/95% NG-air mixtures. Pressure is shown on the left vertical axis, while the equivalence ratio is given on the right vertical axis.

The following experiments were all conducted with $\phi = 1$ but with increasing hydrogen sulfide content. The H_2S fractions in natural gas were 5, 10, 20, 50, 80, 90, and 95%.

Figure 11 shows that, by keeping the H_2S-to-NG ratio constant at 5 : 95 and varying the equivalence ratio, $\phi = 0.72$ and $\phi = 1.26$ give quite similar pressure levels: 0.5 MPa in the obstructed part and 0.3 MPa in the smooth section. The stoichiometric experiment resulted in the fastest pressure rise and a peak pressure of more than 1.3 MPa. A shock wave

of 0.4 MPa was recorded in the smooth section. The rich mixture ($\phi = 1.26$) resulted in strong flame acceleration, 0.5 MPa recorded on channel 4, and a pressure wave in the smooth section.

With the equivalence ratio kept constant at 1 and the H_2S content in the fuel varied from 0% to 100%, the pressure did not change much except for some spikes, as seen in Figure 12. The pressure is shown on the left vertical axis, and the H_2S content in the fuel is shown on the right vertical axis. Time is

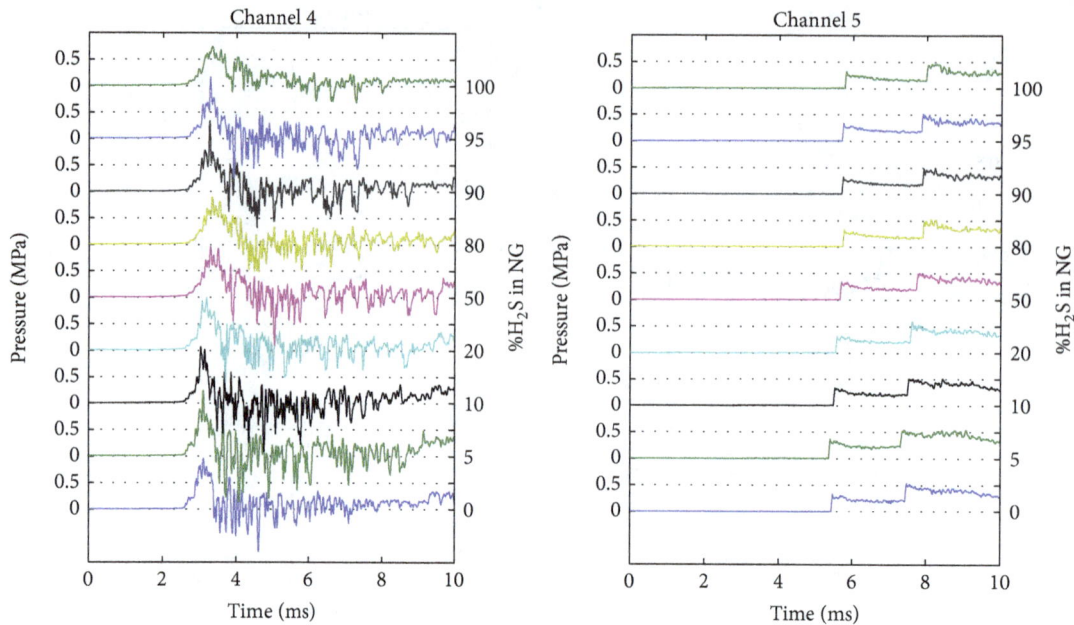

FIGURE 12: From bottom: tests 23, 31, 34, 35, 36, 37, 39, 40, and 28. Comparison of pressure records from channels 4 and 5. The mixture varies from pure natural gas in the fuel (bottom) to pure H$_2$S in the fuel (top). All experiments are stoichiometric mixtures.

shown on the horizontal axis. The pressure in the obstructed part was recorded between 0.8 and 1 MPa, and the shock propagating in the smooth section was about 0.3 to 0.35 MPa and reflected at 0.5 MPa.

5. Discussion

The experimental study for pure natural gas and air showed that the flame propagated fast when the equivalence ratio was lower than 1.27, producing strong deflagrations in the experimental setup. The pressure results showed that the rate of energy release increased as the flame propagated through the square pipe. The richest natural gas mixture investigated was $\phi = 1.27$, and that mixture resulted in a slow pressure rise believed to be due to a slow burning velocity of the flame.

The explosion pressures for lean H$_2$S-air were slightly lower than the pressures for lean NG-air. The lower explosion pressures were to some extent a result of the lower expansion ratio of the H$_2$S-air flame compared with the other fuels. The expansion ratio ($\sigma = \rho_u/\rho_b$) of H$_2$S is about 6.6 while it is 7.6 for NG. This results in a lower flame speed, less turbulence, and, therefore, a lower pressure rise.

By comparing the H$_2$S-air mixtures with mixtures of natural gas and air, as shown in Figure 10 and Figure 8, it was observed experimentally that natural gas and H$_2$S result in a fast flame for $\phi = 0.72$. On the rich side ($\phi = 1.27$), the hydrogen sulfide accelerated as a fast flame while the natural gas was slow. This was expected due to the wider flammability region of H$_2$S [10] compared with NG.

The experiments with stoichiometric H$_2$S-NG-air showed that the flame in the experimental setup produced strong deflagrations with high pressures in the obstructed part of the experimental setup. The pressures seen with

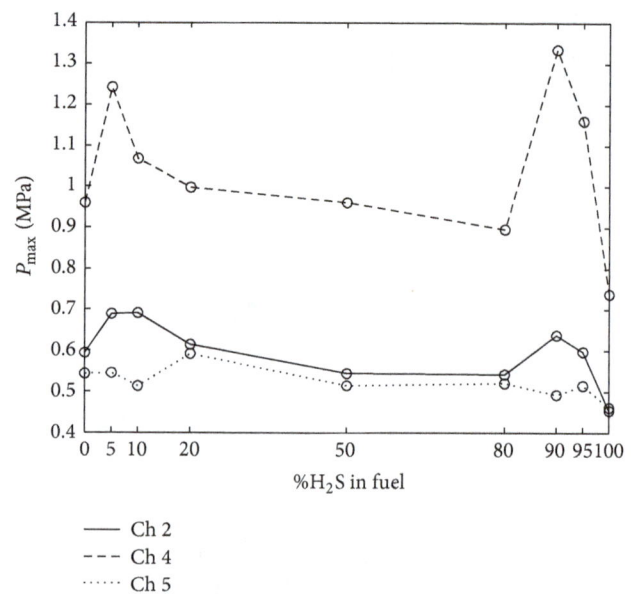

FIGURE 13: Maximum pressure from experiments. The pressure from channels 2, 4, and 5 for various H$_2$S contents in the fuel.

channel 4 in tests with 90% and 95% H$_2$S in the fuel (1.35 and 1.15 MPa) indicate that the compression heating of the reactants caused local ignition in a hot spot.

Comparing the maximum pressure from channels 2, 4, and 5, a trend is observed in Figure 13 in which the maximum pressure decreases as the H$_2$S content in the fuel increases; however, the spikes are also observed when plotting the maximum pressure for three channels when the hydrogen sulfide content was varied. These spikes correspond to 90%

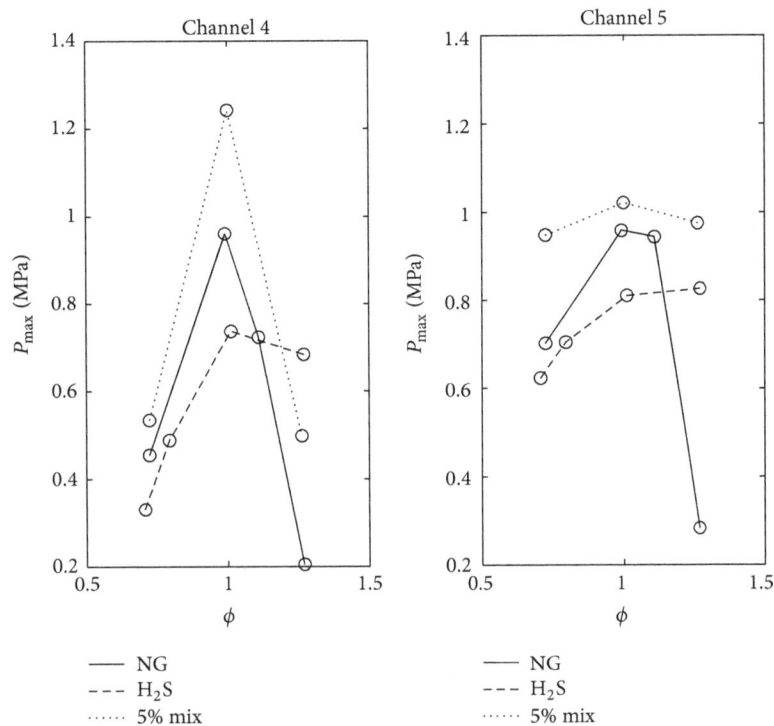

FIGURE 14: Maximum pressure for different equivalence ratios, for pure NG, pure H_2S, and 5% H_2S in NG.

and 95% H_2S in the fuel as well as 5% and 10% H_2S in the fuel.

Compared to the constant volume and constant pressure calculations in Figure 1 it is clear that the pressure spikes originate from different phenomena. One possible explanation could be a more sensitive mixture when small amounts of H_2S are added to natural gas or the opposite. A reduction in chemical induction delay time could lead to local explosions in heated volumes of reactants. These local explosions are very hard to determine even with full view of the channel, but other studies have shown that they are more likely to occur in the obstructed part rather than in the unobstructed parts (Lee [20]).

By comparing Figures 13 and 1, it can be seen that the pressure on channel 4 (section with obstacles) exceeds the constant volume pressure. The equilibrium pressure and the expansion ratio do not explain the spikes seen in Figure 13.

Hot spots and local ignition are closely related to deflagration to detonation transition (DDT), which results in high pressure. No DDT was recorded in these experiments, but the pressure spikes suggest that local explosions could have occurred.

There are always uncertainties when reporting the maximum pressure, since it is measured at one position. Other spikes that may occur in other sections of the experimental setup may be missed by the transducer recording.

By keeping the H_2S content in the fuel constant and changing the equivalence ratio, differences are observed in the combustion. Figure 14 shows the maximum pressure results from the tests with 100% NG, 100% H_2S, and 5% H_2S in NG (mix) for different equivalence ratios.

The addition of 5% H_2S to the natural gas makes the mixture more reactive and, therefore, results in a higher pressure than that with pure NG and pure H_2S. Another notable effect is that the mixture becomes much more insensitive to changes in the equivalence ratio when comparing the maximum pressure from channel 5; that is, it produces higher pressure on both lean and rich sides compared with pure fuels.

A comparison of the pressure in the obstructed section and the smooth section with and without 5% hydrogen sulfide is shown in Figures 15 and 16. Figure 15 shows the stoichiometric case, and the two pressure records from channel 4 and the two pressure records from channel 5 have the same shape and order. This indicates a similar combustion process.

When comparing the explosion pressures with the rich cases (Figure 16), it is seen that there is a major change in the pressure recordings when comparing the same channel. The pure NG burns slowly (a), while the mixed fuel (b) burns much faster and results in a strong pressure wave in the smooth section. This is a significant change caused by the addition of relatively small amounts of hydrogen sulfide to the fuel. There is still more to investigate regarding the combustion of hydrocarbons and sulfur compounds.

These experiments are small/medium scale, and the scale effects of hydrogen sulfide and natural gas explosions are still unknown; however, the presence of hot spots and pressure spikes suggests that DDT might occur if the scale was larger. It was suggested by Moen [15] that the use of a denser obstacle field in experiments would increase the turbulence and flame speed.

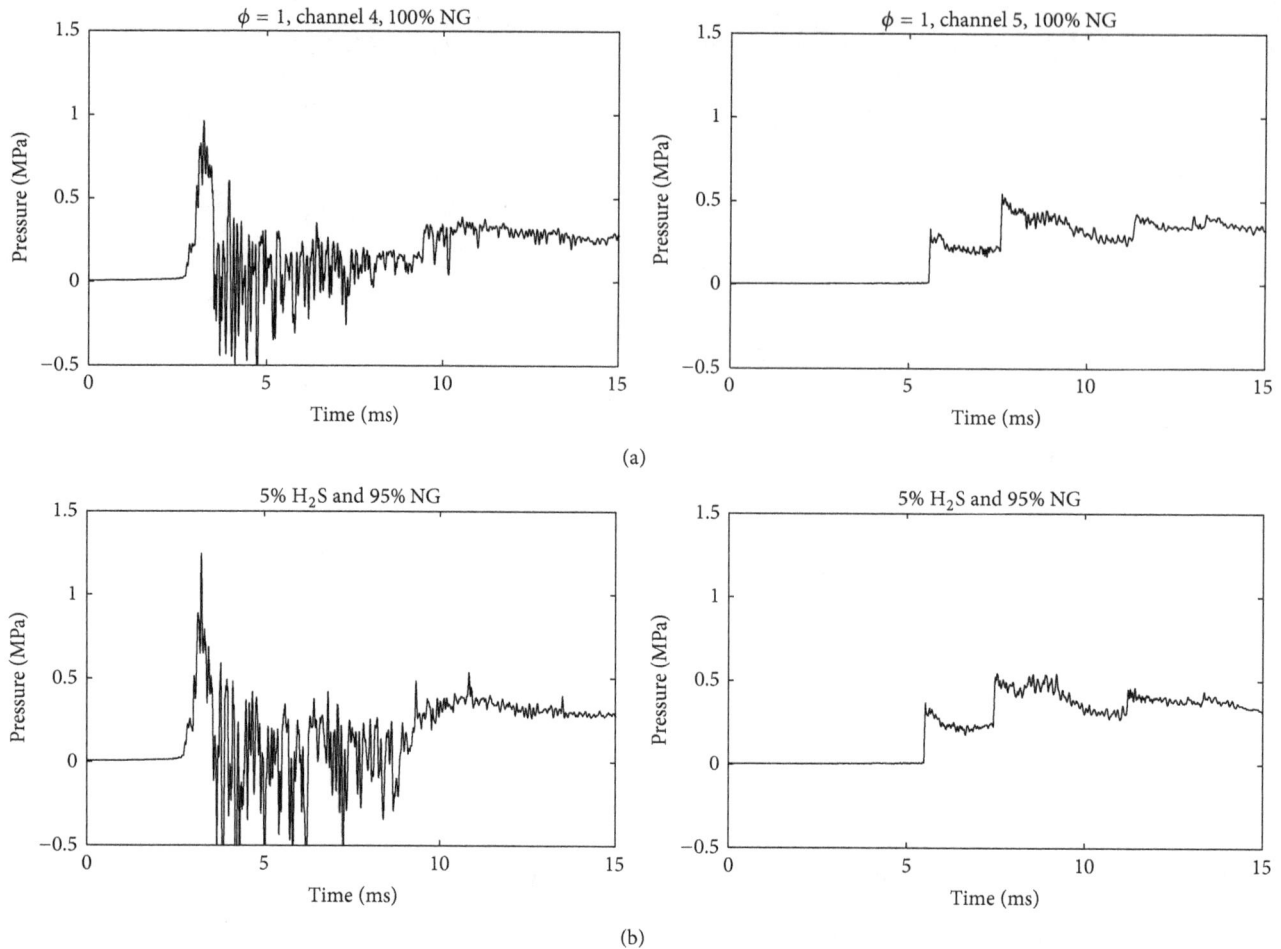

FIGURE 15: Comparison of explosion pressures for $\phi = 1$ in the obstructed section (channel 4) and the smooth section (channel 5). (a) 100% NG and (b) NG with 5% H$_2$S.

6. Conclusion

Only limited data are available in the open literature on H$_2$S-air deflagrations and especially H$_2$S and natural gas mixtures. Data for explosions at conditions supporting strong flame acceleration are lacking. In the present work we have successfully performed such experiments and obtained new and unique experimental data for explosions with hydrogen sulfide and natural gas mixtures. A comparison to pure natural gas is also included.

(i) Pure fuels: hydrogen sulfide has a wide flammability region compared with methane and propane, as shown in the literature. In this study, H$_2$S-air mixtures produced lower explosion pressures at lean-to-stoichiometric compositions relative to natural gas. On the rich side, the H$_2$S-air mixtures produced higher explosion pressures.

(ii) Fuel mixtures at $\phi = 1$: a decrease in the maximum pressure was observed when increasing amounts of hydrogen sulfide were added to the natural gas. There were, however, some maximum pressure spikes observed for 90% and 95% H$_2$S in NG, as well as for

5% and 10% H$_2$S in NG. These spikes could be a result of a local explosion of compressed reactants, but they did not develop into detonations.

(iii) Rich fuel mixtures: rich NG with 5% hydrogen sulfide is more reactive than pure rich NG. When 5% H$_2$S was added to the NG at $\phi = 1$, the result was similar to pure NG but with spikes. When the stoichiometry was changed to $\phi = 1.27$ the result was a fast flame and a strong pressure wave formation in the 5% mixture, while the pure NG had a slow deflagration and a slow and low pressure rise. The 5% mixed fuel also showed decreased sensitivity to changes in the equivalence ratio when the maximum pressures from channel 5 were investigated. These results are important to the process and petroleum industry.

For further work, it is suggested that the experimental results are compared to numerical simulations using commercial and academic software. There is also a need for a thorough study of the laminar properties of H$_2$S-hydrocarbon-air mixtures. Further experimental investigations should be conducted with higher and lower blockage ratios. Larger scale experiments could reveal the possibility of DDT in

(a)

(b)

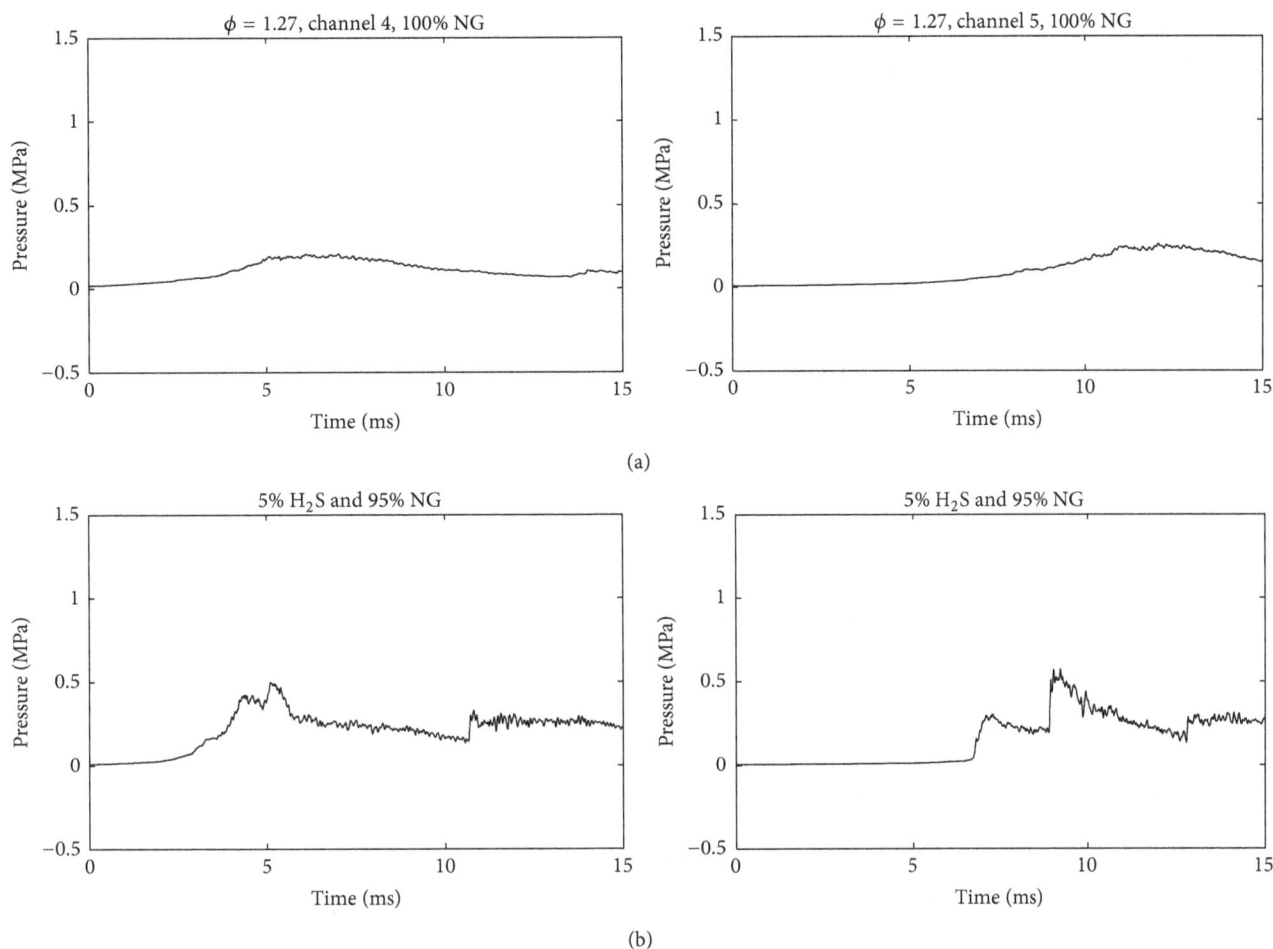

FIGURE 16: Comparison of explosion pressures for $\phi = 1.27$ in the obstructed section (channel 4) and the smooth section (channel 5). (a) 100% NG and (b) NG with 5% H_2S.

H_2S mixtures, and investigations of rich mixtures should be conducted to better understand the effects of added hydrogen sulfide to natural gas.

Conflict of Interests

The authors declare that there is no conflict of interests regarding the publication of this paper.

Acknowledgment

The authors gratefully acknowledge the financial support by Statoil ASA.

References

[1] Z. Jianwen, L. Da, and F. Wenxing, "Analysis of chemical disasters caused by release of hydrogen sulfide-bearing natural gas," *Procedia Engineering*, vol. 26, pp. 1878–1890, 2011.

[2] I. Glassman and R. Yetter, *Combustion*, Academic Press, 4th edition, 2008.

[3] H. Selim, A. Al Shoaibi, and A. K. Gupta, "Effect of H_2S in methane/air flames on sulfur chemistry and products speciation," *Applied Energy*, vol. 88, no. 8, pp. 2593–2600, 2011.

[4] M. U. Alzueta, R. Bilbao, and P. Glarborg, "Inhibition and sensitization of fuel oxidation by SO_2," *Combustion and Flame*, vol. 127, no. 4, pp. 2234–2251, 2001.

[5] J. O. L. Wendt, E. C. Wootan, and T. L. Corley, "Postflame behavior of nitrogenous species in the presence of fuel sulfur I. Rich, moist, $CO/Ar/O_2$ flames," *Combustion and Flame*, vol. 49, no. 1–3, pp. 261–274, 1983.

[6] M. Frenklach, J. H. Lee, J. N. White, and W. C. Gardiner Jr., "Oxidation of hydrogen sulfide," *Combustion and Flame*, vol. 41, pp. 1–16, 1981.

[7] D. S. Chamberlin and D. R. Clarke, "Flame speed of hydrogen sulfide," *Proceedings of the Symposium on Combustion*, vol. 1-2, no. C, pp. 33–35, 1948.

[8] P. F. Kurz, "Influence of hydrogen sulfide on flame speed of propane-air mixtures," *Industrial & Engineering Chemistry*, vol. 45, no. 10, pp. 2361–2366, 1953.

[9] G. J. Gibbs and H. F. Calcote, "Effect of molecular structure on burning velocity," *Journal of Chemical and Engineering Data*, vol. 4, no. 3, pp. 226–237, 1959.

[10] A. P. Bozek and V. Rowe, "Flammable mixture analysis for hazardous area classification," in *Proceedings of the 55th IEEE*

Petroleum and Chemical Industry Technical Conference (PCIC '08), pp. 1–10, September 2008.

[11] R. Pahl and K. Holtappels, "Explosions limits of H_2S/CO_2/air and H_2S/N_2/air," *Chemical Engineering & Technology*, vol. 28, no. 7, pp. 746–749, 2005.

[12] H. Coward and G. Jones, *Limits of Flammability of Gases and Vapors: Bulletin 503*, US Bureau of Mines, Juneau, Alaska, USA, 1952.

[13] A. Sulmistras, I. O. Moen, and A. J. Saber, "Detonations in hydrogen sulphide-air clouds," Suffield Memorandum 1140, Defence Research Establishment Suffield, Alberta, Canada, 1985.

[14] A. J. Saber, A. Sulmistras, I. O. Moen, and P. A. Thibault, "Investigation of the explosion hazard of hydrogen sulphide (Phase I)," Research Report, Defence Research Establishment Suffield, Alberta, Canada, 1985.

[15] I. O. Moen, "Investigation of the explosion hazard of hydrogen sulphide (phase II)," Research Report, Defence Research Establishment Suffield, Alberta, Canada, 1986.

[16] I. O. Moen, A. Sulmistras, B. H. Hjertager, and J. R. Bakke, "Turbulent flame propagation and transition to detonation in large fuel-air clouds," *Symposium (International) on Combustion*, vol. 21, no. 1, pp. 1617–1627, 1988.

[17] J. E. Shepherd, A. Sulmistras, A. J. Saber, and I. O. Moen, "Chemical kinetics and cellular structure of detonations in hydrogen sulfide and air," in *Proceedings of the 10th International Committee on the Dynamics of Explosions and Reactive Systems (ICDERS '85)*, p. 294, Berkeley, Calif, USA, 1985.

[18] L. Vervisch, B. Labegorre, and J. Réveillon, "Hydrogen-sulphur oxy-flame analysis and single-step flame tabulated chemistry," *Fuel*, vol. 83, no. 4-5, pp. 605–614, 2004.

[19] S. R. Turns, *An Introduction to Combustion*, McGraw-Hill, New York, NY, USA, 2nd edition, 2000.

[20] J. H. S. Lee, *The Detonation Phenomena*, Cambridge University Press, New York, NY, USA, 1st edition, 2008.

[21] G. Ciccarelli and S. Dorofeev, "Flame acceleration and transition to detonation in ducts," *Progress in Energy and Combustion Science*, vol. 34, no. 4, pp. 499–550, 2008.

7

Experimental Gasification of Biomass in an Updraft Gasifier with External Recirculation of Pyrolysis Gases

Adi Surjosatyo, Fajri Vidian, and Yulianto Sulistyo Nugroho

Department of Mechanical Engineering, Faculty of Engineering, Universitas Indonesia, UI Campus, Depok 16242, Indonesia

Correspondence should be addressed to Adi Surjosatyo; adisur@eng.ui.ac.id

Academic Editor: Constantine D. Rakopoulos

The updraft gasifier is a simple type of reactor for the gasification of biomass that is easy to operate and has high conversion efficiency, although it produces high levels of tar. This study attempts to observe the performance of a modified updraft gasifier. A modified updraft gasifier that recirculates the pyrolysis gases from drying zone back to the combustion zone and gas outlet at reduction zone was used. In this study, the level of pyrolysis gases that returned to the combustion zone was varied, and as well as measurements of gas composition, lower heating value and tar content. The results showed that an increase in the amount of pyrolysis gases that returned to the combustion zone resulted in a decrease in the amount of tar produced. An increase in the amount of recirculated gases tended to increase the concentrations of H_2 and CH_4 and reduce the concentration of CO with the primary (gasification) air flow held constant. Increasing the primary air flow tended to increase the amount of CO and decrease the amount of H_2. The maximum of lower heating value was 4.9 MJ/m^3.

1. Introduction

The development of industry around the world has resulted in an enormous demand for energy that will continue to rise. However, the supply and the availability of energy from fossil fuels will decrease. Biomass is an environmentally sustainable alternative energy source that is widely available around the world [1]. Using biomass sources such as wood, rice husks, and bagasse, which have the highest energy content, together with the highest-efficiency conversion methods would add a significant amount of energy. The use of biomass for biofuels has reached approximately 9–14% of the total of energy demand worldwide [2].

Gasification is an ecoefficient and sustainable thermochemical conversion method [3] that creates low levels of pollution [4].

Various forms of gasifiers have been developed to meet criteria of being easy to operate, being highly efficient, and producing relatively low amounts of tar. The two most popular types of fixed-bed reactors used are the updraft gasifier and the downdraft gasifier. The updraft gasifier is easy to operate and has quite high conversion efficiency, but this type produces high levels of tar, as high as 0.2 kg/m^3 [5], because the pyrolysis gases containing high levels of tar are extracted directly from the reactor. This increases the load on the gas cleaning system and the level of carcinogenic waste produced [1].

Modifying the reactor is one common method used to reduce the level of tar produced from the gasification. The gasifier could be modified with recirculaton of pyrolysis gas and modification of gas outlet [6].

This study presents a study of a modified updraft gasifier with recirculaton of the pyrolysis gases back to the combustion zone and the gas outlet at the reduction zone. The effects on combustible gas composition, the LHV, and the tar content of the producer gas were investigated.

2. Material and Methods

The fuel used was woody biomass that was cut into pieces of 0.03 m wide and 0.03 m long and had a moisture content of 10.24%. The proximate and ultimate analyses of this fuel are shown in Table 1.

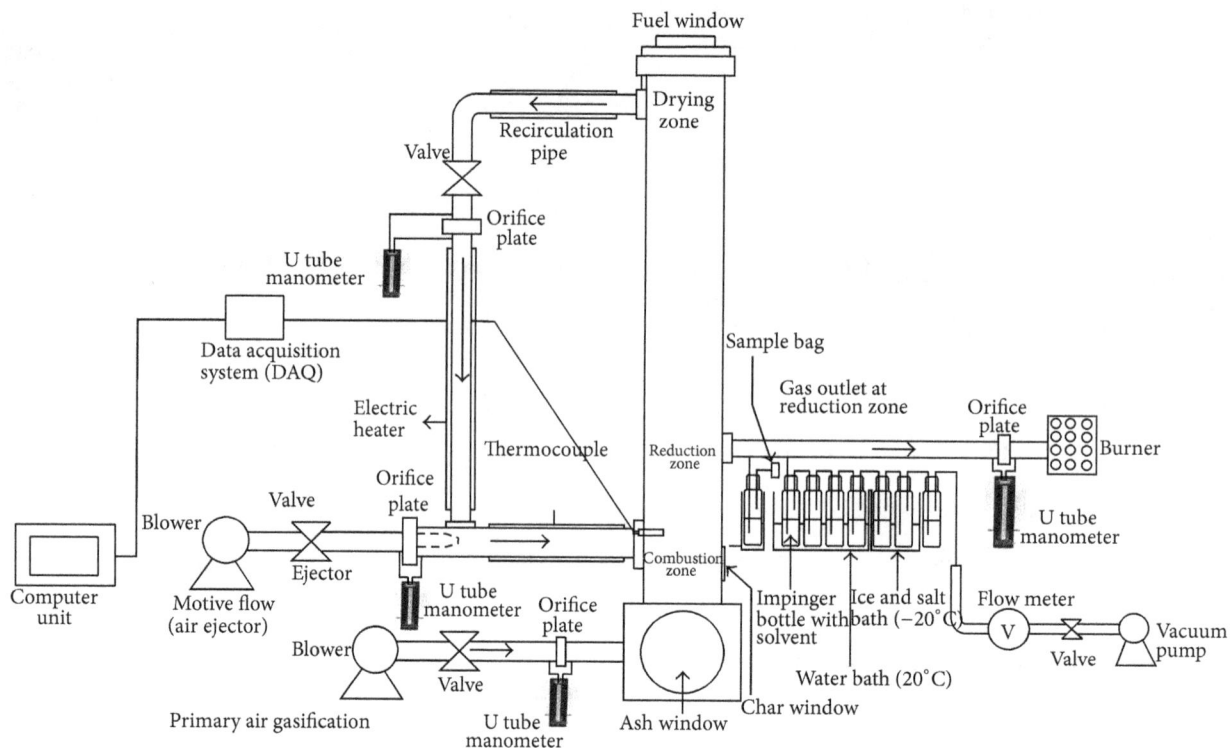

FIGURE 1: Experimental Setup.

TABLE 1: Proximate and ultimate analyses [7].

Proximate and ultimate analyses		
	Unit	Value
Proximate analysis		
Moisture (adb)	%	10.24
Ash	%	2.71
Volatile	%	71.80
Fixed carbon	%	15.25
Ultimate analysis		
Carbon	%	43.33
Hydrogen	%	5.11
Nitrogen	%	Not detected
Sulfur	%	Not detected
Oxygen		38.61
Calorific value	kJ/kg	17025
Density	kg/m^3	640

The gasification process included a gasification reactor with a diameter of 0.22 m and a length of 0.63 m and constructed of stainless steel (SUS 304 [7]), as shown in Figure 1. Type-K thermocouples were placed at the bottom of the reactor wall to measure the temperature inside the combustion zone reactor. Two air supplies were used in this process: the primary air and the motive flow (ejector air). The primary air for combustion was supplied using a blower and the motive flow for driving pyrolysis gas was supplied by a ring blower. The flow rates of the primary air, the motive flow, and the recirculated gas were measured using orifice plate flow meters. The producer gas outlet from the reactor was at a height of 0.13 m above the grate (at reduction zone). Recirculation pipes with a diameter of 0.05 m running from the top (drying zone) to the bottom of the reactor (the combustion zone) were constructed from stainless steel and equipped with control valves to manage the recirculation flow. The pyrolysis gas was driven to the combustion zone by the motive flow (air ejector) in the recirculation pipes. The recirculation pipes were heated using an electric heater to a temperature of 623°K to prevent tar condensation [8]. The tar content in the producer gases was measured using six impinger bottles, five of which were filled with a solvent (isopropanol) and one which was empty, as in [7, 9, 10]. Approximately 0.00005 m^3 (50 mL) of solvent was used to fill each of the five bottles. Subsequently, the solvent containing the tar was vaporised at a temperature of 380°K [11, 12]. The tar that was not vaporised was measured to determine the mass of the tar (kg/m^3). The gas composition was taken using sample tight bags, then it was analyzed using gas chromatography with thermal conductivity detector (TCD).

The ejector was a constant-mixing-area type, and the convergence nozzle had the following dimensions: the diameters of the inlet and outlet air were 0.025 m and 0.0075 m, respectively, and the nozzle exit position (NXP) was −0.03 m before the entrance of mixing chamber [13].

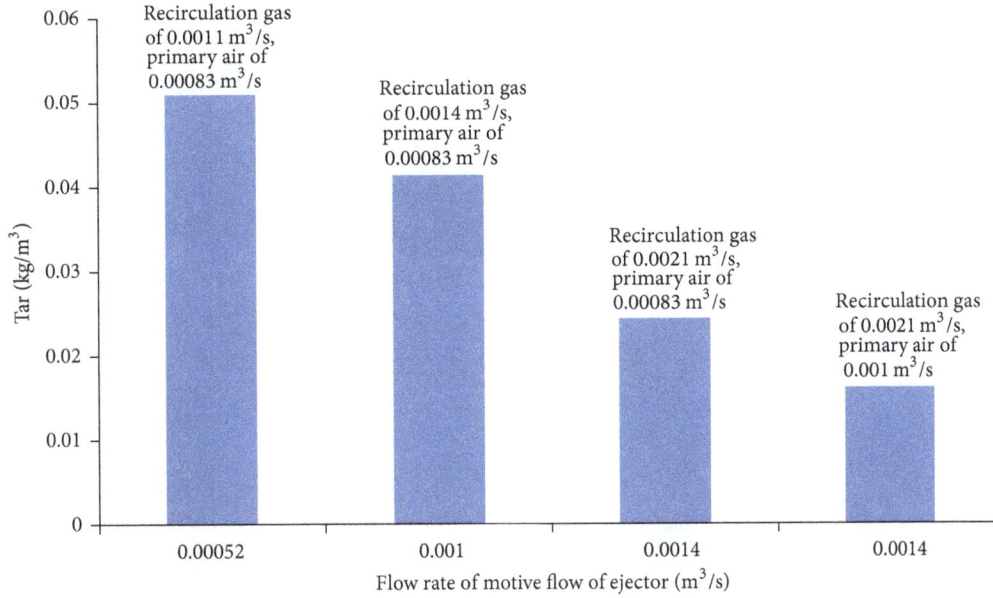

FIGURE 2: Tar content versus motive flow ejector.

The lower heating value (LHV) of producer gas was calculated using calorific value of moles fraction of combustible gas (CO, H_2, CH_4) [14]:

$$\text{LHV}\ (\text{kJ/m}^3) = y_{CO} \cdot 12621 + y_{H_2} \cdot 10779 + y_{CH_4} \cdot 35874, \tag{1}$$

where y_i values are volume fractions of main combustible gas in the producer gas.

The mass of tar was calculated based on differences of mass impinger bottle containing tar and empty impinger bottle:

$$\text{Gravimetric tar}\ (\text{kg/m}^3) = \frac{\text{mass of bottle containing tar (kg)} - \text{mass of empty bottle (kg)}}{\text{flow rate of tar sampling (m}^3/\text{s)} \times \text{time of tar sampling (s)}}. \tag{2}$$

To begin the tests, approximately 0.5 kg of fuel was placed in the reactor and burned until the temperature reached the maximum of 473°K. Next, approximately 6 kg of additional fuel was placed into the reactor until the reactor was nearly full. After 900 seconds of operation, the combustible gases were obtained. The primary air gasification (primary air flow) was varied of 0.00083 m³/s and 0.001 m³/s. The motive flow of ejector (air ejector) was varied of 0.00052 m³/s, 0.001 m³/s, and 0.0014 m³/s. The experiment was carried out at constant primary air gasification and the air ejector was increased, then the air ejector was held constant and the primary air was increased. The gas composition was taken for the sampling at the temperature of combustion zone was stable. Tar samples were taken at the temperature of the combustion zone was stable, while the flow rate of gas sampling was set at 3.3×10^{-5} m³/s and the sampling time was about 240 seconds for each experiment.

3. Results and Discussion

3.1. Tar Content. Figure 2 shows the effect of the flow rate of the recirculated pyrolysis gases on the amount of tar

produced. At recirculated pyrolysis gas flow rates of 0.0011, 0.0014, and 0.0021 m³/s, generated with ejector motive flow rates of 0.00052 m³/s, 0.001 m³/s, and 0.0014 m³/s, respectively, the resulting tar concentrations were approximately 0.051, 0.0414, and 0.0243 kg/m³ respectively, indicating a reduction in the tar content. At a motive flow rate of 0.0014 m³/s, the primary air flow rate was increased to 0.001 m³/s, and the tar concentration was reduced to 0.0161 kg/m³. This reduction in the amount of tar resulted from the cracking (reactions (8)) and reforming (reactions (9) and reactions (10)) process of tar into combustible gases (H_2 and CO) at a temperature of approximately 923–973°K [15, 16]. Figure 5 shows the temperature in the combustion zone for every operating condition above 1073°K for which cracking and reforming tar is possible. The cracking and reforming of the tar will contribute more to H_2 (reactions (8), and (9)) production when the primary air flow rate is constant and the motive flow of the ejector is varied. When the primary gasification was increased, there was a tendency towards an increased contribution of tar reforming to CO production (reactions (10)) as an effect of the increase

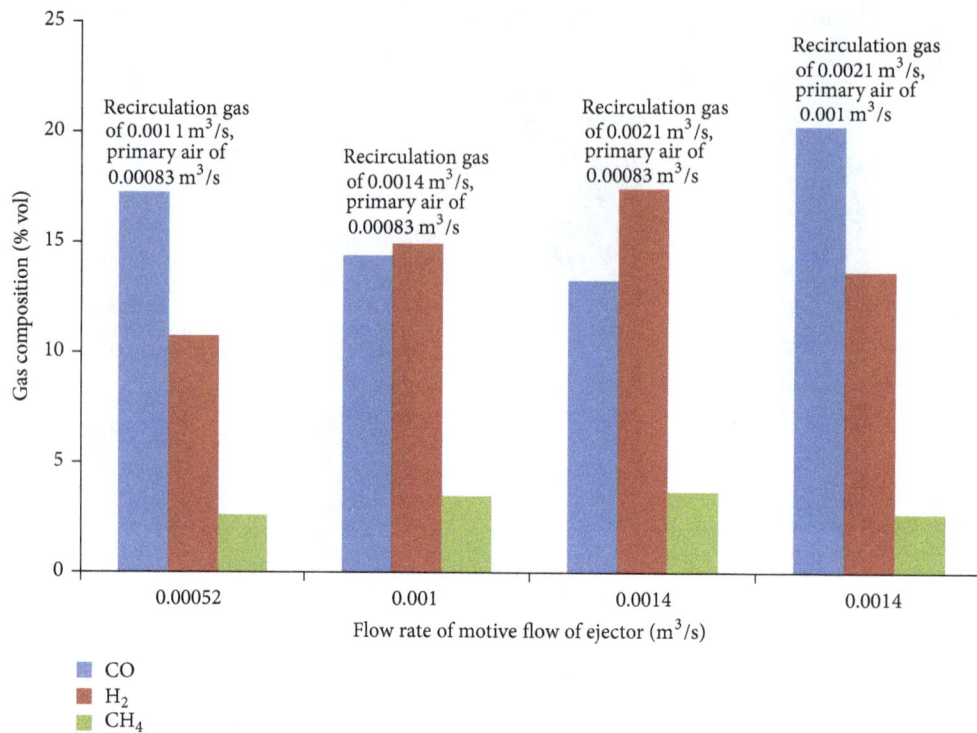

FIGURE 3: Gas concentration versus motive flow ejector.

of combustion reaction (reaction (7)), as shown in Figure 3.

3.2. Gas Compositionzz.

Figure 3 shows the concentrations of the gases resulting from the recirculation of the pyrolysis gases back to the combustion zone.

Increasing the recirculation of the pyrolysis gas from $0.0011\,m^3/s$ to a maximum of $0.0021\,m^3/s$ at a constant primary air flow rate of $0.00083\,m^3/s$ caused the concentration of H_2 to increase from 10.74% to 17.46% and the concentration of CH_4 to increase from 2.62% to 3.7%; however, the concentration of CO decreased from 17.25% to 13.29%. This upward trend was caused by the water vapour (H_2O) present in pyrolysis gases reacting with the C and the CO at high temperatures (above 1073°K) to produce H_2 (reactions (3)) and (reactions (4)). According to Kumar et al. (2009), the water gas reaction should increase at temperatures between 1023°K and 1073°K [17]. The reaction of tar cracking (reactions (8)) and steam reforming of the tar (reactions (9)) at temperatures of approximately 923°K to 973°K [15, 16] would contribute to an increase in H_2. The decrease in the CO concentration due to a decrease in the dominance of the boudouard reaction (reactions (5)) caused a temperature reduction. The increase in the CH_4 level was caused by an increase in the methanisation reaction (reactions (6)) because the pressure increased in the gasifier as an effect of the pressure of the ejector. According to Kaupp and Gross (1981) [18] and Donaj et al. (2011) [19], an increase in the pressure inside the reactor increases the concentration of CH_4.

When the recirculation flow rate was at a maximum of $0.0021\,m^3/s$ and the primary air flow rate was increased to $0.001\,m^3/s$, the concentration of CO increased from 13.29% to 20%, but the concentrations of H_2 and CH_4 decreased from 17.46% to 13.68% and from 3.7% to 2.69%, respectively. Because the increase in the primary air flow rate increases the combustion reaction (reactions (7)) and the temperature inside the reactor, this contributed to a reduction in the reaction that produces H_2 (reaction (3)) and (reaction (4)) and an increase in the boudauard reaction (reaction (5)) [20] and dry reforming of the tar (reaction (10)). The decrease in H_2 contributed to a reduction in CH_4 (reaction (6)) The increased temperature inside the reactor (Figure 5) contributed to the decrease in the CH_4 concentration [20, 21].

Consider gasification, tar cracking, and tar reforming reaction as follows:

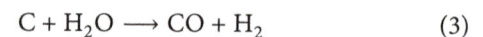

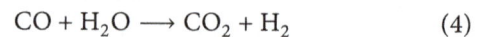

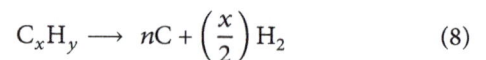

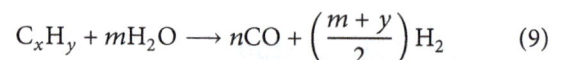

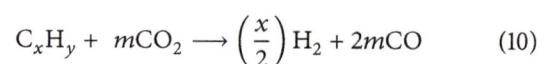

$$C + H_2O \longrightarrow CO + H_2 \tag{3}$$

$$CO + H_2O \longrightarrow CO_2 + H_2 \tag{4}$$

$$C + CO_2 \longrightarrow 2\,CO \tag{5}$$

$$C + H_2 \longrightarrow CH_4 \tag{6}$$

$$C + O_2 \longrightarrow CO_2 \tag{7}$$

$$C_xH_y \longrightarrow nC + \left(\frac{x}{2}\right)H_2 \tag{8}$$

$$C_xH_y + mH_2O \longrightarrow nCO + \left(\frac{m+y}{2}\right)H_2 \tag{9}$$

$$C_xH_y + mCO_2 \longrightarrow \left(\frac{x}{2}\right)H_2 + 2mCO \tag{10}$$

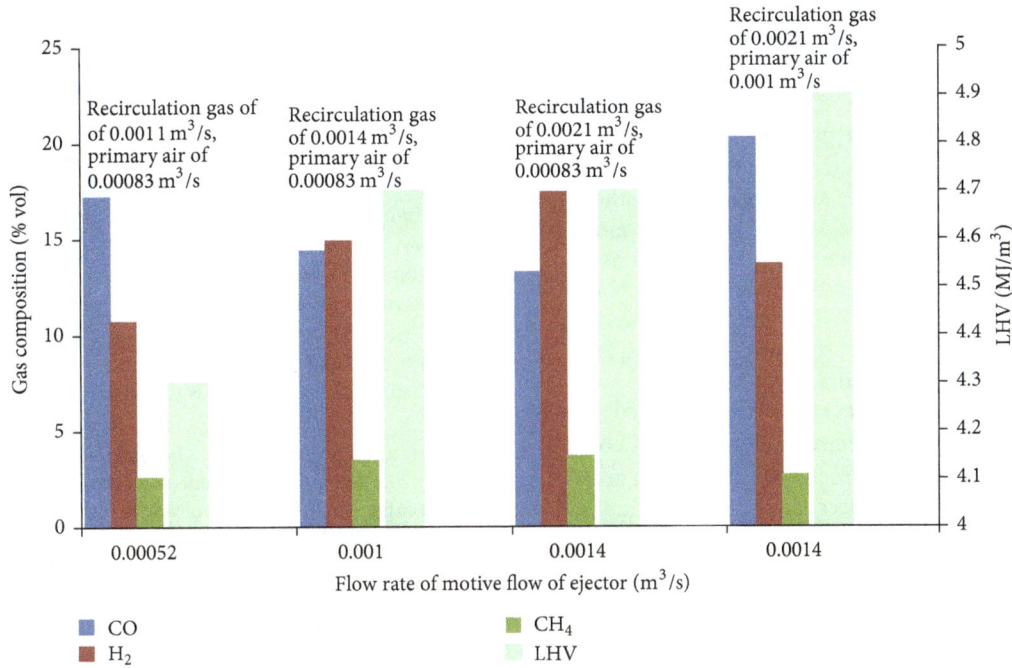

FIGURE 4: Lower heating values of gases versus motive flow ejector.

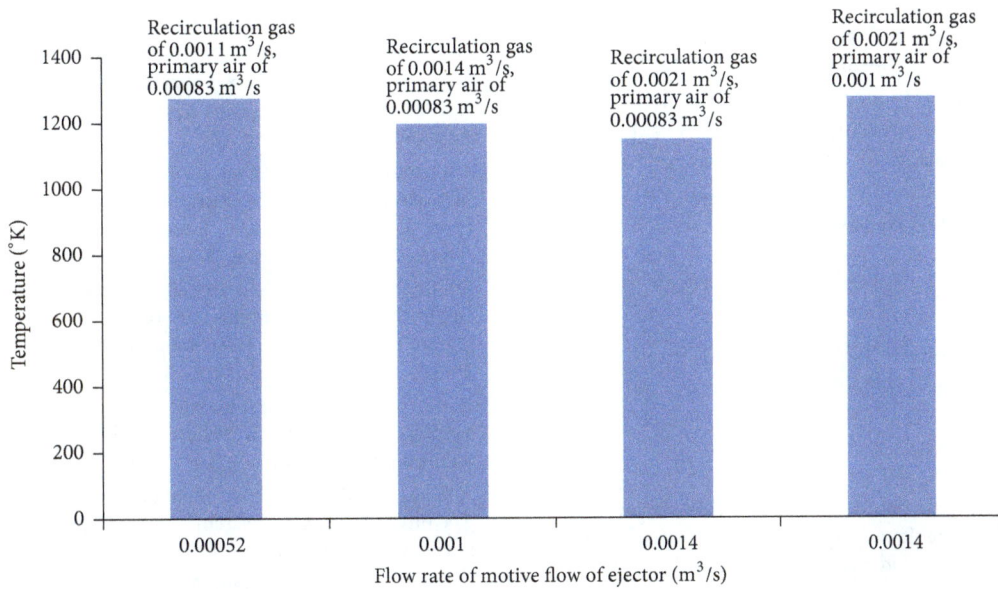

FIGURE 5: Combustion zone temperature versus motive flow ejector.

3.3. *The Lower Heating Value of the Producer Gas.* Figure 4 shows the effect of the flow of the pyrolysis gases in the combustion zone on the lower heating value of the gas.

When the pyrolysis gas recirculation rate was increased from $0.0011\,m^3/s$ to a maximum of $0.0021\,m^3/s$ at a constant primary air flow rate of $0.00083\,m^3/s$, the lower heating value of the gas increased from $4.3\,MJ/m^3$ to $4.7\,MJ/m^3$. This increase resulted from a significant increase in the concentration of H_2 and a moderate decrease in the concentration of CO. Furthermore, there was an increase in the concentration of CH_4, which has a greater heating value than that of the other gases. When the primary air flow rate was increased to $0.001\,m^3/s$ and the pyrolysis gas recirculation rate was held constant at $0.0021\,m^3/s$, the lower heating value increased from $4.7\,MJ/m^3$ to $4.9\,MJ/m^3$ because of the significant increase in the concentration of CO.

3.4. Combustion Zone Temperature. Figure 5 shows the variation in the maximum temperature in the combustion zone with the ejector flow rate. When the pyrolysis gas recirculation was increased from $0.0011\,m^3/s$ to a maximum of $0.0021\,m^3/s$ at a constant primary air flow rate of $0.00083\,m^3/s$, the maximum temperature in the combustion zone decreased from $1273°K$ to $1148°K$ because the endothermic reaction that produces H_2 (reactions (3)) became more dominant. The lower pyrolysis gas temperature led to a need for more heat to increase its temperature, causing a decrease in the temperature of the bed [17]. The temperature inside reactor was constant at over $1148°K$ so that the cracking and steam reforming of the tar were continuous.

When the primary air flow rate was increased to $0.001\,m^3/s$ and the pyrolysis gas recirculation was held constant at $0.0021\,m^3/s$, the temperature in the combustion zone increased from $1148°K$ to $1273°K$ because the increase in the flow rate of the primary air contributed to an increase in the exothermic combustion reaction (reaction (7)).

4. Conclusions

The recirculation of pyrolysis gases from the top of gasifier (drying zone) to the combustion zone and gas outlet from reduction zone in a modified updraft gasifier in this study resulted in maximum lower heating value of $4.9\,MJ/m^3$. Increasing the flow of the pyrolysis gases to the combustion zone tended to reduce the amount of tar produced. The concentration of H_2 tended to increase and the concentration of CO decreased with increasing motive flow rate and constant primary gasification air. Increasing flow rate of the primary gasification air tended to increase the amount of CO and decrease the amount of H_2 produced.

Conflict of Interests

The authors declare that there is no conflict of interests regarding the publication of this paper.

References

[1] E. G. Pereiraa, J. N. da Silvaa, J. L. de Oliveirab, and C. S. Machadoa, "Sustainable energy: a review of gasification technologies," *Renewable and Sustainable Energy Reviews*, vol. 16, pp. 4753–4762, 2012.

[2] K. K. Gupta, A. Rehman, and R. M. Sarviya, "Bio-fuels for the gas turbine: a review," *Renewable and Sustainable Energy Reviews*, vol. 14, no. 9, pp. 2946–2955, 2010.

[3] A. K. Sharma, "Modeling and simulation of a downdraft biomass gasifier 1. Model development and validation," *Energy Conversion and Management*, vol. 52, no. 2, pp. 1386–1396, 2011.

[4] A. F. Kirkels and G. P. J. Verbong, "Biomass gasification: still promising? A 30-year global overview," *Renewable and Sustainable Energy Reviews*, vol. 15, no. 1, pp. 471–481, 2011.

[5] P. Stahlberg, M. Lappi, E. Kurkela, P. Simell, P. Oesch, and M. Nieminen, *Sampling of Contaminants from Product Gases of Biomass Gasifiers*, vol. 5 of *VTT, Technical Research Centre of Finland Espoo*, 1998.

[6] A. Surjosatyo, F. Vidian, and Y. S. Nugroho, "A review on gasifier modification for tar reduction in biomass gasification," *An International Journal of Jurnal Mekanikal*, no. 31, pp. 62–77, 2010.

[7] A. Surjosatyo, F. Vidian, and Y. S. Nugroho, "Performance gasification perbatch rubber wood in conventional updraft gasifier," *Journal of Engineering and Applied Sciences*, vol. 7, no. 8, pp. 494–500, 2012.

[8] P. Gilbert, C. Ryu, V. Sharifi, and J. Swithenbank, "Tar reduction in pyrolysis vapours from biomass over a hot char bed," *Bioresource Technology*, vol. 100, no. 23, pp. 6045–6051, 2009.

[9] J. P. A. -Neeft, H. A. M. Knoef, G. J. Buffinga et al., "Guideline for sampling and analysis of tar and particles in biomass producer gases," Energy Project ERK6-CT1999-20002 (Tar Protocol), 2002.

[10] C. Brage and K. Sjöström, *An Outline of R&D Work Supporting the Tar Guideline*, Department of Chemical Engineering and Technology, Chemical Technology, Royal Institute of Technology (KTH), 2002.

[11] Y. Ueki, T. Torigoe, O. Hirofumi, Y. Ryo, J. H. Kihedu, and I. Naruse, "Gasification characteristics of woody biomass in the packed bed reactor," in *Proceedings of the Combustion Institute*, vol. 33, pp. 1795–1800, 2011.

[12] E. Kurkela, P. Ståhlberg, P. Simell, and J. Leppälahti, "Updraft gasification of peat and biomass," *Biomass*, vol. 19, no. 1-2, pp. 37–46, 1989.

[13] F. Vidian, A. Surjosatyo, and Y. S. Nugroho, "CFD analysis of external recirculation flow at updraft gasifier using ejector," in *AIP Confrences Proceedings*, vol. 1440, pp. 936–941, 2011.

[14] M. Seggiani, S. Vitolo, M. Puccini, and A. Bellini, "Cogasification of sewage sludge in an updraft gasifier," *Fuel*, vol. 93, pp. 486–491, 2012.

[15] T. Damartzis and A. Zabaniotou, "Thermochemical conversion of biomass to second generation biofuels through integrated process design-A review," *Renewable and Sustainable Energy Reviews*, vol. 15, no. 1, pp. 366–378, 2011.

[16] J. Šulc, J. Štojdl, M. Richter et al., "Biomass waste gasification—can be the two stage process suitable for tar reduction and power generation?" *Waste Management*, vol. 32, no. 4, pp. 692–700, 2012.

[17] A. Kumar, D. D. Jones, and M. A. Hanna, "Thermochemical biomass gasification: a review of the current status of the technology," *Energies*, vol. 2, no. 3, pp. 556–581, 2009.

[18] A. Kaupp and J. R. Gross, "State of the art for small (2–50 kW) gas producer engine system," Final Report to USDA, Forest Service Contract No 53-39R-0-141, 1981.

[19] P. Donaj, M. R. Izadpanah, W. Yang, and W. Blasiak, "Effect of pressure drop due to grate-bed resistance on the performance of a downdraft gasifier," *Energy and Fuels*, vol. 25, no. 11, pp. 5366–5377, 2011.

[20] H. Kitzler, C. Pfeifer, and H. Hofbauer, "Pressurized gasification of woody biomass-Variation of parameter," *Fuel Processing Technology*, vol. 92, no. 5, pp. 908–914, 2011.

[21] L. E. Taba, M. F. Irfan, W. A. M. Wan Daud, and M. H. Chakrabarti, "The effect of temperature on various parameters in coal, biomass and CO-gasification: a review," *Renewable and Sustainable Energy Reviews*, vol. 16, pp. 5584–5596, 2012.

8

Large Eddy Simulation of a Bluff Body Stabilized Lean Premixed Flame

A. Andreini, C. Bianchini, and A. Innocenti

Department of Industrial Engineering, University of Florence, Via di Santa Marta 3, 50139 Florence, Italy

Correspondence should be addressed to A. Andreini; antonio.andreini@htc.de.unifi.it

Academic Editor: Eliseo Ranzi

The present study is devoted to verify current capabilities of Large Eddy Simulation (LES) methodology in the modeling of lean premixed flames in the typical turbulent combustion regime of Dry Low NO_x gas turbine combustors. A relatively simple reactive test case, presenting all main aspects of turbulent combustion interaction and flame stabilization of gas turbine lean premixed combustors, was chosen as an affordable test to evaluate the feasibility of the technique also in more complex test cases. A comparison between LES and RANS modeling approach is performed in order to discuss modeling requirements, possible gains, and computational overloads associated with the former. Such comparison comprehends a sensitivity study to mesh refinement and combustion model characteristic constants, computational costs, and robustness of the approach. In order to expand the overview on different methods simulations were performed with both commercial and open-source codes switching from quasi-2D to fully 3D computations.

1. Introduction

The emission reduction, especially of NO_x, has been the major driver for gas turbine development in the last decades. One of the most promising gas turbine combustion technologies to respect the strict legislative limits on pollutant emissions is the adoption of lean premixed flame. In the fields of combustion science and engineering, CFD calculations are now truly competitive with experiments and theory, as a research tool to produce detailed and multiscale information about combustion processes and play a crucial role in the design of environment-friendly devices. In particular, gas turbine combustion modeling, involving the interaction of many complex physical processes such as turbulent mixing and chemical reactions, comprises a wide range of computational and modeling challenges [1]. In this context LES is one of the most promising techniques as it allows a detailed resolution of the flow field and turbulent mixing phenomena.

An axisymmetric bluffbody stabilized flame, reproducing typical lean premixed gas turbine combustor's conditions, has been numerically studied, under adiabatic conditions, with the commercial code ANSYS Fluent vers.14.0, using LES coupled with the progress variable (c) approach closed with Zimont Turbulent Flame Speed Closure (TFC) [2].

Numerical settings, mesh and time step sensitivity analysis are at first performed on a quasi-2D test case, representing a 5-degree slice of the complete geometry. Successively the fully 3D geometry has been simulated varying the combustion model constant controlling the source term in progress variable transport equation (Product Formation Rate, PFR). Finally an improved mesh arrangement in the region near the bluffbody has been generated in order to investigate the influence of mesh refinement on smallest structures wrinkling the flame front and PFR.

LES simulations have also been performed with the open-source CFD code OpenFOAM both using its native modelling for premixed combustion as well as an in-house developed solver with turbulent flame speed closure similar to that available in ANSYS Fluent. This second test permitted separating the discrepancies due to different combustion models by those related to numerics.

Results are compared, in terms of mean velocity and mean temperature, with available experimental data, published LES calculations results found in literature, and RANS simulations performed on the same test case. A computational cost

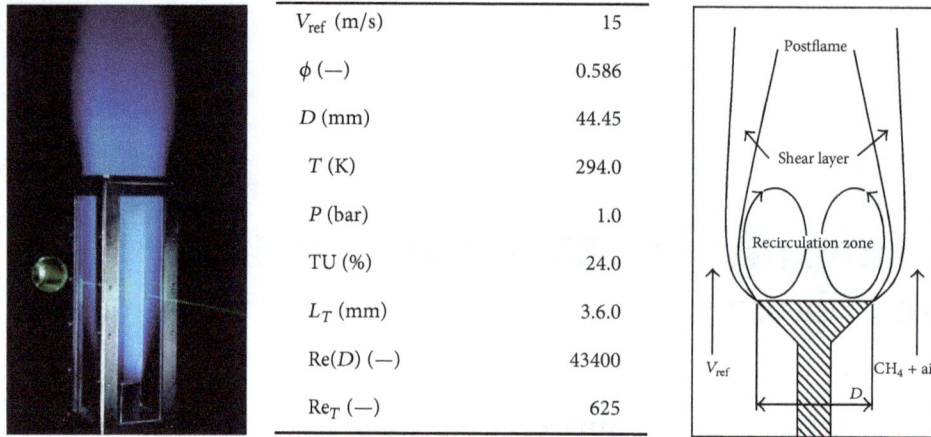

V_{ref} (m/s)	15
ϕ (—)	0.586
D (mm)	44.45
T (K)	294.0
P (bar)	1.0
TU (%)	24.0
L_T (mm)	3.6.0
Re(D) (—)	43400
Re$_T$ (—)	625

FIGURE 1: Vanderbilt combustor and reference conditions.

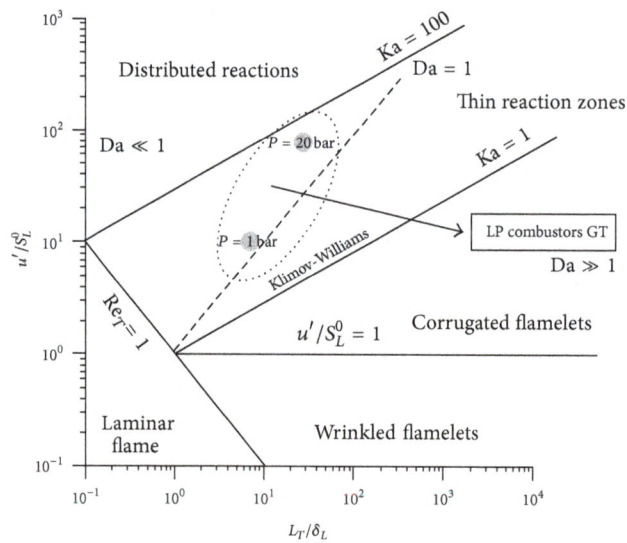

FIGURE 2: Comparison of turbulent combustion regime of VDB and typical lean premixed gas turbine combustors.

TABLE 1: Mesh sensitivity analysis data.

Mesh	Number of elements	β_{max}
A	26727	0.2
B	62160	0.13
C	153276	0.11

TABLE 2: Performed runs.

	RANS Zim-FL	LES-Zim-FL M1-A	LES-Zim-FL M2-A	LES-Zim-FL M2-2A
Mesh elements	915000	915000	2400000	2400000
A constant	0.52	0.52	0.52	1.04

analysis has been finally carried out to provide feasibility guidelines for future works.

1.1. Specific Objectives. The main objective of this work is to test different reactive LES implementations in order to understand their potentials and limitations and to compare them with RANS simulations in terms of mean results, calculation time, and modeling accuracy gains.

2. Turbulent Premixed Flames Modeling

Premixed flames fronts are generally very thin with a thickness δ in the range of $0.1 \div 1$ mm, which is, in many cases, smaller than the filter width Δ used in LES. In a LES context, one approach to model such a flame is to ignore its internal structure and the detailed chemical kinetics and to represent the combustion occurring at the flame front in terms of a progress variable (PV or c), which varies from 0 (fresh reactants) to 1 (burnt gases). Possible definitions of the progress variable can be provided in terms of reduced temperature or reduced fuel mass fraction, given by

$$c = \frac{T - T_u}{T_b - T_u} \qquad c = \frac{Y - Y_u}{Y_b - Y_u}, \tag{1}$$

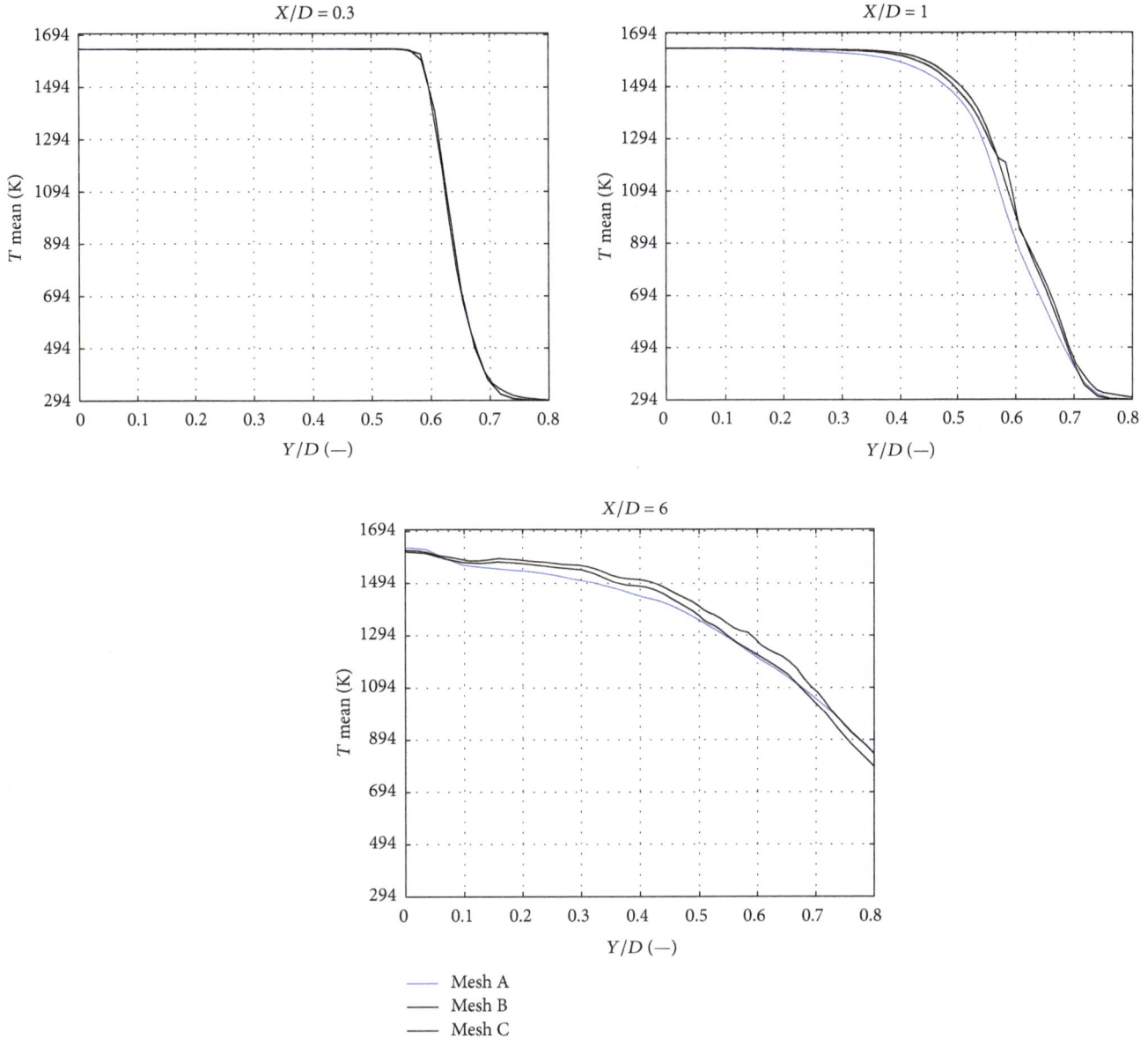

FIGURE 3: Mean temperature profiles resulting from mesh size sensitivity analysis.

where T, T_u, and T_b are, respectively, local, unburnt, and burnt gases temperature, while Y, Y_u, and Y_b are local, unburnt, and burnt species mass fractions. In case of Lewis number $Le = 1$, and if the constant pressure specific heat c_p is retained the same for both states (burnt and unburnt), it is possible to demonstrate that the two definitions coincide [4].

Adopting this approach in a LES context, a Favre-filtered transport equation for c (2) is solved in conjunction with the filtered momentum equations:

$$\frac{\partial \overline{\rho} \tilde{c}}{\partial t} + \frac{\partial \left(\overline{\rho} \tilde{u}_i \tilde{c} \right)}{\partial x_i} = \frac{\partial}{\partial x_i} \left(\left(\overline{\rho D} + \frac{\mu_{sgs}}{Sc_T} \right) \frac{\partial \tilde{c}}{\partial x_i} \right) + \overline{\rho S_c}. \quad (2)$$

In ANSYS Fluent implementation, the subfilter scalar term is modeled by means of gradient based assumption with the turbulent Schmidt number Sc_T set constant to 0.7,

while subgrid scale viscosity is obtained from the standard Smagorinsky model. Source term is modeled using Zimont's TFC and set proportional to the gradient of the filtered progress variable and to a turbulent flame speed S_T that depends on the physical-chemical characteristics of the fuel mixture through its laminar flame speed S_l and on the local turbulence at the subgrid level:

$$\overline{\rho S_c} = \overline{\rho_u} |\nabla \tilde{c}| \overline{S_T} = |\nabla \tilde{c}| \left(A \overline{u'}^{3/4} S_l^{1/2} \chi^{-1/4} l_t^{1/4} \right), \quad (3)$$

where $u' = C_s \Delta |\overline{S}|$, $l_t = C_s \Delta$, $C_s = 0.1$, and $\overline{S}$ the resolved strain tensor and χ the thermal diffusivity. The model constant A is empirical and the suggested value is 0.52 for most hydrocarbon fuels. To take flame stretching into account, the source term for the progress variable is multiplied by a stretch factor G, calculated as a function of

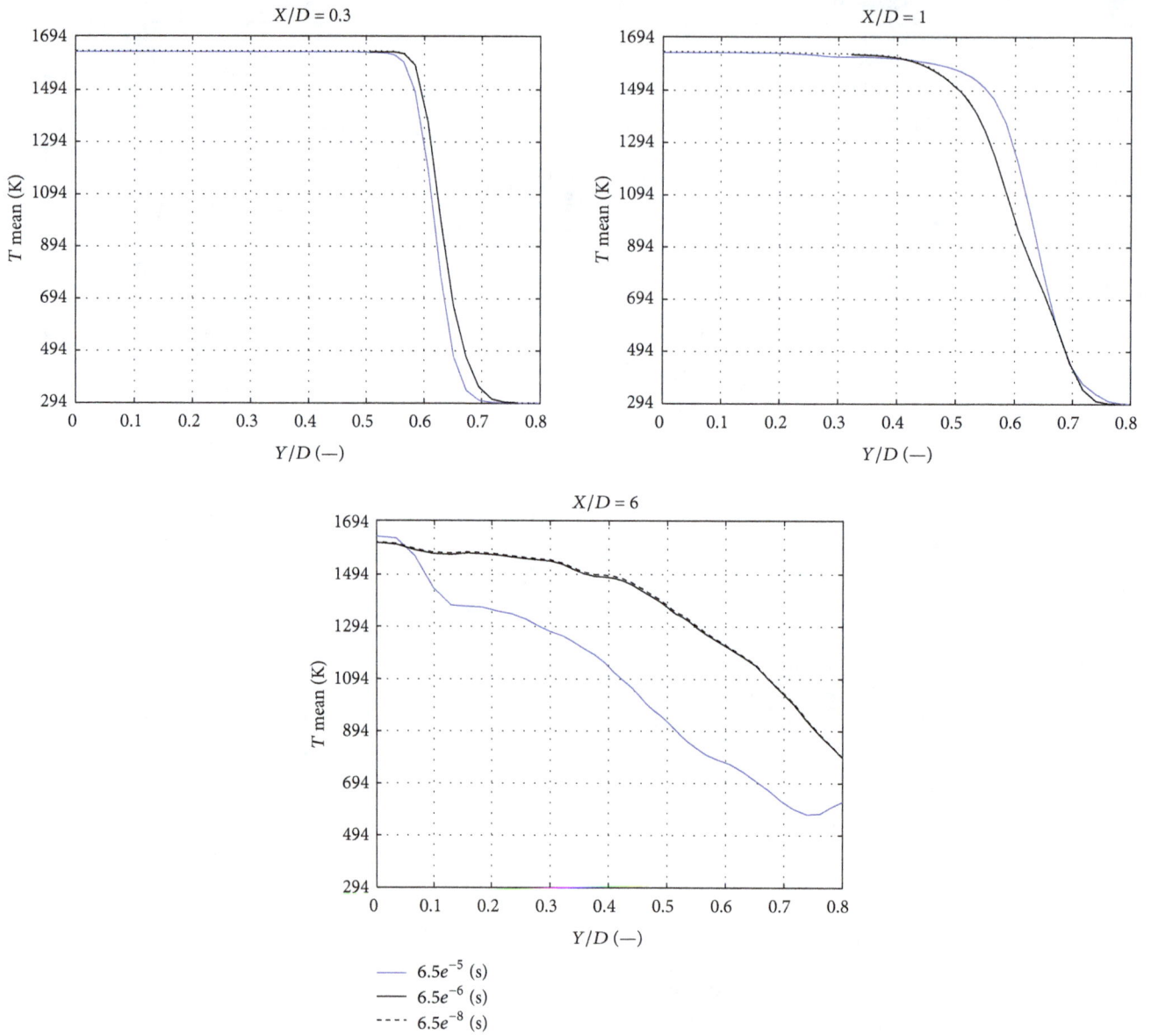

FIGURE 4: Mean temperature profiles resulting from time step size sensitivity analysis.

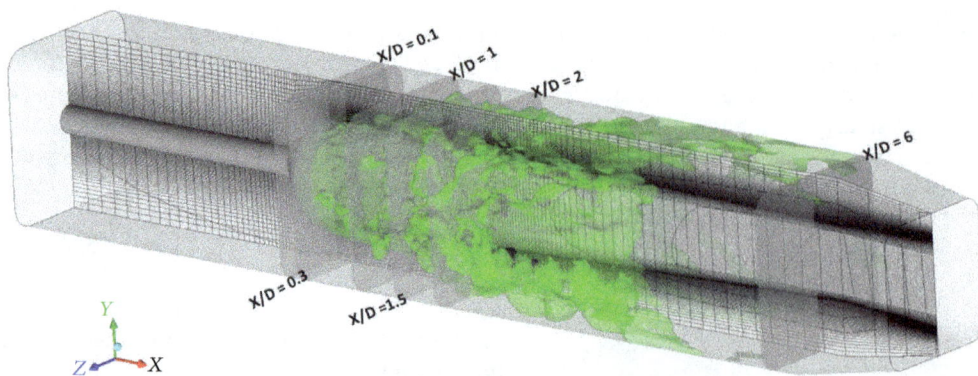

FIGURE 5: Simulated domain, sample planes, and a mesh plane.

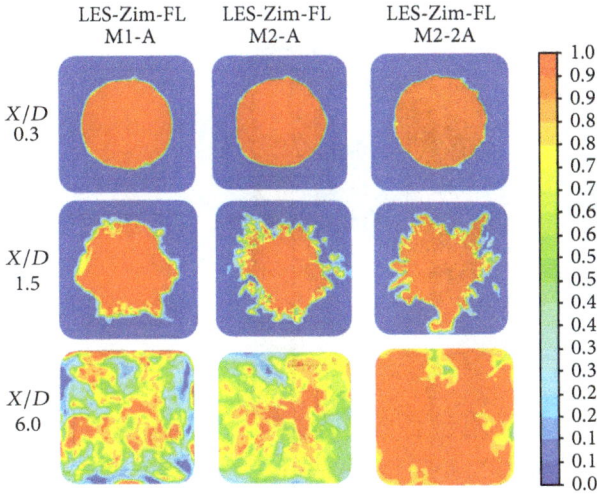

FIGURE 6: Instantaneous progress variable contours at three axial locations.

the turbulent dissipation rate (ε), the subgrid scales properties, and the critical strain rate (g_{cr}) that is by default set to a high value so that no flame stretch occurs [5]. No flame speed reduction has been introduced in the wall proximity to account for wall turbulence damping. RANS formulation of the combustion model has been adopted when performing such simulations together with a RNG k-ε turbulence model.

The OpenFOAM available formulation for flame front propagation within premixed combustion regime is instead based on solving a transport equation for the Favre-filtered regress variable $b = 1 - c$:

$$\frac{\partial \overline{\rho} \widetilde{b}}{\partial t} + \frac{\partial \left(\overline{\rho} \widetilde{u_i} \widetilde{b} \right)}{\partial x_i} = \frac{\partial}{\partial x_i} \left(\left(\overline{\rho D} + \frac{\mu_{sgs}}{Sc_T} \right) \frac{\partial \widetilde{b}}{\partial x_i} \right) - \overline{\rho S_c}. \quad (4)$$

The b-equation source term is modeled as

$$\rho S_c = \rho_u \left| \nabla b \right| \Xi S_l. \quad (5)$$

With turbulent to laminar flame speed ratio Ξ (also referred as Xi) solved by means of a transport equation as proposed in [6]. However, in order to stress the differences only due to the different numerics available in the two codes, the same Zimont closure as proposed in Fluent was also implemented and tested in OpenFOAM, maintaining the same framework based on the regress variable.

Sub-grid fluctuations model is based on a transport equation for the sub-grid turbulent kinetic energy as proposed by Yoshizawa [7] and Menon et al. [8] which is accredited for a better prediction of nonequilibrium zones compared to typical algebraic sgs models.

3. Test Case Description

A classical way of stabilizing a turbulent premixed flame consists in generating a large recirculation where the low velocity zone permits flame anchoring with hot gasses recirculating in proximity of the fresh mixture and continuously igniting the reactive stream. A way of achieving such flow

distribution is using a bluff body inserted in the mixture stream whose low pressure back side guarantees the driving force for the large recirculation bubble. Turbulence enhances mixing downstream of the bluff body, anchoring the flame in an unsteady manner. Combustors with bluff body flame holders, like that analyzed in this work, are characterized by a shear layer where vortices are shed due to Kelvin-Helmholtz instability [9]. This shear layer separates the region of high-speed fresh mixture from the wake region of lower speed (recirculation zone) with hot products.

The test case adopted is an experimental burner studied at Vanderbilt University, Tennessee, USA (VDB), which allows reproducing all the effects of the turbulent mixing and turbulence-combustion kinetic interaction to be observed and investigated at the same time.

Figure 1 shows a picture of the flame realized by VDB combustor. The flame is confined in a square-section duct with quartz walls. A 45° conical bluff body, characterized by a diameter D, is located coaxially to stabilize a premixed methane-air flame.

This work will refer, for calculation comparisons, to Nandula's measurements [10] which include laser Doppler anemometry velocity and turbulence data as well as temperature profiles based on Rayleigh scattering.

Reference values of velocity V_{ref}, turbulent intensity TU, an integral scale length L_T must be considered at the test section, immediately before recirculation zone. Equivalence ratio $\phi = 0.586$, representative of a lean premixed combustor flame, is considered.

In order to point out the capacity of this test case to be representative of turbulent combustion flames of gas turbine lean premixed combustors, an attempt was made to locate its combustion regime in a classical spectral Borghi diagram following the classification proposed by Peters [11]. According to Peters, it is possible to extend the validity of flamelet regime also when Karlovitz number ($Ka = (\delta_L/L_k)^2$) is slightly greater than unity ($Ka < 100$). In this regime, called thin reaction zone, the scale of smallest vortexes (L_k) is still greater than the thickness of laminar flamelet inner layer ($1/10\delta_L$, being δ_L overall laminar flame thickness), which is the reaction zone core of laminar structure. Figure 2 reports the expected locations on the diagram of a typical actual gas turbine lean premixed combustor (operating at 20 bar) and the VDB bluff body (operating at ambient pressure).

4. Domain Definition and Numerical Methods

4.1. Boundary Conditions. At inlet section a constant velocity profile is imposed with free-stream turbulence set to 24% and integral length scale to 3.6 mm in RANS simulation, while LES method employs synthetic turbulence generation methods. ANSYS Fluent implements the so-called "Vortex Method" [5] which is based on a Lagrangian form of the vorticity equation to provide perturbations on a given mean profile. On the other hand, only a quite basic synthetic turbulent generator based on white noise random fluctuation superposition is available in OpenFOAM. Hence, in order to provide realistic flow conditions within the upstream feeding

Progress variable c | PFR

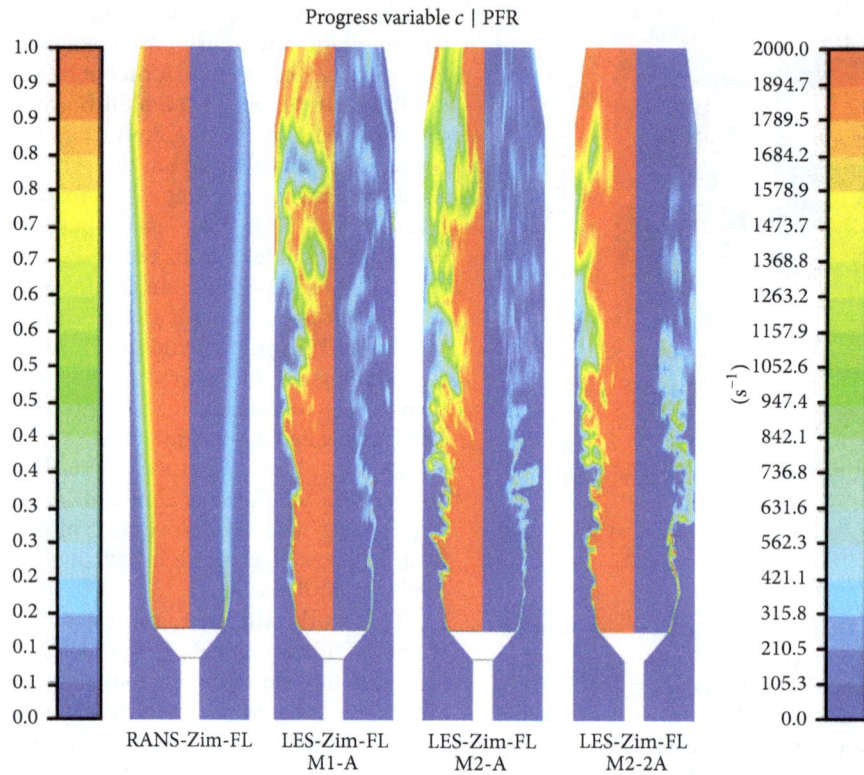

FIGURE 7: Instantaneous progress variable and PFR contours (longitudinal view).

Progress variable c | PRF

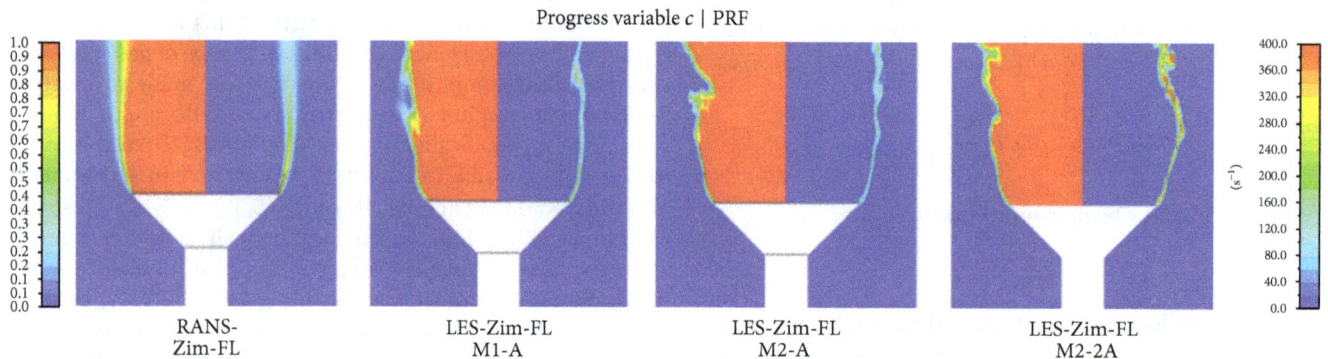

FIGURE 8: Instantaneous progress variable and PRF (bluff body details).

channel, an autorecycling technique [12] was exploited. Such technique consists in the mapping of self-developed turbulent fields back on the inflow introducing feedback mechanism to guarantee prescribed mean profiles. An approach capable of dealing with arbitrarily shaped mesh was implemented and employed in this context. Uniform pressure is assigned on the outlet neglecting all types of reflections of pressure or entropy waves, while bluffbody and duct walls are considered adiabatic realizing no-slip conditions.

4.2. Fluent Numerical Settings. Iterative time advancement is used in Fluent with segregated SIMPLEC algorithm to solve internal iteration pressure-velocity coupling. For pressure equation, a linear discretization is chosen as it leads to less

numerical dissipation, resulting in a more refined flame front reproduction compared with other discretization schemes. A central differencing scheme based on a deferred approach is used for momentum and progress variable equations as it enhances stability respect to the purely linear scheme. The choice of the numerical schemes used has been made after preliminary analysis aimed at defining the best setting in terms of accuracy, robustness, and computational efficiency [13].

The thermophysical properties of the air methane mixture with $\phi = 0.586$ have been determined using CHEMKIN libraries. In particular unburnt mixture density ρ_u at reference temperature of 294 K was found to be 1.115 kg/s, thermal diffusivity 1.0E-5 m^2/s, and adiabatic flame temperature

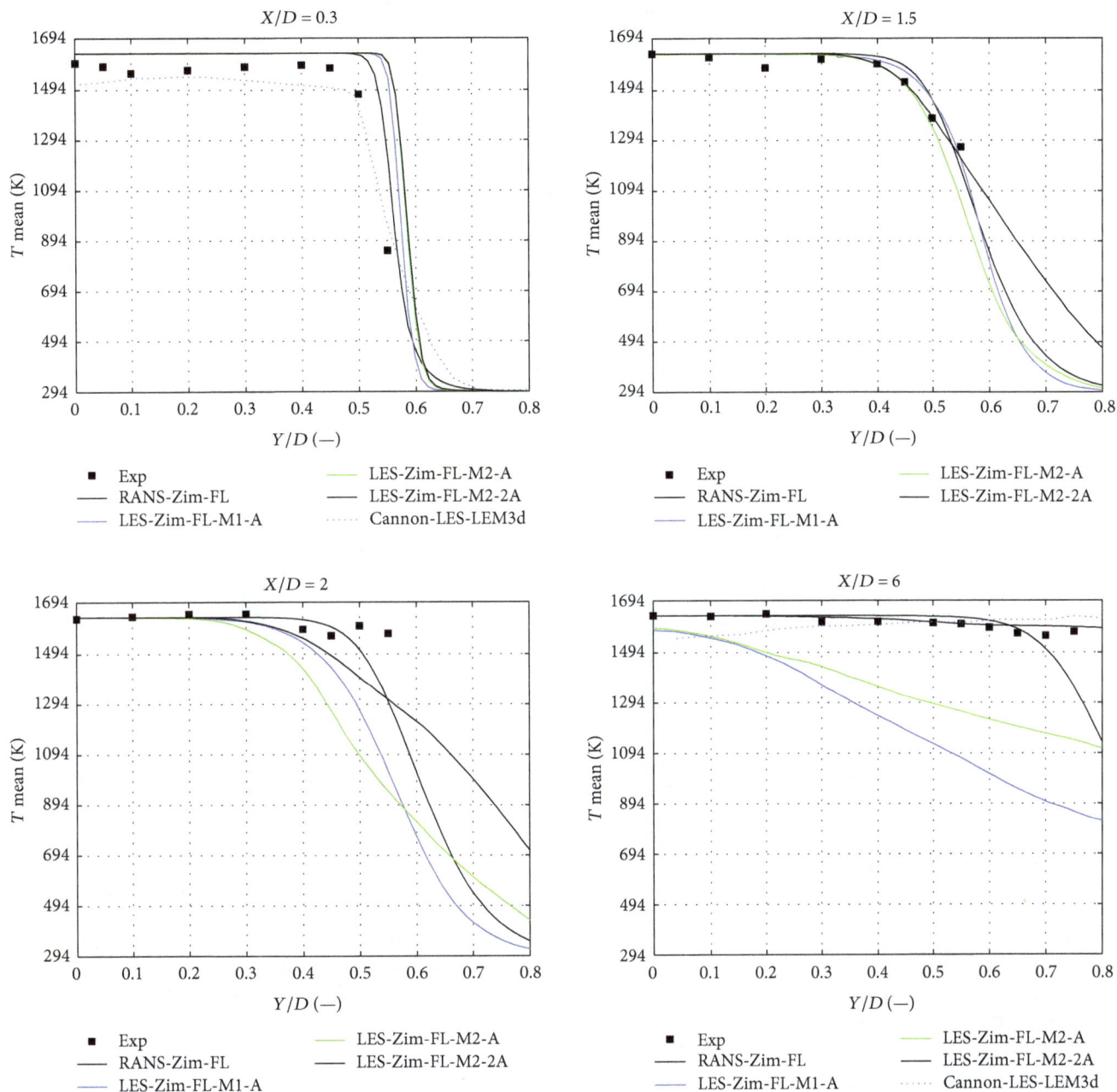

FIGURE 9: Mean temperature profiles at different axial locations.

1640.00 K. For TFC model the unstretched laminar flame speed was assumed equal to 0.11 [m/s].

4.3. OpenFOAM Numerical Settings. Besides the already mentioned combustion and subgrid model, other numerical and modeling strategies differ among Fluent and OpenFOAM. First of all the solver in this case considers nonadiabatic conditions: energy equation is solved, and JANAF table is used to compute thermophysical properties of the mixture. Sutherland law is employed for the transport properties of both burnt and unburnt gases.

The solver follows a classical OpenFOAM segregated method called PIMPLE which is based on a PISO loop nested within a SIMPLE loop to solve the pressure-velocity coupling. Due to the small time step employed in the analysis, the solver uses only 3 internal corrector steps to achieve coupling between continuity and momentum equations and 2 external with high relaxation factors (i.e., >0.8). Both convective and diffusive fluxes are discretized following purely linear schemes to maximize accuracy as necessary for LES, except for the regress variable equation where a filtered scheme is adopted to guarantee stability; time advancement is achieved by means of implicit Euler scheme.

Linear algebra exploits acceleration with the use of Diagonal Incomplete LU factorization to precondition BiConjugate gradient solver. Matrices are solved with a relative tolerance

Mean progress variable c | mean axial velocity

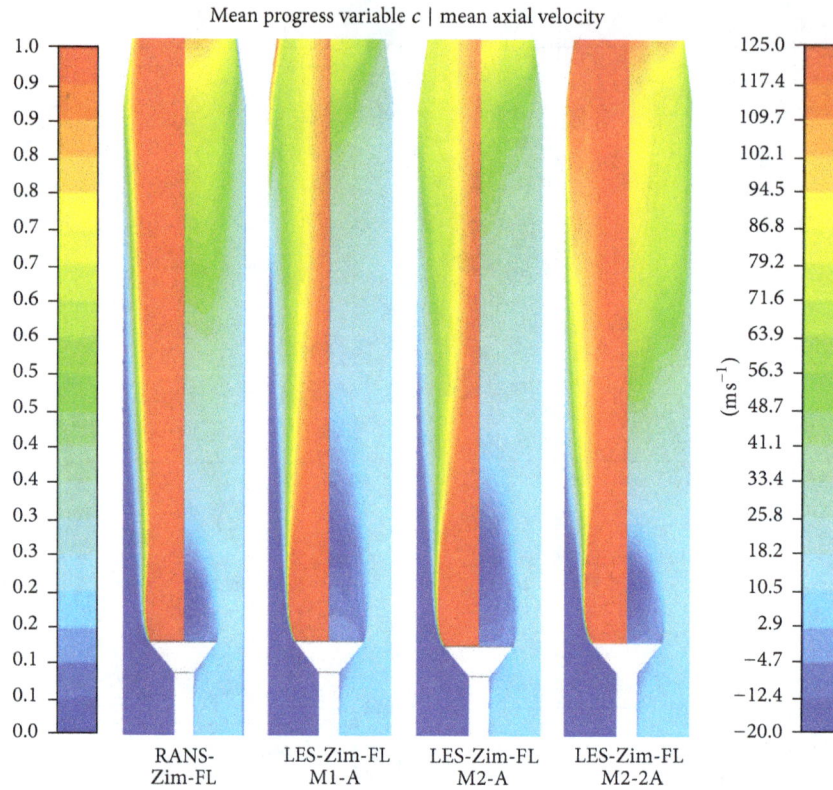

FIGURE 10: Mean progress variable and axial velocity contours (longitudinal view).

of 0.1 for the inner iterations and $1e^{-7}$ absolute tolerance for the last one.

4.4. Computational Mesh.

All finite volume grids have been realized with the commercial tool ICEM-CFD V.13.0. Multi-block structured topology is employed. A progressive refinement of the mesh is realized in the zone where the shear layer is present in order to well resolve the local turbulent structures. Their influence is in fact crucial for an accurate reproduction of the combustion phenomena evolution. A coarsening of the mesh elements is instead accepted in the final part of the domain towards outlet section.

5. Results and Discussion

5.1. Quasi-2D Preliminary Study.

Before simulating the complete 3D test case, it has been considered useful to calibrate the procedure on a simplified and less onerous quasi-2D test case. Even though two-dimensional behaviour should never be assumed in case of direct resolution of turbulent structures (LES or DNS) the axisymmetric geometry allows reducing the problem to an equivalent angular sector (5°).

It should be noticed that tangential fluctuations are considerably less intense and do not produce significant turbulent mixing and combustion enhancement in this case. In the quasi-2D case relevant phenomena are then slightly altered and the main features can still be observed. This means that mesh and time step sensitivity considerations may be retained valid also for the 3D case.

Periodic conditions are imposed on the side boundaries, limiting the tangential fluctuations but substantially reproducing the main flame development. Mesh sensitivity analysis and time step validation have been conducted to establish the optimal numerical setting, which is then applied to the complete 3D calculations.

A criterion for a first estimation of the mesh refinement can be found in [14], according to whom the filter width Δ should be dimensioned relatively to the mean turbulent flame thickness δ_T by

$$\Delta \cong \beta\delta_T, \qquad (6)$$

proposing $\beta \leq 0.09$ for a first estimation. To verify this criteria, it is necessary to define and provide a value for δ_T that is absolutely difficult. It is decided to estimate δ_T a posteriori from RANS solution as follows:

$$\delta_T = \frac{\text{Volume}_{0.1<c<0.9}}{\text{Area}_{c=0.5}}, \qquad (7)$$

that is the ratio between the volume included by the two isosurfaces at progress variable $c = 0.1$ and $c = 0.9$ and the area of the isosurface at $c = 0.5$ bearing in mind that RANS flame front is thicker than the instantaneous LES one.

The maximum values of β for the three meshes are reported in Table 1 together with the number of elements. The coarsest mesh did not allow correctly reproducing the smallest structures wrinkling the flame front if compared with the two more refined meshes. Such effect does not affect

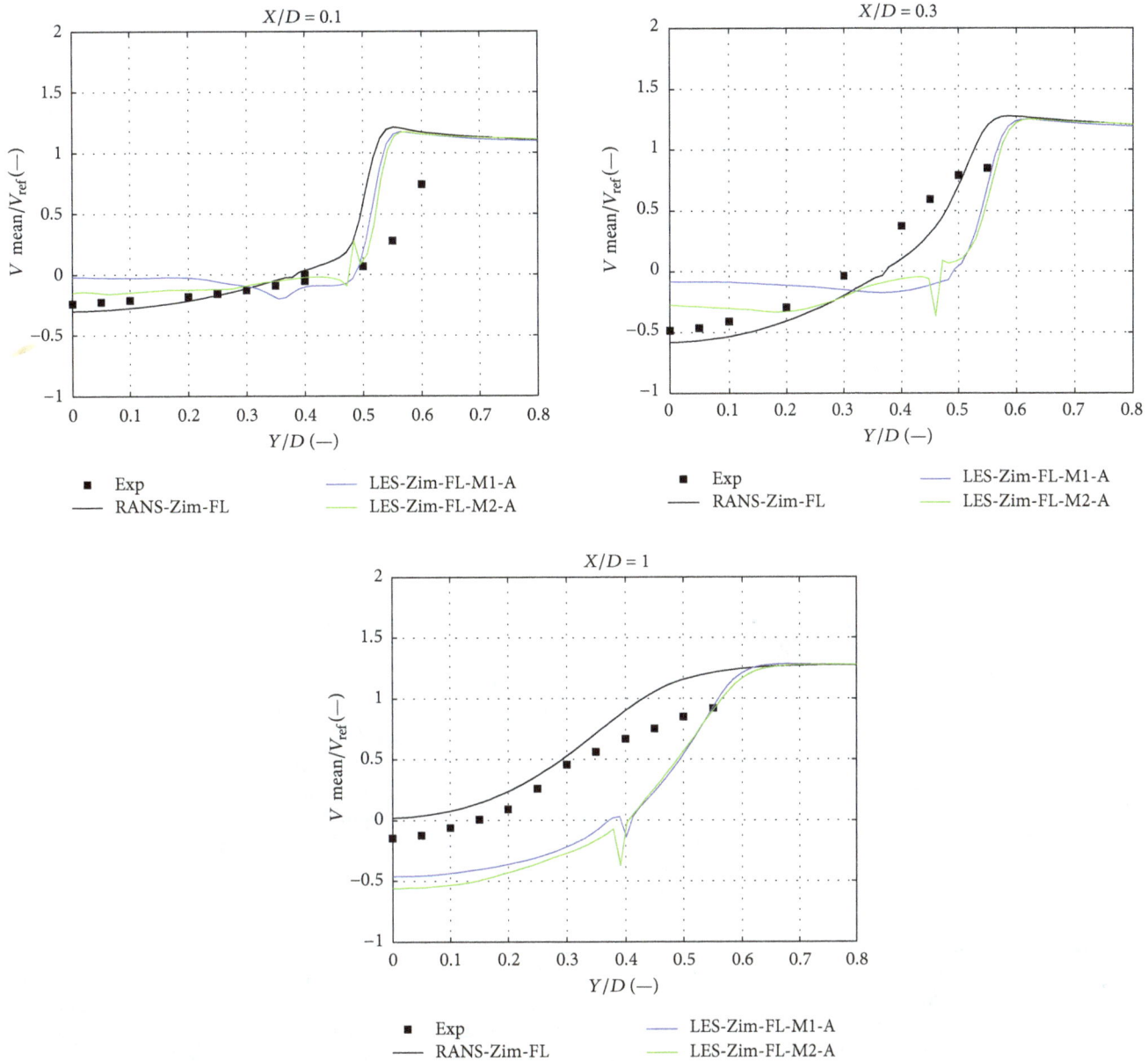

FIGURE 11: Mean axial velocity profiles at different axial locations.

mean temperature profiles as it is possible to observe from Figure 3.

A time step sensitivity analysis has been conducted using mesh B (Table 2) which is the best compromise between calculation time and solution accuracy. The time step needs to be small enough to catch the characteristic time of smallest structures [3]. Three time steps have been chosen ($6.5 e^{-5}$ s, $6.5 e^{-6}$ s, and $6.5 e^{-8}$ s). The longest was found not suited as it did not satisfy what was mentioned just above. This effect can be appreciated even looking at the mean temperature profiles in Figures 3 and 4. From the same figures can be observed how the two time steps $< 1 e^{-5}$ s lead to the same result as they allow catching the smallest structure fluctuations. A time step of $6.5 e^{-6}$ s has been then chosen to continue the study, confirming the findings by other authors [3]. More detailed results on this preliminary quasi-2D study can be found in [8].

5.2. Fully 3D Calculations. After the mentioned preliminary study, fully 3D calculations have been performed. The geometry (Figure 5) reproduces the experimental device. Two levels of mesh refinement have been realized and tested, guaranteeing in both cases a refinement in the region of the shear layer. The coarse mesh was not sufficiently refined to satisfy the above mentioned mesh criteria everywhere in the first part of the flame zone, while the fine one did.

On both meshes, the effects of an increase (doubling) in the constant A in (3) have been evaluated. Table 2 reassumes the setup of the performed runs.

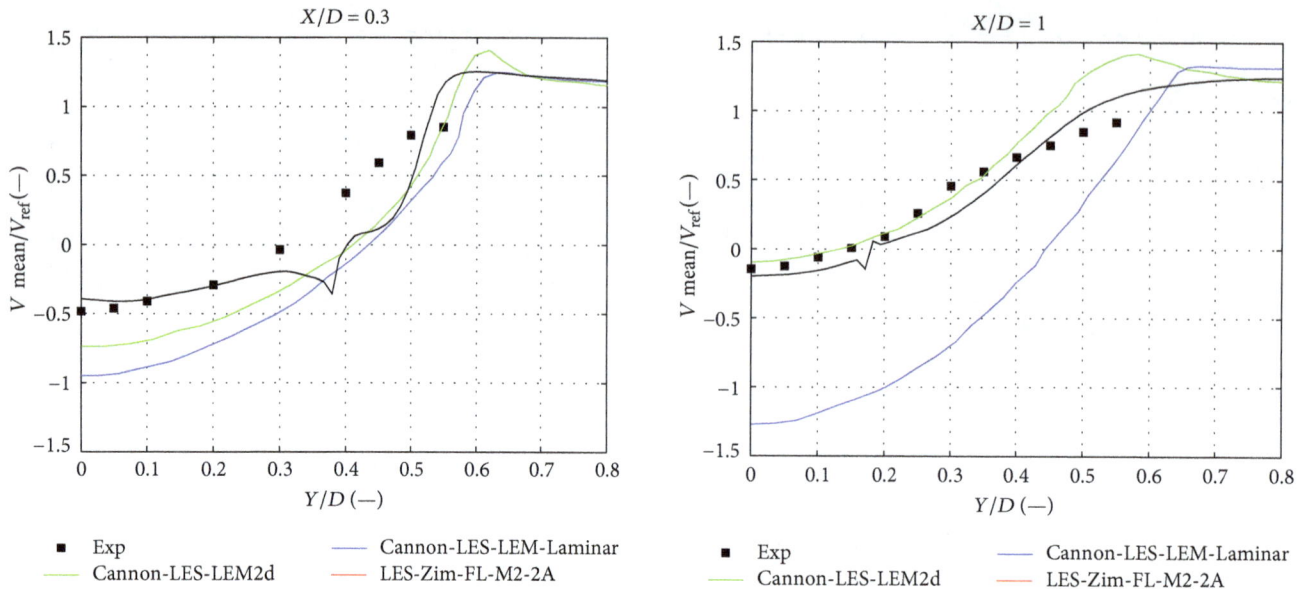

FIGURE 12: Mean axial velocity compared with Cannon et al. results [3].

5.3. Upstream Flow Conditions. Before discussing the flame behavior downstream the bluff body, it is due to underline some aerodynamics features observed in the feeding duct. In particular, turbulence intensity levels obtained in the bluff body throat section for the performed simulations are compared against experimental one.

All the simulations predict lower free turbulence intensity levels than the measured 24%: RANS simulation reaches a TU of 13% while Fluent and OpenFOAM LES show a reduction of the provided inlet value up to 9% and 11%, respectively. Such TU is computed for LES sampling and averaging velocity fluctuations to evaluate time-averaged resolved turbulent kinetic energy and adding the subgrid contribution. The low turbulence levels reached in all the cases might have a direct influence on the solution leading to local underestimation of reaction rates or missing quenching effects induced by flame stretch due to locally high turbulence levels in the cases when such effect is included in the simulation, that is, LES-Zim-FL-M2-2A.

5.4. Flame Structure. Results obtained with Fluent LES are presented below describing the main features of the predicted flame. The influence of a mesh refinement as well as the mentioned change in the combustion submodel constant A on the turbulence flame interaction and on the local burning rate is described together with a possible explanation about the resulting behavior and the differences emerged in the comparisons with the experimental data.

From progress variable contours at three different axial locations in Figure 6 emerges how the flame remains confined in the central part of the domain for the major part of the combustor length, while the external squared duct does not significantly affect the flame shape. The effect of the increased resolution cannot be appreciated in the first part of the burner, close to the bluff body, but becomes more evident downstream at X/D larger than 1. The flame is more wrinkled

due to the small turbulent structure which were filtered out by the coarse mesh.

This behaviour affects the PFR that is increased close to the bluffbody, see Figure 8, as well as in the central part of the combustor, Figure 7. Such a change is not sufficient to complete combustion at the outlet nor to significantly change the mean progress variable contours. From mean temperature profiles in Figure 9 a slight difference is observed at $X/D = 6$ section but the level predicted underestimates anyhow the measured temperature.

In all these cases the recirculation length is overpredicted (refer to Figures 10 and 11) with respect to RANS simulation and experimental results.

Changing the combustion model constant A to 1.04 allows an excellent agreement with experimental results in terms of mean temperature and velocity profiles as depicted in Figures 9 and 11. The flow field and the recirculation length are well predicted. Such a change in A constant boosts the combustion since the very first part of the bluff body end and the flame tends to move upstream the bluff body itself. To avoid flame flashback, critical strain rate limit is set to a lower value ($20000\,s^{-1}$), thus introducing local flame quenching in the regions of high stretch. The flame is enlarged and PFR increased (Figure 8). The result is to change the shear layer interactions between the fresh mixture and burnt gasses intensifying both turbulent mixing and chemical reactions. Being the equations coupled, an increased reaction turns into higher velocities in the second part of the domain and a shortening of the recirculation zone.

From instantaneous progress variable contours in Figure 7, it is possible to see how changing the model constant does not alter the flame wrinkling. The flame remains confined in the central part of the duct maintaining a hot core for the greatest part of the combustor length. Unlike the other LES results towards the outlet, at $X/D = 6$ section, fully burnt gasses are now obtained.

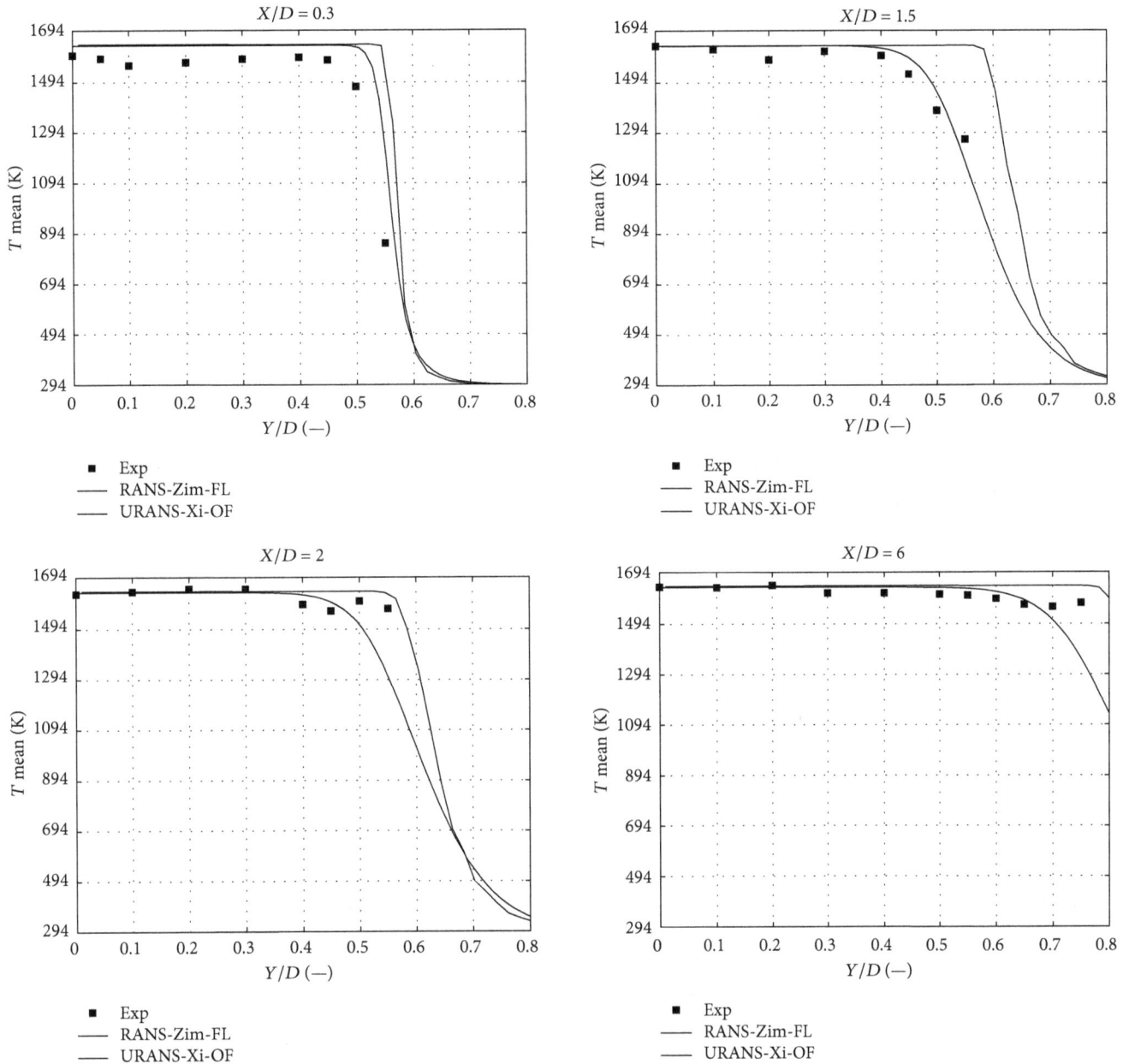

FIGURE 13: Mean temperature profiles with OpenFOAM (URANS).

Looking at mean progress variable contours predictions with LES-Zim-FL-M2-2A lead to a significantly different c levels when compared to other LES simulations and the results are closer to the RANS one. With RANS the flame does not attach to the duct wall but in the converging part, while in LES-Zim-FL-M2-2A mean progress variable contours show that flame anchoring is achieved on the straight duct.

Temperature profiles from simulations performed by Cannon et al. [3] are included for reference in Figure 9. It refers to a Large Eddy Simulation coupled with a linear eddy model (LES-LEM), used to describe the sub-grid chemistry and turbulence-chemistry interactions, using a single-step mechanism, performed on a fully 3D test case. Figure 9 shows how the LES-LEM nicely predicts temperature profiles even though they are available only at $X/D = 0.3$ and 6, whereas also case LES-Zim-FL-M2-2A presents a quite good agreement with measurements. The effect of LEM is to enhance the filtered reaction rate due to sub-grid turbulent fluctuations, substantially acting in the same way of the modified A constant in Zimont model. This may suggest that a much more refined grid or a different subgrid scale model ought to be used when performing LES simulation in order to catch the smallest scale effect on reaction rates. Both these solutions have the drawback of a large increase in calculation time.

A general overprediction of temperature levels in the recirculation zone by the computational models is found in all the studied cases (Figure 9 $X/D = 0.3$). This effect is mainly

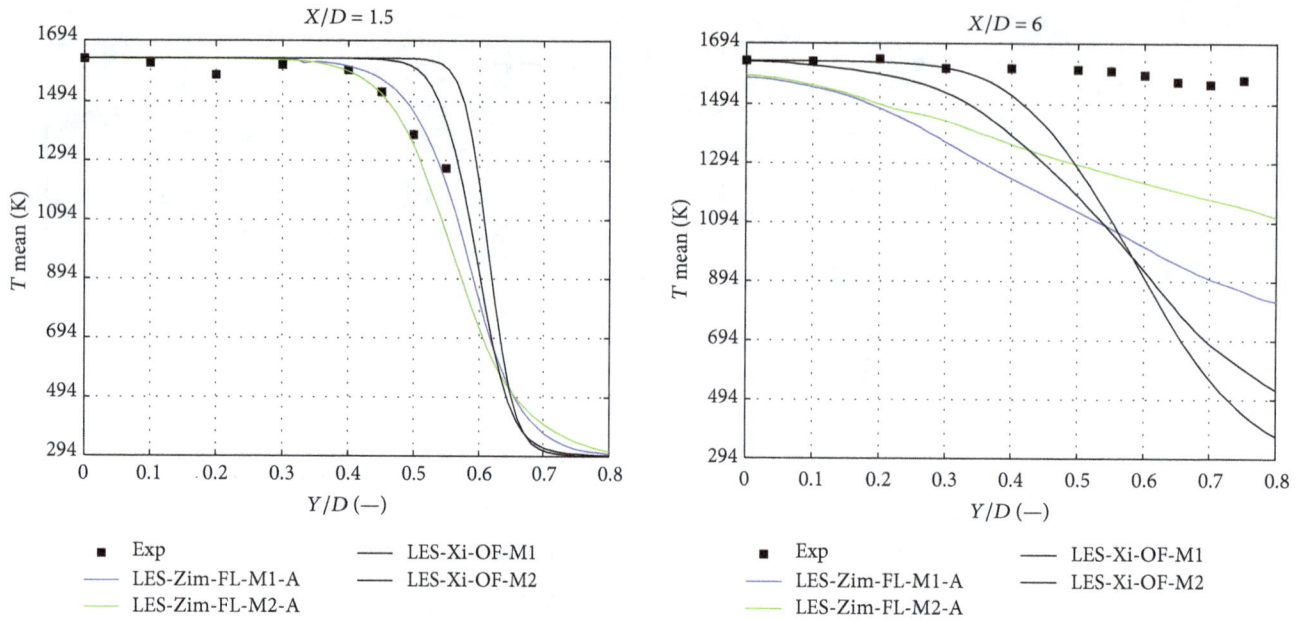

FIGURE 14: Mean temperature profiles with OpenFOAM and Fluent (LES).

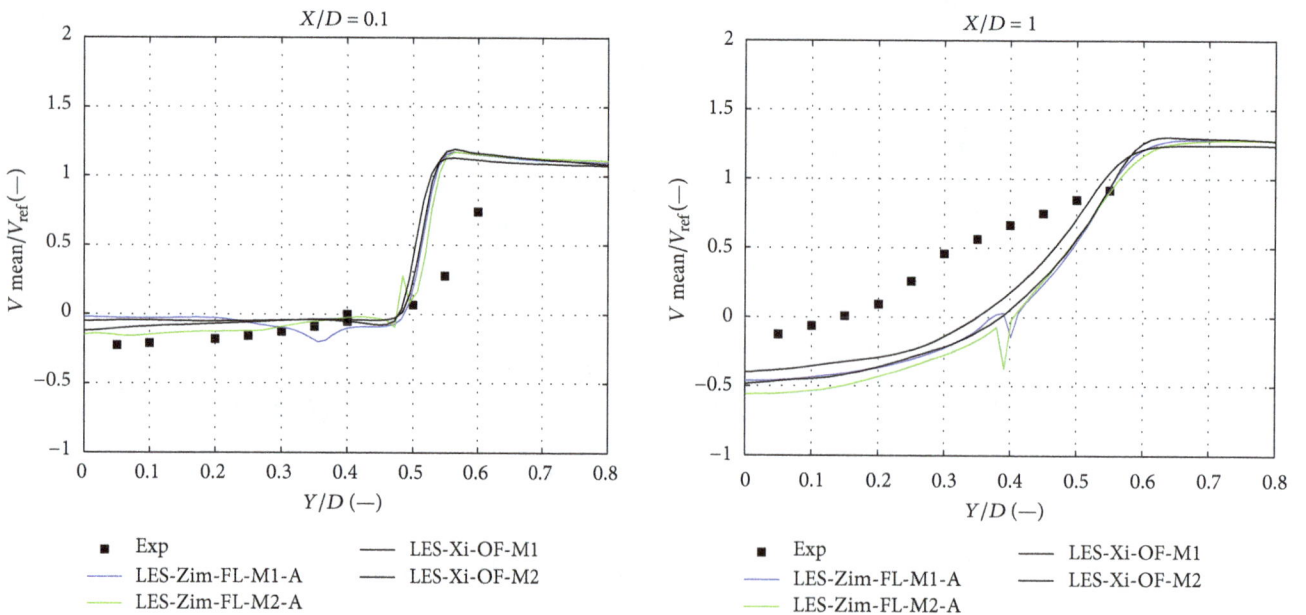

FIGURE 15: Mean axial velocity profiles with OpenFOAM and Fluent (LES).

due to the thermal boundary condition on the bluffbody wall which in the computations is considered adiabatic. Nandula [10], in fact, estimates a heat loss at the bluffbody walls of around 6%. A fair inappropriate prediction of aerodynamic stretch may also locally influence the accuracy [15]: this is confirmed by recalling that implemented TFC model does not account for flame stretch giving an explanation to the underestimated mean flame brush thickness and to the higher values of flame temperature predicted by the model.

Another common aspect, highlighted in Figure 10, is the overprediction of the flame length respect to experiments.

As already pointed out, at $X/D = 6$ only LES-Zim-FL-M2-2A succeeds in achieving a completely developed and burnt flame while the reaction is not complete in all the other cases. This discrepancy also affects the upstream sections. Looking at temperature profiles at $X/D = 1.5$ and $X/D = 3$ in Figure 9, the observed flame brush is thinner, with sharp gradient, characteristic of a delayed profile which has not conveyed the fresh gases entrainment and diffusion. Such effect tends in fact to reduce when the total flame length is well predicted.

The regions more sensitive to an increase in turbulence levels are the reaction zones downstream of the recirculation

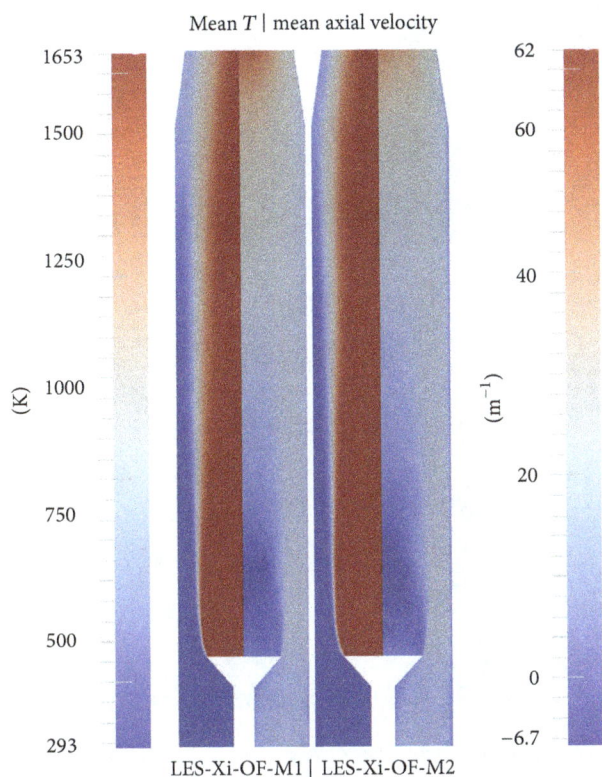

FIGURE 16: Mean temperature and axial velocity with OpenFOAM (LES).

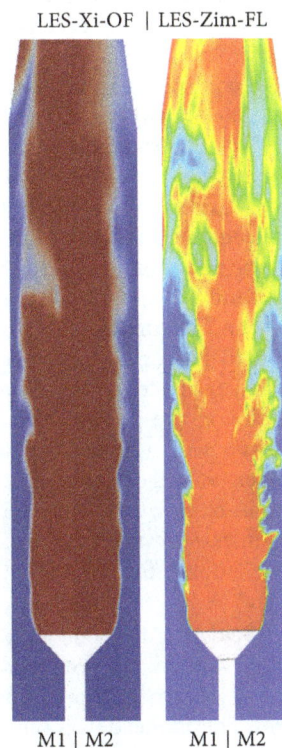

FIGURE 17: Instantaneous temperature field with OpenFOAM and Fluent (LES).

zone. In fact the flame is not supported by the hot products of the recirculation zone, but it behaves as a freely propagating flame in that region. The measured strain rates are lower and a thin flame (shear layer) begins its growth. This shear layer (reaction zone) thickens as flame travels downstream due to entrainment of fresh reactants and large-scale coherent structures fold around the flame edge [10]. The effect of an increase in the modeling constant A in (3) is to partially compensate the lower turbulence level (lower u') thus leading to profiles closer to the experimental ones.

In the axial velocity plots shown in Figure 12, two results from Cannon et al. [16] are presented together with experimental and LES-Zim-FL-M2-2A results. In this study they used a laminar chemistry LES model and a LES-LEM with 1-step mechanism, both performed on a 2D test case. LES-LEM of Cannon shows a good agreement with experimental results while the LES with laminar chemistry overpredicts the recirculation zone length. This confirms how an underpredicted reaction rate leads to a different flow field with a longer recirculation zone and how increasing the model constant may lead to satisfying results as well as a more onerous subgrid scale modeling approach.

5.5. OpenFOAM LES Results with Xi Closure. In order to generate a realistic initial condition and provide more comparable references for the LES results, URANS was also performed. Comparing results and measurements in Figure 13, it may be observed that the temperature is overpredicted in the first part of flame while fits well experimental data in the downstream sections. Towards the outlet, the combustion is complete and the flame tends to reach the ducts wall slightly before than in Fluent RANS solution.

LES simulation results, presented in Figure 14, suggest the same trend highlighted by the RANS calculations that is the quite sharp mean flame front close to the bluff body positioned at higher radii compared to both experiments and calculations with Fluent and the incomplete level of burnt gases even at the end of the flame tube. This incomplete combustion is however partially reduced with the finer mesh showing temperature profiles more in agreement with the experiments at the outflow section than the equivalent computations performed with Fluent.

Also the retarded closure of the recirculation bubble compared to measurements is recorded for the OpenFOAM computations as for runs with unmodified Zimont constant, see Figure 15.

These findings may also be noted in Figure 16, where the time-averaged values of temperature and axial velocity are shown on a longitudinal plane.

However, it is important to verify the behavior of the flame also from a qualitative point of view, looking at the flame front shape to describe the regime of reaction. This can be done observing the instantaneous temperature fields in Figure 17 which depicts a less wrinkled and less diffused flame front compared to Fluent.

5.6. OpenFOAM LES Results with Zimont Closure. A comparison between ANSYS Fluent LES and OpenFOAM native

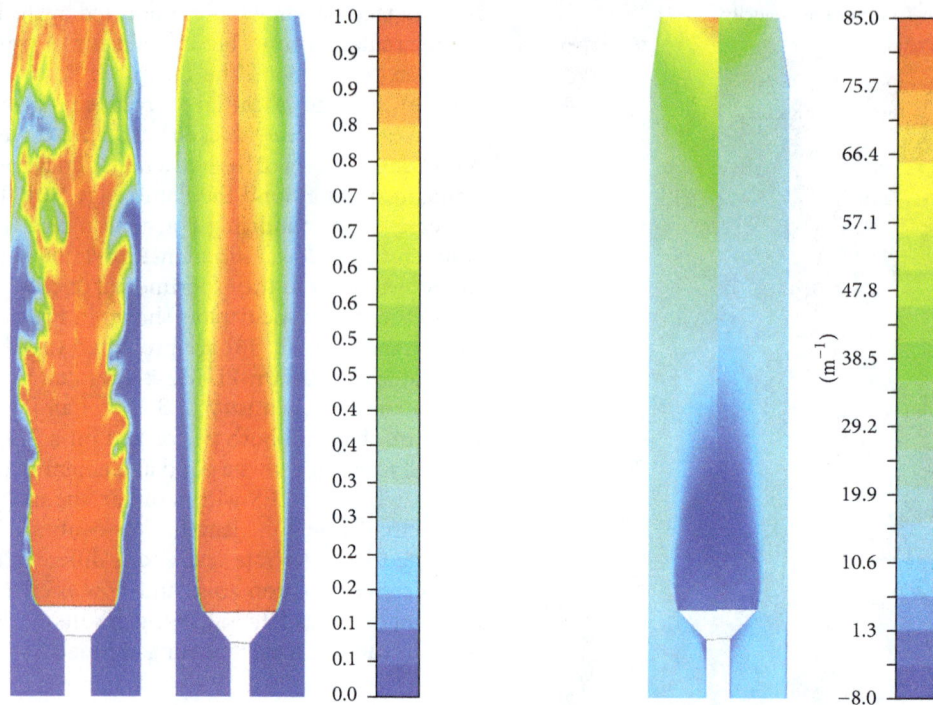

FIGURE 18: Instantaneous and mean progress variable and mean axial velocity contours for Fluent LES-ZIM-FL-M1-A and Zimont OF on mesh M1.

solver XiFoam LES was the initial aim of the work. However, obtained solutions predict different flame behaviors and turbulence-chemistry interactions. Having used different combustion submodels, it is not possible to know if obtained discrepancies are due to numerical aspects or to the combustion submodel itself. For a direct evaluation of the effects of the numerics implemented in the two codes, Zimont turbulent closure has been implemented in OpenFOAM to provide a solver nominally equivalent to that used in Fluent. Results of this analysis are presented here, comparing the two solutions in terms of predicted flame features, mean temperature, and mean velocity profiles, highlighting the main numerical aspects which might influence the solution.

When Zimont closure is used in OpenFOAM, obtained results are in line with Fluent ones as demonstrated by the test performed on M1. Looking at Figure 18, it is possible to observe that the two flames show the same features: flame front wrinkling is characterized by equivalent length scales with large vortices entraining the hot core for $X/D > 2$. Some discrepancies however emerge between the two cases: from mean progress variable contours it is evident that the Zimont OF M1 flame is not able to attach the duct wall. This effect may be due to the different near wall treatment implemented in the subgrid model which results in a lower velocity close to the wall for Fluent calculation which helps stabilize the flame to the wall. The missed wall attachment of the flame in the final part of the combustor induces an intermittent flame detachment in OpenFOAM. The main differences in mean progress variable contours can be appreciated in the central part of the combustor, where the mean flame front angle is higher in LES-ZIM-FL-M1-A and even more from

the point where this latter simulation predicts an attached flame on the wall. The effect is visible also in the temperature profiles in Figure 20 at the last sampled section $X/D = 6$ where the reaction progress, and in turn temperature which is calculated according to (1), does not reach the same values predicted by Fluent LES.

From both mean c contours and temperature profiles, it is possible to see that OpenFOAM model predicts a thicker flame brush. This is probably due to the filtered linear discretization scheme exploited for regress variable b equation which is an extension of the purely linear scheme with a filter for high-frequency ringing [17]. This filtering operation acts in the same direction of a mesh coarsening (see i.e., M1-M2 differences) and has an impact on local turbulent flame speed (S_t) values and in turn on the flame brush; see Figure 19. The reaction to be completed (progress variable from 0 to 1) needs a larger space (brush). Looking at PFR contours in Figure 19, being the product between S_t and c gradient, results in lower peak values, due to smoothed c gradients but with nonzero values over a larger zone. Reaction progress variations are in fact spread over a longer distance. From temperature profiles shown in Figure 20, a general agreement is, however, found for all the other sections.

The resulting flow field is consistent with the one predicted by Fluent. Higher velocity are predicted by the latter (Figure 18); the effect is due to the previously discussed phenomena of intermittent flame detachment which lead to intermittent density variation, gas expansion, and mean velocity. From profiles in Figure 21, a general agreement is observed in the velocity field predicted by the two codes.

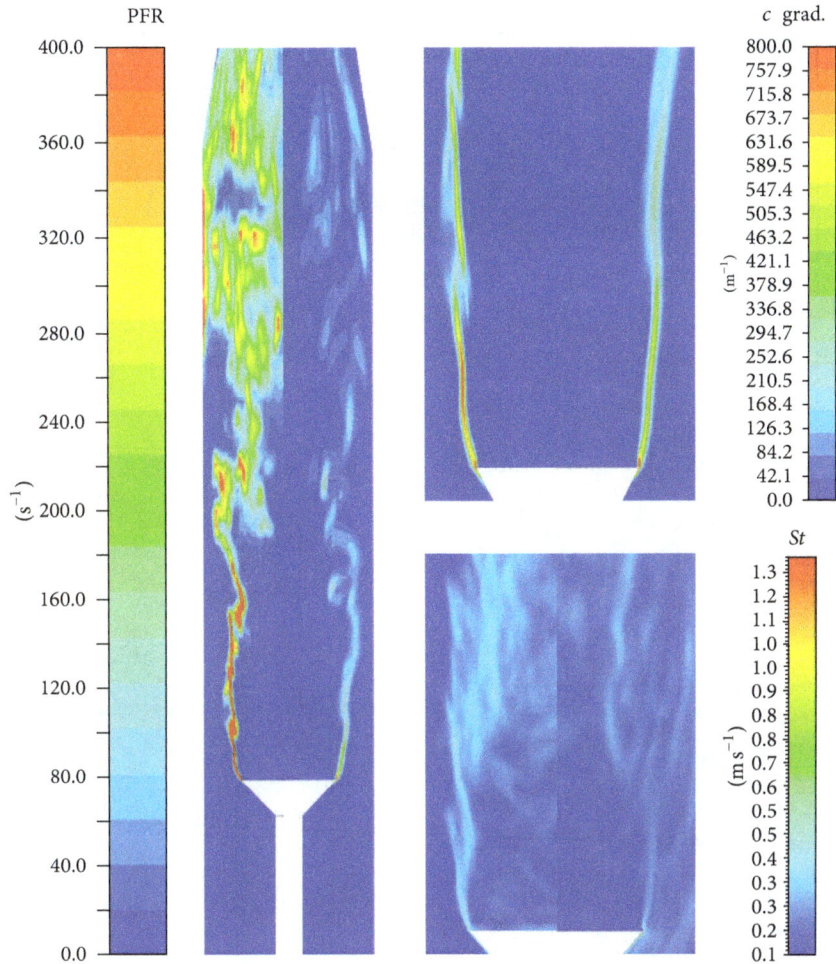

FIGURE 19: Product formation rate, turbulent flame speed, and progress variable gradient contours.

5.7. Calculation Time Considerations. In order to provide a guideline for future works, it has been retained useful to evaluate and compare calculation times between the two codes and to provide an estimate of the computational overhead connected with LES compared to RANS and its increase due to the need of a larger number of time steps and a more refined mesh level.

In this case where RANS can exploit 2D assumptions the additional computational resources needed for LES become relatively enormous: almost 100 times.

Comparison of CPU time is hence conducted for LES only on simulations performed on the coarsest mesh (915000 elements). The global computational time is computed as function of the time needed to complete a time step and the total number of time steps employed by the solution to stabilize. In this analysis, the employed time step is equal for all computations so the total number of iterations to cover the same amount of physical time is the same. However, depending on the obtained regime of flame, it is necessary to extend the averaging time to stabilize mean quantities on different time periods.

In terms of computational efficiency per single time step, OpenFOAM overwhelms Fluent with a time ratio of approximately 0.15 due to the different time advancement algorithm which minimizes the number of inner iterations. This ratio is almost independent by the type of combustion modeling (closure following [6] or [2]) employed. The additional time related to the use of M2 is approximately 30% for OpenFOAM and 150% for Fluent. However, as already hinted, the total number of iterations needed to reach smooth and stable time-averaged profile is quite different between the considered cases. Computations with Zimont closure showed that only 8000 time steps were necessary for Fluent while more than 90000 time steps were computed with OpenFOAM. This large difference is principally due to the low frequency of flame detachment from the wall.

6. Conclusions

LES simulations of a bluff body stabilized lean premixed flame have been performed using both a commercial and an open-source CFD code. Preliminary analysis conducted on a quasi-2D test permitted assessing the influence of local mesh resolution and computational settings as well as providing a reference RANS solution which was exploited to evaluate the effective computational cost increase and the physical insight enhancement associated with LES compared to standard modeling practices.

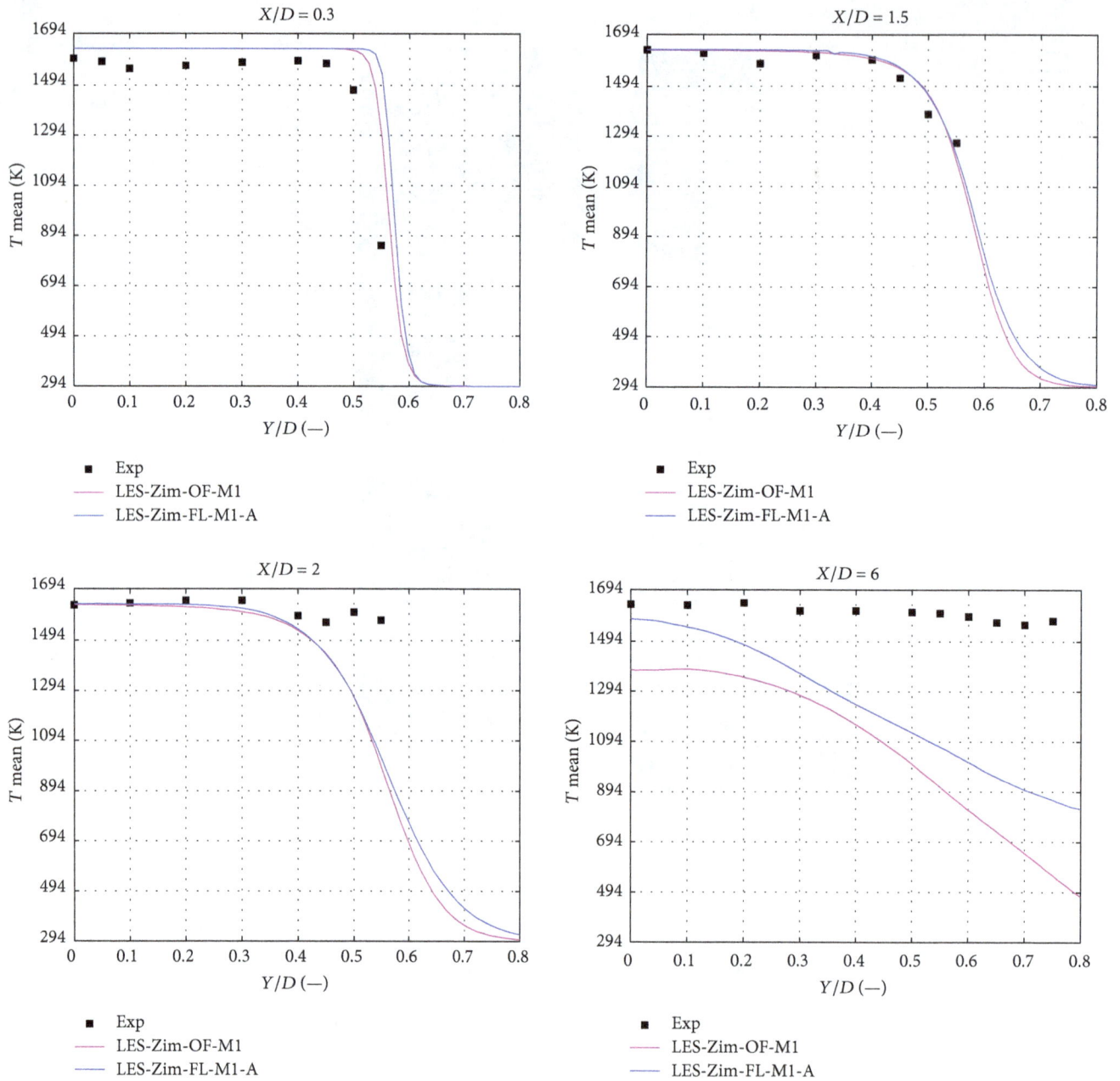

FIGURE 20: Temperature profiles for Fluent LES-Zim-FL-M1-A, experimental results, and LES-Zim-OF-M1.

LES performed with Zimont turbulent flame speed closure show that if the grid clustering is based on aerodynamic and RANS flame thickness considerations, it is necessary to tune model constant A to achieve a complete combustion at the exit of the burner. Solutions obtained with Fluent and OpenFOAM using nominally same models are equivalent in terms of mean profiles at least up to $X/D < 2$. Main differences are limited to the near wall flame behavior and the flame brush thickness which are, respectively, due to the employed wall functions and the advective interpolation scheme for the progress variable equation.

Native OpenFOAM closure based on Xi transport equation predicts an even thinner flame brush associated with a very low wrinkling resulting in limited interaction between fresh and burnt gases downstream the bluffbody and in reduced combustion progress in the region at larger radii for $X/D > 3$.

In terms of computational resources, OpenFOAM requires a much smaller time to complete a single time step; however in the only case of equivalent physical modeling performed (M1 with Zimont closure), the higher number of time step necessary to achieve converged mean solution, due to a more unsteady behavior of the flame, resulted in equivalent global computational efforts.

RANS calculations succeed in predicting mean values with substantially equivalent accuracy and considerably

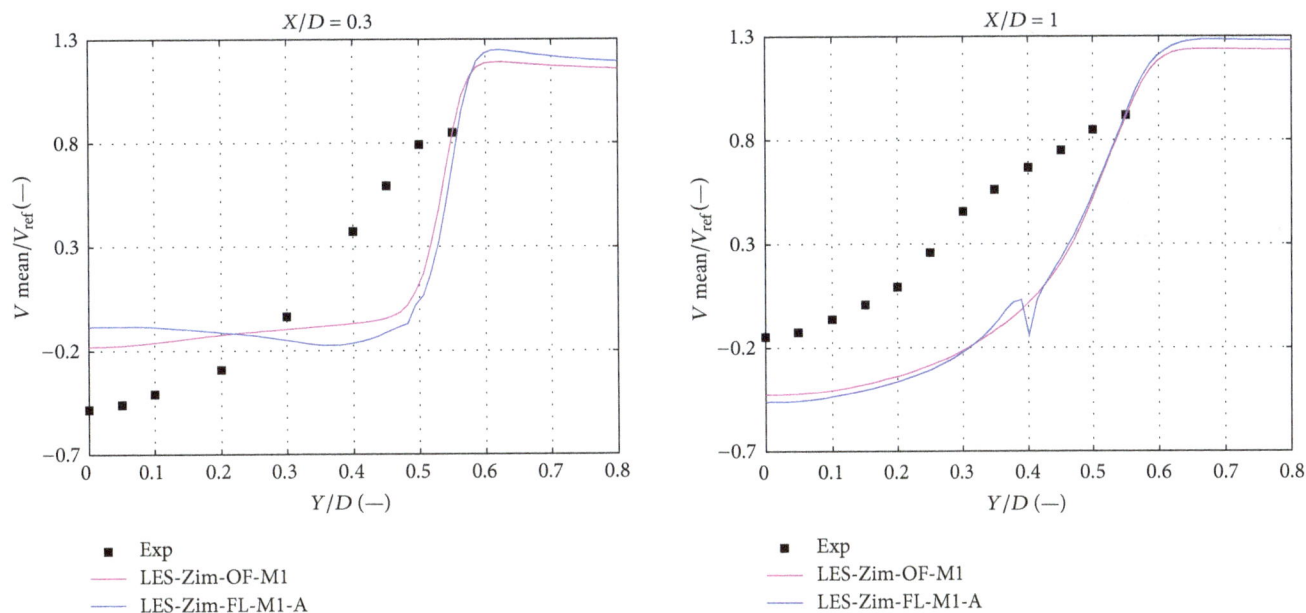

FIGURE 21: Velocity profiles for Fluent LES-ZIM-FL-M1-A, experimental results, and Zimont OF.

lower calculation time 1%. LES potential thus may be identified in providing more accurate and detailed resolution of the field of motion and better capture the turbulence-chemistry interaction. This is however reflected in more accurate prediction of the flame shape only after a long tuning process of the computational setup.

The large computational resources required for LES would be compensated by really improved, qualitatively and quantitatively, results, only when applied to complex cases, such as swirled flames or complete combustor geometry where RANS models fail in accurately predicting the flame stabilization process as well as the flame anchoring.

As far as RANS simulations are concerned, more accurate results on the studied test case might be obtained exploiting a turbulent combustion model able to properly predict the flame length and which takes into account chemical species as well as nonequilibrium effects on these latter, due to the interaction with turbulent flow field. Possible models to be used are G-equation model, which has been shown to provide good agreement with experimental data [15], or flamelet generated manifolds in both RANS and LES context, being their LES implementation straightforward.

Conflict of Interests

The authors declare that there is no conflict of interests regarding the publication of this paper.

References

[1] V. Raman and H. Pitsch, "Large-Eddy Simulation of a bluff-body-stabilized non-premixed flame using a recursive filter-refinement procedure," Combustion and Flame, vol. 142, no. 4, pp. 329–347, 2005.

[2] V. Zimont, W. Polifke, M. Bettelini, and W. Weisenstein, "An efficient computational model for premixed turbulent combustion at high Reynolds numbers based on a turbulent flame speed closure," Journal of Engineering for Gas Turbines and Power, vol. 120, no. 3, pp. 526–532, 1998.

[3] S. Cannon, B. Zuo, V. Adumitroaie, and C. Smith, Linear Eddy Subgrid Modeling of a lean Premixed methane-Air Combustion, CFDRC Report 8321/8, 2002.

[4] T. Poinsot and D. Veynante, Theoretical and Numerical Combustion, Edwards, 2001.

[5] Fluent Vers. 14. 0 Theory Guide.

[6] H. G. Weller, G. Tabor, A. D. Gosman, and C. Fureby, "Application of a flame-wrinkling LES combustion model to a turbulent mixing layer," Symposium (International) on Combustion, vol. 27, no. 1, pp. 899–907, 1998.

[7] A. Yoshizawa, "Bridging between eddy-viscosity-type and second-order models using a two-scale DIA," in Proceedings of 9th Symposium on Turbulent Shear Flows,, 1993.

[8] S. Menon, P. K. Yeung, and W. W. Kim, "Effect of subgrid models on the computed interscale energy transfer in isotropic turbulence," AIAA Paper, pp. 2394–2378, 1994.

[9] H. Lugt, Introduction To Vortex Theory, Potomac, Vortex Flow Press, 1996.

[10] S. P. Nandula, Lean Premixed Flame Structure in Intense Turbulence Rayleight/Raman/LIF Measurements and Modeling [Ph.D. thesis], Faculty of the Graduate School of Vanderbilt University.

[11] N. Peters, Turbulent Combustion, Monographs on Mechanics, Cambridge, Mass, USA, 2000.

[12] M. H. Baba-Ahmadi and G. Tabor, "Inlet conditions for LES using mapping and feedback control," Computers and Fluids, vol. 38, no. 6, pp. 1299–1311, 2009.

[13] A. Innocenti, Large Eddy Simulation of a Bluff Body Stabilized Lean Premixed Flame [M.S. thesis], University of Florence, 2011.

[14] L. Vervisch, P. Domingo, G. Lodato, and D. Veynante, "Scalar energy fluctuations in Large-Eddy Simulation of turbulent

flames: statistical budgets and mesh quality criterion," *Combustion and Flame*, vol. 157, no. 4, pp. 778–789, 2010.

[15] M. Cerutti, A. Andreini, B. Facchini, and L. Mangani, "Modeling of turbulent combustion and radiative heat transfer in a object-oriented CFD code for gas turbine application," in *Proceedings of the ASME Turbo Expo*, pp. 809–822, June 2008.

[16] S. Cannon, V. Adumitroaie, and C. Smith, *LES Software for the Design of Low Emission Combustion Systems For Vision 21 Plants*, CFDRC Report 8321/7, 2002.

[17] OpenCFD, OpenFOAM User Guide.

9

Comparative Numerical Study of Four Biodiesel Surrogates for Application on Diesel 0D Phenomenological Modeling

Claude Valery Ngayihi Abbe,[1,2] Raidandi Danwe,[1,3] and Robert Nzengwa[1,2]

[1]National Advanced School of Engineering, University of Yaounde, P.O. Box 337, Yaounde, Cameroon
[2]Faculty of Industrial Engineering, University of Douala, P.O. Box 2701, Douala, Cameroon
[3]Higher Institute of the Sahel, University of Maroua, P.O. Box 46, Maroua, Cameroon

Correspondence should be addressed to Claude Valery Ngayihi Abbe; ngayihiclaude@yahoo.fr

Academic Editor: Yiguang Ju

To meet more stringent norms and standards concerning engine performances and emissions, engine manufacturers need to develop new technologies enhancing the nonpolluting properties of the fuels. In that sense, the testing and development of alternative fuels such as biodiesel are of great importance. Fuel testing is nowadays a matter of experimental and numerical work. Researches on diesel engine's fuel involve the use of surrogates, for which the combustion mechanisms are well known and relatively similar to the investigated fuel. Biodiesel, due to its complex molecular configuration, is still the subject of numerous investigations in that area. This study presents the comparison of four biodiesel surrogates, methyl-butanoate, ethyl-butyrate, methyl-decanoate, and methyl-9-decenoate, in a 0D phenomenological combustion model. They were investigated for in-cylinder pressure, thermal efficiency, and NO_x emissions. Experiments were performed on a six-cylinder turbocharged DI diesel engine fuelled by methyl ester (MEB) and ethyl ester (EEB) biodiesel from wasted frying oil. Results showed that, among the four surrogates, methyl butanoate presented better results for all the studied parameters. In-cylinder pressure and thermal efficiency were predicted with good accuracy by the four surrogates. NO_x emissions were well predicted for methyl butanoate but for the other three gave approximation errors over 50%.

1. Introduction

Extensive studies regarding biodiesel combustion as an alternative for conventional diesel fuel have been performed recently [1, 2]. Experimental studies are more and more accompanied by numerical studies to better understand the physical phenomena involved in the combustion process in diesel engines when fuelled by biodiesel [3–5]. Numerical studies in diesel combustion are often performed using surrogates for which combustion kinetic mechanisms are well established and are comparable to those of investigated fuels [6–9].

Biodiesel combustion is often found difficult to model, due to the diversity of its sources and also the complexity molecular structure of biodiesel components which consist of saturated and unsaturated fatty acid [10]. This makes the modeling procedure for such type of fuel complex. It is widely accepted that numerical simulations provide a useful tool for engine conception and optimization. It is therefore important to identify suitable biodiesel surrogates that can be used for simulation purpose.

Most of the numerical works in literature present the use of methyl butanoate and methyl decanoate [11–14] as surrogates for biodiesel fuel. These works are mostly about 1D and 3D CFD detailed combustion kinetic modeling. As it was stated in previous researches CFD modeling presents the disadvantage of a high computer cost [15]. This is where 0D phenomenological modeling can be effective, because it is less time-consuming.

Investigations on different biodiesel surrogates for 0D modeling are somehow scarce. Galle et al. [16] performed a numerical analysis of a simplified spray model for different biodiesel surrogates. Their work showed that the choice of the surrogates is of high importance, and the fuel properties are highly influencing the model performance. Stagni et al. [17] proposed a reduced kinetic model for biodiesel fuel for application on 0D model. They showed that the mechanism, when applied, permits a considerable time saving in computation.

Som et al. [18] performed four different chemical kinetic models for 0D and 3D simulation; the kinetic models were mainly based on mixture of two different set of surrogates: (a) methyl decanoate, methyl 9 decenoate, and n-heptane and (b) methyl butanoate and n-heptane. Only species mole fraction simulations were performed using 0D modeling. All the mechanisms performed very well against experimental data but at the expense of computer cost.

The objective of this study is to compare 4 different methyl and ethyl ester surrogates in strictly 0D phenomenological modeling for biodiesel combustion simulation purpose. The surrogates will be compared according to computed maximum pressure, NO_x emission, and thermal efficiency.

2. Material and Methods

2.1. 0D Model Governing Equations. The model used in this study was validated in a previous study [15]; we are presenting here some fundamental relationships for a better understanding of the computing methodology.

2.1.1. Spray Submodel. Injection and spray are modeled using the theory described by Lyshevsky and Razleytsev [19–22]. The diesel jet is assumed to be constituted of two phases, an initial and a base, which are separated by a transitory characteristic length which is computed as follows:

$$L_b = C_s d_n W_e^{0.25} M^{0.4} \rho_l^{-0.6}, \tag{1}$$

where $t_b = l_b^2 / B_s$ is the characteristic time (s), C_s is a weighing coefficient, d_n is injector nozzle diameter, W_e is the Webber number, and M is an dimensional criterion [22]. Webber number is computed by

$$W_e = \frac{U_0^2 \rho_f d_c}{\sigma_f}, \tag{2}$$

where U_0 is the spray initial velocity given by

$$U_0 = \frac{24 q_c RPM}{0.75 \rho_f \pi d_c^2 i_c \varphi_{\text{inj}}}; \tag{3}$$

q_c is the fuel mass flow rate in Kg/cycle; ρ_f is the fuel density kg/m³; d_c is the injector nozzle diameter in mm; φ_{inj} is the injection duration in crank angle degrees; i_c is the number of injector holes.

2.1.2. Evaporation Submodel. The injected fuel is scattered into fine droplets after the transition length; these droplet are evaporating at a given rate which is proportional to their size [20, 21]. The fuel spray is divided in two zones: In the first zone, droplets are concentrated behind the front flame area. In the second zone, droplets have reached the front flame and the evaporation is turbulent.

The size of the pulverized droplets is evaluated for each zone using the Sauter mean diameter, calculated by

$$d_{32} = 1.7 d_n M^{0.0733} \left(We \cdot \rho\right)^{-0.266}, \tag{4}$$

where $\rho = \rho_a / \rho_l$ is the ratio of density between air and liquid fuel.

Droplets are assumed to evaporate following the d^2 law [23] for which the evaporation rate for a given droplet is

$$d_i^2 = d_0^2 - Kt, \tag{5}$$

where d_0 represents the droplet diameter after breakup (see (4)) and d_i the diameter of the same droplet at time t. K is the evaporation constant:

$$K = 10^6 \cdot \frac{4 N_u D_p p_s}{\rho_l}, \tag{6}$$

where N_u is the Nusselt number for diffusion processes, D_p is the diffusion constant for fuel vapor, and p_s is the fuel saturated vapor pressure.

Combustion duration, which is going to be used as an entry data for heat release rate calculation, is calculated in seconds by

$$\tau_{\text{ev}} = \frac{A_z}{\left(b_{it} \lambda^{0,6}\right)}, \tag{7}$$

where $b_{it} = K/d_{32}^2$ is the relative evaporation constant and λ the excess air coefficient. A_z is a fitting weighing coefficient where its value varies from 4 to 12 [21].

2.1.3. Ignition Delay and Heat Release Submodels. Ignition delay is computed using the following correlation [24]:

$$\text{ID} = \left[0.36 + 0.22 \overline{U_p}\right] \exp \left[E_a \cdot \left(\frac{1}{RT_{\text{im}} \varepsilon^{n_c-1}} - \frac{1}{17190}\right)\right.$$
$$\left. + \left(\frac{21.2}{P_{\text{im}} \varepsilon^{n_c} - 12.4}\right)^{0.63}\right], \tag{8}$$

where E_a is the fuel activation energy which is computed by $E_a = 1310000/(CN + 25)$, CN is the fuel cetane number, $\overline{U_p}$ is the engine piston speed, T_{im} and P_{im} are, respectively, the temperature and pressure at the intake manifold, ε is the engine's compression ratio, and n_c is the polytropic exponent for compression.

The fuel burnt ratio at any crank angle $x = f(\varphi)$ is computed using a double Wiebe function [25, 26] by

$$x = \beta x_p + (1 - \beta) x_d, \tag{9}$$

where x_p and x_d represent the fuel burnt ratio for premixed and diffusion combustion phases and β represents the fraction of injected fuel during premixed phase.

For each combustion phase we can write

$$x_i = 1 - \exp \left[-a_i \left(\frac{\varphi - \varphi_{\text{comb}}}{\Delta \varphi}\right)^{m_i+1}\right], \tag{10}$$

where a_i and m_i are Wiebe weighing coefficients; φ_{comb} is the start of ignition corresponding angle; $\Delta\varphi$ is the combustion duration in degrees of rotation.

The heat release rate is computed by derivation of x about φ.

2.1.4. Thermodynamic Combustion Modeling.

The governing chemical reaction used in the study is given by

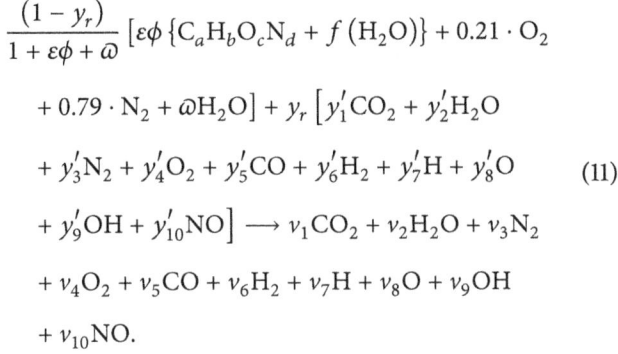

$$\frac{(1 - y_r)}{1 + \varepsilon\phi + \varpi} \left[\varepsilon\phi \left\{ C_aH_bO_cN_d + f\left(H_2O\right) \right\} + 0.21 \cdot O_2 \right.$$

$$+ 0.79 \cdot N_2 + \varpi H_2O \left] + y_r \left[y_1'CO_2 + y_2'H_2O \right. \right.$$

$$+ y_3'N_2 + y_4'O_2 + y_5'CO + y_6'H_2 + y_7'H + y_8'O \tag{11}$$

$$+ y_9'OH + y_{10}'NO \left] \longrightarrow v_1CO_2 + v_2H_2O + v_3N_2 \right.$$

$$+ v_4O_2 + v_5CO + v_6H_2 + v_7H + v_8O + v_9OH$$

$$+ v_{10}NO.$$

The mass fraction of each species is calculated based on equilibrium assumption [27]. The following equations were exploited to determine the thermodynamic state of each surrogate of chemical formula $C_aH_bO_cN_d$, such as specific heat, enthalpy, and entropy depending on the temperature of the combusting mixture. Consider

$$\frac{c_p}{R} = a_1 + a_2T + a_3T^2 + a_4T^3 + a_5\frac{1}{T^2};$$

$$\frac{h}{RT} = a_1 + \frac{a_2}{2}T + \frac{a_3}{4}T^2 + \frac{a_4}{4}T^3 + \frac{a_5}{5}\frac{1}{T^2} + a_6\frac{1}{T}; \tag{12}$$

$$\frac{s}{R} = a_1\ln T + a_2T + \frac{a_3}{2}T^2 + \frac{a_4}{3}T^3 + \frac{a_5}{4}\frac{1}{T^2} + a_7.$$

The fitting coefficients a_1 to a_7 are the first seven Chemkin NASA coefficients of specific species in the therm.dat Chemkin file. The reduced thermodynamic data are incorporated into the code and are in compliance with the temperature range that our study is covering.

The thermodynamic tables used for air properties about temperature are taken from the Chemkin data base for temperature range of 300 to 5000 K [28].

The thermodynamic state of the burning mixture is calculated at each time step as a function of each species mass fraction issued from combustion products.

The molar enthalpy of the mixture is then calculated by

$$\overline{h} = \sum y_i\overline{h}_i + \overline{h}_f y_f, \quad \text{(kJ/kmol)}; \tag{13}$$

its molar entropy is calculated by

$$\overline{s} = \left(s_f - \ln y_f \right) y_f + \sum \left(s_i - \ln y_i \right) y_i, \quad \text{kJ/kmolK}; \tag{14}$$

the heat capacity of the mixture is given by

$$\overline{c}_p = c_{p_f} y_f + \sum c_{pi} y_i. \tag{15}$$

Specific volume and internal energy of the mixture are computed by

$$u = h - RT,$$

$$v = R\frac{T}{P}, \tag{16}$$

where R is the universal gas constant and h_f, s_f, c_{pf} are, respectively, the enthalpy, entropy, and specific of the fuel (surrogate) involved in reaction (11).

The above determined parameters with equilibrium hypothesis permit the determination of in-cylinder burning mixture pressure and temperature. The following three equations give relationships for the internal energy, specific volume, entropy, enthalpy, and specific heat with temperature and pressure as functions of the crank angle [24, 25]:

$$\frac{du}{d\theta} = \left(c_P - \frac{pv}{T}\frac{\partial\ln v}{\partial\ln T} \right)\frac{dT}{d\varphi}$$

$$- \left(v\left(\frac{\partial\ln v}{\partial\ln T} + \frac{\partial\ln v}{\partial\ln P} \right) \right)\frac{dp}{d\varphi}; \tag{17}$$

$$\frac{dv}{d\theta} = \frac{v}{T}\frac{\partial\ln v}{\partial\ln T}\frac{dT}{d\varphi} - \frac{v}{p}\frac{\partial\ln v}{\partial\ln p}\frac{dp}{d\varphi}; \tag{18}$$

$$\frac{ds}{d\theta} = \left(\frac{c_P}{T} \right)\frac{dT}{d\varphi} - \frac{v}{T}\frac{\partial\ln v}{\partial\ln T}\frac{dp}{d\varphi}, \tag{19}$$

with $(\partial h/\partial T)_P = c_P$.

Considering the in-cylinder burning gases made of two zones (unburned and burned), variations of in-cylinder pressure and temperature about the crank angle for unburned and burned gases are given by the following relations:

$$\frac{dp}{d\varphi} = \frac{A + B + C}{D + E}; \tag{20}$$

$$\frac{dT_b}{d\varphi} = \frac{-h\left(\pi b^2/2 + 4V/b \right)x^{1/2}T_b - T_w}{vmc_{Pb}x}$$

$$+ \frac{v_b}{c_{Pb}}\left(\frac{\partial\ln v_b}{\partial\ln T_b} \right)\left(\frac{dp}{d\varphi} \right) \tag{21}$$

$$+ \frac{h_{cu} - h_{cb}}{xc_{Pb}}\left[\frac{dx}{d\varphi} - x - x^2\frac{C}{\omega} \right];$$

$$\frac{dT_u}{d\varphi} = \frac{-h\left(\pi b^2/2 + 4V/b \right)x^{1/2}T_u - T_w}{vmc_{Pu}x}$$

$$+ \frac{v_u}{c_{Pu}}\left(\frac{\partial\ln v_u}{\partial\ln T_u} \right)\left(\frac{dp}{d\varphi} \right) \tag{22}$$

$$+ \frac{h_{cu} - h_{cb}}{xc_u}\left[\frac{dx}{d\varphi} - x - x^2\frac{C}{\omega} \right],$$

with

$$A = \frac{1}{m}\left(\frac{dV}{d\varphi} + \frac{VC_{\text{blowby}}}{\omega}\right);$$

$$B = h\frac{\left(dV/d\varphi + VC_{\text{blowby}}/\omega\right)}{\omega m}\left[\frac{v_b}{c_{Pb}}\frac{\partial \ln v_b}{\partial \ln T_b}x^{1/2}\right.$$
$$\left. \cdot \frac{T_b - T_w}{T_b} + \frac{v_u}{c_{Pu}}\frac{\partial \ln v_u}{\partial \ln T_u}\left(1 - x^{1/2}\right)\frac{T_u - T_w}{T_u}\right];$$

$$C = -v_b - v_u\frac{dx}{d\varphi} - v_b\frac{\partial \ln v_b}{\partial \ln T_b}\frac{h_{cu} - h_{cb}}{c_{Pb}T_b}\left[\frac{dx}{d\varphi}\right. \tag{23}$$
$$\left. - \frac{\left(x - x^2\right)C_{\text{blowby}}}{\omega}\right];$$

$$D = x\left[\frac{v_b^2}{c_{Pb}T_b}\left(\frac{\partial \ln v_u}{\partial \ln T_u}\right)^2 + \frac{v_b}{P}\frac{\partial \ln v_b}{\partial \ln P}\right];$$

$$E = 1 - x\left[\frac{v_u^2}{c_{Pu}T_u}\left(\frac{\partial \ln v_u}{\partial \ln T_u}\right)^2 + \frac{v_u}{p}\frac{\partial \ln v_u}{\partial \ln p}\right],$$

where h_{cb} and h_{cu} are heat transfer coefficients to the wall, respectively, for burned and unburned zones [29]. x is the fraction of burned fuel. m is the fuel mass, v is the specific volume, ω is the engine speed in rad/s, and C_{blowby} is the blow by coefficient.

Nonlinear equations (17)–(22) are simultaneously solved using Runge-Kutta method (Dormand-Prince).

2.1.5. Nitric Oxides Emission Submodels. NO_x emissions are computed using the extended Zeldovich [30, 31] where NO_x formation rate is computed as

$$\frac{d[\text{NO}]}{dt} = \frac{2C_1\left[1 - \left([\text{NO}]/[\text{NO}]_e\right)\right]}{1 + \left[[\text{NO}]/[\text{NO}]_e\right]C_2}, \tag{24}$$

where $C_1 = K_{f,1}[\text{O}]_e[\text{N}_2]_e$ and

$$C_2 = \frac{C_1}{K_{r,2}[\text{NO}]_e[\text{O}]_e K_{r,3}[\text{NO}]_e[\text{H}]_e}; \tag{25}$$

$[X]$ is the concentration of a given species X and $[X]_e$ the concentration of the same species at equilibrium. $K_{f,i}$ and $K_{r,2}$, respectively, represent the forward and reverse rate of each reaction involved in the NO_x formation mechanism.

2.2. Biodiesel Surrogates. The surrogates used in this study are, respectively, methyl butanoate, ethyl butyrate, methyl decanoate, and methyl 9 decenoate. These four surrogates were chosen firstly because of the high amount of work on their combustion kinetics and study as biodiesel surrogates [33–36] and secondly for the wildly availability of their validated reduced mechanisms data. The thermodynamic data for each surrogate were taken, respectively, from the work of Liu et al. [37] for methyl butanoate, Herbinet et al. for methyl decanoate [38], Luo et al. [39] for methyl 9 decenoate, and the work of Goos et al. [40] for methyl butyrate.

TABLE 1: Engine specification [32].

Engine	6 liters, Ford cargo
Type	Direct injection, turbocharged
Number of cylinders	6
Bore × stroke (mm)	104.0–114.9
Compression ratio	16.4 : 1
Maximum power (kW)	136 at 2400 rpm
Maximum brake torque (Nm)	650 at 1400 rpm
Injection pump	In-line type
Injection opening pressure	197 bar

TABLE 2: Fuel properties [32].

Property	Methyl ester biodiesel	Ethyl ester biodiesel
Density (15°C)	884.3 kg/m^3	883.4 kg/m^3
Viscosity (40°C)	4.5 mm^2/s	4.9 mm^2/s
Lower heating value	37.33 kJ/kg	37.550 kJ/kg
Cetane number	54.9	53.5

TABLE 3: Values of start of injection angle [32].

Fuel	Start of injection (BTDC, °CA)		
	1100 rpm	1400 rpm	1700 rpm
Methyl ester biodiesel	14	13	12
Ethyl ester biodiesel	14	13.25	12.25

2.3. Engine Specification and Characteristic of the Biodiesel. The experimental analysis against which our study is about was performed by Sanli et al. [32]. Combustion and emission characteristics of a six-cylinder turbocharged DI diesel engine were investigated for methyl ester biodiesel and ethyl ester biodiesel from wasted frying oil. The characteristics of the engine are given in Table 1 and that of biodiesel fuels is given in Table 2.

The numerical study was performed for three different engine speeds for each surrogate: 1100 rpm, 1400 rpm, and 1700 rpm. The injection timing for each biodiesel was in accordance with the measured start of injection angle during experimental setup; the values are given in Table 3. The measured crank angle at which the fuel line reached the injector needle opening pressure was taken as the injection timing angle or start of injection (SOI).

The biodiesel surrogates were numerically investigated for three parameters:

(i) Maximum pressure.

(ii) NO_x emission.

(iii) Thermal efficiency.

3. Results and Discussions

3.1. Evaluation of Thermodynamic Properties of Biodiesel Surrogates. A first evaluation was performed for each surrogate according to (12). Equations (13) to (15) permit the computation of the mixture enthalpy, specific heat, and entropy as a function of the surrogate used. These values are used for

TABLE 4: Comparative experimental and simulated results for methyl ester biodiesel and each surrogate at 1100 rpm.

RPM			MEB			
1100	P_{max} (Mpa)	Error	NO_x (ppm)	Error	Thermal eff	Error
Experimental	**9.68**		**1400**		**39.00%**	
(1) Methyl butanoate	9.79	1.15%	1408	0.57%	38.06%	02.41%
(2) Ethyl butyrate	9.10	5.97%	101.2	92.77%	33.07%	15.21%
(3) Methyl decanoate	9.58	0.98%	532	62.00%	34.87%	10.59%
(4) Methyl 9 decenoate	9.14	5.51%	142.2	89.84%	33.43%	14.28%

TABLE 5: Comparative experimental and simulated results for methyl ester biodiesel and each surrogate at 1400 rpm.

RPM			MEB			
1400	P_{max} (Mpa)	Error	NO_x (ppm)	Error	Thermal eff	Error
Experimental	**9.89**		**1550**		**42%**	
(1) Methyl butanoate	9.79	0.96%	1511	2.52%	41.82%	0.43%
(2) Ethyl butyrate	9.09	8.06%	108	93.03%	36.22%	13.76%
(3) Methyl decanoate	9.57	3.22%	578	62.71%	38.43%	8.50%
(4) Methyl 9 decenoate	9.13	7.61%	152.2	90.18%	36.63%	12.79%

TABLE 6: Comparative experimental and simulated results for methyl ester biodiesel and each surrogate at 1700 rpm.

RPM			MEB			
1700	P_{max} (Mpa)	Error	NO_x (ppm)	Error	Thermal eff	Error
Experimental	**9.75**		**1525**		**43%**	
(1) Methyl butanoate	9.91	1.69%	1782	16.85%	43.66%	1.53%
(2) Ethyl butyrate	9.20	5.57%	128.3	91.59%	37.75%	12.21%
(3) Methyl decanoate	9.70	0.49%	686.6	54.98%	40.16%	6.60%
(4) Methyl 9 decenoate	9.25	5.09%	180	88.20%	38.18%	11.21%

the computation of the in-cylinder pressure and temperature; (21) and (22) show that in-cylinder temperature for each zone is linearly proportional to the mixture enthalpy and inversely proportional to the mixture specific heat. The variation of the mixture's entropy (19) is a nonlinear function of the temperature variation as well as the variation of in-cylinder pressure. Thus, the variation of surrogate's properties values such as enthalpy, entropy, and specific heat as a function of temperature could be a good indicator of its suitability for our 0D modeling.

Figure 1 presents the variation of enthalpy against temperature for each surrogate. The curve shapes are similar and the residual values are higher for methyl butanoate and methyl 9 decenoate, with maximum difference of 36.59 kJ/kmol. Concerning entropy and specific heat (Figures 2 and 3), we also notice a similar curve shape with higher residual values between methyl butanoate and methyl decanoate, maximum values which are, respectively, of 112 kJ/kmolK and 50.1 kJ/kmolK for entropy and specific heat. The residuals presented in the plots are the differences between each computed parameter for each surrogate for a given temperature. The plot presents the higher residual values. These results show that some notable differences shall be expected in terms of thermodynamic computed values of combusting species during the simulation for each surrogate, which is the subject of the discussion in the next sections.

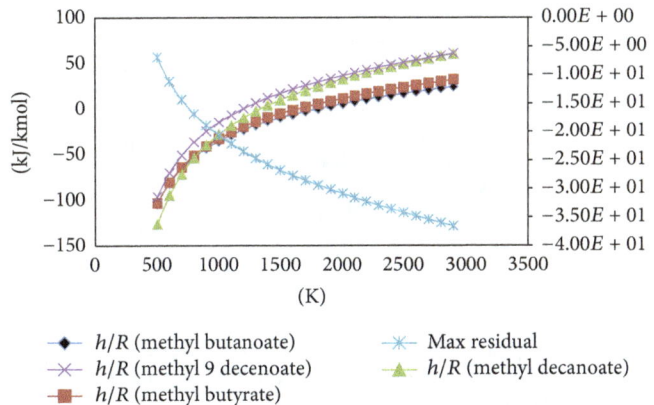

FIGURE 1: Variation of enthalpy against temperature for biodiesel surrogates.

3.2. Biodiesel Surrogates Numerical Investigation. Maximum pressure, thermal, and efficiency and NO_x emission were computed for methyl ester biodiesel and ethyl ester biodiesel from waste fried oil. The four mentioned biodiesel surrogates thermodynamic data were used in the simulation at 1100, 1400, and 1700 rpm. Results of simulations and subsequent error evaluations are displayed in Tables 4, 5, 6, 7, 8, and 9, respectively, at 1100, 1400, and 1700 rpm. Errors are evaluated as relative errors between measured and computed values.

TABLE 7: Comparative experimental and simulated results for ethyl ester biodiesel and each surrogate at 1100 rpm.

RPM	EEB					
1100	P_{max} (Mpa)	Error	NO_x (ppm)	Error	Thermal eff	Error
Experimental	**9.86**		**1410**		**41%**	
(1) Methyl butanoate	9.86	0.06%	1420	0.71%	38.04%	7.22%
(2) Ethyl butyrate	9.16	7.02%	102.3	92.74%	33.07%	19.34%
(3) Methyl decanoate	9.65	2.09%	537	61.91%	34.85%	15.00%
(4) Methyl 9 decenoate	9.21	6.56%	143.9	89.79%	33.42%	18.49%

TABLE 8: Comparative experimental and simulated results for ethyl ester biodiesel and each surrogate at 1400 rpm.

RPM	EEB					
1400	P_{max} (Mpa)	Error	NO_x (ppm)	Error	Thermal eff	Error
Experimental	**9.78**		**1510**		**42.50%**	
(1) Methyl butanoate	9.77	0.04%	1507	0.20%	41.81%	1.62%
(2) Ethyl butyrate	9.35	4.40%	131.6	91.28%	35.32%	16.89%
(3) Methyl decanoate	9.84	0.63%	692.3	54.15%	37.42%	11.95%
(4) Methyl 9 decenoate	9.39	3.90%	184.9	87.75%	35.70%	16.00%

TABLE 9: Comparative experimental and simulated results for ethyl ester biodiesel and each surrogate at 1700 rpm.

RPM	EEB					
1700	P_{max} (Mpa)	Error	NO_x (ppm)	Error	Thermal eff	Error
Experimental	**9.7**		**1500**		**43.50%**	
(1) Methyl butanoate	9.96	2.73%	1659	10.60%	44.29%	1.82%
(2) Ethyl butyrate	9.47	2.34%	158	89.47%	36.93%	15.10%
(3) Methyl decanoate	9.98	2.89%	831	44.60%	39.24%	9.79%
(4) Methyl 9 decenoate	9.52	1.86%	221.9	85.21%	37.35%	14.14%

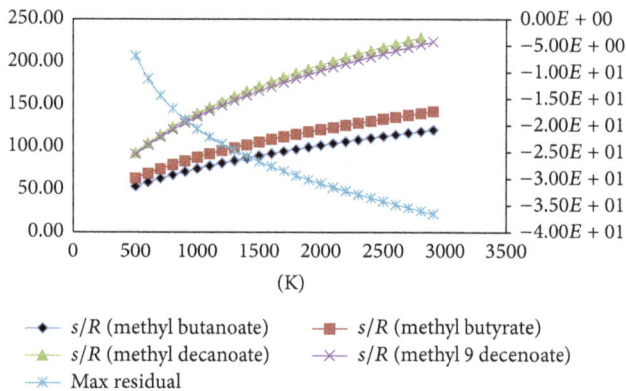

FIGURE 2: Variation of entropy against temperature for biodiesel surrogates.

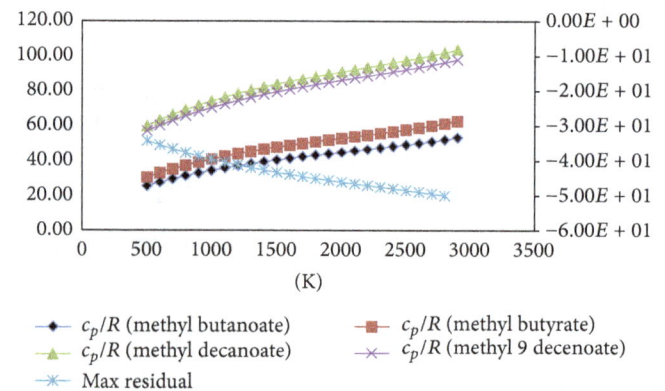

FIGURE 3: Variation of specific against temperature for biodiesel surrogates.

3.2.1. Methyl Ester Biodiesel Numerical Investigation

Maximum in-Cylinder Pressure. Maximum pressure is fairly well predicted for each surrogate. However methyl butanoate shows a better prediction rate of an average of 1.27% accuracy across the different rpm. The other surrogates present an average accuracy of 6.53%, 1.56%, and 6.07% for ethyl butyrate, methyl decanoate, and methyl 9 decenoate, respectively. In Figure 4 it can be seen that experimental and simulated

pressure traces closely match for each biodiesel surrogates compared to MEB experimental data. Better pressure simulations are achieved by methyl butanoate and methyl decanoate compared to the other two.

Thermal Efficiency. Thermal efficiency provides a good insight of fuel heat input to mechanical energy output during combustion cycle, especially when evaluating alternative fuel [41, 42]. The simulation results show that methyl butanoate

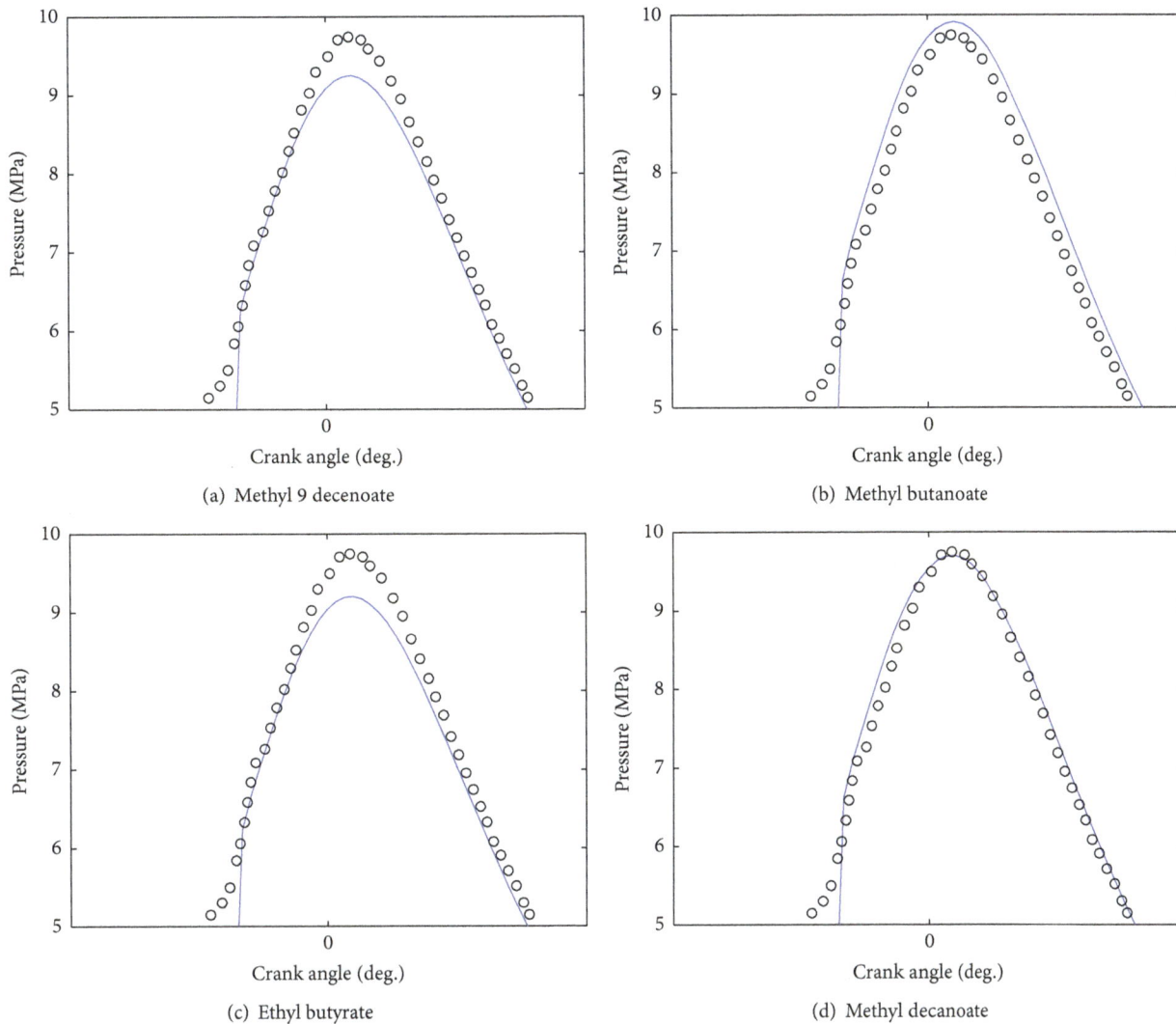

(a) Methyl 9 decenoate

(b) Methyl butanoate

(c) Ethyl butyrate

(d) Methyl decanoate

FIGURE 4: Experimental and simulated pressure trace at 1700 RPM for MEB for each surrogate.

gives the better approximation with an average error of 1.46%. Ethyl butyrate, methyl decanoate, and methyl 9 decenoate present average errors below 15%, which are acceptable for engine simulation. Methyl decanoate presents an approximation of average 8.56%, followed by methyl 9 decenoate at 12.76% and ethyl butyrate at 13.76%.

NO_x Emissions. NO_x emissions are simulated using the extended Zeldovich mechanism [30, 43]; the simulation results present a good agreement with experimental results from methyl butanoate with an average of 6.65% error across the three regimes. Methyl decanoate, methyl 9 decenoate, and ethyl butyrate however poorly predict NO_x emissions with respective average errors of 59.9%, 92.46%, and 89.49%.

3.2.2. Ethyl Ester Biodiesel Numerical Investigation

Maximum in-Cylinder Pressure. Maximum pressure was calculated through simulation in the 0D phenomenological model, as well as in the case of methyl ester biodiesel.

Maximum pressure is also fairly well predicted for each surrogate. Methyl butanoate shows a better prediction rate of an average of 0.94% accuracy across the different rpm, which is slightly better than during the methyl ester biodiesel simulation. Simulations present an average accuracy of 4.59%, 1.87%, and 4.10% for ethyl butyrate, methyl decanoate, and methyl 9 decenoate, respectively. Figure 5 shows that experimental and simulated pressure traces closely match for each biodiesel surrogate compared to EEB experimental data. Better pressure simulations are achieved by methyl butanoate and methyl decanoate compared to the other two, just as in the case of MEB.

Thermal Efficiency. The simulation results present a better approximation for methyl butanoate with an average error of 3.55%. Ethyl butyrate, methyl decanoate, and methyl 9 decenoate present average errors close and over 15%, which is below acceptable for engine simulation. Methyl decanoate presents an approximation of average 12.25%, followed by methyl 9 decenoate at 16.21% and ethyl butyrate at 17.11%.

(a) Methyl 9 decenoate

(b) Methyl butanoate

(c) Ethyl butyrate

(d) Methyl decanoate

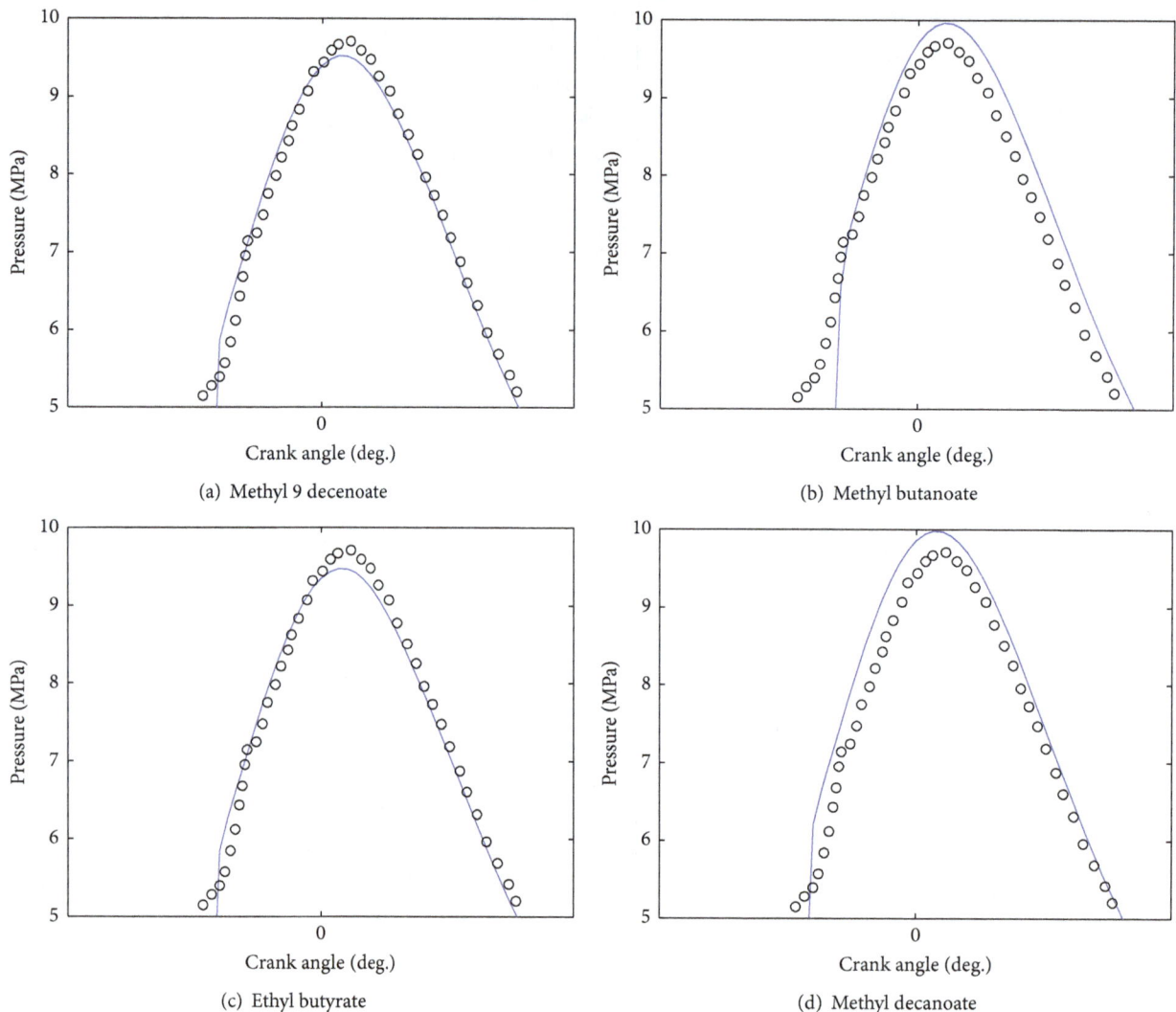

FIGURE 5: Experimental and simulated pressure trace at 1700 RPM for EEB for each surrogate.

NO$_x$ Emissions. As it was the case for MEB, the simulation results present a good agreement with experimental results from methyl butanoate with an average of 3.84% error across the three regimes, which is an improvement compared to MEB simulation. Methyl decanoate, methyl 9 decenoate, and ethyl butyrate, just as it was the case for MEB, poorly predict NO$_x$ emissions with respective average errors of 53.56%, 87.59%, and 91.17%.

3.3. Discussion. The approximations results are similar, whether we are investigating methyl ester biodiesel or ethyl ester biodiesel. For each case, pressure trace and thermal efficiency are fairly well predicted. Ethyl butyrate and methyl 9 decenoate present the worst results in terms of simulation precision.

The overall trend observed from the various simulation results is that methyl butanoate gives better predictions of engine parameters compared to other surrogates; this is in accordance with its high rate of usage as a surrogate for biodiesel in 3D and 1D simulation studies [4, 5, 44]. However,

this result is somehow contradictory with comments from Brakora et al. [45] and Hakka et al. [33] that say that methyl butanoate is not a particular good surrogate for biodiesel. While this was explained by the fact that methyl butanoate possesses a short alkylic chain and therefore cannot adequately capture ignition delay and species history, it should be noted that most of these findings were done under 3D or experimental investigations. Our 0D model only uses Chemkin Nasa coefficients to compute in-cylinder thermodynamic parameters; the ignition delay is computed as a function of the investigated biodiesel cetane number. The reaction kinetic in our model is therefore independent of the surrogate thermodynamic data.

The three other surrogates present overall good approximation for thermal efficiency and maximum pressure. Methyl decanoate presents better simulation approximation errors than ethyl butyrate and methyl 9 decenoate; that trend is interesting since one would have expected a closer approximation value for ethyl butyrate since its enthalpy, heat capacity, and entropy curves are closer to the ones for methyl butanoate. However, it has been noticed in the literature that methyl decanoate has been extensively used as a surrogate for biodiesel combustion simulation as compared to ethyl butyrate and methyl 9 decanoate [13, 46, 47].

Another trend observed is the poorly predicted NO_x emission simulation for methyl decanoate, ethyl butyrate, and methyl 9 decenoate. For these three surrogates, NO_x emissions are underpredicted compared to experimental results. An earlier study, performed on Kiva-2 showed that surrogates with lesser content of oxygen tend to predict lower NO_x emission concentration [48]. In our case, each surrogate presents 2 atoms of oxygen in its formula; methyl butanoate presents a chemical formula of $C_5H_{10}O_2$, ethyl butyrate of $C_6H_{12}O_2$, methyl decanoate of $C_{11}H_{22}O_2$, and methyl 9 decenoate $C_{11}H_{20}O_2$. The stoichiometric air/fuel ratio which is calculated by $L = ((8/3)a + 8 \cdot b - c)/0.232$ for each surrogate of formula $C_aH_bO_cN_d$ will be, respectively, 8.788, 9.4928, 11.37, and 11.24. According to experimental parameters, the same mass of injected fuel was considered for each surrogate in the simulation. The model takes into account the molar fraction of each species contained in the surrogate. The composition of the burning mixture will therefore be dependable of the surrogate composition. The model will therefore predict a richer mixture during combustion for methyl butanoate, while others will be leaner. That could influence the calculation of oxygen content in the burning mixture and therefore will impact the computation of NO_x emission through the extended Zeldovich mechanism. Another area of investigation could be the extension of the thermodynamic data for each surrogate by integration of more chain reactions into the reduced chemical mechanical mechanism. That could lead to an increase of computation time.

4. Conclusion

Four biodiesel surrogates were investigated for their performance in 0D phenomenological combustion modeling. The surrogated thermodynamic data were those of butanoate, ethyl butyrate, methyl decanoate, and methyl 9 decenoate. Simulations were compared against experimental data gathered from the combustion of ethyl ester and methyl ester biodiesel from waste cooking oil in a 6-cylinder DI diesel engine at engine speed varying to 1100, 1400, and 1700 rpm. Each biodiesel surrogate was investigated for in-cylinder pressure, maximum pressure, thermal efficiency, and NO_x emissions. After the analysis of the simulation, the following conclusions could be derived:

(i) Enthalpy, entropy, and heat capacity of the surrogates follow similar curve shapes; the highest differences were identified for methyl butanoate and methyl decanoate concerning entropy and heat capacity and between methyl butanoate and methyl 9 decenoate for enthalpy.

(ii) Each surrogate fairly well predicted maximum pressure and thermal efficiency for each engine regime and each biodiesel.

(iii) Methyl butanoate showed the best accuracy for all three simulated parameters, 1.27% and 0.94% for

maximum pressure, 1.46% and 3.55% for thermal efficiency, and 6.65% and 3.84% for NO_x emissions for MEB and EEB, respectively, followed by methyl decanoate, methyl 9 decenoate, and ethyl butyrate.

(iv) Out of the four surrogates, only methyl butanoate, could well predict NO_x emissions at 6.65% and 3.84% for MEB and EEB, respectively; the other three surrogates presented heavy average errors higher than 50%.

Conflict of Interests

The authors declare that there is no conflict of interests regarding the publication of this paper.

Acknowledgment

The authors are grateful to the Ministry of Higher Education of Cameroon for financing this research.

References

[1] S. A. Basha, K. R. Gopal, and S. Jebaraj, "A review on biodiesel production, combustion, emissions and performance," *Renewable and Sustainable Energy Reviews*, vol. 13, no. 6-7, pp. 1628–1634, 2009.

[2] N. Usta, E. Öztürk, Ö. Can et al., "Combustion of bioDiesel fuel produced from hazelnut soapstock/waste sunflower oil mixture in a Diesel engine," *Energy Conversion and Management*, vol. 46, no. 5, pp. 741–755, 2005.

[3] X. Wang, K. Li, and W. Su, "Experimental and numerical investigations on internal flow characteristics of diesel nozzle under real fuel injection conditions," *Experimental Thermal and Fluid Science*, vol. 42, pp. 204–211, 2012.

[4] J. Yang, M. Johansson, C. Naik et al., "3D CFD modeling of a biodiesel-fueled diesel engine based on a detailed chemical mechanism," SAE Technical Paper 2012-01-0151, SAE International, 2012.

[5] S. Som and D. E. Longman, "Numerical study comparing the combustion and emission characteristics of biodiesel to petrodiesel," *Energy and Fuels*, vol. 25, no. 4, pp. 1373–1386, 2011.

[6] N. Samec, B. Kegl, and R. W. Dibble, "Numerical and experimental study of water/oil emulsified fuel combustion in a diesel engine," *Fuel*, vol. 81, no. 16, pp. 2035–2044, 2002.

[7] H. J. Curran, P. Gaffuri, W. J. Pitz, and C. K. Westbrook, "A comprehensive modeling study of n-heptane oxidation," *Combustion and Flame*, vol. 114, no. 1-2, pp. 149–177, 1998.

[8] S. Hong, M. S. Wooldridge, H. G. Im, D. N. Assanis, and H. Pitsch, "Development and application of a comprehensive soot model for 3D CFD reacting flow studies in a diesel engine," *Combustion and Flame*, vol. 143, no. 1-2, pp. 11–26, 2005.

[9] G. Vourliotakis, D. I. Kolaitis, and M. A. Founti, "Development and parametric evaluation of a tabulated chemistry tool for the simulation of n-heptane low-temperature oxidation and autoignition phenomena," *Journal of Combustion*, vol. 2014, Article ID 237049, 13 pages, 2014.

[10] L. Coniglio, H. Bennadji, P. A. Glaude, O. Herbinet, and F. Billaud, "Combustion chemical kinetics of biodiesel and related compounds (methyl and ethyl esters): experiments

and modeling—advances and future refinements," *Progress in Energy and Combustion Science*, vol. 39, no. 4, pp. 340–382, 2013.

[11] S. Um and S. W. Park, "Modeling effect of the biodiesel mixing ratio on combustion and emission characteristics using a reduced mechanism of methyl butanoate," *Fuel*, vol. 89, no. 7, pp. 1415–1421, 2010.

[12] J. Y. W. Lai, K. C. Lin, and A. Violi, "Biodiesel combustion: advances in chemical kinetic modeling," *Progress in Energy and Combustion Science*, vol. 37, no. 1, pp. 1–14, 2011.

[13] P. Diévart, S. H. Won, S. Dooley, F. L. Dryer, and Y. Ju, "A kinetic model for methyl decanoate combustion," *Combustion and Flame*, vol. 159, no. 5, pp. 1793–1805, 2012.

[14] V. I. Golovitchev and J. Yang, "Construction of combustion models for rapeseed methyl ester bio-diesel fuel for internal combustion engine applications," *Biotechnology Advances*, vol. 27, no. 5, pp. 641–655, 2009.

[15] C. V. Ngayihi Abbe, R. Nzengwa, R. Danwe, Z. M. Ayissi, and M. Obonou, "A study on the 0D phenomenological model for diesel engine simulation: application to combustion of Neem methyl esther biodiesel," *Energy Conversion and Management*, vol. 89, pp. 568–576, 2015.

[16] J. Galle, R. Verschaeren, and S. Verhelst, "The behavior of a simplified spray model for different diesel and bio-diesel surrogates," SAE Technical Paper 2015-01-0950, 2015.

[17] A. Stagni, C. Saggese, M. Bissoli et al., "Reduced kinetic model of biodiesel fuel combustion," *Chemical Engineering Transactions*, vol. 37, pp. 877–882, 2014.

[18] S. Som, W. Liu, and D. E. Longman, "Comparison of different chemical kinetic models for biodiesel combustion," in *Proceedings of the Internal Combustion Engine Division Fall Technical Conference (ASME '13)*, American Society of Mechanical Engineers, Dearborn, Mich, USA, October 2013.

[19] A. S. Lyshevsky, *Fuel Atomization in Marine Diesels*, Machgiz, Leningrad, Russia, 1971 (Russian).

[20] N. F. Razleytsev, *Combustion Simulation and Optimization in Diesels*, Vischa Shkola, Kharkiv, Ukraine, 1980 (Russian).

[21] F. I. Abramchuk, A. P. Marchenko, N. F. Razlejtsev, and E. I. Tretiak, *Modern Diesel Engines: Increase of Fuel Economy and Durability*, Technika, Kiev, Ukraine, 1992.

[22] A. S. Kuleshov, "Multi-zone DI diesel spray combustion model for thermodynamic simulation of engine with PCCI and high EGR level," *SAE International Journal of Engines*, vol. 2, no. 1, pp. 1811–1834, 2009.

[23] D. B. Spalding, "The combustion of liquid fuels," in *Proceedings of the Fourth Symposium (International) on Combustion and Detonation Waves*, The Combustion Institute, Cambridge, Mass, USA, 1953.

[24] Y. Aghav, V. Thatte, M. Kumar et al., "Predicting ignition delay and HC emission for DI diesel engine encompassing EGR and oxygenated fuels," SAE Technical Paper 2008-28-0050, SAE, 2008.

[25] N. Watson, A. Pilley, and M. Marzouk, "A combustion correlation for diesel engine simulation," SAE Technical Paper 800029, 1980.

[26] O. Grondin, *Modélisation du moteur à allumage par compression dans la perspective du contrôle et du diagnostic [Ph.D. thesis]*, Université de Rouen, Rouen, France, 2004.

[27] C. R. Ferguson and T. K. Allan, *Internal Combustion Engines: Applied Thermosciences*, John Wiley & Sons, Hoboken, NJ, USA, 2nd edition, 2000.

[28] R. J. Kee, F. M. Rupley, and J. A. Miller, "The chemkin thermodynamic data base," Tech. Rep., 1990.

[29] G. Woschni, "A universally applicable equation for the instantaneous heat transfer coefficient in the internal conbustion engine," SAE Technical Paper 670931, 1967.

[30] J. B. Heywood, *Internal Combustion Engine Fundamentals*, McGraw-Hill, New York, NY, USA, 1988.

[31] G. A. Lavoie, J. B. Heywood, and J. C. Keck, "Experimental and theoretical study of nitric oxide formation in internal combustion engines," *Combustion Science and Technology*, vol. 1, no. 4, pp. 313–326, 1970.

[32] H. Sanli, M. Canakcib, E. Alptekinb, A. Turkcanb, and A. N. Ozsezenb, "Effects of waste frying oil based methyl and ethyl ester biodiesel fuels on the performance, combustion and emission characteristics of a DI diesel engine," *Fuel*, vol. 159, pp. 179–187, 2015.

[33] M. H. Hakka, P.-A. Glaude, O. Herbinet, and F. Battin-Leclerc, "Experimental study of the oxidation of large surrogates for diesel and biodiesel fuels," *Combustion and Flame*, vol. 156, no. 11, pp. 2129–2144, 2009.

[34] W. Wang and M. A. Oehlschlaeger, "A shock tube study of methyl decanoate autoignition at elevated pressures," *Combustion and Flame*, vol. 159, no. 2, pp. 476–481, 2012.

[35] S. Zhang, L. J. Broadbelt, I. P. Androulakis, and M. G. Ierapetritou, "Comparison of biodiesel performance based on HCCI engine simulation using detailed mechanism with on-the-fly reduction," *Energy & Fuels*, vol. 26, no. 2, pp. 976–983, 2012.

[36] A. M. El-Nahas, M. V. Navarro, J. M. Simmie et al., "Enthalpies of formation, bond dissociation energies and reaction paths for the decomposition of model biofuels: ethyl propanoate and methyl butanoate," *The Journal of Physical Chemistry A*, vol. 111, no. 19, pp. 3727–3739, 2007.

[37] W. Liu, R. Sivaramakrishnan, M. J. Davis, S. Som, D. E. Longman, and T. F. Lu, "Development of a reduced biodiesel surrogate model for compression ignition engine modeling," *Proceedings of the Combustion Institute*, vol. 34, no. 1, pp. 401–409, 2013.

[38] O. Herbinet, W. J. Pitz, and C. K. Westbrook, "Detailed chemical kinetic mechanism for the oxidation of biodiesel fuels blend surrogate," *Combustion and Flame*, vol. 157, no. 5, pp. 893–908, 2010.

[39] Z. Luo, M. Plomer, T. Lu et al., "A reduced mechanism for biodiesel surrogates for compression ignition engine applications," *Fuel*, vol. 99, pp. 143–153, 2012.

[40] E. Goos, A. Burcat, and B. Ruscic, "Transfer of extended third millennium ideal gas and condensed phase thermochemical database," in *Extended Third Millennium Ideal Gas and Condensed Phase Thermochemical Database*, Engineering FOA, Haifa, Israel, 2009.

[41] G. Labeckas and S. Slavinskas, "The effect of rapeseed oil methyl ester on direct injection Diesel engine performance and exhaust emissions," *Energy Conversion and Management*, vol. 47, no. 13-14, pp. 1954–1967, 2006.

[42] C. D. Rakopoulos, K. A. Antonopoulos, D. C. Rakopoulos, D. T. Hountalas, and E. G. Giakoumis, "Comparative performance and emissions study of a direct injection Diesel engine using blends of Diesel fuel with vegetable oils or bio-diesels of various origins," *Energy Conversion and Management*, vol. 47, no. 18-19, pp. 3272–3287, 2006.

[43] G. A. Lavoie and P. N. Blumberg, "A Fundamental model for predicting fuel consumption, NO_x and HC emissions of the

conventional spark-ignition engine," *Combustion Science and Technology*, vol. 21, no. 5-6, pp. 225–258, 1980.

[44] I. Uryga-Bugajska, M. Pourkashanian, D. Borman, E. Catalanotti, and C. W. Wilson, "Theoretical investigation of the performance of alternative aviation fuels in an aero-engine combustion chamber," *Proceedings of the Institution of Mechanical Engineers, Part G: Journal of Aerospace Engineering*, vol. 225, no. 8, pp. 874–885, 2011.

[45] J. L. Brakora, Y. Ra, and R. D. Reitz, "Combustion model for biodiesel-fueled engine simulations using realistic chemistry and physical properties," *SAE International Journal of Engines*, vol. 4, no. 1, pp. 931–947, 2011.

[46] K. Seshadri, T. Lu, O. Herbinet et al., "Experimental and kinetic modeling study of extinction and ignition of methyl decanoate in laminar non-premixed flows," *Proceedings of the Combustion Institute*, vol. 32, no. 1, pp. 1067–1074, 2009.

[47] S. M. Sarathy, M. J. Thomson, W. J. Pitz, and T. Lu, "An experimental and kinetic modeling study of methyl decanoate combustion," *Proceedings of the Combustion Institute*, vol. 33, no. 1, pp. 399–405, 2011.

[48] C. Y. Choi and R. D. Reitz, "Modeling the effects of oxygenated fuels and split injections on DI diesel engine performance and emissions," *Combustion Science and Technology*, vol. 159, no. 1, pp. 169–198, 2000.

10

Testing Vegetation Flammability:
The Problem of Extremely Low Ignition Frequency
and Overall Flammability Score

Zorica Kauf,[1] **Andreas Fangmeier,**[1] **Roman Rosavec,**[2] **and Željko Španjol**[2]

[1] *Institute of Landscape and Plant Ecology, University of Hohenheim, August-von-Hartmann Straße 3, 70599 Stuttgart, Germany*
[2] *Department of Forest Ecology and Silviculture, Faculty of Forestry, University of Zagreb, Svetošimunska 25, 10002 Zagreb, Croatia*

Correspondence should be addressed to Zorica Kauf; zorica.kauf@uni-hohenheim.de

Academic Editor: Michael A. Delichatsios

In the recent decades changes in fire regimes led to higher vulnerability of fire prone ecosystems, with vegetation being the only component influencing fire regime which can be managed in order to reduce probability of extreme fire events. For these management practices to be effective reliable information on the vegetation flammability is being crucial. Epiradiator based testing methods are one of the methods commonly used to investigate vegetation flammability and decrease in ignition frequency is always interpreted as a decrease in flammability. Furthermore, gathered information is often combined into a single flammability score. Here we present results of leaf litter testing which, together with previously conducted research on similar materials, show that material with very low ignition frequency under certain testing conditions can be extremely flammable if testing conditions are slightly changed. Additionally, our results indicate that combining measured information into one single flammability score, even though sometimes useful, is not always meaningful and should be performed with caution.

1. Introduction

In recent decades land use coupled with climate change led to changes in the fire regime making fire prone ecosystems more vulnerable to wildfires [1, 2], with further shift towards more devastating fire regimes being predicted in the future [3, 4]. Fire regime is the result of complex interactions between ignitions, weather, topography, and vegetation acting as fuel [5–7]. Even though vegetation is rarely the most influential factor, it is the only one that can be managed in order to reduce the probability of occurrence of extreme wildfires [8]. To achieve this goal correct and reliable information on vegetation flammability is being crucial, becoming one of the essential components of fire risk assessment and management planning [9, 10]. Even though there are numerous lists of species based on their flammability [11–13], the complexity of vegetation flammability makes such a ranking challenging and resulting lists unreliable and possibly misleading [10].

Flammability is comprised of (i) ignitibility—the fuel ignition delay once exposed to heat, (ii) sustainability—the measure of how well a fire will continue to burn with or without the heat source, (iii) combustibility—the reflection of the rapidity with which a fire burns [14], and (iv) consumability—the proportion of mass or volume consumed by fire (Martin et al. 1994). There is currently no validated method of integrating all flammability components into one single index of plant flammability [15]. Furthermore, vegetation fuels are highly variable; their flammability changes with genotype, age, season, location, and material tested [8, 16].

Due to rising importance of reliable vegetation flammability information numerous researchers worked on quantifying it with studies ranging from field burning experiments [17, 18] and burning tables/benches [19–22] to individual leaf testing [16, 23]. In this effort a high number of testing procedures and measured parameters were associated with vegetation flammability. Nevertheless, results of laboratory scale testing as well as vegetation flammability as a concept were challenged due to the discrepancy between laboratory testing results, field testing, and modelling outputs [24]. Simultaneously, they were defended as an opportunity to better understand the influence of fire as a selective evolutionary force [25] and being a useful tool for providing basic information for assessing fire risk despite the inability to directly transfer testing results to a bigger scale [26].

Epiradiator based methods are often criticized for their low reproducibility [27] and low heat flux used [24]. Nevertheless, due to low technical demands and presumably straight-forward interpretation they were often used [20, 28–35]. Up to now they are providing valuable information and are contributing to our better understanding of vegetation flammability [26, 36, 37].

Most of the authors refer to Valette's [38] work as to the reference epiradiator based vegetation flammability testing method. This method attributes flammability score ranging from 0 (the lowest flammability) to 5 (extreme flammability) based on combined information on average ignition delay (average time elapsed between placing a sample on the epiradiator surface and appearance of flame) and ignition frequency (percentage of tests in which flame occurred). Decrease in ignition delay leads to increase in flammability score, whereas decrease in ignition frequency leads to its decrease. If ignition frequency is lower than 50%, material can be assigned only 0 and 1 flammability score regardless of ignition delay. Even when alternative interpretation of data is given [26, 32] or alternative testing method is used [39], its final intention is to give a single flammability score.

Comparing our results to those previously published, we tried to answer two questions.

(i) Does low ignition frequency guarantee low flammability?

(ii) Is it always meaningful to combine gathered information into one single flammability score?

2. Materials and Methods

2.1. Samples Gathering and Storing. Sampling was performed in the coastal region of Croatia. Leaf litter was gathered on May 15, 2010, in the Trstenik area ($42°54'$N, $17°23'$E) on the southwest coast of the Pelješac peninsula, where samples of *Arbutus unedo* L. (strawberry tree), *Ceratonia siliqua* L. (carob), *Laurus nobilis* L. (laurel), *Olea europaea* L. (olive), *Pinus halepensis* Mill. (aleppo pine), *Pistacia lentiscus* L. (mastic), *Pittosporum tobira* Thunb. (Japanese mock orange), and *Quercus ilex* L. (holm oak) were collected.

Samples of *P. tobira* and *P. lentiscus* originated from two and four individual plants, respectively. All the other samples were gathered randomly across the sampling sites and were composed of material originated from more than ten individuals. Only whole leaves and bigger leaf fragments which could be ambiguously identified as belonging to species of interest were gathered; thus only the upper leaf litter layer was sampled. Material was stored in open paper bags, on a storage table, at room temperature and humidity until further processing. Position of samples was occasionally changed.

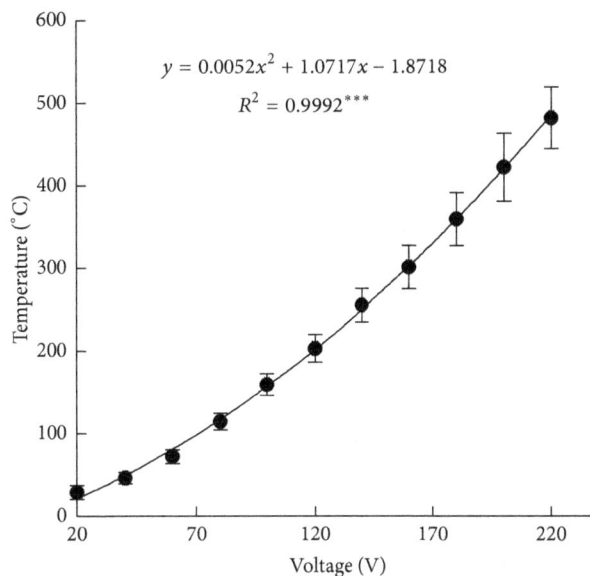

FIGURE 1: Relationship between input voltage and epiradiator surface temperature. Mean and standard deviation of 10 measurements per voltage input.

2.2. Flammability Testing. For flammability testing, a 500 W epiradiator with 10 cm diameter radiant disc and nominal surface temperature of 420°C was used as a heat source. It was connected to a variable voltage transformer allowing us to reduce and control surface temperature. While determining the voltage-temperature curve, it was observed that the actual temperature was higher than nominal temperature and varied substantially across the epiradiator surface (Figure 1). This was confirmed on three separated epiradiators, one being completely new. Thus, in order to ensure uniform start temperature, the same epiradiator (type 534 Rc2, Quartz Saint-Gobain) was used for all tests with its surface temperature being constantly measured and monitored at fixed point using a K-type temperature probe (GES 900, Greisinger) connected to a digital thermometer (GMH 3210, Greisinger). Additionally, during the process of determining the appropriate surface temperature-material amount combination, it was demonstrated that if the surface temperature rise during the test is substantial, the temperature does not drop back to the initial start temperature, regardless of the waiting time. To give an example for this effect: when testing 5 grams of leaf litter material at a start surface temperature of 500°C and waiting time between two consecutive tests sufficiently long for surface temperature to stabilise, after three successive tests the surface temperature stabilised at 590°C. Under these conditions, the ignition delay dropped from 5.65 seconds for the first test to 1.05 seconds for the third test, indicating that monitoring and stabilising the surface temperature is crucial for reducing systematic errors.

In order to capture the ignitability component of leaf litter flammability we aimed to achieve a higher ignition frequency than Petriccione et al. [34] and a longer ignition delay than Ormeño et al. [20]. Preliminary testing demonstrated that an increase in both epiradiator surface temperature and amount of material tested leads to an increase in ignition frequency; higher temperature results in shorter ignition delay and bigger amount of material tested in longer ignition delay.

After extensive pretesting we chose to test 3.0 ± 0.1 gram sample material at $400 \pm 5°C$ epiradiator surface temperature, respectively. This combination led to 100% ignition frequency with relatively long ignition delay.

The chosen amount of test material formed a layer on the epiradiator surface for all the tested samples ensuring similar heat exposure despite varying surface temperature. After placement on the epiradiator surface samples surrounded the temperature probe, allowing us to use the same probe for measuring initial temperature of the epiradiator surface and temperature at the lower side of the leaf litter during a test. Samples were not additionally oven dried before flammability testing as final moisture content was attributed to intrinsic characteristics of materials governing their ability to retain moisture [40, 41].

The horizontal pilot flame was positioned 4.5 cm above the epiradiator surface. Four parameters were measured: (i) ignition delay (ID)—the time elapsed between placing a sample on the epiradiator surface and appearance of a flame, (ii) flame extinguish time (FET)—time elapsed between placing a sample on the epiradiator surface and end of the flaming combustion, (iii) ignition temperature (IT)—temperature measured at the moment of ignition, and (iv) max temperature (MT)—maximal temperature reached during the test. Flame residence time (FRT) was calculated as the difference between FET and ID, representing the duration of flaming combustion. ID is considered to be a measure of ignitability and FRT a measure of sustainability component of flammability [20, 36, 42]. FET is taken as an additional sustainability measure. IT was interpreted in relation to ID; MT, as it was measured at the bottom of the sample, could not be interpreted in light of flammability components, but provided limited information on heat transfer to the soil during wildfires. Flame intensity was not determined. Time variables were measured using a stopwatch.

Flammability testing was performed on September 5 and 6, 2010, with each sample being tested five times, once in every replication. Samples were tested in random order. Testing was performed in a closed room under a simple chamber, minimizing disturbance due to external air movement. The chamber was opened at the top to allow natural air convection and partially at one side to allow for sample manipulation.

2.3. Physical Measurements. Moisture content (MC), specific leas area (SLA), average area (AA), and average mass (AM) of the single particle are the physical parameters measured and reported. During flammability testing one or two subsamples per species, equal in size to the test samples, were taken. Their dry mass was determined by reweighting them after drying for 24 hours at $85°C$ and MC was expressed on dry weight basis [43]. After drying fragile leaf litter was soaked in warm tap water and flattened with cloths flatiron before its area was measured with a portable area meter. SLA was calculated by dividing leaf litter area by the corresponding dry mass; AM and AA were calculated dividing dry mass and

area of the sample by the corresponding number of particles. An exception was made for *P. halepensis*, as needles were fragmented and the number of particles was difficult to count and it was estimated based on an even smaller subsample and set at 250.

2.4. Statistical Analysis. IBM SPSS Statistics 21 software was used for statistical analysis of gathered data. As all flammability parameters were normally distributed and had homogeneous variances on the species level one-way ANOVA was performed, followed by Duncan post hoc test in order to determine differences between species based on single parameters.

Regression analysis was performed to determine if any of the physical parameters influenced flammability and to what extent different flammability parameters were related. Overall significance of the regression analysis was checked by ANOVA; regression coefficients were tested with t-test against null hypothesis. A relationship between flammability parameters was considered important in order to give appropriate interpretation to collected data, whereas relationships between physical parameters and flammability were reported but are not discussed in detail here.

In order to account for combined influence of all measured parameters on fire behaviour hierarchical cluster analysis was performed using squared Euclidean distances of standardised values (0-1) of measured flammability characteristics (ID, IT, FET, and MT) to determine within-group linkage and try to attribute an overall flammability score to leaf litters of different species.

3. Results and Discussion

When choosing sample mass-temperature combination for our testing procedure we were aware of the argument that masses larger than 1 gram should not be tested with epiradiator based methods as properties such as fuel height could influence the results [20] as well as criticism directed towards the laboratory flammability tests and their use of low heat fluxes [24]. Nevertheless, we consider our testing combination to be acceptable as Anderson [14] stated that there is a clear relationship between sample height and ignition when holding sample density and specific heat constant. If mass is held constant, difference in heights and densities will be present between materials, with both influencing flammability, but also being governed by particle geometry—an inherent characteristic of the material. Furthermore, Fernandes [24] stated that heat fluxes measured during wildland fires are several orders of magnitude higher than that used in laboratory studies, yet still ignition sources do not need to have big heat fluxes. If ignition source has low heat flux (e.g., cigarette butt, sparks, firebrands, not completely extinguished grilling amber, etc.), flammability together with other material properties can have a relevant role in determining whether or not ignition occurs and how fast initial combustion process proceeds, as shown in research made by Ganteaume et al. on reconstructed [22] and undisturbed [39] leaf litters tested with standardised firebrand.

3.1. Means, ANOVA, and Regression Analysis. All five flammability parameters showed statistically significant differences between tested species (Table 1), with the highest level of significance obtained for ignition related parameters: ID and IT. *P. lentiscus* and *P. halepensis* had the shortest ID and were considered the most ignitable of the tested species, whereas *C. siliqua* had by far the longest average ID and was the least ignitable species in this study. *C. siliqua* was followed by *Q. ilex*, *O. europaea*, and *A. unedo*, which formed a group of species with intermediate ignitability as all of the members were significantly different from *P. halepensis*. *L. nobilis* and *P. tobira* were designated as fairly ignitable species as they did not show any significant difference in comparison to *P. halepensis* but had significantly longer ID when compared to *P. lentiscus*. ID and IT grouping results were similar, as could be expected for two parameters measuring the same flammability component and had significant positive linear regression (Figure 2(a)). Nevertheless, the two species *L. nobilis* and *O. europaea* showed substantial difference in their ordering when comparing ID and IT (Table 1). Furthermore, same species showed a deviation from ID-IT regression line with *L. nobilis* having higher and *O. europaea* lower IT in comparison with species with similar ID (Figure 2(a)). These results suggest that in the same time period *L. nobilis* releases more energy and *O. europaea* less energy in comparison with other species involved in this study.

When comparing our ignitability results to those previously published it was observed that *P. lentiscus*, the species with the shortest ID and the highest ignitability in our study, had no positive flammability tests and was identified as species with "null flammability" (ignition frequency = 0) when 1 gram of leaf litter was tested at the epiradiator temperature of 250°C [34]. These contrary results together with the fact that in the research performed by Petriccione et al. [34] leaf litter of four out of fourteen species was considered to have "null flammability" challenge data interpretation with respect to the correlation between low ignition frequency and low flammability.

When taken into consideration that leaf litter is among the most flammable vegetation fraction [44], it is unreasonable for such a big number of species as reported by Petriccione et al. [34] to be "nonflammable"; as small, disturbed samples especially were tested and thus oxygen limitation due to dense packing of the material is highly unlikely.

Vegetation combustion does not start with the appearance of a flame; rather, materials undergo thermal degradation before the start of flaming combustion [45, 46] and complete combustion is possible without flames ever appearing [44, 47]. Therefore, interpretation of negative tests should be reconsidered and testing combinations which do not yield sufficient number of positive tests should be regarded as inconclusive as they do not guarantee low fire hazard. Lack of ignition tells us that critical mass flux of fuel vapours for piloted ignition was not reached, but it does not tell us the reason. A very small amount of highly ignitable material tested at low heat flux can lead to a lack of ignition as release of fuel vapours starts at low temperature and low release rates shortly upon positioning of the sample on the heater surface. As pyrolysis proceeds, mass flux of fuel vapours will increase, but the possibility of complete combustion before a critical mass flux of fuel vapours is reached cannot be excluded. In the same time heat resistant material tested under similar conditions can be completely combusted before a critical mass flux of fuel vapours is reached. In this case due to very slow pyrolysis and slow increase in fuel vapour mass flux. We suggest that tests with extremely low ignition frequency to be repeated under different testing conditions (e.g., higher fuel load and/or heater temperature/external heat flux) or characteristics other than the appearance of a flame are monitored as well [46, 48]. Depending on the purpose of the research and the material tested, different combinations might be appropriate with extra caution being necessary when interpreting and comparing results derived from different testing procedures.

When discussing appropriate fuel load-heater temperature (external heat flux) it should be noted that at low heat fluxes glowing combustion precedes ignition and is necessary for augmenting the radiation heating and enabling ignition [49]. Furthermore, with decreasing heat flux a bigger amount of material needs to be combusted in order for ignition to occur [50]. Thus, to avoid complete preignition combustion of a sample an increased amount of material should be considered at low heat fluxes. Epiradiator testing of leaf litter samples bigger than 1 gram is justified as leaf litter naturally occurs in layers, presenting more or less continuous fuel patches, which even in very fragmented state have mass bigger than 1 gram and a fuel load higher than that corresponding to testing 1 gram at epiradiator with 10-centimetre diameter [18, 22, 51, 52].

Fewer differences with lower level of significance ($P > 0.001$) were found between species when examining sustainability related parameters. Regardless of the parameter used *L. nobilis* and *A. unedo* were shown to be the species with the lowest sustainability, whereas *P. tobira*, *O. europaea*, and *P. halepensis* were those species with the highest sustainability. Both parameters placed *P. lentiscus* and *Q. ilex* in-between these groups, but in case of FRT *P. lentiscus* was significantly different from the species with the lowest sustainability, whereas that was the case for *Q. ilex* with regard to flame extinguish time. Low sustainability of *L. nobilis* and high sustainability of *O. europaea* confirm that the former has fast energy release rate and the latter has slow energy release rate, as previously indicated by ID-IT results.

C. siliqua was the species that changed its order the most when comparing FRT and FET results. It was among the species with the lowest sustainability regarding FRT (ranking third) and among the species with the highest sustainability (ranking sixth) based on FET.

While similarities in grouping of species based on two different sustainability factors can be explained through their positive linear regression (Figure 2(b)), we were also interested in finding an explanation for differences in species ranking. When taking into consideration that a longer heating process results in an increase of ignition mass loss and ignition delay [53] it can be expected that larger differences in ignition delay will result in larger differences in masses still available for combustion at the moment of ignition. Thus, in cases where ID is relatively long and significantly different

Table 1: Mean values, standard deviations, results of ANOVA, and Duncan post hoc test for measured flammability parameters. Different letters indicating statistically significant differences at $P < 0.05$ level of probability.

	ID (s)			IT (°C)			FET (s)			FRT (s)			MT (°C)		
	Mean	St. dev.		Mean	St. dev.		Mean	St. dev.		Mean	St. dev.		Mean	St. dev.	
A. unedo	10.064	1.076	cde	402.8	2.59	abc	41.232	3.445	ab	31.168	3.637	ab	548.6	18.34	a
C. siliqua	14.758	1.481	f	419.2	13.14	d	47.458	2.977	bc	32.700	1.568	abc	556.2	15.45	a
L. nobilis	8.584	0.694	bcd	412.4	12.80	cd	37.084	3.359	ab	28.500	3.971	a	588.2	22.95	b
O. europaea	10.894	2.438	de	397.4	1.14	ab	49.622	7.970	c	38.728	6.616	bcd	557.4	20.38	a
P. halepensis	7.072	2.268	ab	396.6	8.08	ab	46.210	7.803	bc	39.138	7.018	cd	550.0	15.73	a
P. lentiscus	5.688	1.748	a	390.8	11.69	a	44.358	6.431	abc	38.670	7.134	bcd	545.2	9.76	a
P. tobira	8.172	1.997	bcd	398.2	5.93	ab	50.304	5.901	c	42.132	6.649	d	571.6	19.81	ab
Q. ilex	11.344	1.892	e	409.4	8.41	bcd	45.600	4.591	bc	34.256	4.469	abc	561.0	20.54	a
F	12.597			5.561			3.023			3.689			2.993		
P	<0.001			<0.001			0.015			0.005			0.016		

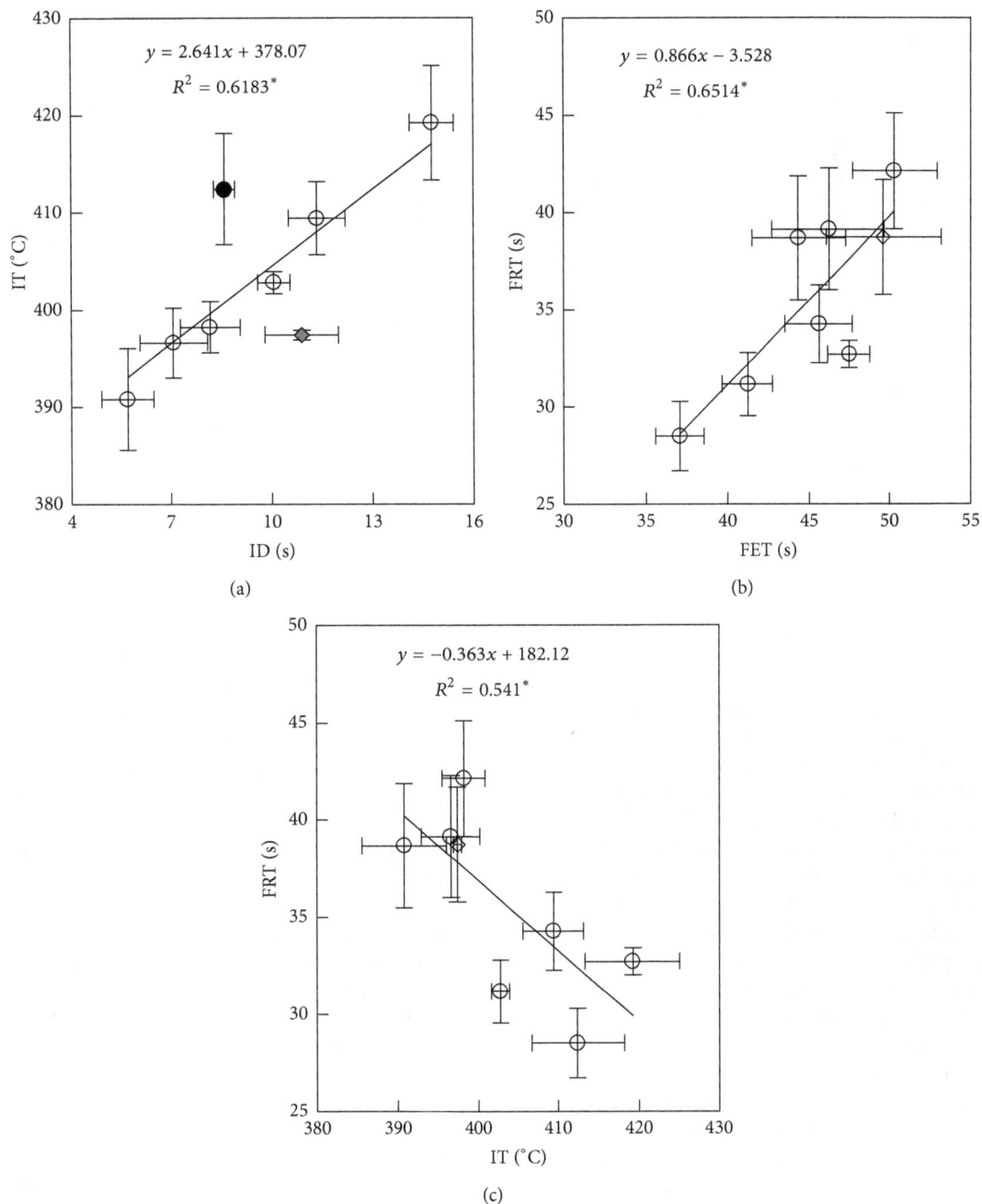

FIGURE 2: Significant liner regressions between flammability parameters. Points indicate mean values and error bars represent standard error. (a) Linear regression between ignitability related parameters ignition delay (ID) and ignition temperature (IT), with black data point indicating position of *L. nobilis* and grey data point indicating position of *O. europaea*. (b) Linear regression between sustainability related parameters flame extinguish time (FET) and flame residence time (FRT). (c) Liner regression between ignitability parameter ignition temperature (IT) and sustainability parameter flame residence time (FRT).

FRT might be distorted as a measure of sustainability and capture a portion of ignitability as well. This explanation is further reinforced by the significant negative linear regression between IT and FRT (Figure 2(c)), as mass needs to be combusted in order for temperature to increase. In our study the average ID of the species with the highest value was more than 2.5 times longer than that of the species with the shortest ID; thus differences in masses still available for

combustion at the moment of ignition are very likely even when possible variations in energy release rate are taken into consideration. Nevertheless, in studies where ignition delays were short and there were no significant differences between them [20] the duration of flaming combustion might still be an appropriate measure of sustainability. In order to fully understand the influence of prolonged ID on FRT it might be useful to consider performing two series of epiradiator

TABLE 2: Physical parameters of tested leaf litter.

	SLA (cm^2/g)	AM (g)	AA (cm^2)	MC (%)
A. unedo	64.89	0.171	11.07	11.72
C. siliqua	72.19	0.232	16.75	13.03
L. nobilis	95.37	0.237	22.60	10.83
O. europaea	50.78	0.087	4.40	10.16
P. halepensis	33.43	0.011	0.36	11.47
P. lentiscus	61.15	0.023	1.43	9.55
P. tobira	74.51	0.095	7.05	9.77
Q. ilex	60.75	0.176	10.73	10.22

TABLE 3: Significant linear regressions between physical and flammability parameters.

	Linear regression models $y = a + bx$				
y	a	b	x	r^2	P
SLA	522.121	0.587	MT	0.564	0.032
AM	41.387	−44.403	FRT	0.693	0.010
AA	40.479	−0.518	FRT	0.709	0.009
AA	393.198	1.092	IT	0.766	0.004

tests for each material: one with prolonged ignition delay in order to capture the ignitability portion of flammability and another with extremely short ignition delay in order to capture sustainability while reducing the influence of differences in masses at the moment of ignition.

Out of the measured flammability parameters MT was the one that showed the least differences between species. It identified *L. nobilis* as the species that reached the highest maximal temperature. *P. tobira* did not significantly differ from any other species regarding maximal temperature reached. All the other species had significantly lower maximal temperature than *L. nobilis*, but no significant differences were found among them.

MC was the only physical parameter measured that had no significant influence on any of the flammability parameters (Table 2.). This can be attributed to the relatively small range of values with the minimum achieved for *P. lentiscus* being 9.55% and the maximum for *C. siliqua* being 13.03%. SLA showed a positive linear regression with MT (Table 3), whereas both AM and AA showed a negative relationship with FRT. AA also influenced IT, with an increase in single particle area leading to a higher ignition temperature.

3.2. Hierarchical Cluster Analysis. Based on hierarchical cluster analysis performed on four measured flammability parameters (ID, IT, FET, and MT) five separate clusters could be distinguished (Figure 3). *P. halepensis* and *P. lentiscus* formed the first cluster. On the single parameter bases they had the highest ignitability and intermediate to high sustainability in comparison to the other species tested (Table 1). A second cluster was formed by *O. europaea* and *P. tobira*, two species with the longest FET, which followed species from the first flammability group based on their IT and had a short to intermediate ID, indicating high

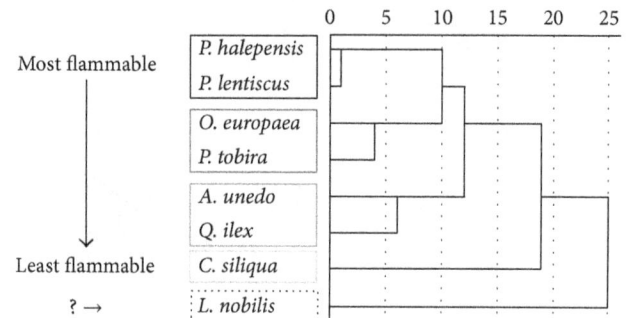

FIGURE 3: Hierarchical cluster analysis based on standardized measured flammability parameters: ignition delay (ID), ignition temperature (IT), flame extinguish time (FET), and maximal temperature (MT). Distances presented in the dendrogram are automatically rescaled by the SPSS software and thus the presented scale does not show the calculated distances.

sustainability and intermediate to high ignitability. *A. unedo* and *Q. ilex*, species from the third cluster, had intermediate to low both ignitability and sustainability. *C. siliqua* formed a fourth cluster with its relatively high sustainability when assessed by FET, the highest IT, and by far the longest ID—indicating species with remarkably low ignitability. All aforementioned species showed similar behaviour and it was possible to compare their flammability based on combined information and, if considered appropriately, attribute an overall flammability score with decreasing flammability from the first to the fourth cluster. Nevertheless, *L. nobilis* results were difficult to compare to the rest of the species. The most pronounced difference between *L. nobilis* and other species included in this study was its comparably faster energy release, which was shown through the shortest FET and FRT, highest MT, and higher IT in comparison to species with similar ID (Figure 2(a)). Combining this data we could conclude that *L. nobilis* was the least sustainable and intermediate to high ignitable species; but where does this information put *L. nobilis* in comparison to other species included in this research? Figure 3 shows that *L. nobilis* differs from any other species, but it does not tell us why and how this difference influences *L. nobilis* overall flammability score. This example raises a further question: can one single score describe the flammability of a material?

In our study we measured four flammability parameters, which could be attributed to two out of four components of vegetation flammability on small standardised samples of eight different leaf litters and were able to meaningfully compare seven species based on combined information. Ganteaume et al. [26], while performing epiradiator based testing of live vegetation, tested eight different materials and were able to attribute meaningful flammability scores to all the materials based on hierarchical cluster analysis that took into account ignition frequency, ignition delay, and flame residence time. Although the results of their analysis were meaningful and useful for the given situation, the included parameters did not account for all the flammability components.

In a different study which tested flammability of undisturbed leaf litters Ganteaumne et al. [39] also performed hierarchical cluster analysis based on four parameters (ignition frequency: IF, time to ignite: TTI, flaming duration: FD, and number of sides reached by the flame/flame spread: S). Even though we acknowledge the remarkable quality of the gathered data we disagree with its interpretation. In our opinion the clusters showed different relationship between flammability parameters which could not be transferred into single and comparable flammability score, similar to our own observations. For instance, in the mentioned work, the cluster containing *Prunus laurocerasus* L. was considered to be more flammable than that containing *Cupressus sempervirens* L. Nevertheless, when comparing values of the measured flammability parameters, only one out of four suggested this relationship between the species in question.

Hierarchical cluster analysis groups elements based on similarity of the measured parameters, maximising within group similarities and between group dissimilarities. Thus it can be expected that species in the same cluster will have similar fire behaviour and that the distance between clusters corresponds to differences in fire behaviour. Nevertheless, it does not tell us anything about the overall flammability score. Thus, attributing an overall flammability score depends strongly on the conclusions of the respective researcher and can only be done by taking into consideration the values of the input parameters on which the analysis is based.

If clustering or any other ranking based on combined information of different measured parameters is made, input parameters, limitations of the used method, and components of flammability taken into consideration should always be kept in mind when assessing output results. Furthermore, it can be expected that with every additional parameter and/or element (material, species, etc.) included into the analysis a meaningful interpretation of the combined results will be more difficult and more of the original information on the fire behaviour of material will be lost. Therefore, even though flammability assessment based on combined information can sometimes be useful, it should be performed with caution and should not be insisted upon.

4. Conclusions

The results presented here suggest that epiradiator testing with extremely low ignition frequency can occur in extremely flammable fuels if an inappropriate fuel load-heater temperature (external heat flux) combination is applied. The importance of these findings lies in the fact that wrongly identifying extremely flammable fuels as having low flammability can lead to misapprehension of vegetation characteristics influencing flammability and management practices with potentially adverse effects on the environment. It is our suggestion to treat tests with extremely low ignition frequency as inconclusive and retest materials yielding these results under different testing conditions or monitor characteristics other than the appearance of flame as well.

Performed clustering analysis showed that, even though sometimes useful, combining all the gathered vegetation flammability information into one single flammability score is not always meaningful. It should be performed with extra caution and should not always be insisted upon. In our study combining information of four measured parameters, which describe two out of four flammability components, allowed us to attribute comparable flammability scores to seven out of eight species included in this study. Nevertheless, in the case of *L. nobilis* we were unable to do so. As such, in this study, we preferred an explanatory comparison of species. Taking into consideration all the parameters in their original state allowed us to compare species based on their ignitability, sustainability, and energy release rates.

Conflict of Interests

The authors declare that there is no conflict of interests regarding the publication of this paper.

Acknowledgments

The authors wish to thank Mr. Vilim Pantlik for the technical support. They are also very grateful for the financial support provided by the "Rudolf und Helene Glaser-Stiftung."

References

[1] A. Badia, D. Saurí, R. Cerdan, and J. Llurdés, "Causality and management of forest fires in Mediterranean environments: an example from Catalonia," *Environmental Hazards*, vol. 4, no. 1, pp. 23–32, 2002.

[2] J. González-Olabarria and T. Pukkala, "Integrating fire risk considerations in landscape-level forest planning," *Forest Ecology and Management*, vol. 261, no. 2, pp. 278–287, 2011.

[3] J. Piñol, J. Terradas, and F. Lloret, "Climate warming, wildfire hazard, and wildfire occurrence in coastal Eastern Spain," *Climatic Change*, vol. 38, no. 3, pp. 345–357, 1998.

[4] R. Marzano, E. Lingua, and M. Garbarino, "Post-fire effects and short-term regeneration dynamics following highseverity crown fires in a Mediterranean forest," *IForest*, vol. 5, no. 3, pp. 93–100, 2012.

[5] R. C. Rothermel, "How to predict the spread and intensity of forest and range fires," USDA General Technical Report INT-143, Intermountain Forest and Range Experiment Station, Ogden, Utah, USA, 1983.

[6] F. Moreira, O. Viedma, M. Arianoutsou et al., "Landscape—wildfire interactions in southern Europe: implications for landscape management," *Journal of Environmental Management*, vol. 92, no. 10, pp. 2389–2402, 2011.

[7] P. M. Fernandes, "Fire-smart management of forest landscapes in the Mediterranean basin under global change," *Landscape and Urban Planning*, vol. 110, no. 1, pp. 175–182, 2013.

[8] L. G. Fogarty, *A Flammability Guide for Some Common New Zealand Native Tree and Shrub Species*, Forest Research Bulletin no. 197, Forest and Rural Fire Scientific and Technical Series Report no. 6, New Zealand Fire Service Commission and National Rural Fire Authority, Wellington, New Zealand, 2001.

[9] A. P. Dimitrakopoulos, "Thermogravimetric analysis of Mediterranean plant species," *Journal of Analytical and Applied Pyrolysis*, vol. 60, no. 2, pp. 123–130, 2001.

[10] R. H. White and W. C. Zipperer, "Testing and classification of individual plants for fire behaviour: plant selection for the

wildlandurban interface," *International Journal of Wildland Fire*, vol. 19, no. 2, pp. 213–227, 2010.

[11] S. Fitzgerald and A. J. Waldo, *Fire-Resistant Plants for Oregon Home Landscapes*, Forest Resource Note No. 6, Oregon State University Extentsion Service, Salem, Ore, USA, 2001.

[12] L. Lorenson and K. Callahan, *Firewise Plants for Western Nevada County*, Fire Safe Council of Nevada County, Grass Valley, Nev, USA, 2010.

[13] C. Lippi and M. Kuypers, *Making Your Landscape More Resistant to Wildfires*, Institute of Food and Agricultural Sciences, University of Florida, Bunnell, Fla, USA, 1998.

[14] H. E. Anderson, "Forest fuel ignitibility," *Fire Technology*, vol. 6, no. 4, pp. 312–319, 1970.

[15] A. L. Behm, A. J. Long, M. C. Monroe, C. K. Randall, W. C. Zipperer, and L. A. Hermansen-Baez, *Fire in the Wildland-Urban Interface: Preparing a Firewise Plant List for WUI Residents*, Circular 1453, Institute of Food and Agricultural Sciences, University of Florida and Southern Center for Wildland-Urban Interface Research and Information, Southern Research Station, USDA Forest Service, Gainesville, Fla, USA, 2011.

[16] A. M. Gill and P. H. R. Moor, "Ignitability of leaves of Australian plants," A Contract Report to the Australian Flora Fundation, Centre for Plant Biodiversity Research, CSIRO Plant Industry, Canberra, Australia, 1996.

[17] C. D. Grant, W. A. Loneragan, J. M. Koch, and D. T. Bell, "Fuel characteristics, vegetation structure and fire behaviour of 11–15 year-old rehabilitated bauxite mines in Western Australia," *Australian Forestry*, vol. 60, no. 3, pp. 147–157, 1997.

[18] D. P. Ellair and W. J. Platt, "Fuel composition influences fire characteristics and understorey hardwoods in pine savanna," *Journal of Ecology*, vol. 101, no. 1, pp. 192–201, 2013.

[19] F. Morandini, Y. Perez-Ramirez, V. Tihay, P. Santoni, and T. Barboni, "Radiant, convective and heat release characterization of vegetation fire," *International Journal of Thermal Sciences*, vol. 70, pp. 83–91, 2013.

[20] E. Ormeño, B. Céspedes, I. A. Sánchez et al., "The relationship between terpenes and flammability of leaf litter," *Forest Ecology and Management*, vol. 257, no. 2, pp. 471–482, 2009.

[21] P. Santoni, F. Morandini, and T. Barboni, "Determination of fireline intensity by oxygen consumption calorimetry," *Journal of Thermal Analysis and Calorimetry*, vol. 104, no. 3, pp. 1005–1015, 2011.

[22] A. Ganteaume, C. Lampin-Maillet, M. Guijarro et al., "Spot fires: fuel bed flammability and capability of firebrands to ignite fuel beds," *International Journal of Wildland Fire*, vol. 18, no. 8, pp. 951–969, 2009.

[23] J. D. Engstrom, J. K. Butler, S. G. Smith, L. L. Baxter, T. H. Fletcher, and D. R. Weise, "Ignition behavior of live California chaparral leaves," *Combustion Science and Technology*, vol. 176, no. 9, pp. 1577–1591, 2004.

[24] P. M. Fernandes and M. G. Cruz, "Plant flammability experiments offer limited insight into vegetation-fire dynamics interactions," *New Phytologist*, vol. 194, no. 3, pp. 606–609, 2012.

[25] J. G. Pausas and B. Moreira, "Flammability as a biological concept," *New Phytologist*, vol. 194, no. 3, pp. 610–613, 2012.

[26] A. Ganteaume, M. Jappiot, C. Lampin, M. Guijarro, and C. Hernando, "Flammability of some ornamental species in wildland-urban interfaces in southeastern France: laboratory assessment at particle level," *Environmental Management*, vol. 52, no. 2, pp. 467–480, 2013.

[27] S. Liodakis, T. Kakardakis, S. Tzortzakou, and V. Tsapara, "How to measure the particle ignitability of forest species by TG and LOI," *Thermochimica Acta*, vol. 477, no. 1-2, pp. 16–20, 2008.

[28] G. A. Alessio, J. Peñuelas, M. De Lillis, and J. Llusià, "Implications of foliar terpene content and hydration on leaf flammability of *Quercus ilex* and *Pinus halepensis*," *Plant Biology*, vol. 10, no. 1, pp. 123–128, 2008.

[29] D. Alexandrian and E. Rigolot, "Sensibilité du pin d'Alep à l'incendie," *Forêt Méditerranéenne*, vol. 13, no. 3, pp. 185–198, 1992.

[30] P. Delabraze and J. Valette, ""The use of fire in silviculture ," in *Proceedings of the Symposium on Dynamics and Management of Mediterranean-Type Ecosystems; June 22–26, 1981; San Diego, CA, pp. 475–482*," General Technical Report PSW-GTR-058, USDA Forest Service, Pacific Southwest Forest and Range Experiment Station, Berkeley, Calif, USA, 1982.

[31] A. Ganteaume, M. Guijarro, M. Jappiot et al., "Laboratory characterization of firebrands involved in spot fires," *Annals of Forest Science*, vol. 68, no. 3, pp. 531–541, 2011.

[32] M. Hachmi, A. Sesbou, H. Benjelloun, N. El Handouz, and F. Bouanane, "A simple technique to estimate the flammability index of moroccan forest fuels," *Journal of Combustion*, vol. 2011, Article ID 263531, 11 pages, 2011.

[33] C. Moro, "Inflammabilité, matériels et méthodes," INRA's methodology, Internal report, Version 26/10/94, modified 27/01/04, French Mediterranean Forest Research Institute, INRA, Avignon, France, 2004.

[34] M. Petriccione, C. Moro, and F. A. Rutigliano, "Preliminary studies on litter flammability in Mediterranean region," in *Proceedings of the 5th International Conference on Forest Fire Research*, D. X. Viegas, Ed., ADAI, Figueira da Foz , Portugal, 2006.

[35] S. Saura-Mas, S. Paula, J. G. Pausas, and F. Lloret, "Fuel loading and flammability in the Mediterranean Basin woody species with different post-fire regenerative strategies," *International Journal of Wildland Fire*, vol. 19, no. 6, pp. 783–794, 2010.

[36] M. Blackhall, E. Raffaele, and T. T. Veblen, "Is foliar flammability of woody species related to time since fire and herbivory in Northwest Patagonia, Argentina?" *Journal of Vegetation Science*, vol. 23, no. 5, pp. 931–941, 2012.

[37] J. G. Pausas, G. A. Alessio, B. Moreira, and G. Corcobado, "Fires enhance flammability in Ulex parviflorus," *New Phytologist*, vol. 193, no. 1, pp. 18–23, 2012.

[38] J. Valette, "Inflammabilités des espèces forestières méditerranéennes. Conséquences sur la combustibilité des formations forestières," *Revue Forestière Française*, vol. 42, pp. 76–92, 1990.

[39] A. Ganteaume, M. Jappiot, and C. Lampin, "Assessing the flammability of surface fuels beneath ornamental vegetation in wildland-urban interfaces in Provence (south-eastern France)," *International Journal of Wildland Fire*, vol. 22, no. 3, pp. 333–342, 2013.

[40] J. K. Kreye, J. M. Varner, J. K. Hiers, and J. Mola, "Toward a mechanism for eastern North American forest mesophication: differential litter drying across 17 species," *Ecological Applications*, vol. 23, no. 8, pp. 1976–1986, 2013.

[41] H. E. Anderson, "Predicting equilibrium moisture content of some foliar forest litter in the northern Rocky Mountains," USDA Forest Services Research Paper INT-429, Intermountain Research Station, Ogden, Utah, USA, 1990.

[42] A. Ganteaume, J. Marielle, L. Corinne, C. Thomas, and B. Laurent, "Effects of vegetation type and fire regime on flammability

of undisturbed litter in Southeastern France," *Forest Ecology and Management*, vol. 261, no. 12, pp. 2223–2231, 2011.

[43] N. C. Turner, "Techniques and experimental approaches for the measurement of plant water status," *Plant and Soil*, vol. 58, no. 1–3, pp. 339–366, 1981.

[44] L. F. DeBano, D. G. Neary, and P. F. Ffolliott, *Fire's Effects on Ecosystems*, John Wiley & Sons, New York, NY, USA, 1998.

[45] L. Trabaud, "Inflammabilité et combustibilité des principales espèces des garrigues de la région méditerranéennes," *Oecologia Plantarum*, vol. 11, no. 2, pp. 117–136, 1976.

[46] M. Bernard and N. Nimour, "Inflammabilité des vegetaux Méditerranéens et feux de forêts: rôle de l'eau sur exothermicité de leur réaction," *Science Technique Technologie*, vol. 26, pp. 24–30, 1993.

[47] P. Delabraze and J. Valette, "Inflammabilité et combustibilité de la végétation forestière méditerranéenne," *Revue Forestière Française*, vol. 26, pp. 171–177, 1974.

[48] G. Massari and A. Leopaldi, "Leaf flammability in Mediterranean species," *Plant Biosystems*, vol. 132, no. 1, pp. 29–38, 1998.

[49] J. G. Quintiere, *Fundamentals of Fire Phenomena*, John Wiley & Sons, Chichester, UK, 2006.

[50] A. Atreya, "Ignition of fires," *Philosophical Transactions of the Royal Society A: Mathematical, Physical and Engineering Sciences*, vol. 356, no. 1748, pp. 2787–2813, 1998.

[51] T. Curt, A. Schaffhauser, L. Borgniet et al., "Litter flammability in oak woodlands and shrublands of southeastern France," *Forest Ecology and Management*, vol. 261, no. 12, pp. 2214–2222, 2011.

[52] A. L. Behm, M. L. Duryea, A. J. Long, and W. C. Zipperer, "Flammability of native understory species in pine flatwood and hardwood hammock ecosystems and implications for the wildland-urban interface," *International Journal of Wildland Fire*, vol. 13, no. 3, pp. 355–365, 2004.

[53] J. L. Consalvi, F. Nmira, A. Fuentes, P. Mindykowski, and B. Porterie, "Numerical study of piloted ignition of forest fuel layer," *Proceedings of the Combustion Institute*, vol. 33, no. 2, pp. 2641–2648, 2011.

11

Consistent Conditional Moment Closure Modelling of a Lifted Turbulent Jet Flame Using the Presumed β-PDF Approach

Ahmad El Sayed and Roydon A. Fraser

Department of Mechanical and Mechatronics Engineering, University of Waterloo, 200 University Avenue West, Waterloo, ON, Canada N2L 3G1

Correspondence should be addressed to Ahmad El Sayed; aselsaye@uwaterloo.ca

Academic Editor: Ishwar K. Puri

A lifted H_2/N_2 turbulent jet flame issuing into a vitiated coflow is investigated using the conditional moment closure. The conditional velocity (CV) and the conditional scalar dissipation rate (CSDR) submodels are chosen such that they are fully consistent with the moments of the presumed β probability density function (PDF). The CV is modelled using the PDF-gradient diffusion model. Two CSDR submodels based on the double integration of the homogeneous and inhomogeneous mixture fraction PDF transport equations are implemented. The effect of CSDR modelling is investigated over a range of coflow temperatures (T_c) and the stabilisation mechanism is determined from the analysis of the transport budgets and the history of radical build-up ahead of the stabilisation height. For all T_c, the balance between chemistry, axial convection, and micromixing, and the absence of axial diffusion upstream of the stabilisation height indicate that the flame is stabilized by autoignition. This conclusion is confirmed from the rapid build-up of HO_2 ahead of H, O, and OH. The inhomogeneous CSDR modelling yields higher dissipation levels at the most reactive mixture fraction, which results in longer ignition delays and larger liftoff heights. The effect of the spurious sources arising from homogeneous modelling is found to be small but nonnegligible, mostly notably within the flame zone.

1. Introduction

The investigation of stabilisation mechanisms in lifted flames is an active area in combustion research. A substantial number of experimental and numerical studies have been dedicated to the understanding and analysis of the process by which lifted flames are stabilised. Several theories have been proposed in the literature. Vanquickenborne and van Tiggelen [1] suggest the premixed flame propagation stabilisation theory. Their experimental findings indicate that the fuel and the oxidiser are premixed ahead of the base of lifted diffusion flames and that stabilisation takes place at stoichiometric locations where the local mean velocity and the turbulent burning velocity of a stoichiometric premixed flame are equal. Peters and Williams [2] suggest that stabilisation is achieved by the local quenching of diffusion flamelets due to excessive straining. Cabra et al. [3] conduct experimental measurements for a lifted turbulent H_2/N_2 jet flame issuing into a coaxial vitiated coflow consisting of the hot combustion products of a lean premixed H_2/air flame. The coflow conditions allow for the possibility of stabilisation by means of autoignition. It is postulated that autoignition takes place as inert fuel parcels are convected downstream and become well mixed with the hot coflow. However, premixed flame propagation remains another possible stabilisation mechanism.

The Cabra flame is attractive for the validation of combustion models due to its simple configuration and well-defined boundary conditions. It is extensively investigated using PDF methods [3–6]. Masri et al. [4] use the composition PDF approach coupled with the k-ε turbulence model and conclude that flame is largely controlled by chemical kinetics. They identify key reactions and show that the predicted liftoff height and the composition at the base of the flame are highly sensitive to the rate parameters of these reactions. Gordon et al. [6] use the same computational methodology in order to determine the underlying stabilisation mechanism. They develop two numerical indicators for the distinction between stabilisation via autoignition and premixed flame propagation. These indicators are based on the analysis of the transport budgets of convection, diffusion, and reaction and the history of radical build-up ahead of the reaction zone. A convection-reaction balance without significant contribution from axial diffusion, along with an early and rapid build-up of HO_2 ahead of the remaining radicals, indicate that stabilisation occurs via autoignition, whereas an axial diffusion-convection balance with negligible reaction accompanied by a simultaneous build-up of all radicals points to stabilisation by means of premixed flame propagation. With the aid of these indicators, Gordon et al. [6] conclude that autoignition is the controlling mechanism for a wide range of coflow temperatures. Cao et al. [5] employ the joint velocity-turbulence frequency-composition PDF method and perform a detailed parametric study using several mixing models and chemical kinetic mechanisms, over a wide range of jet and coflow conditions. They show that although the modelling of mixing affects the prediction of liftoff height, the flame is mainly controlled by chemical kinetics. They also demonstrate that the liftoff height increases with increasing jet and coflow velocities and decreases with increasing jet and coflow temperatures. In all of the PDF studies described above, autoignition was found to be the controlling stabilisation mechanism. In the context of CMC, Stankovic and Merci [7] and Navarro-Martinez and Kronenburg [8] investigate this flame using Large Eddy Simulation (LES) with a first-order closure for the conditional chemical source. The main difference between the two LES-CMC studies lies in the treatment of the CSDR. The first employs the Amplitude Mapping Closure (AMC) [9], which produces symmetric profiles in mixture fraction space. The second employs a different conditioning technique described in [10], which allows for skewed profiles with the possibility of local minima and maxima in mixture fraction space. In both sets of calculations the coflow temperature ranges between 1020 and 1080 K. The first authors conclude that stabilisation is controlled by autoignition for all coflow temperature. The second authors draw the same conclusions for high coflow temperatures; however, they report a transition to flame stabilization by premixed flame propagation as the coflow temperature is reduced. The different conclusions are most likely attributed to the treatment of the CSDR and, to a lesser extent, to the employed numerical methods, discretization schemes, and grid resolutions in physical and mixture fraction spaces. Patwardhan et al. [11] also use the first-order CMC to study this flame. They employ the standard k-ε model for the computation of the turbulence and mixing fields. To a certain degree, their conclusions in regard to the controlling stabilisation mechanism are in line with those of Navarro-Martinez and Kronenburg [8]. Premixed flame propagations are found to be the controlling stabilisation mechanism of the lifted flame for low and intermediate coflow temperatures (1025 and 1045 K), whereas the flame is stabilised by autoignition for high coflow temperatures (1080 K). To close the CMC equations, they employ the β-distribution for the PDF, the linear velocity model [12] for the CV, and the AMC for the CSDR. The assumption of homogeneous Gaussian turbulence is inherent to the adopted CV and CSDR closures. The

linear velocity model is exact only if the velocity and the mixture fraction are jointly Gaussian [13, 14], which is in general not the case for inhomogeneous flows. Though being consistent with the first moment of the PDF, this CV model is inconsistent with the second moment [15]. As for the AMC, this closure is a particular solution of the homogeneous PDF transport equation, wherein the employed mapping function is Gaussian. Patwardhan et al. [11] use a β-distribution instead. Further, the fact that the AMC does not account for inhomogeneous effects leads to an inconsistent CMC implementation. When such closure is employed, the integration of the PDF-weighted CMC equations over the mixture fractions space yields the unconditional equations in addition to extra spurious source terms [14, 16]. This term may lead to inaccurate results and misleading conclusions, depending on the magnitudes of these sources. As such, the CV and CSDR closures employed in [11] do not provide a fully consistent CMC implementation.

In this work, the lifted hydrogen jet flame of Cabra et al. [3] is revisited and investigated using the two-dimensional first-order CMC and a modified version of the k-ε model. The conditional submodels required for the closure of the CMC equations are selected such that they are fully consistent with the moments of the presumed PDF. Additionally, a transport equation is solved for the Favre-averaged scalar dissipation rate rather employing the traditional "equal time scales" algebraic approach. The simulations are set up such that the settings of Patwardhan et al. [11] are reproduced as closely as possible. This study is aimed at determining (1) the effect of micromixing on liftoff via the modelling of the CSDR, (2) the response of the flame to changes in the coflow temperature, (3) the controlling stabilisation mechanism, (4) the influence of the spurious sources, and (5) the impact of chemical kinetics.

2. Experimental Configuration

The flame under investigation is the lifted turbulent jet flame of Cabra et al. [3]. A schematic of the experimental setup is shown in Figure 1 and the flow conditions are summarized in Table 1. The burner consists of a central H_2/N_2 jet surrounded with a coaxial flow composed of the hot combustion products of a lean premixed H_2/Air flame stabilized on a perforated disk. The perforated disk is surrounded by an exit collar for the purpose of delaying the entrainment of ambient air into the coflow. The nozzle exit is placed above the surface of the disk in order to allow the fuel to exit into the coflow with a uniform composition. The experimental criterion for the determination of the liftoff height (H_{exp}) is taken as the first location where the mass fraction of OH reaches 600 ppm. The measured height is equal to 10 nozzle diameters.

3. Mathematical Model

3.1. Conditional Moment Closure. The modelling of the turbulence-chemistry interactions is carried out using the first-order CMC [14]. In CMC, reactive scalars such as the species mass fractions and the temperature are conditionally

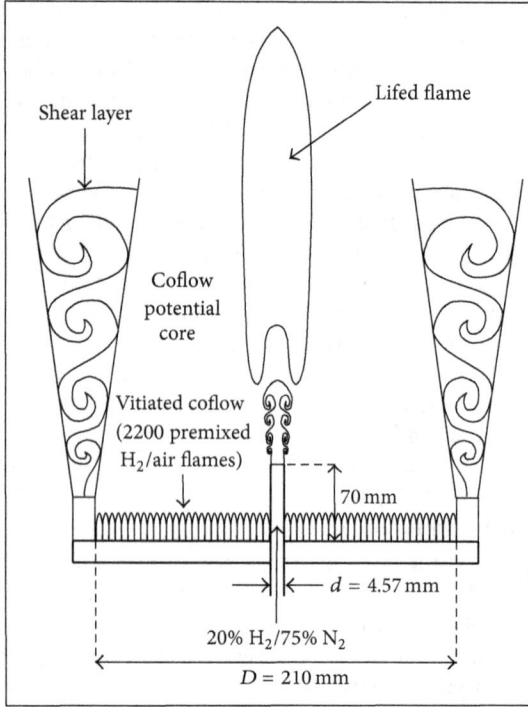

FIGURE 1: The lifted H_2/N_2 flame. Reproduced from [3].

TABLE 1: Conditions of the jet and coflow in the lifted flame of Cabra et al. [3].

	Jet		Coflow
d [mm]	4.57	D [mm]	210
T_j [K]	305	T_c [K]	1045
U_j [m/s]	107	U_c [m/s]	3.5
X_{H_2}	0.2537	X_{H_2}	0.0005
X_{O_2}	0.0021	X_{O_2}	0.1474
X_{N_2}	0.7427	X_{N_2}	0.7534
X_{H_2O}	0.0015	X_{H_2O}	0.0989
ξ_{st}	0.474	ϕ	0.25

Nomenclature: d: jet diameter; D: coflow diameter; T: temperature; U: velocity; X: mole fraction; ξ: mixture fraction; ϕ: equivalence ratio. j: jet; c: coflow; st: stoichiometric.

where

$$e_{Q_\kappa} = \frac{\partial}{\partial x_i}\left(\rho D \frac{\partial Q_\kappa}{\partial x_i}\right) + \left\langle \rho D \frac{\partial \xi}{\partial x_i} \frac{\partial}{\partial \xi}\left(\frac{\partial Q_\kappa}{\partial x_i}\right) \mid \eta \right\rangle, \quad (4)$$

$$e_{y_\kappa''} = -\left\langle \rho \frac{\partial y_\kappa''}{\partial t} + \rho u_i'' \frac{\partial y_\kappa''}{\partial x_i} - \frac{\partial}{\partial x_i}\left(\rho D \frac{\partial y_\kappa''}{\partial x_i}\right) \mid \eta \right\rangle, \quad (5)$$

$$\chi = 2D \frac{\partial \xi}{\partial x_i} \frac{\partial \xi}{\partial x_i} \quad (6)$$

is the scalar dissipation rate. The species and mixture fraction diffusivities in (3)–(6) are assumed to be equal to the molecular diffusivity, D, and the Lewis number is set to unity. In (3), $\langle u_i \mid \eta \rangle$ is the CV, $\langle \chi \mid \eta \rangle$ is the CSDR, and $\langle \dot{\omega}_\kappa \mid \eta \rangle$ is the conditional chemical source. The expressions of e_Q (see (4)) and $e_{y_\kappa''}$ (see (5)) are unclosed. They are modelled using the primary closure hypothesis. Klimenko and Bilger [14] show that, given finite Schmidt numbers, all the terms of e_Q scale as the inverse of the Reynolds number (Re). Therefore, this term is neglected in high Re applications. By neglecting the conditional density and diffusivity fluctuations, they further show that and $e_{y_\kappa''}$ reduces to

$$e_{y_\kappa''} = -\frac{\langle \rho \mid \eta \rangle}{\bar{\rho}\tilde{P}(\eta)} \frac{\partial}{\partial x_i}\left[\bar{\rho}\left\langle u_i'' y_\kappa'' \mid \eta \right\rangle \tilde{P}(\eta)\right], \quad (7)$$

where $\tilde{P}(\eta)$ is the Favre PDF of mixture fraction and $\langle u_i'' y_\kappa'' \mid \eta \rangle$ is the conditional turbulent flux. Using these approximations, the conditional species transport equation takes the form:

$$\frac{\partial Q_\kappa}{\partial t} = -\langle u_i \mid \eta \rangle \frac{\partial Q_\kappa}{\partial x_i} - \frac{1}{\bar{\rho}\tilde{P}(\eta)} \frac{\partial}{\partial x_i}\left[\bar{\rho}\left\langle u_i'' y_\kappa'' \mid \eta \right\rangle \tilde{P}(\eta)\right]$$
$$+ \frac{\langle \chi \mid \eta \rangle}{2} \frac{\partial^2 Q_\kappa}{\partial \eta^2} + \frac{\langle \dot{\omega}_\kappa \mid \eta \rangle}{\langle \rho \eta \rangle}. \quad (8)$$

The left-hand side term of (8) is the rate of change of the conditional species mass fraction. The first term on the

averaged with respect to the mixture fraction, ξ, and their conditional transport equations are solved. The conditional average of the mass fraction of a species κ is defined as

$$Q_\kappa(\eta, \mathbf{x}, t) = \left\langle Y_\kappa(\mathbf{x}, t) \mid \xi(\mathbf{x}, t) = \eta \right\rangle = \left\langle Y_\kappa \mid \eta \right\rangle, \quad (1)$$

where Y_κ is the mass fraction of the species, $\langle \cdot \mid \cdot \rangle$ denotes the conditional average of the quantity to the left of the vertical bar subject to the quantity to its right, and η is a sample variable of ξ, with $0 \leq \eta \leq 1$. Following the decomposition approach [14], Y_κ is written as the sum of its conditional average Q_κ and a fluctuation y_κ'':

$$Y_\kappa(\mathbf{x}, t) = Q_\kappa(\eta, \mathbf{x}, t) + y_\kappa''(\eta, \mathbf{x}, t) \quad (2)$$

such that $\langle y_\kappa'' \mid \eta \rangle = 0$. Substitution of (2) into the transport equation of species κ, followed by the conditional averaging of the resulting expression, leads to [14]:

$$\frac{\partial Q_\kappa}{\partial t} = -\langle u_i \mid \eta \rangle \frac{\partial Q_\kappa}{\partial x_i} + \frac{\langle \chi \mid \eta \rangle}{2} \frac{\partial^2 Q_\kappa}{\partial \eta^2} + \frac{\langle \dot{\omega}_\kappa \mid \eta \rangle}{\langle \rho \mid \eta \rangle}$$
$$+ \frac{e_{Q_\kappa} + e_{y_\kappa''}}{\langle \rho \mid \eta \rangle}, \quad (3)$$

right-hand side (r.h.s.) represents transport by means of convection, the second corresponds to transport by conditional turbulent fluxes (diffusion in physical space), the third accounts for micromixing (diffusion in mixture fraction space), and the fourth is the chemical source. The conditional temperature equation is derived in a similar fashion and is given by

$$\frac{\partial Q_T}{\partial t} = \underbrace{-\langle u_i \mid \eta \rangle \frac{\partial Q_T}{\partial x_i}}_{T_{C,x} \text{ and } T_{C,y}} \underbrace{- \frac{1}{\bar{\rho}\tilde{P}(\eta)} \frac{\partial}{\partial x_i} \left[\bar{\rho} \langle u_i'' T'' \mid \eta \rangle \tilde{P}(\eta) \right]}_{T_{D,x} \text{ and } T_{D,y}}$$

$$\underbrace{+ \frac{\langle \chi \mid \eta \rangle}{2} \left\{ \frac{\partial^2 Q_T}{\partial \eta^2} + \frac{1}{\langle c_p \mid \eta \rangle} \right.}_{} $$
$$\left. \times \left[\frac{\partial \langle c_p \mid \eta \rangle}{\partial \eta} + \sum_{\kappa=1}^{N} \left(\langle c_{p,\kappa} \mid \eta \rangle \frac{\partial Q_\kappa}{\partial \eta} \right) \right] \right.$$
$$\left. \times \frac{\partial Q_T}{\partial \eta} \right\}$$
$$\underbrace{}_{T_{MM}}$$

$$+ \frac{1}{\langle c_p \mid \eta \rangle} \left\langle \frac{1}{\rho} \frac{\partial p}{\partial t} \mid \eta \right\rangle$$

$$\underbrace{- \frac{\langle \dot{\omega}_h \mid \eta \rangle}{\langle \rho \mid \eta \rangle \langle c_p \mid \eta \rangle}}_{T_{CS}} \underbrace{- \frac{\langle \dot{\omega}_r \mid \eta \rangle}{\langle \rho \mid \eta \rangle \langle c_p \mid \eta \rangle}}_{T_{RS}},$$

$$(9)$$

where $\langle \dot{\omega}_h \mid \eta \rangle = \sum_{\kappa=1}^{N} \langle h_\kappa \mid \eta \rangle \langle \dot{\omega}_\kappa \mid \eta \rangle$ is the chemical source term, N is the total number of species in the mixture, and $\langle \dot{\omega}_r \mid \eta \rangle$ is the conditional radiative source term. The unsteady pressure term in (9) is neglected due to the fact that the investigated flame is open to the atmosphere, and hence temporal pressure changes are expected very small.

The unconditional (Favre) averages of the reactive scalars are calculated by integrating their conditional values weighted by $\tilde{P}(\eta)$ over the mixture fraction space:

$$\tilde{\phi} = \int_0^1 \langle \phi \mid \eta \rangle \tilde{P}(\eta) \, d\eta, \quad \phi = Y_\kappa, T. \quad (10)$$

3.2. Conditional Submodels.
The quantities $\tilde{P}(\eta)$, $\langle u_i'' y_\kappa'' \mid \eta \rangle$, $\langle u_i'' T'' \mid \eta \rangle$, $\langle u_i \mid \eta \rangle$, $\langle \chi \mid \eta \rangle$, $\langle \dot{\omega}_\kappa \mid \eta \rangle$, and $\langle \dot{\omega}_r \mid \eta \rangle$ in (8) and (9) are unclosed and require further modelling. The submodels employed in this study are discussed in this section.

3.2.1. Probability Density Function.
The PDF is presumed using the β-distribution. The Favre β-PDF is given by

$$\tilde{P}(\eta; v, w) = \frac{\eta^{v-1}(1-\eta)^{w-1}}{B(v, w)}, \quad (11)$$

where the parameters v and w are related to the mean and variance of the mixture fraction by $v = \gamma \tilde{\xi}$ and $w = \gamma(1 - \tilde{\xi})$ with $\gamma = [\tilde{\xi}(1 - \tilde{\xi})/\widetilde{\xi''^2} - 1] \geq 0$, and $B(v, w) = \int_0^1 \eta^{v-1}(1 - \eta)^{w-1} d\eta$ is the beta function.

3.2.2. Conditional Turbulent Fluxes.
The conditional turbulent fluxes are modelled using the gradient diffusion assumption:

$$\langle \phi'' u_i'' \mid \eta \rangle = -D_t \frac{\partial \langle \phi \mid \eta \rangle}{\partial x_i}, \quad \phi = Y_\kappa, T, \quad (12)$$

where

$$D_t = \frac{C_\mu}{Sc_t} \frac{\tilde{k}^2}{\tilde{\varepsilon}} \quad (13)$$

is the turbulent diffusivity. The constant C_μ is equal to 0.09 and the turbulent Schmidt number Sc_t is set to 0.7 following Jones and Whitelaw [17]. Richardson and Mastorakos [18] suggest adding a correction to (12) in order to account for counter-gradient transport effects. When applied to a lifted flame, they conclude that the inclusion of counter-gradient effects leads to a slight increase in liftoff height and to a decrease in flame thickness. This extension is not employed in the current study. However, it is worth investigating in future studies.

3.2.3. Conditional Velocity.
The PDF-gradient model proposed by Pope [19] is employed to model the CV fluctuations. The resulting expression for the CV takes the form:

$$\langle u_i \mid \eta \rangle = \tilde{u} - \frac{D_t}{\tilde{P}(\eta)} \frac{\partial \tilde{P}(\eta)}{\partial x_i}$$

$$= \tilde{u}_i - D_t \left\{ \frac{\partial \ln \left[\tilde{P}(\eta) \right]}{\partial \tilde{\xi}} \frac{\partial \tilde{\xi}}{\partial x_i} + \frac{\partial \ln \left[\tilde{P}(\eta) \right]}{\partial \widetilde{\xi''^2}} \frac{\partial \widetilde{\xi''^2}}{\partial x_i} \right\}. \quad (14)$$

As $\tilde{P}(\eta)$ tends to zero, $\langle u_i \mid \eta \rangle$ diverges to $\pm\infty$, depending on the sing of the gradient of $\tilde{P}(\eta)$. This behavior is of minor importance, because low-probability events have a negligible effect on mixing [20]. One important feature of the PDF-gradient model is its consistency with both the first and second moments of the mixture fraction [15, 20]. This does not hold for the commonly used linear model [12], which is only consistent with the first moment [15].

3.2.4. Conditional Scalar Dissipation Rate

Girimaji's Model. Girimaji [21] derives a model for $\langle \chi \mid \eta \rangle$ by doubly integrating the homogeneous mixture fraction PDF transport equation. His formulation is based on the observation that a presumed β-PDF is capable of accurately characterizing the evolution of the scalar PDF over all stages of two-scalar, constant-density mixing in statistically stationary, isotropic turbulence. The model is given by:

$$\langle \chi \eta \rangle = -2\tilde{\chi} \frac{\tilde{\xi}(1 - \tilde{\xi})}{\left(\widetilde{\xi''^2} \right)^2} \frac{I(\eta)}{\tilde{P}(\eta)}, \quad (15)$$

where $\tilde{\chi}$ is the Favre-averaged scalar dissipation rate (discussed in more detail in Section 3.3) and $I(\eta)$ is given by the integral:

$$
\begin{aligned}
I(\eta) &= \int_0^\eta \left\{ \tilde{\xi} \left[\ln \eta' - \int_0^1 \ln \eta'' \tilde{P}(\eta'') \, d\eta'' \right] + (1 - \tilde{\xi}) \right. \\
&\quad \left. \times \left[\ln(1 - \eta') - \int_0^1 \ln(1 - \eta'') \tilde{P}(\eta'') \, d\eta'' \right] \right\} \\
&\quad \times \tilde{P}(\eta')(\eta - \eta') \, d\eta'.
\end{aligned}
\tag{16}
$$

As mentioned above, Girimaji's model is valid for homogeneous turbulence. Therefore, its usage in the CMC of inhomogeneous flows is inconsistent and results in spurious source terms.

Mortensen's Model. In a similar fashion, Mortensen [22] derives an expression for $\langle \chi \mid \eta \rangle$ by doubly integrating the mixture fraction PDF transport equation. However, in his derivation, the inhomogeneous terms are retained. The CV fluctuations appearing in the transport equation are modelled using the PDF gradient model of Pope [19] and the PDF is presumed using a functional form described by the vector of mixture fraction moments. When the β-PDF is employed, this model reads as follows [23]:

$$
\begin{aligned}
\langle \chi \mid \eta \rangle &= \frac{2}{\tilde{P}(\eta)} \left\{ - \frac{\partial II(\eta)}{\partial \widetilde{\xi''^2}} S_{\widetilde{\xi''^2}} \right. \\
&\quad + D_t \left[\frac{\partial^2 II(\eta)}{\partial \widetilde{\xi''^2} \partial \widetilde{\xi''^2}} \frac{\partial \widetilde{\xi''^2}}{\partial x_i} \frac{\partial \widetilde{\xi''^2}}{\partial x_i} + \frac{\partial^2 II(\eta)}{\partial \tilde{\xi} \partial \tilde{\xi}} \right. \\
&\quad \left. \left. \times \frac{\partial \tilde{\xi}}{\partial x_i} \frac{\partial \tilde{\xi}}{\partial x_i} + 2 \frac{\partial^2 II(\eta)}{\partial \tilde{\xi} \partial \widetilde{\xi''^2}} \frac{\partial \tilde{\xi}}{\partial x_i} \frac{\partial \widetilde{\xi''^2}}{\partial x_i} \right] \right\},
\end{aligned}
\tag{17}
$$

where

$$
II(\eta; v, w) = \int_0^\eta I(\eta'; v, w) \, d\eta',
\tag{18}
$$

$$
I(\eta; v, w) = \int_0^\eta \tilde{P}(\eta; v, w) \, d\eta = \frac{\int_0^\eta \eta^{v-1}(1-\eta)^{w-1} \, d\eta}{B(v, w)}
\tag{19}
$$

is the incomplete β-function. By applying some identities and integral properties of the incomplete β-function, Mortensen [23] shows that $II(\eta; v, w)$ simplifies to

$$
II(\eta; v, w) = (\eta - \tilde{\xi}) I(\eta; v, w) + \widetilde{\xi''^2} \tilde{P}(\eta; v+1, w+1).
\tag{20}
$$

The term $S_{\widetilde{\xi''^2}}$ in (17) represents the source term of the mixture fraction variance transport equation (discussed in more detail in Section 3.3). It is given by

$$
S_{\widetilde{\xi''^2}} = 2D_t \frac{\partial \tilde{\xi}}{\partial x_i} \frac{\partial \tilde{\xi}}{\partial x_i} - \tilde{\chi}.
\tag{21}
$$

Mortensen's model ensures a fully consistent CMC implementation as it accounts for the inhomogeneous terms of the PDF transport equation and employs a consistent closure for the CV fluctuations. As such, this model does not yield any spurious sources. It is important to note that Girimaji's model is the exact equivalent of the homogeneous portion of Mortensen's, $\langle \chi \mid \eta \rangle = 2[\tilde{\chi}/\tilde{P}(\eta)] \partial II(\eta)/\partial \widetilde{\xi''^2}$.

3.2.5. Conditional Chemical Source. The conditional chemical source term, $\langle \dot{\omega}_\kappa \mid \eta \rangle$, is modelled using a first-order closure. In this closure, it is assumed that the conditional fluctuations about the conditional averages of the reactive scalars are small. Accordingly, $\langle \dot{\omega}_\kappa \mid \eta \rangle$ is modelled as a function of the conditional density, mass fractions, and temperature:

$$
\begin{aligned}
\langle \dot{\omega}_\kappa(\rho, T, \mathbf{Y}) \mid \eta \rangle &\approx \langle \dot{\omega}_\kappa(\langle \rho \mid \eta \rangle, \langle T \mid \eta \rangle, \langle \mathbf{Y} \mid \eta \rangle) \mid \eta \rangle \\
&= \dot{\omega}_\kappa(\langle \rho \mid \eta \rangle, Q_T, \mathbf{Q}),
\end{aligned}
\tag{22}
$$

where $\mathbf{Y} = \{ Y_\kappa \mid \kappa = 1, 2, \ldots, N \}$ and $\mathbf{Q} = \langle \mathbf{Y} \mid \eta \rangle$. This closure has been successfully applied in the CMC of lifted flames [7, 8, 11, 24].

3.2.6. Conditional Radiative Source. The conditional radiative source term is modelled using the optically thin assumption with H_2O being the predominantly participating species. To a first order approximation,

$$
\begin{aligned}
\langle \dot{\omega}_r \mid \eta \rangle &= 4\sigma p_{H_2O} a_{p,H_2O} \left(\langle T \mid \eta \rangle^4 - T_b^4 \right) \\
&= 4\sigma p_{H_2O} a_{p,H_2O} \left(Q_T^4 - T_b^4 \right),
\end{aligned}
\tag{23}
$$

where $\sigma = 5.669 \times 10^8 \, \mathrm{W/m^2 K^4}$ is the Stefan-Boltzmann constant, p_{H_2O} and a_{p,H_2O} are the partial pressure, and the Planck mean absorption coefficient of H_2O, respectively, and T_b is the background temperature (300 K). A Curve fit for a_{p,H_2O} is obtained from [25].

3.3. Turbulent Flow Field Calculations. The Favre averages of the velocity, mixture fraction mean and variance, turbulence kinetic energy, and turbulence eddy dissipation are obtained from an inert flow calculation using the commercial Computational Fluid Dynamics (CFD) software FLUENT [26]. This approach was successfully applied in the CMC calculations of lifted flames performed by Kim and Mastorakos [24]. They justify the validity of this simplification by the fact that the flow remains frozen until the stabilisation height. Nevertheless, they note that density changes caused by the flame are non-negligible and may affect the flow field.

A two-dimensional axisymmetric computational domain is employed to perform the inert flow calculations. The domain extends 55 nozzle diameters above the nozzle exit in the axial direction and 15 diameters in the radial direction with the origin of the domain placed at the centre of the nozzle. The thickness of the wall of the nozzle is neglected [3, 27]. A quadratic unstructured mesh is generated using ANSYS ICEM CFD [28]. It consists of $1,000 \times 265$ nodes (axial $\times$ radial). The k-ε model is employed to perform the calculations. Default model constants are employed, except for $C_{\varepsilon 1}$ which is modified from the standard value of 1.44 to 1.6 in order to improve the prediction of the spreading rate of the jet [29]. The Semi-Implicit Method for Pressure-Linked Equations (SIMPLE) algorithm is used for pressure-velocity coupling. The PREssure STaggering Option (PRESTO) pressure interpolation scheme is employed to compute the pressure at cell faces. Spatial discretization is performed using the second-order upwind scheme. The inlet velocity boundary condition is specified using the 1/7 power law with the centreline velocity set to 5/4 times the experimentally estimated mean value of 107 m/s as in [27]. The coflow velocity is assumed to be uniform and is set to 3.5 m/s. The turbulence intensity is taken to be 5% at both the inlet and the coflow. The boundary conditions of the temperature and species mass fractions are specified following Table 1. Transport equations for the mean and variance of the mixture fraction are added to the solver. These are given by

$$\bar{\rho}\frac{\partial \tilde{\xi}}{\partial t} + \bar{\rho}\tilde{u}_i\frac{\partial \tilde{\xi}}{\partial x_i} = -\frac{\partial \left(\overline{\rho u_j'' \xi''}\right)}{\partial x_i}, \tag{24}$$

$$\bar{\rho}\frac{\partial \widetilde{\xi''^2}}{\partial t} + \bar{\rho}\tilde{u}_i\frac{\partial \widetilde{\xi''^2}}{\partial x_i} = -\frac{\partial \left(\overline{\rho u_j'' \xi''^2}\right)}{\partial x_i} - 2\overline{\rho u_i'' \xi''}\frac{\partial \tilde{\xi}}{\partial x_i} - \bar{\rho}\tilde{\chi}. \tag{25}$$

The gradient diffusion assumption is employed to close the turbulent fluxes as $\overline{u_i'' \xi''} = -D_t\partial\tilde{\xi}/\partial x_i$ and $\overline{u_i'' \xi''^2} = -D_t\partial\widetilde{\xi''^2}/\partial x_i$. The last term on the r.h.s. of (25), $\tilde{\chi}$, is the Favre-averaged scalar dissipation rate. It is a standard practice to model this term as

$$\tilde{\chi} = C_\chi\frac{\tilde{\varepsilon}}{\tilde{k}}\widetilde{\xi''^2}, \tag{26}$$

where C_χ is a constant set to 2. In this algebraic model, it is assumed that the flow time $(\tilde{k}/\tilde{\varepsilon})$ is proportional to the time scale of scalar turbulence $(\widetilde{\xi''^2}/\tilde{\chi})$, with C_χ being the proportionality constant (the time scale ratio). This modelling approach was adopted in the early stages of this study. However, it was abandoned later because the calculation yielded very short liftoff heights for moderate and low T_c and nearly attached flames for sufficiently high T_c. This behavior was not observed in the lifted flames simulations of Patwardhan et al. [11] and Kim and Mastorakos [24].

Alternatively, the transport equation derived for $\tilde{\chi}$ by Jones and Musonge [30] is solved. This equation reads

$$\bar{\rho}\frac{\partial \tilde{\chi}}{\partial t} + \bar{\rho}\tilde{u}_i\frac{\partial \tilde{\chi}}{\partial x_i} = \frac{\partial}{\partial x_j}\left(C_s\overline{\rho u_j'' u_l''}\frac{\tilde{k}}{\tilde{\varepsilon}}\frac{\partial \tilde{\chi}}{\partial x_l}\right)$$
$$- C_1\bar{\rho}\frac{\tilde{\chi}^2}{2\widetilde{\xi''^2}} - C_2\bar{\rho}\frac{\tilde{\varepsilon}}{2\tilde{k}}\tilde{\chi} - 2C_3\bar{\rho}\frac{\tilde{\varepsilon}}{\tilde{k}}\overline{u_i'' \xi''}\frac{\partial \tilde{\xi}}{\partial x_i}$$
$$- C_4\bar{\rho}\frac{\tilde{\chi}}{\tilde{k}}\overline{u_i'' u_j''}\frac{\partial \tilde{u}_i}{\partial x_j}, \tag{27}$$

where $C_s = 0.22$, $C_1 = 2$, $C_2 = 2(C_{\varepsilon 2} - 1) = 1.84$, $C_3 = 1.70$, and $C_4 = C_{\varepsilon 1} = 1.6$ are model constants. The turbulent flux $\overline{u_i'' \xi''}$ is modelled using the gradient diffusion assumption. For consistency with the Boussinesq hypothesis, the production of turbulence kinetic energy appearing in the last term on the r.h.s of (27) is modelled as

$$-\bar{\rho}\overline{u_i'' u_j''}\frac{\partial \tilde{u}_i}{\partial x_j} = \mu_t S^2, \tag{28}$$

where μ_t is the turbulent viscosity and S is the modulus of the mean strain rate tensor. The tensor $C_s\overline{\rho u_i'' u_l''}\tilde{k}/\tilde{\varepsilon}$ appearing in the first term on the r.h.s. of (27) represents the anisotropic diffusivity. Initial trials showed numerical instabilities and suffered from severe convergence issues when accounting for anisotropic effects. Therefore, for simplicity, the diffusivity was assumed to be isotropic and standard modelling was employed (μ_t/Sc_t used instead of $C_s\overline{\rho u_j'' u_l''}\tilde{k}/\tilde{\varepsilon}$). The inclusion of the transport equation of $\tilde{\chi}$ (see (27)) in the context of CMC was first employed in the counterflow flames calculations of Kim and Mastorakos [31]. The authors solve an additional transport equation for $\overline{u_i'' \xi''}$ rather than resorting to gradient diffusion modelling.

3.4. The CMC Implementation. A smaller physical domain is employed in the CMC calculations for computational efficiency. The chosen domain is 30 nozzle diameters in length and 5 diameters in width. The mesh consists of 70×40 (axial $\times$ radial) nonuniformly distributed nodes, with the mesh density being highest near the experimentally measured stabilization height $(10\,d)$ [3]. The mixture fraction grid consists of 80 points. These resolutions ensure grid independence. The CMC equations are discretized using the finite difference method. The first-order derivative appearing in the convective term is discretized using the second-order upwind difference scheme with the kappa flux limiter of Koren [32], following Patwardhan et al. [11]. The second-order derivatives in physical and mixture fraction spaces are discretized using the second-order central difference scheme. The steady-state solution is obtained from the transient CMC transport equations, (8) and (9), by relaxation of the time step, for which a value of 10^{-5} s is used. A three-step fractional method (Strang operator splitting [33]) is implemented in order to treat the stiff chemical source term separately.

The first fractional step accounts for convection and diffusion in physical space over the first half of the time step. The second step handles the chemical source and micromixing in mixture fractions space over the whole time step. The third and last step accounts for convection and diffusion in physical space over the second half of the time step. Each step uses the solution of the previous one as initial conditions. Splitting errors are negligible due to the usage of a small time step. The Ordinary Differential Equation (ODE) solver VODPK [34–36] is used to solve the system of equations. The ODEs in the first and third steps are nonstiff and hence solved with the implicit Adams method. Conversely, the ODEs in the second step are stiff and therefore treated with the Backward Differentiation Formulas (BDF). The boundary conditions for Q_κ and Q_T at the inlet and the coflow are specified using the inert mixing solution. The composition of the fuel and coflow streams is obtained from the experiments [3]. Zero-gradient boundary conditions are specified at the remaining boundaries. The solution is initialized by igniting some nodes above the experimentally measured stabilization height with a steady laminar flamelet computed with a strain rate of $1000 \, \mathrm{s}^{-1}$. The Inert mixing solution is set elsewhere. The Favre-averaged velocity, mixture fraction mean and variance, turbulence kinetic energy, and turbulence eddy dissipation are transferred from FLUENT to the CMC solver by applying bilinear interpolation. This interpolation scheme is accurate enough given the high resolution of the CFD grid. The integrals appearing in Girimaji's model are computed using the quadrature integration package QUADPACK [37]. In Mortensen's model, the incomplete beta-function $I(\eta; v, w)$ (see (19)) is calculated using the method of continued fractions [38] and the partial derivatives of $II(\eta; v, w)$ (see (20)) with respect to the mean and variance of the mixture fraction are computed using Ridders' method of polynomial extrapolations [38, 39]. Due to the division by the PDF in the PDF-gradient and Mortensen's models, numerical instabilities may arise at low probabilities. To resolve this issue, $\langle u_i \mid \eta \rangle$ is set equal to its unconditional value, $\tilde{u}_i$, and $\langle \chi \mid \eta \rangle$ is set to zero when η falls outside $\tilde{\xi} \pm 7(\widetilde{\xi''^2})^{1/2}$. Additionally, in the event where Moretnsen's model yields unphysical negative values, $\langle \chi \mid \eta \rangle$ is set to zero.

The chemical kinetics mechanisms developed by Mueller et al. [40], Li et al. [41], Burke et al. [42], and Ó Conaire et al. [43] are considered in this study. The corresponding numbers of species and reactions are summarised in Table 2. All four mechanisms have a comparable level of detail, with those of Mueller and Burke being the least and most detailed, respectively. The Mueller mechanism is the most extensively used throughout this work. It is provided in Table 3.

4. Results and Discussion

4.1. Flow Field Results. The radial profiles of the Favre-averaged axial velocity are compared to the experimental measurements of Kent [44] in Figure 2(a). The predictions are in very good agreement with the experimental data up to $x/d = 11$. The measurements are slightly underpredicted at $x/d = 14$ for $y/d > 0.5$. No velocity measurements are

TABLE 2: Number of species and reactions in the considered chemical kinetics mechanisms.

Mechanism	Species	Reactions	Reference
Mueller	9	21	[40]
Li	13	25	[41]
Burke	13	27	[42]
Ó Conaire	10	21	[43]

TABLE 3: Chemical kinetics mechanism of Mueller et al. [40].

R1	$H + O_2 \rightleftharpoons O + OH$
R2	$O + H_2 \rightleftharpoons H + OH$
R3	$H_2 + OH \rightleftharpoons H_2O + H$
R4	$O + H_2O \rightleftharpoons OH + OH$
R5	$H_2 + M \rightleftharpoons H + H + M$
R6	$O + O + M \rightleftharpoons O_2 + M$
R7	$O + H + M \rightleftharpoons OH + M$
R8	$H + OH + M \rightleftharpoons H_2O + M$
R9	$H + O_2 \,(+M) \rightleftharpoons HO_2 \,(+M)$
R10	$HO_2 + H \rightleftharpoons H_2 + O_2$
R11	$HO_2 + H \rightleftharpoons OH + OH$
R12	$HO_2 + O \rightleftharpoons O_2 + OH$
R13	$HO_2 + OH \rightleftharpoons H_2O + O_2$
R14	$HO_2 + HO_2 \rightleftharpoons H_2O_2 + O_2{}^\dagger$
R15	$H_2O_2 \,(+M) \rightleftharpoons OH + OH \,(+M)$
R16	$H_2O_2 + H \rightleftharpoons H_2O + OH$
R17	$H_2O_2 + H \rightleftharpoons HO_2 + H_2$
R18	$H_2O_2 + O \rightleftharpoons OH + HO_2$
R19	$H_2O_2 + OH \rightleftharpoons HO_2 + H_2O^\dagger$

†R14 and R19 are expressed as the sum of two rate expressions. Hence, the mechanism has a total of 21 reactions.

available at $x/d = 26$. The profile at this location is provided for completeness. The radial profiles of the Favre-averaged mixture fraction mean are displayed in Figure 2(b) along with the experimental data of Cabra et al. [3]. The predictions are in very good agreement with the experiments at $x/d = 1, 8, 9, 10,$ and 11. Slight underpredictions are obtained at $x/d = 14$ and 26 for $0 \leq y/d \leq 1$. Figure 2(c) shows a comparison of the radial Favre-averaged mixture fraction variance profiles with the experimental measurements. A very good agreement is obtained at all axial locations except at $x/d = 26$. The discrepancies at this location are attributed to the underprediction of the mixture fraction mean (bottom pane in Figure 2(b)). The results obtained with the algebraic modelling of $\tilde{\chi}$ (see (26)) tend to be underpredictive near the centreline and overpredictive around the peaks of the experimental data (not shown here). Therefore, it is concluded that the alternative approach of incorporating an additional transport equation for $\tilde{\chi}$ (see (27)) provides more accurate mixture fraction variance predictions. Overall, the results presented in Figure 2 show that the inert flow calculations employed in this study provide reliable flow and mixing fields for the CMC calculations.

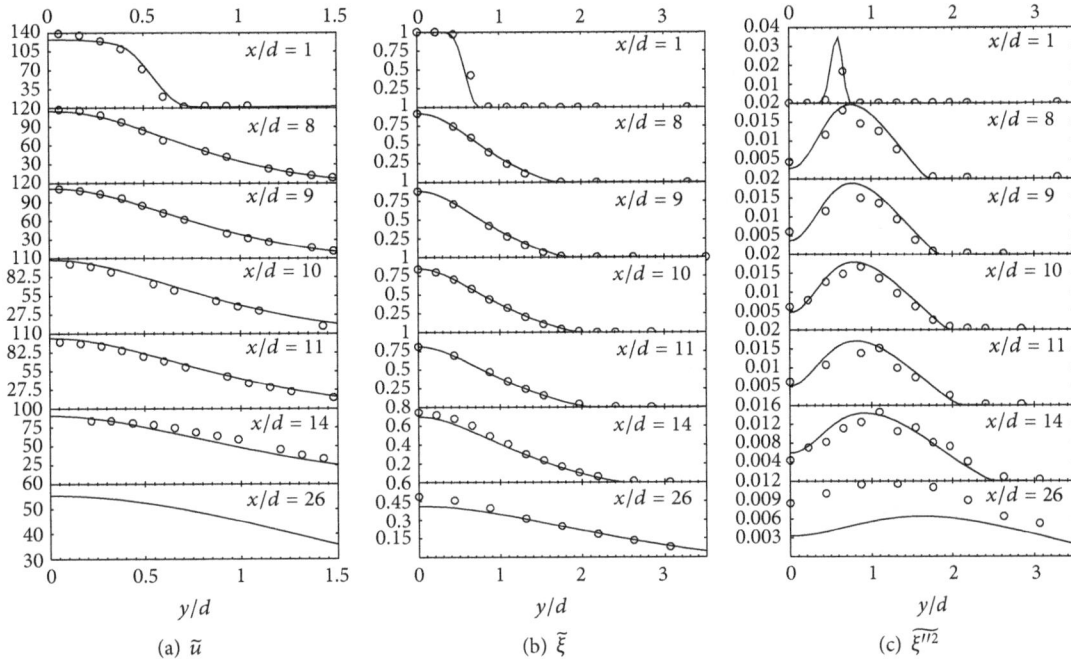

FIGURE 2: Radial profiles of (a) $\tilde{u}$, (b) $\tilde{\xi}$, and (c) $\widetilde{\xi''^2}$. Lines, numerical results; symbols, experimental data ($\tilde{u}$ obtained from [44] and $\tilde{\xi}$ and $\widetilde{\xi''^2}$ obtained from [3]). No velocity measurements are available at $x/d = 26$.

The centreline evolution of the time scale ratio, C_χ, is shown in Figure 3. Here, C_χ is computed using (26) with $\tilde{\chi}$ obtained from the solution of (27). In the vicinity of the nozzle, C_χ is well above the commonly used constant value of 2. Away from the nozzle, C_χ decays gradually and tends asymptotically to 2.76. This suggests that the usage of (26) with $C_\chi = 2$ would generally underpredict the centreline scalar dissipation levels, especially near the nozzle area. The trend obtained in Figure 3 is in line with the experimental observations of Markides [45] who performs scalar dissipation measurements in autoignitive nitrogen-diluted fuel jets injected into coflowing heated air, a configuration similar to the one under investigation. His results show that the centreline values of C_χ decay from high levels near the injector to a constant value around 2 at downstream locations. Accordingly, in a subsequent CMC study in the same context, Markides et al. [46] tune C_χ in (26) in the vicinity of the injector in order to emulate the decay of the time scale ratio and set it to 2 downstream. In the absence of scalar dissipation measurements, as in the present case, such treatment is *ad hoc* and tedious: C_χ must be tuned by trial and error until a reasonable agreement between the predicted and measured $\widetilde{\xi''^2}$ profiles is achieved, and most likely, a nonconstant functional form for C_χ may be necessary. This process is easily avoided by means of solving (27).

4.2. Effect of the CSDR Modelling for Various Coflow Temperatures.
The effect of the modelling of the CSDR is investigated using the homogeneous model of Girimaji [21] and the inhomogeneous model of Mortensen [22]. The reader is reminded that the latter model degenerates to the

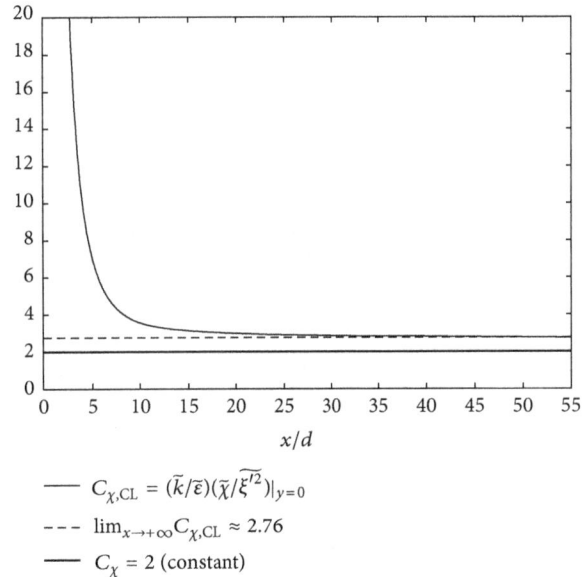

FIGURE 3: Evolution of the time scale ratio along the centreline.

former when the inhomogeneous terms in (17) are discarded. Calculations are performed for three coflow temperatures. In addition to the experimentally reported value $T_c = 1045$ K [3], two other temperatures are investigated: $T_c = 1030$ and 1060 K. Given the experimental uncertainty of 3% in the temperature measurements [3], these two values lie within the experimental error. The chemical kinetics mechanism of Mueller et al. [40] is employed throughout this section.

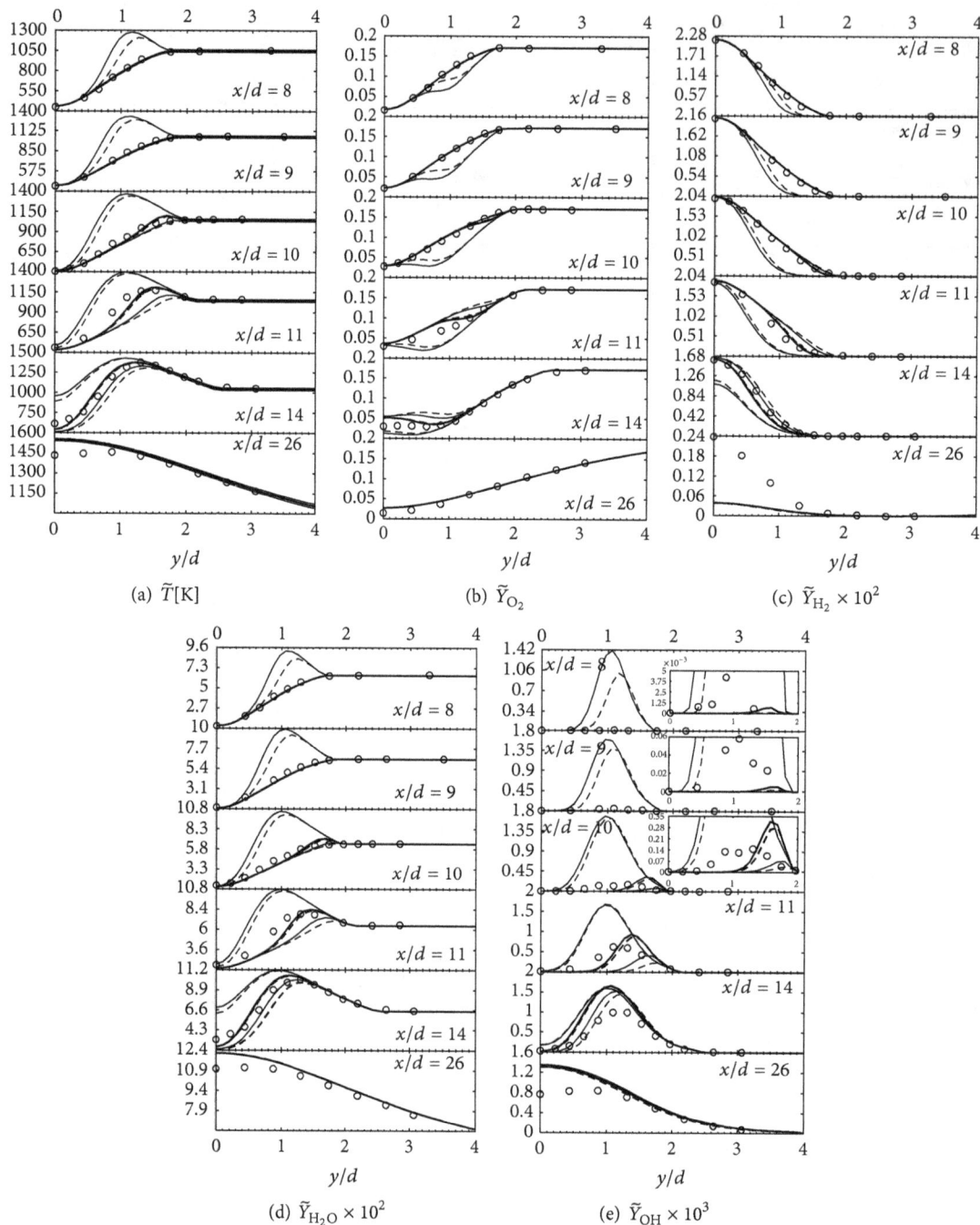

FIGURE 4: Radial profiles of (a) $\tilde{T}$, (b) $\tilde{Y}_{O_2}$, (c) $\tilde{Y}_{H_2}$ ($\times 10^2$), (d) $\tilde{Y}_{H_2O}$ ($\times 10^2$) (e), and $\tilde{Y}_{OH}$ ($\times 10^3$) using Girimaji's and Mortensen's models for the modelling of the CSDR. In (a)–(e): solid lines, Girimaji's model; dashed lines, Mortensen's model; symbols, experimental data [3]. Colours: black (thin), T_c = 1030 K; black (thick), T_c = 1045 K; grey, T_c = 1060 K.

4.2.1. Results in Physical Space. The radial profiles of the Favre-averaged reactive scalars are shown in Figures 4(a)–4(e) at the axial locations x/d = 8, 9, 10, 11, 14, and 26. For T_c = 1045 K (thick black lines), the results obtained using the two CSDR models are almost identical, except for some minor differences at x/d = 10 and 11. When T_c is decreased to 1030 K (thin black lines), the predictions remain in good agreement with the experiments up to x/d = 10 and the two CSDR

models yield very similar results. Farther downstream, the temperature and products mass fractions are underpredicted and the reactants mass fractions are overpredicted, most notably at x/d = 11 and 14. The influence of the CSDR modelling becomes more apparent at these axial locations. Overall, the results remain in reasonable agreement with the experiments. When T_c is increased to 1060 K (grey lines), the CSDR models yield distinct results for x/d ≤ 14. Using

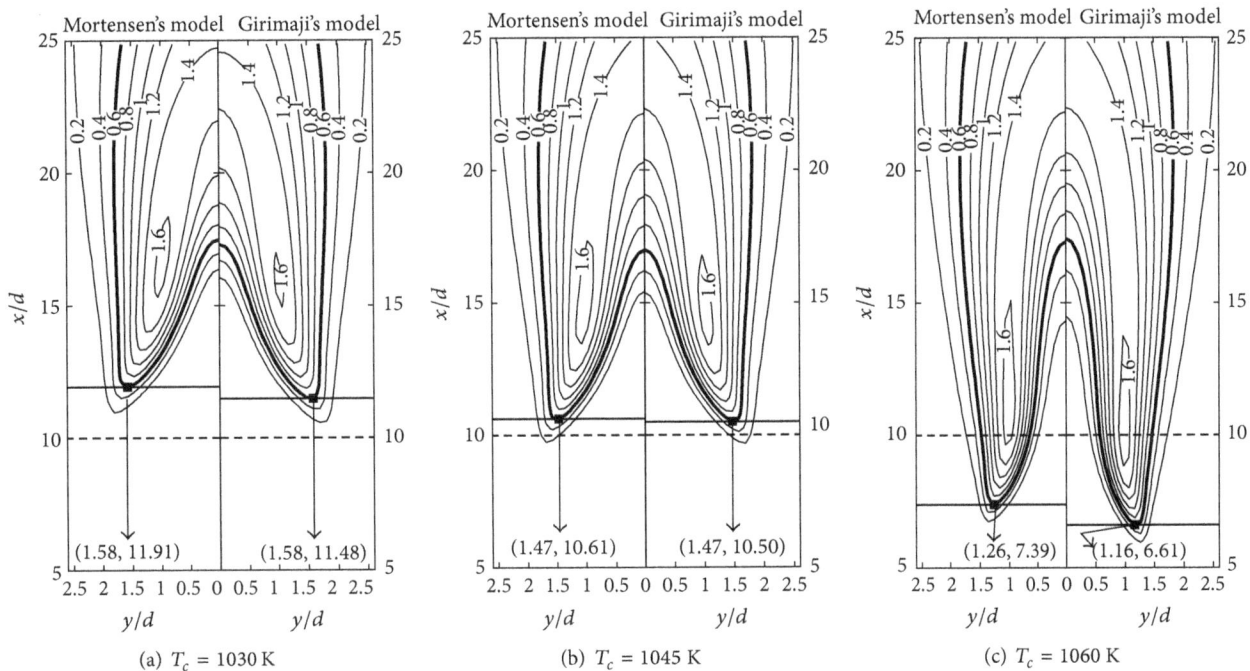

FIGURE 5: Contours of $\widetilde{Y}_{OH}$ ($\times 10^3$): (a) $T_c = 1030\,\text{K}$, (b) $T_c = 1045\,\text{K}$, and (c) $T_c = 1060\,\text{K}$ obtained using Mortensen's model (left pane) and Girimaji's model (right pane). The thick contour corresponds to 600 ppm. The solid and dashed horizontal lines correspond to the numerical and experimental liftoff heights, respectively.

both models, the temperature and products mass fractions are grossly overpredicted and the reactants mass fractions are underpredicted. All three coflow temperatures result in very similar profiles at $x/d = 26$, irrespective of the CSDR model. As a general observation in Figure 4, the usage of Girimaji's model always yields lower temperatures and product mass fractions and higher reactant mass fractions when compared to Mortensen's, which leads to a shorter liftoff height. This is attributed to the fact that Girimaji's model results in earlier ignition delays and hence shorter ignition kernels for all coflow temperatures. This phenomenon will be discussed shortly in more detail. Before doing so, the liftoff heights obtained from the cases considered in Figure 4 are first presented. The contours of $\widetilde{Y}_{OH}$ ($\times 10^{-3}$) are shown in Figure 5(a) for $T_c = 1030\,\text{K}$, Figure 5(b) for $T_c = 1045\,\text{K}$, and Figure 5(c) for $T_c = 1060\,\text{K}$. In each subfigure, the left and right panes correspond to the results obtained with Mortensen's and Girimaji's models, respectively. As shown in Figure 5(a), stabilization takes place at slightly distinct locations due to the very small $\widetilde{Y}_{OH}$ differentials (see the radial profiles in Figure 4(e) for $T_c = 1030\,\text{K}$). Mortensen's model (left) results in $H/d = 11.91$ (19.1% in excess of H_{exp}/d) while Girimaji's model (right) gives $H/d = 11.48$ (14.8% in excess of H_{exp}/d). The same scenario is observed in Figure 5(b) where $T_c = 1045\,\text{K}$. However, in this case, the location of stabilisation is better predicted. Mortensen's model (left) yields $H/d = 10.61$ (6.1% in excess of H_{exp}/d) while Girimaji's model (right) results in $H/d = 10.50$ (5% in excess of H_{exp}/d). The advantage of the inclusion of the effects of inhomogeneity Mortensen's model is judged to be insignificant for these

two coflow temperatures. Conversely, the differences in the stabilisation locations are more substantial for $T_c = 1060\,\text{K}$, as displayed in Figure 5(c). This is due to the much larger $\widetilde{Y}_{OH}$ differentials (see the radial profiles in Figure 4(e) for $T_c = 1060\,\text{K}$). Mortensen's model (left) yields $H/d = 7.39$ (26.1% below H_{exp}/d) while Girimaji's model (right) gives $H/d = 6.61$ (33.9% below H_{exp}/d). As for the radial location of stabilisation (the radial distance from the stabilisation point to the centreline normalised d, denoted here as W/d), the values obtained using the two CSDR models for $T_c = 1030\,\text{K}$ and $1045\,\text{K}$ are the same (1.58 and 1.47, resp.). As for $T_c = 1060\,\text{K}$, the two models yield slightly different values (Mortensen's, 1.26; Girimaji's, 1.16). It is obvious from the above results that the liftoff height becomes smaller and that flame base becomes narrower as T_c is decreased. This is due to the fact that at lower T_c, the mixture can autoignite in areas closer to the nozzle which are characterised by high scalar dissipation rate levels. This behaviour will be addressed in more detail in Section 4.2.3.

4.2.2. Stabilisation Mechanism. The numerical indicators developed by Gordon et al. [6] for the distinction between stabilisation by autoignition and premixed flame propagation are employed here to determine the underlying controlling mechanism. For this purpose, the transport budget of the temperature and the history of radical build-up ahead of the stabilisation height are analysed.

Budgets in Mixture Fraction Space. Figure 6 shows the transport budget of the steady-state conditional temperature

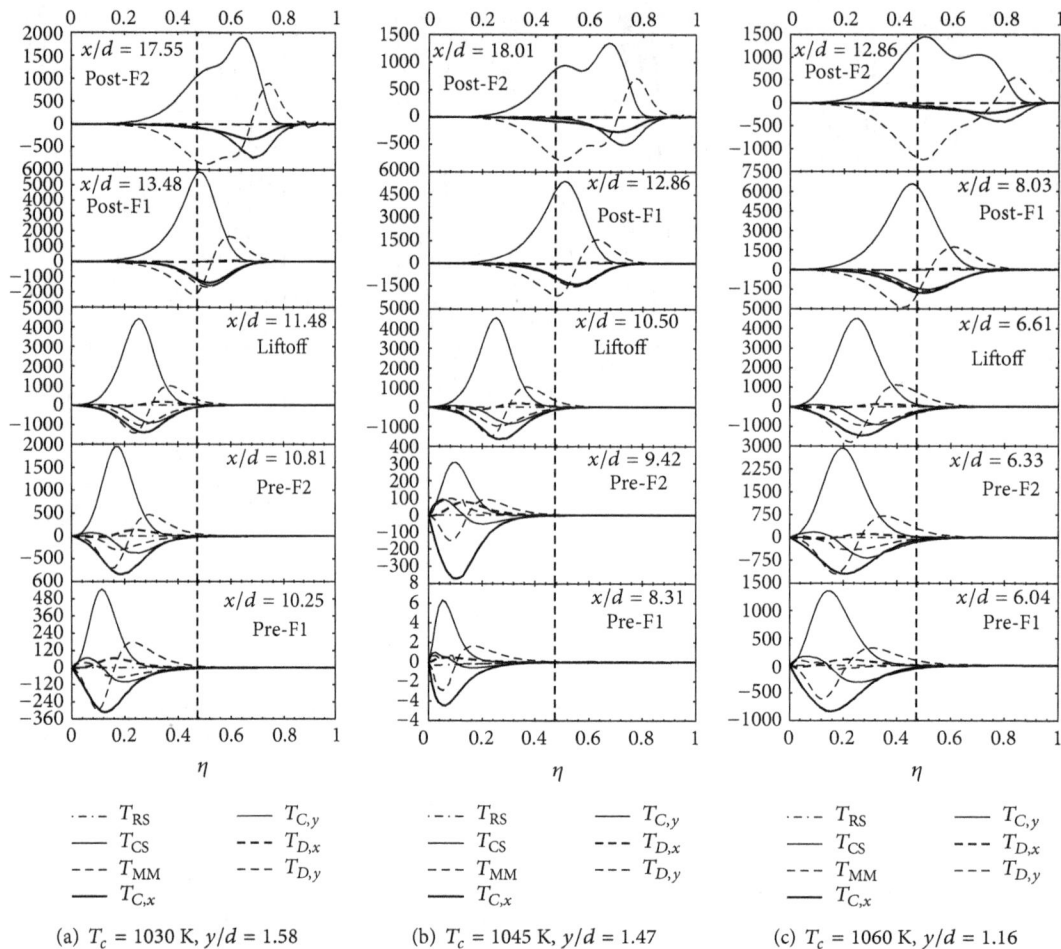

FIGURE 6: Transport budget of the steady-state Q_T equation (r.h.s. terms of (9)) at several axial locations around the stabilisation height: (a) $T_c = 1030$ K ($y/d = 1.58$), (b) $T_c = 1045$ K ($y/d = 1.47$), and (c) $T_c = 1060$ K ($y/d = 1.16$). T_{RS}, radiative source; T_{CS}, chemical source; T_{MM}, micromixing; $T_{C,x}$, axial convection; $T_{C,y}$, radial convection; $T_{D,x}$, axial diffusion; $T_{D,y}$, radial diffusion. The vertical dashed line corresponds to the location of the stoichiometric mixture fraction. All terms are scaled down by a factor of 10^3 and the units are K/s. The CSDR is modelled using Girimaji's model.

equation, that is the individual contributions of the r.h.s. terms of (9). Only the results obtained using Girimaji's model are presented here. Mortensen's model produces similar trends. For each T_c, the budgets are displayed at several axial locations around the corresponding liftoff height. The radial locations are held fixed and set to the W/d value obtained with each T_c. As such, the chosen coordinates cover the preflame, liftoff, and postflame regions. Before analysing the balance in these budgets, the axial evolution of the chemical source term, T_{CS}, is first examined. As shown in Figures 6(a), 6(b), and 6(c), common trends are observed in the evolution of this term for all three coflow temperatures. Below the stabilisation height (first two locations in the preflame regions, Pre-F1 and Pre-F2), as x/d increases, T_{CS} shifts from very lean to less lean mixtures and its amplitude increases dramatically. At liftoff (middle panes), T_{CS} peaks at $\eta \approx 0.24$ with a significantly higher amplitude. Within approximately two to three nozzle diameters downstream the stabilisation height (first locations in the postflame regions, Post-F1) the

peak of T_{CS} increases further and occurs around the stoichiometric mixture fraction. Further downstream (second locations in the postflame regions, Post-F2), the magnitude of T_c decreases substantially from the Post-F1 locations. The reaction zone becomes wider and presents two peaks. The first occurs around the stoichiometric mixture fraction and the second in rich mixtures. The second peak is most likely attributed to the propagation of a rich reaction zone towards stoichiometric mixtures [24]. All of the above trends are consistent with the direct numerical simulation Yoo et al. [47]. As in the evolution of T_{CS}, common trends are observed in the axial variation of the remaining terms contributing to the r.h.s. of (9). In the preflame region (locations Pre-F1 and Pre-F2), there is a clear balance in lean mixtures between T_{CS}, the axial convection term, $T_{C,x}$, and micromixing, T_{MM}. The axial and radial diffusion terms, $T_{D,x}$ and $T_{D,y}$, and the radial convection term, $T_{C,y}$, are nonnegligible but have little contribution to the overall budget. This balance suggests that stabilisation by premixed flame propagation wherein $T_{D,x}$

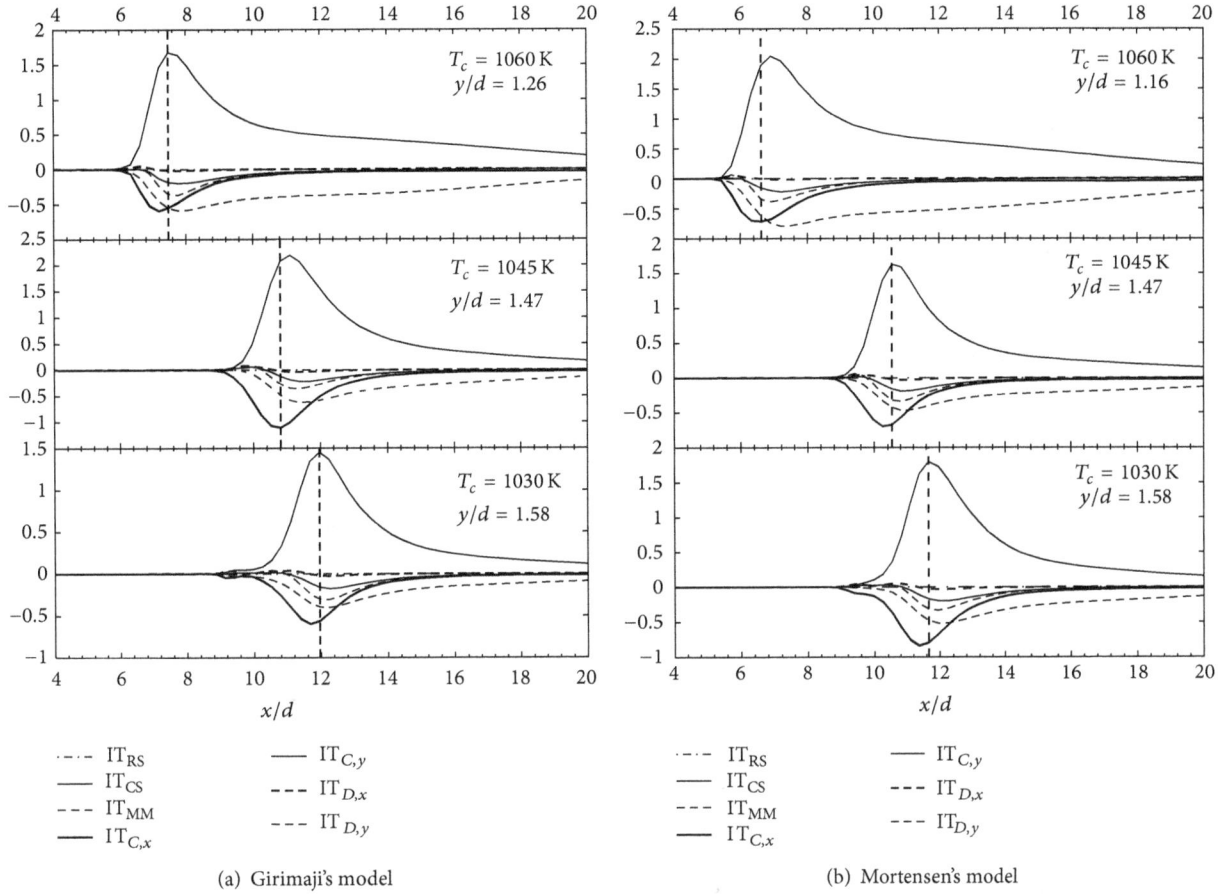

(a) Girimaji's model

(b) Mortensen's model

FIGURE 7: Axial profiles of the Integrated Transport (IT) budget of the steady-state Q_T equation (integrated r.h.s. terms of (9)): (a) Girimaji's model and (b) Mortensen's model. The bottom, middle, and top panes correspond to $T_c = 1030$ K, $T_c = 1045$ K, and $T_c = 1060$ K, respectively. The IT subscripts follow the same notation used in Figure 6. All terms are scaled down by a factor of 10^6 and the units are K/s.

balances $T_{C,x}$ in the preheat zone with negligible T_{CS} is not the case. Instead, stabilisation occurs via autoignition as can be seen from the T_{CS}-$T_{C,x}$-T_{MM} balance. At liftoff, the balance shifts to $\eta \approx 0.24$. The role of $T_{C,x}$ in balancing T_{CS} diminishes but this term remains a major contributor to the overall balance, and T_{MM} prevails. As for the remaining terms, $T_{D,x}$ becomes much weaker while $T_{D,y}$ and $T_{C,y}$ remain important. In the postflame regions (locations Post-F1 and Post-F2), the role of $T_{c,x}$ diminishes further, T_{MM} becomes more dominant, and $T_{D,x}$ is virtually null. The contributions of $T_{D,y}$ and $T_{C,y}$ are as significant as that of $T_{C,x}$ at Post-F1 and supersede it at Post-F2. As T_{CS} increases in mixture fraction space, the contributions of $T_{D,y}$, $T_{C,y}$, and $T_{C,x}$ remain smaller than that of T_{MM}, which acts as the major heat sink. In contrast, as T_{CS} decays, T_{MM} increases and acts as a source (notice its positive contribution), and both of these terms are counterbalanced by $T_{D,y}$, $T_{C,y}$, and $T_{C,x}$. Thus, beyond the stabilisation height, the flame budgets indicate the structure of a nonpremixed flame, which is largely characterised by a T_{CS}-T_{MM} balance. To be noted that the radiative source, T_{RS}, is negligible at all locations for all coflow temperatures due to the fact that hydrogen in the fuel stream is highly diluted with nitrogen.

Therefore, this term can be safely neglected without loss of accuracy.

Budgets in Physical Space. The fact that autoignition is the controlling stabilisation mechanism can be further confirmed by analysing the transport budget of Q_T in physical space. To do so, the PDF-weighted integration of the r.h.s. terms of (9) is performed over the mixture fractions space. This yields IT_{RS}, IT_{CS}, IT_{MM}, $IT_{C,x}$, $IT_{C,y}$, $IT_{D,x}$, and $IT_{D,y}$ ("IT" stands for "Integrated Transport"). The axial profiles of these terms are plotted in Figure 7(a) for Girimaji's model and in Figure 7(b) for Mortensen's model at the respective W/d locations obtained with each T_c. The parallels between the observation made in Figure 6 and this figure are evident. IT_{CS} is mostly balanced by $IT_{C,x}$ and IT_{MM} in the preflame regions, with the former being more dominant. The remaining terms are small but nonnegligible. The absence of $IT_{D,x}$ eliminates the possibility of stabilisation by premixed flame propagation and the current balance indicates that autoignition is the controlling mechanism. Right ahead of liftoff, $IT_{C,y}$ and $IT_{D,y}$ emerge and IT_{MM} becomes more important. These three terms start to dominate at the expense of $IT_{C,x}$, which remains

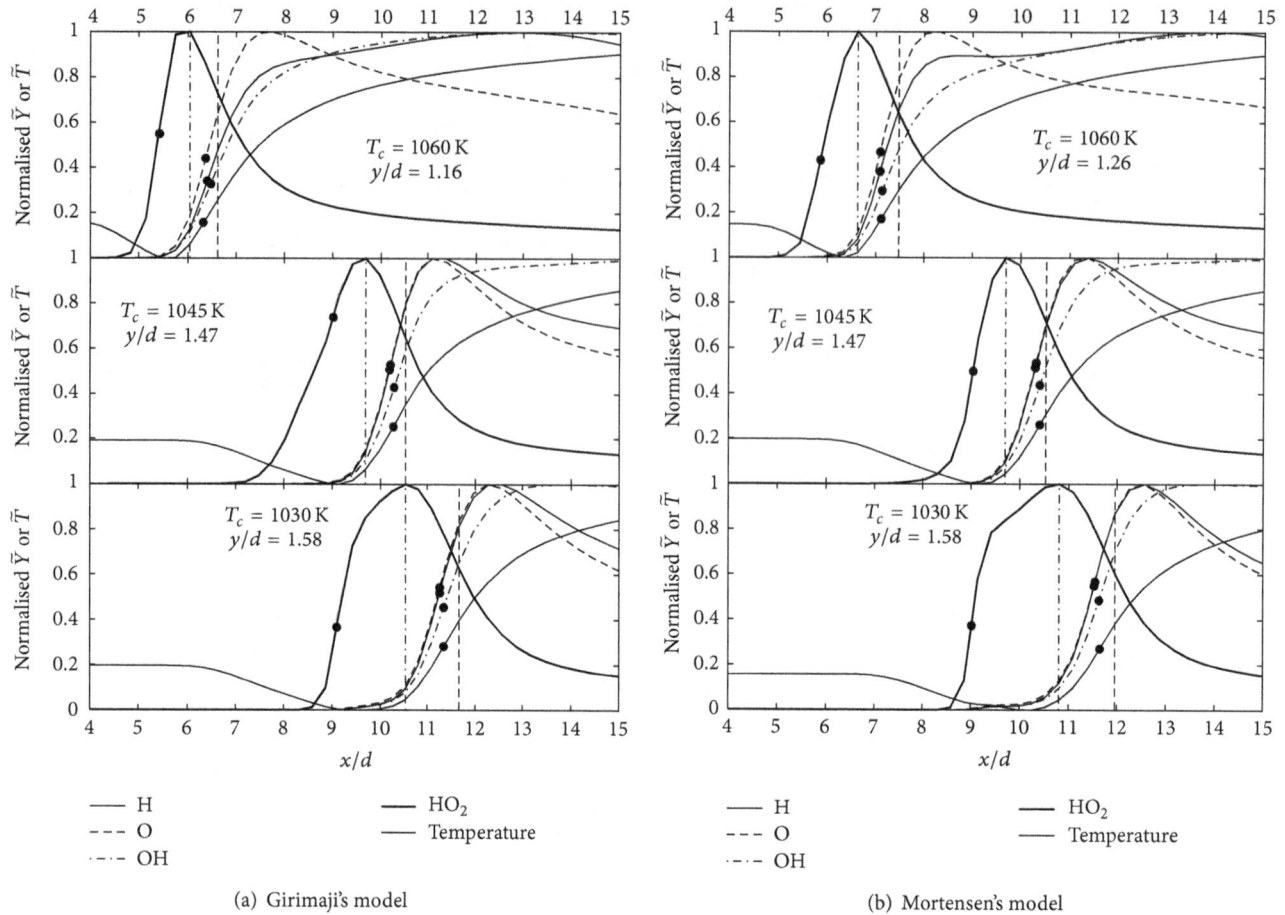

(a) Girimaji's model   (b) Mortensen's model

FIGURE 8: Axial profiles of the normalised Favre-averaged temperature and mass fractions of H, O, OH, and HO_2 obtained at the radial locations of stabilisation corresponding to each T_c: (a) Girimaji's model and (b) Mortensen's model. The vertical dash-dotted and dashed lines correspond to the axial locations of maximum HO_2 and liftoff height, respectively. The circles indicate the locations of the maximum slopes.

the main heat sink. Downstream of the stabilisation height, $IT_{C,x}$, $IT_{C,y}$, and $IT_{D,y}$ diminish gradually and IT_{MM} becomes more important. Further downstream IT_{CS} is mostly counterbalanced by IT_{MM}, with smaller contributions $IT_{C,x}$, $IT_{C,y}$, and $IT_{D,y}$, which indicates the structure of a nonpremixed flame.

Radical History ahead of the Stabilisation Height. The transport budgets of the temperature in mixture fraction and physical spaces are both indicative of stabilisation via autoignition. In order to ascertain this conclusion, the history of radical build-up ahead of the reaction zone is now investigated. Figure 8 displays the axial profiles of the normalised Favre-averaged temperature $((\widetilde{T} - \widetilde{T}_{\min})/(\widetilde{T}_{\max} - \widetilde{T}_{\min}))$ and mass fractions $(\widetilde{Y}/\widetilde{Y}_{\max})$ of H, O, OH, and HO_2. The subscripts "min" and "max" denote the minimum and maximum values of the reactive scalars at $y/d = W/d$. The results are reported for Girimaji's and Mortensen's models in Figures 8(a) and 8(b), respectively. Common trends are observed ahead of the stabilisation height for the different combinations of T_c values and CSDR models.

(1) The production of the intermediate HO_2 initiates upstream of the axial locations where the radicals H, O, and OH emerge.

(2) H, O, and OH start building up before the consumption of HO_2 begins. As shown, the mass fractions of these three species start to increase before HO_2 reaches its peak.

(3) HO_2 builds up rapidly prior to the runaway of H, O, and OH, as can be seen from the axial locations of the maximum slopes.

Therefore, radicals do not build up simultaneously as in premixed flame propagation, but rather HO_2 acts as a precursor to the production of H, O, and OH as in autoignition scenarios [6]. This confirms that the flame is stabilised by autoignition.

4.2.3. Impact of CSDR Modelling on Autoignition. To better understand the autoignition process, standalone zero-dimensional CMC (0DCMC) calculations are performed for each T_c. In these calculations, the convective and (physical space) diffusive terms in (8) and (9) are turned off and the CSDR is modelled using the AMC [9]. This CSDR model is parametrised by its maximum value $\chi_\circ$ and it is given by

$$\langle \chi \mid \eta \rangle = \chi_\circ \exp \left\{ -2 \left[\text{erf}^{-1} \left(2\eta - 1 \right) \right]^2 \right\} = \chi_\circ G(\eta), \quad (29)$$

where erf^{-1} is the inverse error function. The resulting homogeneous CMC equations are identical to those of the unsteady laminar flamelet model (with unity Lewis number) [48]. For a given $\chi_\circ$, the solution evolves from the inert mixing solution at $t = 0$ up to the moment when autoignition occurs, $t = \tau$. Autoignition is declared at the moment when the maximum OH mass fraction reaches 2×10^{-4} at any point in mixture fraction space, following Stanković and Merci [49]. Figure 9 displays the ignition delay (τ) as a function of $\chi_\circ$. These results are obtained using the mechanism of Mueller et al. [40] for $T_c = 1030$ K, 1045 K, and 1060 K. It is evident from this figure that, for a given coflow temperature, τ reaches an asymptotic limit as $\chi_\circ$ is gradually increased. This means that the occurrence of autoignition becomes impossible beyond a critical value of $\chi_\circ$, denoted here by $\chi_{\circ,c}$. The approximate values of $\chi_{\circ,c}$ are 227 s^{-1}, 482 s^{-1}, and 991 s^{-1} for $T_c = 1030$ K, 1045 K, and 1060 K, respectively. Therefore, $\chi_{\circ,c}$ increases with increasing T_c. In order to link these findings to the flow field of the flame under investigation, $\chi_\circ$ is first obtained by integrating (29) weighted by $\widetilde{P}(\eta)$ over the mixture fraction space:

$$\chi_\circ = \frac{\int_0^1 \langle \chi \mid \eta \rangle \widetilde{P}(\eta) \, d\eta}{\int_0^1 G(\eta) \widetilde{P}(\eta) \, d\eta} = \frac{\widetilde{\chi}}{\int_0^1 G(\eta) \widetilde{P}(\eta) \, d\eta} \quad (30)$$

which requires the usage of the readily available quantities $\widetilde{\chi}$, $\widetilde{\xi}$, and $\widetilde{\xi''^2}$. Figure 10 shows the Contours of $\chi_\circ$ computed from (30) for $T_c = 1045$ K. The contours for $T_c = 1030$ K and 1060 K are virtually the same and are not included here for brevity. The contour of Figure 10 will be used in all of the subsequent analysis irrespective of the value of T_c. The thick contours represent the levels 227 s^{-1}, 482 s^{-1}, and 991 s^{-1} which correspond to the $\chi_{\circ,c}$ values determined in Figure 9. Also shown are the ignition kernel locations obtained from the different combinations of CSDR models and coflow temperatures. As expected, for all three coflow temperatures, ignition occurs at locations where $\chi_\circ$ is lower that $\chi_{\circ,c}$ (the circles, squares, and triangles lie outside the 227 s^{-1}, 482 s^{-1}, and 991 s^{-1} contour levels, resp.). Since $\widetilde{\chi}$ decays away from the nozzle, autoignition does not take place until $\chi_\circ$ (which depends on $\widetilde{\chi}$ via (30)) falls below $\chi_{\circ,c}$. This explains why increasing T_c results in earlier ignition and shorter liftoff from a mixing point of view.

In another set of 0DCMC calculations, the "most reactive" mixture fraction, η_{mr}, is computed for each T_c by setting $\chi_\circ$ in (29) to 10^{-20} s^{-1} (homogeneous reactor). The results are summarised in Table 4. Consistent with the findings of Stanković and Merci [49], η_{mr} decreases with increasing T_c. The objective behind determining η_{mr} is to compare the local CSDR values obtained using Mortensen's and Girimaji's models at $\eta = \eta_{\text{mr}}$, $\langle \chi \mid \eta_{\text{mr}} \rangle^M$ and $\langle \chi \mid \eta_{\text{mr}} \rangle^G$, respectively.

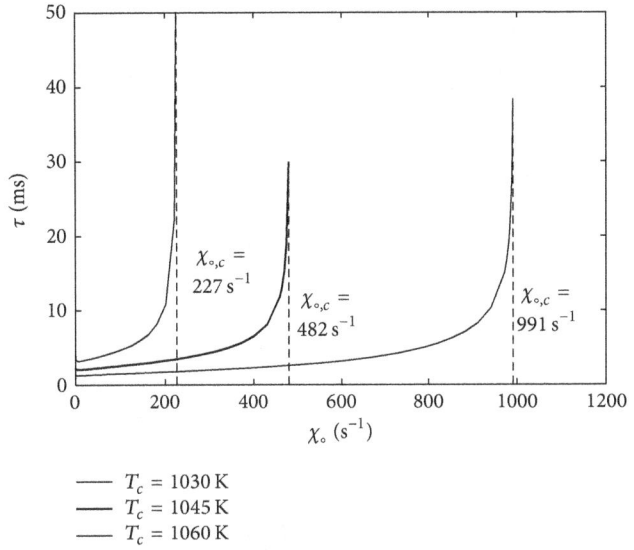

FIGURE 9: Ignition delays as a function of the peak scalar dissipation rate obtained using the mechanism of Mueller et al. [40] for $T_c = 1030$, 1045, and 1060 K. The vertical dashed lines represent the asymptotic limit beyond which ignition cannot occur ($\chi_\circ = \chi_{\circ,c}$).

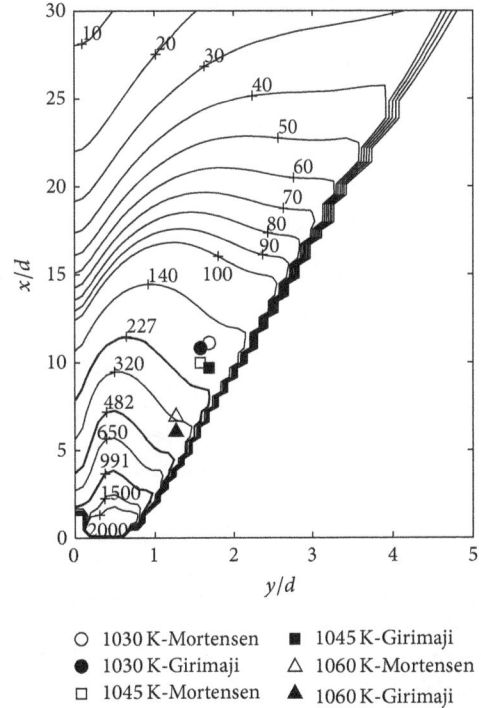

FIGURE 10: Contours of $\chi_\circ$ (see (30)). The thick contours correspond to $\chi_\circ = \chi_{\circ,c}$ for $T_c = 1030$ K (227 s^{-1}), 1045 K (482 s^{-1}), and 1060 K (991 s^{-1}). The symbols represent the ignition locations.

As shown in Figure 11, for all given coflow temperatures and radial locations, the axial profiles of $\langle \chi \mid \eta_{\text{mr}} \rangle^M$ and $\langle \chi \mid \eta_{\text{mr}} \rangle^G$ reveal that Mortensen's models generally result in higher scalar dissipation rate levels and therefore yield longer ignition delays and higher liftoff heights. Also displayed in

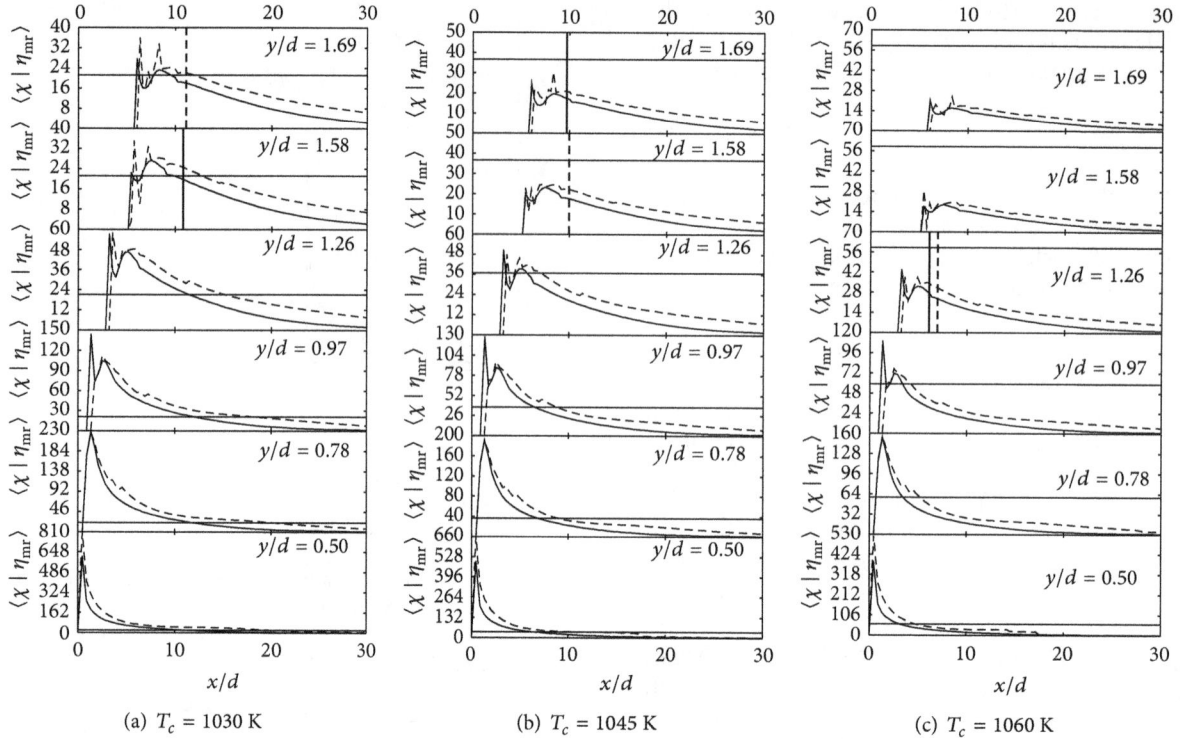

FIGURE 11: Axial profiles of $\langle \chi \mid \eta_{\mathrm{mr}} \rangle$ at several radial locations. (a) $T_c = 1030$ K, (b) $T_c = 1045$ K, and (c) $T_c = 1060$ K. Solid black lines, Girimaji's model; dashed black lines, Mortensen's model. The vertical black lines indicate the radial locations where ignition occurs. The grey horizontal lines represent the value of the CSDR computed using the AMC model parametrised by $\chi_{\circ,c}$ at $\eta = \eta_{\mathrm{mr}}$ ($\langle \chi \mid \eta_{\mathrm{mr}} \rangle_c^{\mathrm{AMC}} = \chi_{\circ,c} \exp\{-2[\mathrm{erf}^{-1}(2\eta_{\mathrm{mr}} - 1)]^2\}$).

Figure 11 are the critical CSDR values computed using the AMC model parametrised by $\chi_{\circ,c}$ at $\eta = \eta_{\mathrm{mr}}$; that is, $\langle \chi \mid \eta_{\mathrm{mr}} \rangle_c^{\mathrm{AMC}} = \chi_{\circ,c} \exp\{-2[\mathrm{erf}^{-1}(2\eta_{\mathrm{mr}} - 1)]^2\}$ (horizontal grey lines). These values are also provided in Table 4 for all coflow temperatures. As expected, $\langle \chi \mid \eta_{\mathrm{mr}} \rangle_c^{\mathrm{AMC}}$ increases with increasing T_c. Figure 11(a) ($T_c = 1030$ K) shows that ignition cannot occur as long as $\langle \chi \mid \eta_{\mathrm{mr}} \rangle^M$ or $\langle \chi \mid \eta_{\mathrm{mr}} \rangle^G$ is below $\langle \chi \mid \eta_{\mathrm{mr}} \rangle_c^{\mathrm{AMC}}$. Indeed, when Mortensen's model is used, ignition happens at the first axial and radial locations where $\langle \chi \mid \eta_{\mathrm{mr}} \rangle^M$ falls below $\langle \chi \mid \eta_{\mathrm{mr}} \rangle_c^{\mathrm{AMC}}$. This takes place at $x/d = 11.09$ and $y/d = 1.69$ (first pane from the top in Figure 11(a)). A similar scenario is observed when Girimaji's model is used. Ignition occurs at the first axial and radial locations where $\langle \chi \mid \eta_{\mathrm{mr}} \rangle^G$ falls below $\langle \chi \mid \eta_{\mathrm{mr}} \rangle_c^{\mathrm{AMC}}$, which happens at $x/d = 10.81$ and $y/d = 1.58$ (second pane from the top in Figure 11(a)). Similar observation can be made in Figure 11(b) ($T_c = 1045$ K) and Figure 11(c) ($T_c = 1060$ K). However, in these two cases, although $\langle \chi \mid \eta_{\mathrm{mr}} \rangle^M$ and $\langle \chi \mid \eta_{\mathrm{mr}} \rangle^G$ fall quickly below $\langle \chi \mid \eta_{\mathrm{mr}} \rangle_c^{\mathrm{AMC}}$ at y/d locations smaller than the ones where ignition is indicated, ignition occurs at the first axial location with lowest $\langle \chi \mid \eta_{\mathrm{mr}} \rangle$. As shown in Figure 11(b) ($T_c = 1045$ K), ignition takes place at $x/d = 9.97$ and $y/d = 1.58$ using Mortensen's model and at $x/d = 9.69$ and $y/d = 1.69$ using Girimaji's model. Similarly, in Figure 11(c) ($T_c = 1060$ K), ignition happens at $x/d = 6.90$ and $y/d = 1.26$ using Mortensen's model and at $x/d = 6.04$ and $y/d = 1.26$ using Girimaji's model.

TABLE 4: The most reactive mixture fraction, η_{mr}, and the CSDR values obtained from the AMC parametrised by $\chi_{\circ,c}$ at η_{mr}.

T_c [K]	η_{mr}	$\langle \chi \mid \eta_{\mathrm{mr}} \rangle_c^{\mathrm{AMC}\dagger}$
1030	0.0620	21.27
1045	0.0543	36.65
1060	0.0465	59.02

$\dagger \langle \chi \mid \eta_{\mathrm{mr}} \rangle_c^{\mathrm{AMC}} = \chi_{\circ,c} \exp\{-2[\mathrm{erf}^{-1}(2\eta_{\mathrm{mr}} - 1)]^2\}$ [s^{-1}].

4.2.4. Effect of Spurious Sources.

Ideally, the integration of the PDF-weighted CMC equations over η-space should yield the unconditional set of equations without any additional source terms [14]. This outcome is guaranteed when fully consistent CSDR models (e.g., the inhomogeneous model of Mortensen) are employed. To be noted that in order for an inhomogeneous CSDR model to be fully consistent, the CV model employed in the derivation should be consistent with the moments of the presumed PDF. This condition is satisfied in Moretnsen's model due to the usage of the PDF-gradient model. An example of an inhomogeneous yet inconsistent CSDR closure is the one proposed in [50]. The discrepancy arises from the modelling of the CV using the linear velocity model, which is not consistent with the second moment of the presumed β-PDF.

When inconsistent CSDR models are employed, the integration of the CMC equations results in spurious (or false) source terms. In this study, this occurs when the CSDR

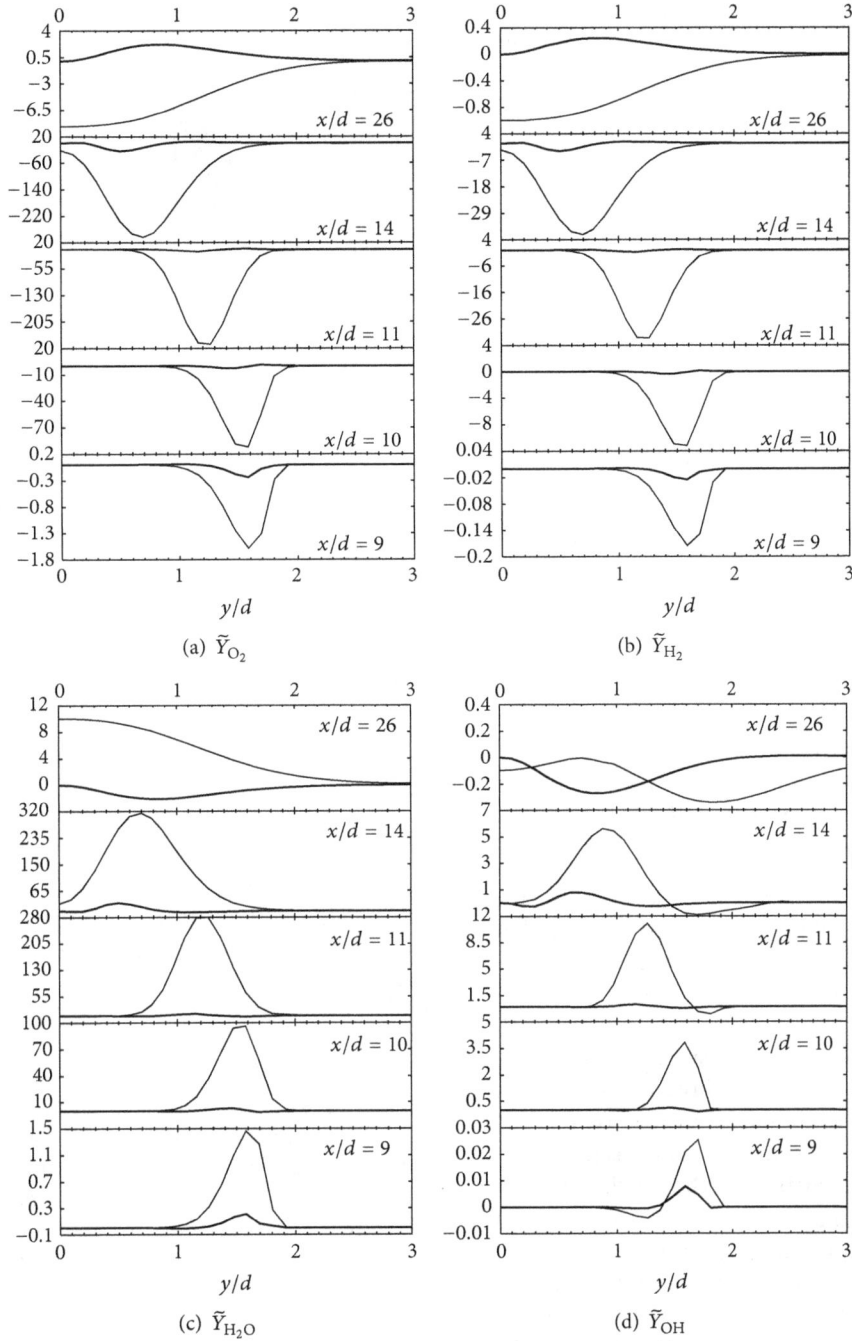

FIGURE 12: Radial profiles of the chemical and spurious sources for $T_c = 1045\,\text{K}$: (a) O_2, (b) H_2, (c) H_2O, and (d) OH. Thin lines, chemical sources; thick lines, spurious sources. The units are s^{-1}.

model of Girimaji is used. The spurious source associated with a species κ is calculated through [14, 16]:

$$\widetilde{S}_\kappa = \frac{1}{2} \int_0^1 \left[\underbrace{\langle \chi \mid \eta \rangle^G}_{\substack{\text{Inonsistent} \\ \text{CSDR model}}} - \underbrace{\langle \chi \mid \eta \rangle^M}_{\substack{\text{Consistent} \\ \text{CSDR model}}} \right] \frac{\partial^2 Q_{\kappa,G}}{\partial \eta^2} \widetilde{P}(\eta)\, d\eta, \quad (31)$$

where $Q_{\kappa,G}$ is the conditional mass fraction of species κ obtained in the inconsistent realisation (using Girimaji's model).

Figure 12 shows the radial profiles of the mean chemical sources (IT_{CS}) and spurious sources of O_2, H_2, H_2O, and OH for $T_c = 1045\,\text{K}$. In comparison to the mean chemical sources, the magnitudes of the spurious sources are small but nonnegligible, in particular within the flame zone ($x/d = 14$ and 26). The effect of the spurious sources ahead of the predicted stabilisation heights ($x/d = 9$ and 10) is not sufficiently large to change the nature of the stabilisation mechanism, as the flame is stabilised by autoignition using both Mortensen's and Girimaji's models. However, the fact that Girimaji's model yields a relatively earlier ignition and a smaller liftoff height indicates that the spurious sources are influential. Although this is not the case in the calculations of the current flame, inconsistent CMC implementations may lead erroneous results and conclusions.

4.3. Effect of Chemical Kinetics.

The mechanism of Mueller et al. [40] has been used in all of the results reported so far. The sensitivity of the flame to chemical kinetics is now assessed by testing the performance of the mechanisms of Li et al. [41], Burke et al. [42], and Ó Conaire et al. [43]. The coflow temperature is set to the experimentally reported value of 1045 K.

4.3.1. Comparison of the Chemical Kinetic Mechanisms

0DCMC A Priori Analysis. 0DCMC calculations are first performed using the three mechanisms in order to determine the corresponding $\chi_{\circ,c}$ values. The results are shown in Figure 13, along with those obtained with the Mueller mechanism in Section 4.2.3 for T_c = 1045 K. By inspecting the calculated $\chi_{\circ,c}$ values and referring to Figure 10, it can be postulated that the Li mechanism would yield the shortest ignition kernel and smallest liftoff height followed by the Mueller mechanism, then the Ó Conaire mechanism, and finally the Burke mechanism. This analysis provides a preliminary qualitative description of the sensitivity of liftoff height to the considered chemical kinetics.

Results in Physical Space. To ascertain these findings, the flame is simulated using the three newly considered mechanisms. The CSDR is closed using Girimaji's model in all of the following calculations. The results are displayed in Figure 14 at the axial locations x/d = 9, 10, 11, 14, and 26, along with those obtained using Mueller mechanism in Section 4.2.1. Unlike the Burke mechanism, the other three kinetic schemes are capable of reproducing the experimental trends. The results of the Mueller and Ó Conaire mechanisms show best agreement with the experimental data. The Mueller mechanism yields better predictions at x/d = 9 and 10, while Ó Conaire's results in better agreement with the measurements at x/d = 11. The differences between the two are small at x/d = 14 and 26. When the Li mechanism is employed, the temperature rises at early axial locations owing to the ability of the mixture to ignite at high scalar dissipation rate levels ($\chi_{\circ,c}^{Li}$ = 562 s^{-1}). This effect propagates downstream and results in the deviation of the numerical predictions from the experiments. The performance of the Burke mechanisms is unsatisfactory, as the mixture fails to achieve early ignition ($\chi_{\circ,c}^{Burke}$ = 169 s^{-1}) and remains inert up to approximately x/d = 14. The contours of $\widetilde{Y}_{OH}$ ($\times 10^3$) are displayed in Figure 15. As predicted from the 0DCMC calculations, the Li and Burke mechanisms yield the smallest (H/d = 9.73) and largest (H/d = 16.11) liftoff heights, as shown in Figures 15(a) and 15(b), respectively. Also being consistent with the 0DCMC findings, the liftoff heights obtained using the Mueller and Ó Conaire mechanisms fall between those of the Li and Burke

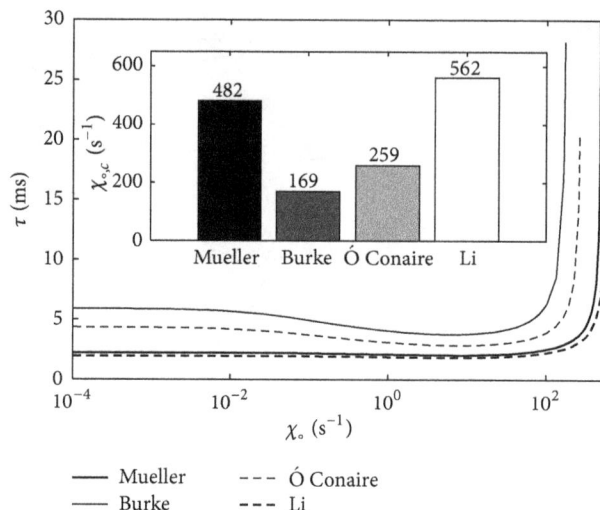

FIGURE 13: Ignition delay as a function of the peak scalar dissipation rate $\chi_\circ$ obtained using the mechanisms of Mueller et al. [53], Ó Conaire et al. [43], Li et al. [41], and Burke et al. [42] for T_c = 1045 K. The inset shows the critical values of $\chi_\circ$ above which ignition cannot occur.

mechanisms, as demonstrated in Figures 15(c) and 15(d). However, the Mueller mechanism results in a slightly larger liftoff height (H/d = 10.50) in comparison to the Ó Conaire mechanism (H/d = 10.26), which contradicts with the 0DCMC predictions. Apart from the Burke mechanism, the liftoff heights computed from the remaining mechanisms are similar and are in close agreement with the experimental value H_{exp}/d = 10. As for the radial location of liftoff, the Mueller, Li, and Ó Conaire mechanisms result is W/d = 1.47, whereas Burke's yields a wider flame base with W/d = 1.92. As in the Mueller mechanism calculations, the flame is also stabilised by autoignition when the Ó Conaire, Li, and Burke mechanisms are employed. This can be clearly seen in Figure 16 where H, O, OH, and HO$_2$ exhibit the same trends observed in Figure 8(a).

Results in Mixture Fraction Space. For completeness, the conditional profiles are shown in Figure 17. The calculation of the conditional data from the experimental scatter is described in the Appendix. The numerical results are reported at x/d = 9, 10, 11, 14, and 26, and y/d = 1.47 (W/d obtained with all mechanisms except Burke's). Overall, the results of the Mueller and Ó Conaire mechanisms are in reasonable agreement with the experiments, especially within mixture fraction intervals where the PDF is large (see Figure 17(f)). This explains the good agreement these two mechanisms yield with the measurements in physical space (Figure 14). The Li mechanism predicts the highest OH levels up to x/d = 11 (Figure 17(e)) which leads to the occurrence of the earliest ignition and consequently results in the smallest liftoff height among the considered

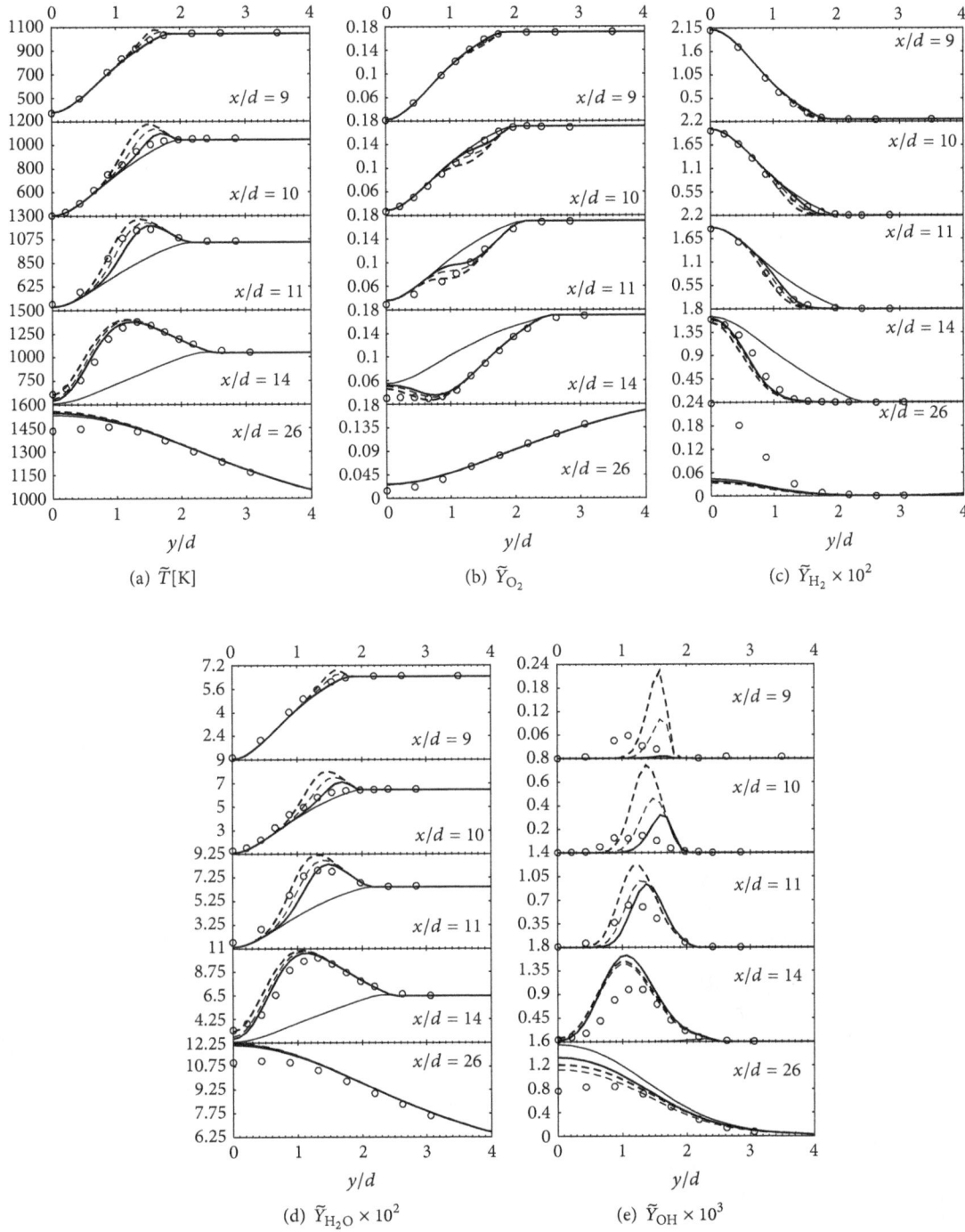

FIGURE 14: Radial profiles of (a) $\tilde{T}$, (b) $\tilde{Y}_{O_2}$, (c) $\tilde{Y}_{H_2}$ ($\times 10^2$), (d) $\tilde{Y}_{H_2O}$ ($\times 10^2$), and (e) $\tilde{Y}_{OH}$ ($\times 10^3$) using different chemical kinetic mechanisms. In (a)–(e): thick solid lines, Mueller et al. [40] mechanism; thick dashed lines, Li et al. [41] mechanism; thin solid lines, Burke et al. [42] mechanism; thin dashed lines, Ó Conaire et al. [43] mechanism; Symbols, experimental data [3]. The CSDR is modelled using Girimaji's model and T_c = 1045 K.

kinetic schemes. In turn, this early ignition produces higher than expected temperature levels, an effect which propagates downstream resulting in the deviation of the conditional predictions from the experiments. In the three mechanisms discussed up to this point, the peak of Q_T in Figure 17(a)

shifts gradually from lean to stoichiometric mixtures as x/d increases, which is consistent with the axial evolution of the conditional heat release term T_{CS} in Figure 6. The profiles of the Burke mechanism remain very close to the inert mixing solution up to x/d = 14. This behaviour results in the

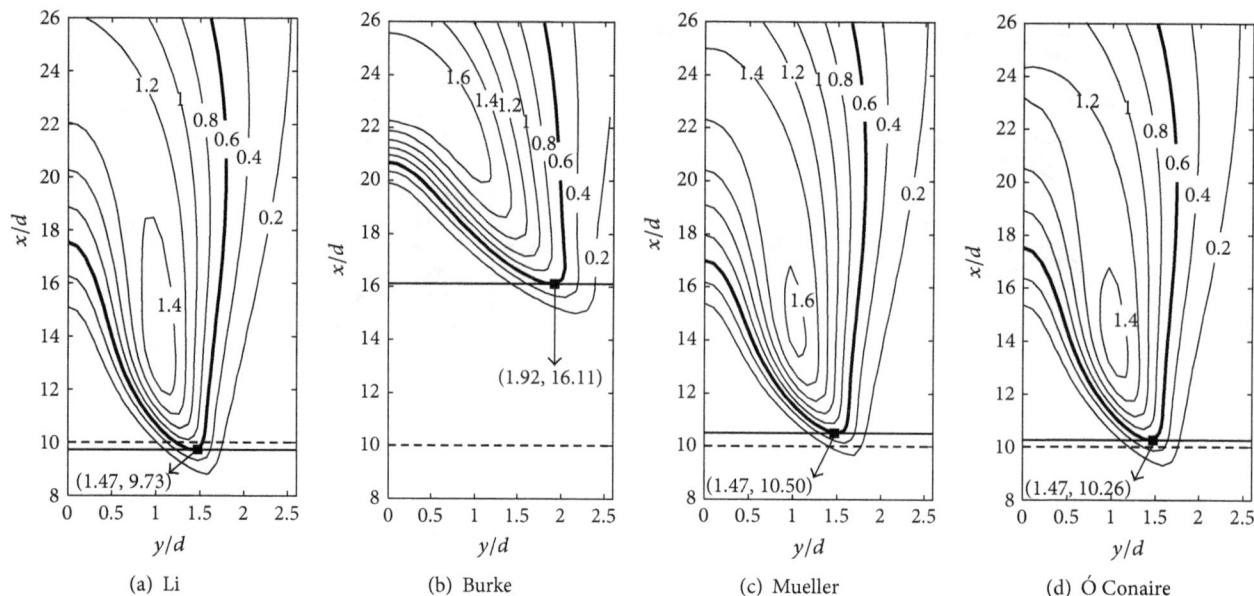

FIGURE 15: Contours of $\tilde{Y}_{OH}$ ($\times 10^3$) obtained using the mechanisms of (a) Li et al. [41], (b) Burke et al. [42], (c) Mueller et al. [40], and (d) Ó Conaire et al. [43]. The thick contours correspond to 600 ppm. The solid and dashed horizontal lines correspond to the numerical and experimental liftoff heights, respectively. The CSDR is modelled using Girimaji's model and $T_c = 1045$ K.

most delayed ignition and leads to the largest liftoff height. The conditional profiles at $y/d = 1.92$ where liftoff occurs undergo very similar trends (not shown here).

4.3.2. Diagnosis of the Burke Mechanism. In order to diagnose the unsatisfactory performance of the Burke mechanism, the rate parameters of two of the key reactions are investigated. The recombination reaction

$$H + O_2 (+M) \rightleftharpoons HO_2 (+M) \qquad (R9)$$

and the chain-branching reaction

$$H + O_2 \rightleftharpoons O + OH \qquad (R1)$$

are two of the most important reactions in H_2 oxidation chemistry. (R9) and (R1) (labelled here as in [40–43]) compete for H atoms and therefore control the overall branching ratio and determine the second explosion limit in homogeneous H_2/O_2 systems [42]. The treatment of (R9) and (R1) in the Li and Burke mechanisms differs significantly from the Ó Conaire and Mueller mechanisms. The rate parameters of (R9) are provided in Table 5. The high-pressure limit rate parameters are identical in the Ó Conaire, Mueller, and Li mechanisms [51]. Burke's mechanism uses an updated set of parameters [52]. In the low-pressure limit, the Ó Conaire and Mueller mechanisms use the same parameters [53] except for the third-body efficiency factors of H_2 and H_2O. The Li mechanism employs a completely different set of parameters [54]. The Burke parameters differ from Li's by the broadening factor (decreased by a factor of 1.6 and equal to the ones employed in the Ó Conaire and Mueller mechanisms) and the

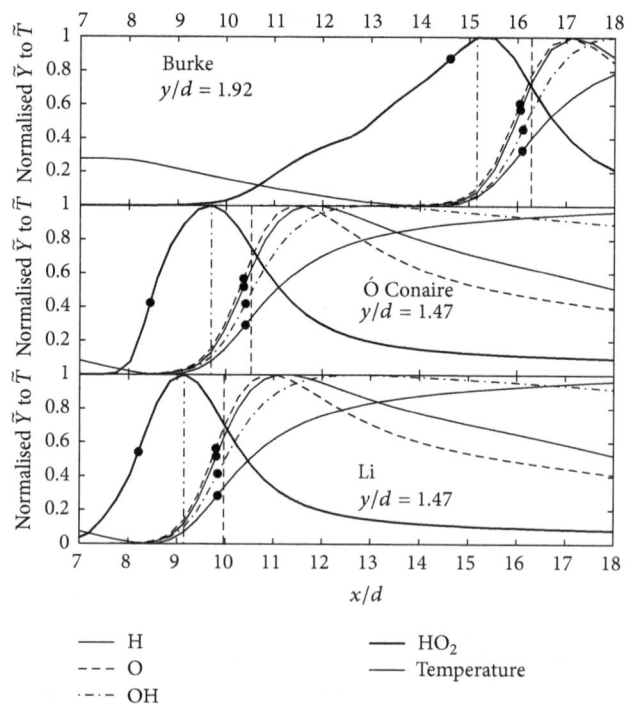

FIGURE 16: Axial profiles of the normalised Favre-averaged temperature and mass fractions of H, O, OH, and HO_2 obtained at the radial locations of liftoff using the mechanisms of (from top to bottom) Burke et al. [42], Ó Conaire et al. [43], and Li et al. [41]. The vertical dash-dotted and dashed lines correspond to the axial locations of maximum HO_2 and liftoff, respectively. The circles indicate the locations of the maximum slopes. The CSDR is modelled using Girimaji's model and $T_c = 1045$ K.

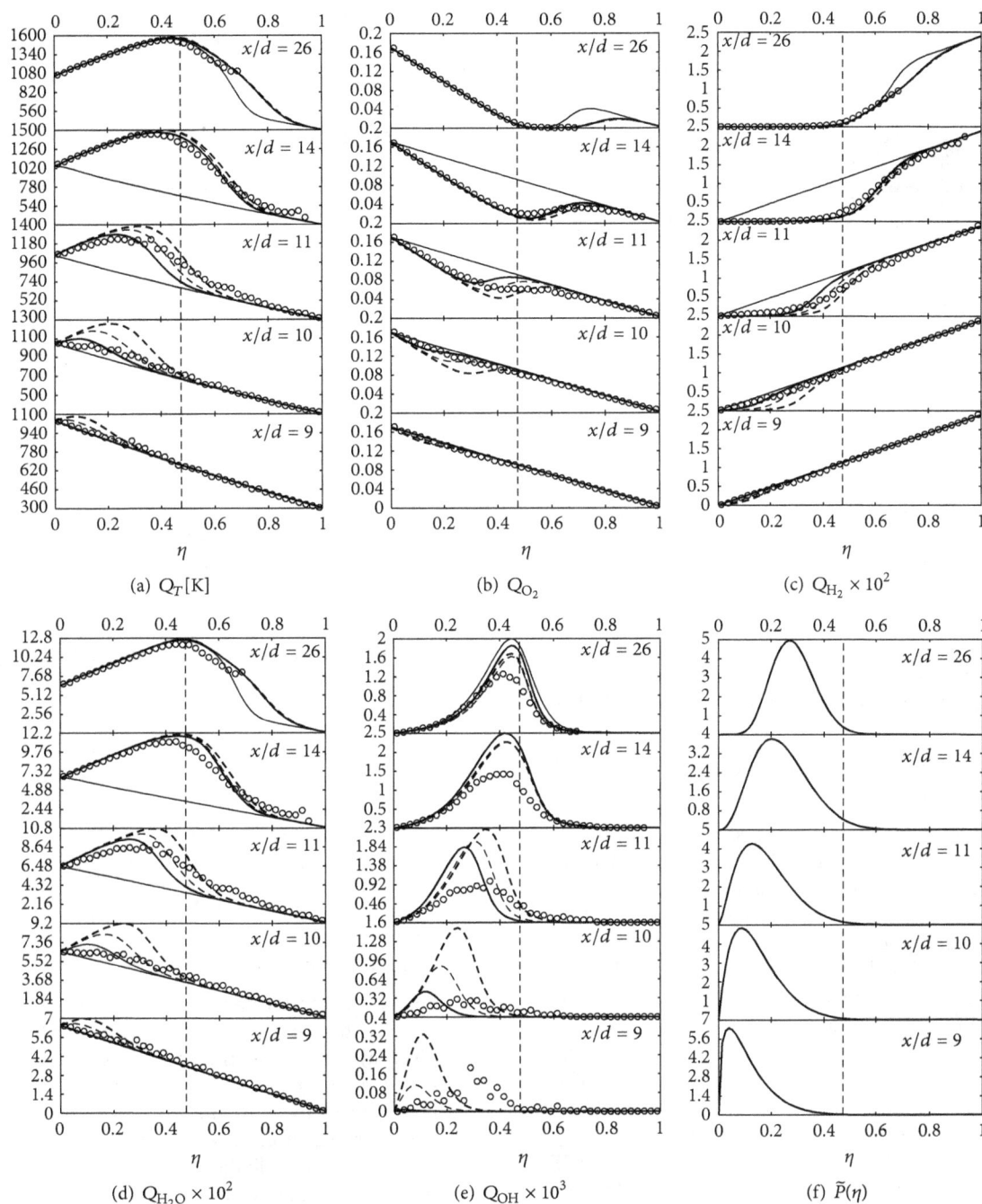

FIGURE 17: Conditional profiles at $x/d = 9, 10, 11, 14$ and 26 and $y/d = 1.47$ using different chemical kinetic mechanisms: (a) Q_T, (b) Q_{O_2}, (c) Q_{H_2} ($\times 10^2$), (d) Q_{H_2O} ($\times 10^2$), (e) Q_{OH} ($\times 10^3$), and (f) $\tilde{P}(\eta)$. In (a)–(e): thick solid lines, Mueller et al. [40]; thick dashed lines, Li et al. [41]; thin solid lines, Burke et al. [42]; thin dashed lines, Ó Conaire et al. [43]; symbols, experimental data (see the Appendix). The vertical dashed line indicates the location of the stoichiometric mixture fraction. The CSDR is modelled using Girimaji's model and $T_c = 1045$ K.

third-body efficiency factor for H_2O (increased by a factor of 1.3) [42]. As for (R1), the corresponding rate parameters are listed in Table 6. The parameters employed in the Ó Conaire and Mueller mechanisms are identical [55]. Two distinct sets of updated parameters are employed in the mechanisms of Li [56] and Burke [57]. In light of the above description, the variability of the predictions in Figure 14 and the close

agreement between the results of the Mueller and Ó Conaire mechanisms may be explained in part by the treatment of (R9) and (R1). This can be demonstrated, for instance, by adjusting the rate parameters of these two reactions in the Burke mechanism, in an attempt to improve its predictions. Since this mechanism is an updated version of Li's, two separate experiments are carried out as follows.

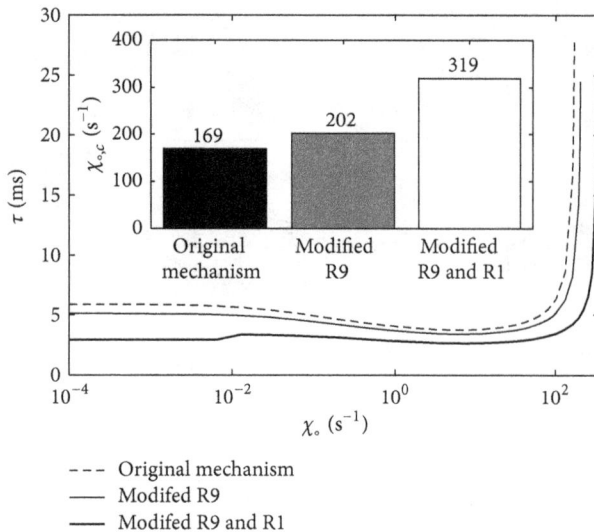

--- Original mechanism
— Modifed R9
— Modifed R9 and R1

FIGURE 18: Ignition delay as a function of the peak scalar dissipation rate $\chi_\circ$ obtained using the original and modified mechanisms of Burke et al. [42]. The inset shows the critical values of $\chi_\circ$ above which ignition cannot occur.

(1) The parameters of (R9) are substituted by those of the Li mechanism, with the parameters of (R1) left unchanged.

(2) The parameters of both (R9) and (R1) are substituted by those of the Li mechanism.

The preliminary 0DCMC calculations shown in Figure 18 reveal that $\chi_{\circ,c}$ increases significantly when these modifications are applied. Therefore, the mixture is now anticipated to ignite earlier and the liftoff height is expected to decrease. Two additional CMC runs are performed using these modifications while keeping the same settings for the CSDR and the coflow temperature. The results are shown in Figure 19 at x/d = 10, 11, and 14. It is evident that the modification of (R9) alone brings a substantial improvement over the rate parameters of the original mechanism. The temperature (Figure 19(a)) and the species mass fractions (Figures 19(b)–19(e)) are much better predicted, especially at x/d = 11 and 14. This further proves the importance of (R9) as a precursor to autoignition reactions. Better agreement with the experimental data is obtained when both (R9) and (R1) are modified. As shown in Figure 19(f), the liftoff height decreases by 26.69% when (R9) is modified and by 29.24% when both (R9) and (R1) are modified. In both cases the computed liftoff heights are substantially closer to the experimentally measured value. Figure 19(f) also shows a remarkable decrease in W/d as liftoff occurs closer to the centreline. With the modified settings, both the axial and radial positions of liftoff become in line with those obtained using the Ó Conaire, Mueller, and Li mechanisms. There should be further room for improvement by applying additional modifications to the rate parameters of other key reactions. However, extreme care must be taken while doing so because chemical kinetic mechanisms are usually optimised as a whole, rather than on an individual reaction

basis. A sensitivity analysis is beyond the scope of this work and was not performed here. The modifications applied above are *ad hoc*.

5. Conclusions

A lifted H_2/N_2 jet flame issuing into a vitiated coflow was investigated using the first-order CMC. The flow and mixing fields were obtained using a modified version of the k-ε model. The calculations included an additional transport equation for the Favre-averaged scalar dissipation rate. Two formulations were implemented for the CSDR, the homogeneous model of Girimaji, and the inhomogeneous model of Mortensen. The CV was modelled using the PDF-gradient model. Calculations were performed for three coflow temperatures, 1030, 1045, and 1060 K, and four chemical kinetic mechanisms were assessed. In light of the previous results, the following conclusions are drawn.

(1) The modification of the constant $C_{\varepsilon 1}$ in the k-ε model from the standard value of 1.42 to 1.6 improves the predictions of the spreading rate of the jet.

(2) The solution of a transport equation for the scalar dissipation rate in lieu of using the traditional algebraic modelling approach provides a more reliable mixing field.

(3) The flame is very sensitive to small changes in the coflow temperature. A variation of roughly ±1.44% in the experimentally reported value of 1045 K results in substantial changes in the predictions. T_c = 1045 K yields best agreement with the experiments. The calculated liftoff height at the 1030 K level remains close to the experimentally measured value. A drastic decrease occurs when the coflow temperature is increased to 1060 K.

(4) For all coflow temperatures and using both CSDR models, the transport budgets in mixture fraction and physical spaces and the history of radical build-up ahead of the reaction zone indicate that the flame is stabilised by autoignition rather than premixed flame propagation. These findings fully agree with previous PDF calculations [4–6], with the CMC results of Stankovic and Merci [7], and in part with those of Navarro-Martinez and Kronenburg [8] and Patwardhan et al. [11].

(5) Standalone zero-dimensional CMC calculation indicates that the mixture is capable of igniting at higher scalar dissipation levels as the coflow temperature is increased. This provides an explanation for the occurrence of ignition at locations closer to the nozzle exit and the decrease in liftoff height with increasing coflow temperature.

(6) In comparison to Mortensen's model, Girimaji's model always results in lower CSDR at the "most reactive" mixture fraction and therefore results in earlier ignition and smaller liftoff heights for all coflow temperatures.

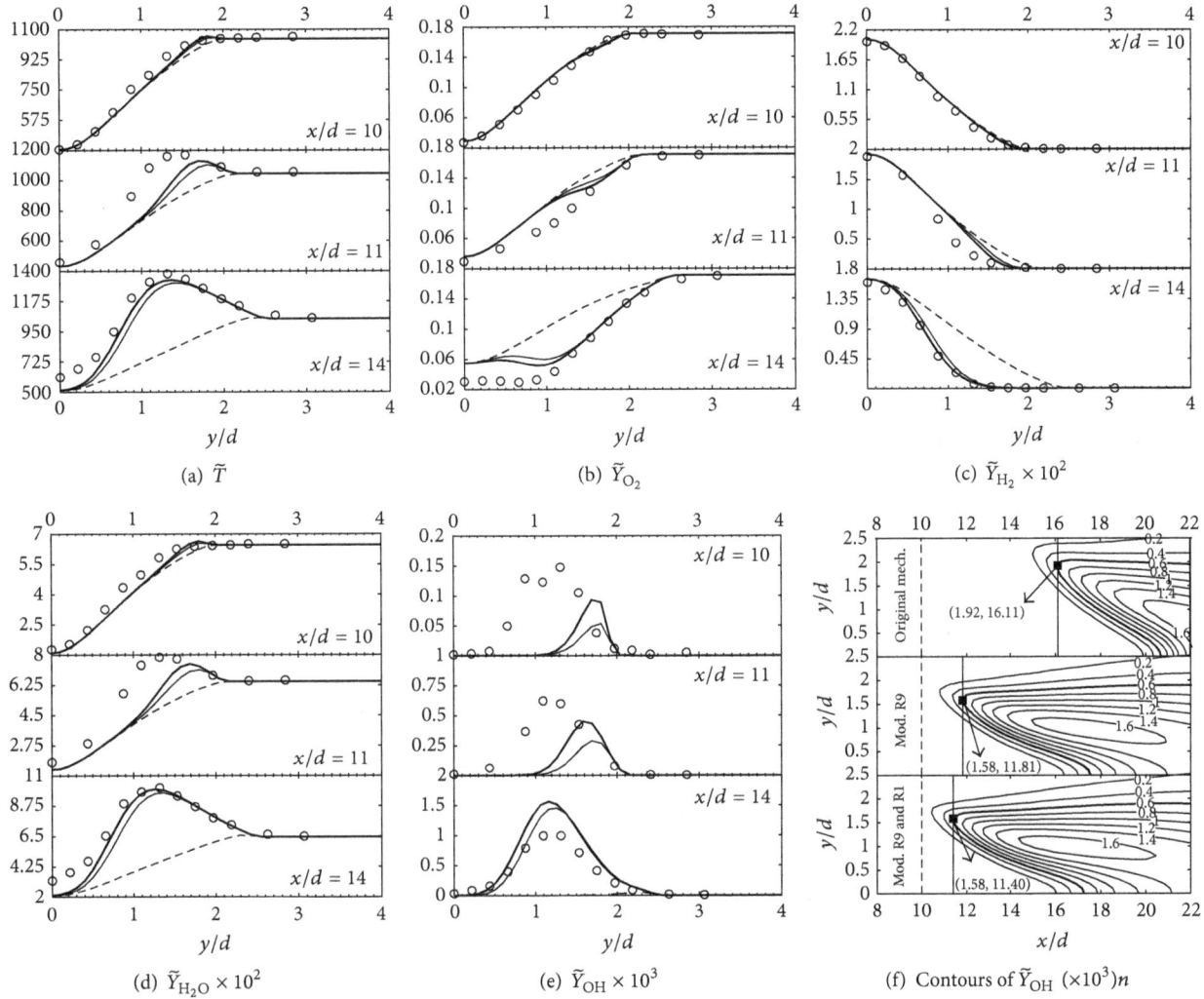

(a) $\tilde{T}$

(b) $\tilde{Y}_{O_2}$

(c) $\tilde{Y}_{H_2} \times 10^2$

(d) $\tilde{Y}_{H_2O} \times 10^2$

(e) $\tilde{Y}_{OH} \times 10^3$

(f) Contours of $\tilde{Y}_{OH}$ ($\times 10^3$)n

FIGURE 19: Radial profiles obtained using the original and modified versions of the Burke mechanism: (a) $\tilde{T}$, (b) $\tilde{Y}_{O_2}$, (c) $\tilde{Y}_{H_2}$ ($\times 10^2$), (d) $\tilde{Y}_{H_2O}$ ($\times 10^2$), and (e) $\tilde{Y}_{OH}$ ($\times 10^3$). In (a)–(e): dashed lines, original mechanism; thin solid lines, modified (R9); thick solid lines, modified (R9) and (R1); symbols, experimental data [3]. Subfigure (f) shows the contours of $\tilde{Y}_{OH}$ ($\times 10^3$). The thick contour corresponds to 600 ppm. The solid and dashed horizontal lines correspond to the numerical and experimental liftoff heights, respectively. The CSDR is modelled using Girimaji's model and $T_c = 1045$ K.

TABLE 5: Rate parameters of reaction (R9), H + O$_2$ (+M) $\rightleftharpoons$ HO$_2$ (+M).

		Ó Conaire	Mueller	Li	Burke
High-pressure limit	A	1.475×10^{12}	1.475×10^{12}	1.475×10^{12}	4.65084×10^{12}
	n	0.6	0.6	0.6	0.44
	E_a	0	0	0	0
Low-pressure limit	A	3.482×10^{16}	3.482×10^{16}	6.366×10^{20}	6.366×10^{20}
	n	-0.411	-0.411	-1.72	-1.72
	E_a	-1.115×10^3	-1.115×10^3	5.248×10^2	5.248×10^2
	F_c	0.5	0.5	0.8	0.5
	ε_{H_2}	1.3	2.5	2	2
	ε_{H_2O}	14	12	11	14
	ε_{O_2}	1	1	0.78	0.78

Units are cm^3-mole-sec-kcal-K; $k = AT^n \exp(-E_a/RT)$.
Nomenclature: A: preexponential coefficient; n: temperature exponent; E_a: activation energy; F_c: broadening factor; ε: third-body efficiency factor.

TABLE 6: Rate parameters of reaction (R1), $H + O_2 \rightleftharpoons O + OH$.

	Ó Conaire	Mueller	Li	Burke
A	1.915×10^{14}	1.915×10^{14}	3.547×10^{15}	1.040×10^{14}
n	0	0	-0.406	0
E_a	1.6439×10^4	1.6439×10^4	1.6599×10^4	1.5286×10^4

Same units and nomenclature as in Table 5.

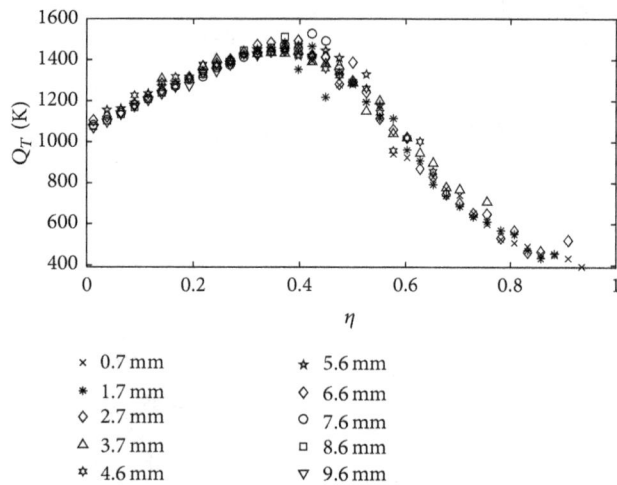

(a) Radial Q_T profiles

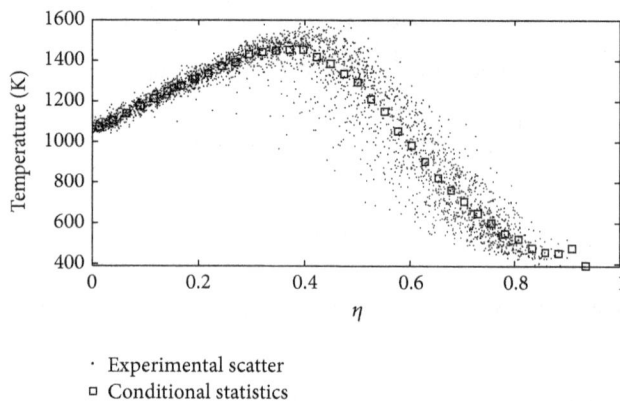

· Experimental scatter
□ Conditional statistics

(b) Q_T from all samples

FIGURE 20: Computation of the conditional temperature from the experimental scatter at $x/d = 14$: (a) profiles obtained from measurements at fixed radii (b) profile obtained from all samples superimposed over the experimental scatter.

(7) The spurious source terms resulting from the modelling of the CSDR are in general small but nonnegligible, mostly notably within the flame zone.

(8) The flame shows high sensitivity to chemical kinetics. The results obtained with the Mueller, Li, and Ó Conaire mechanisms reproduce the experimental trends with varying degrees of accuracy. The predictions are in general in good agreement with the experimental measurements, particularly those of the Mueller and Ó Conaire mechanisms. The three kinetic schemes predict very similar liftoff heights.

Conversely, the performance of the Burke mechanism is unsatisfactory due to the inability of the mixture to achieve early ignition. The temperature and species mass fractions are not well predicted and the liftoff height is grossly overestimated. The modification of the rate parameters of some of the key reactions improves the predictions substantially.

Further improvements are possible by modelling the conditional chemical sources with a second-order closure and by inclusion of counter-gradient and differential diffusion effects.

Appendix

Calculation of the Conditional Experimental Data

For each of the axial locations $x/d = 1, 8, 9, 10, 11, 14,$ and 26, measurements are reported at several radial locations. For a given axial location, the data is conditionally averaged at each radius as described in [14].

(1) The range of the mixture fraction is subdivided into 40 bins.

(2) The data is sorted into the bins.

(3) The average of the data in each bin is taken to be the conditional average at the centre of the bin.

As an example, the radial profiles of the experimental conditional temperature at $x/d = 14$ are shown in Figure 20(a). The weak radial dependence of the conditional data is obvious. Therefore, Steps 1–3 are repeated by using all of the experimental scatter at $x/d = 14$, irrespective of the radial location. The results are displayed in Figure 20(b).

Conflict of Interests

The authors declare that there is no conflict of interests regarding the publication of this paper.

Acknowledgments

This project is funded by the Ministry of Training, Colleges and Universities of Ontario through the OGS program, and the Natural Sciences and Engineering Research Council of Canada. The authors are grateful for their financial support.

References

[1] L. Vanquickenborne and A. van Tiggelen, "The stabilization mechanism of lifted diffusion flames," *Combustion and Flame*, vol. 10, no. 1, pp. 59–69, 1966.

[2] N. Peters and F. A. Williams, "Lift-off characteristics of turbulent jet diffusion flames," *AIAA journal*, vol. 21, no. 3, pp. 423–429, 1983.

[3] R. Cabra, T. Myhrvold, J. Y. Chen, R. W. Dibble, A. N. Karpetis, and R. S. Barlow, "Simultaneous laser raman-rayleigh-lif measurements and numerical modeling results of a lifted

turbulent H_2/N_2 jet flame in a vitiated coflow," *Proceedings of the Combustion Institute*, vol. 29, no. 2, pp. 1881–1888, 2002.

[4] A. R. Masri, R. Cao, S. B. Pope, and G. M. Goldin, "PDF calculations of turbulent lifted flames of H_2/N_2 fuel issuing into a vitiated co-flow," *Combustion Theory and Modelling*, vol. 8, no. 1, pp. 1–22, 2004.

[5] R. R. Cao, S. B. Pope, and A. R. Masri, "Turbulent lifted flames in a vitiated coflow investigated using joint PDF calculations," *Combustion and Flame*, vol. 142, no. 4, pp. 438–453, 2005.

[6] R. L. Gordon, A. R. Masri, S. B. Pope, and G. M. Goldin, "A numerical study of auto-ignition in turbulent lifted flames issuing into a vitiated co-flow," *Combustion Theory and Modelling*, vol. 11, no. 3, pp. 351–376, 2007.

[7] I. Stankovic and B. Merci, "LES-CMC simulations of turbulent hydrogen flame in a vitiated air co-flow," in *Proceedings of the 7th Mediterranean Combustion Symposium*, September 2011.

[8] S. Navarro-Martinez and A. Kronenburg, "Flame stabilization mechanisms in lifted flames," *Flow, Turbulence and Combustion*, vol. 87, no. 2-3, pp. 377–406, 2011.

[9] E. E. O'Brien and T.-L. Jiang, "The conditional dissipation rate of an initially binary scalar in homogeneous turbulence," *Physics of Fluids A*, vol. 3, no. 12, pp. 3121–3123, 1991.

[10] S. Navarro-Martinez, A. Kronenburg, and F. Di Mare, "Conditional moment closure for large eddy simulations," *Flow, Turbulence and Combustion*, vol. 75, no. 1-4, pp. 245–274, 2005.

[11] S. S. Patwardhan, S. De, K. N. Lakshmisha, and B. N. Raghunandan, "CMC simulations of lifted turbulent jet flame in a vitiated coflow," *Proceedings of the Combustion Institute*, vol. 32, no. 2, pp. 1705–1712, 2009.

[12] V. R. Kuznetsov and V. A. Sabel'nikov, *Turbulence and Combustion*, Hemisphere Publishing, New York, NY, USA, 1990.

[13] R. W. Bilger, "Some aspects of scalar dissipation," *Flow, Turbulence and Combustion*, vol. 72, no. 2-4, pp. 93–114, 2004.

[14] A. Y. Klimenko and R. W. Bilger, "Conditional moment closure for turbulent combustion," *Progress in Energy and Combustion Science*, vol. 25, no. 6, pp. 595–687, 1999.

[15] M. J. Cleary, *CMC Modelling of Enclosure Fires [Ph.D. thesis]*, University of Sydney, Sydney, Australia, 2004.

[16] K. Tsai and R. O. Fox, "Modeling multiple reactive scalar mixing with the generalized IEM model," *Physics of Fluids*, vol. 7, no. 11, pp. 2820–2830, 1995.

[17] W. P. Jones and J. H. Whitelaw, "Calculation methods for reacting turbulent flows: a review," *Combustion and Flame*, vol. 48, pp. 1–26, 1982.

[18] E. S. Richardson and E. Mastorakos, "Simulations of non-premixed edge-flame propagation in turbulent non-premixed jets," in *Proceedings of the 3rd European Combustion Meeting (ECM '07)*, vol. 3, pp. 1–6, April 2007.

[19] S. B. Pope, "The probability approach to the modelling of turbulent reacting flows," *Combustion and Flame*, vol. 27, pp. 299–312, 1976.

[20] S. M. de Bruyn Kops and M. Mortensen, "Conditional mixing statistics in a self-siminar scalar mixing layer," *Physics of Fluids*, vol. 17, no. 9, Article ID 095107, pp. 1–11, 2005.

[21] S. S. Girimaji, "On the modeling of scalar diffusion in isotropic turbulence," *Physics of Fluids A*, vol. 4, no. 11, pp. 2529–2537, 1992.

[22] M. Mortensen, "Consistent modeling of scalar mixing for presumed, multiple parameter probability density functions," *Physics of Fluids*, vol. 17, no. 1, Article ID 018106, p. 4, 2005.

[23] M. Mortensen, "Implementation of a conditional moment closure for mixing sensitive reactions," *Chemical Engineering Science*, vol. 59, no. 24, pp. 5709–5723, 2004.

[24] I. S. Kim and E. Mastorakos, "Simulations of turbulent lifted jet flames with two-dimensional conditional moment closure," *Combustion and Flame*, vol. 30, no. 1, pp. 911–918, 2005.

[25] http://www.sandia.gov/TNF/radiation.html .

[26] ANSYS Inc., "ANSYS FlUENT 12. 1," 2009.

[27] R. Cabra, *Turbulent Jet Flames Into a Vitiated Coflow [Ph.D. thesis]*, University of California, Berkeley, Calif, USA, 2004.

[28] ANSYS Inc., "ANSYS ICEM CFD," 2009.

[29] S. B. Pope, "An explanation of the turbulent roundjet/plane-jet anomaly," *AIAA Journal*, vol. 16, no. 3, pp. 279–281, 1978.

[30] W. P. Jones and P. Musonge, "Closure of the Reynolds stress and scalar flux equations," *Physics of Fluids*, vol. 31, no. 12, pp. 3589–3604, 1988.

[31] I. S. Kim and E. Mastorakos, "Simulations of turbulent non-premixed counterflow flames with first-order conditional moment closure," *Flow, Turbulence and Combustion*, vol. 76, no. 2, pp. 133–162, 2006.

[32] B. Koren, "A robust upwind discretization method for advection, diffusion and source terms," in *Numerical Methods For Advection-Diffusion Problems*, C. B. Vreugdenhil and B. Koren, Eds., vol. 45 of *Notes on Numerical Fluid Mechanics*, pp. 117–138, Vieweg, Braunschweig, Germany, 1993.

[33] G. Strang, "On the construction and comparison of difference schemes," *SIAM Journal on Numerical Analysis*, vol. 5, no. 3, pp. 506–517, 1968.

[34] P. N. Brown, G. D. Byrne, and A. C. Hindmarsh, "VODE: a Variable-Coefficient ODE solver," *SIAM Journal on Scientific and Statistical Computing*, vol. 10, no. 5, pp. 1038–1051, 1988.

[35] P. N. Brown and A. C. Hindmarsh, "Reduced storage matrix methods in stiff ODE systems," *Applied Mathematics and Computation*, vol. 31, pp. 40–91, 1989.

[36] G. D. Byrne, "Pragmatic experiments with Krylov methods in the stiff ODE setting," in *Computational Ordinary Differential Equations*, J. Cash and I. Gladwell, Eds., pp. 323–356, Oxford University Press, Oxford, UK, 1992.

[37] R. Piessens, E. de Doncker-Kapenga, and C. W. Ueberhuber, *Quadpack: A Subroutine Package for Automatic Integration*, Springer Series in Computational Mathematics, Springer, Berlin, Germany, 1983.

[38] W. H. Press, S. A. Teukolsky, W. T. Vetterling, and B. P. Flannery, *Numerical Recipes in Fortran 77: The Art of Scientific Computing*, Cambridge University Press, Cambridge, UK, 2nd edition, 1992.

[39] C. J. F. Ridders, "Accurate computation of $F'(x)F''(x)$," *Advances in Engineering Software*, vol. 4, no. 5, pp. 75–76, 1982.

[40] M. A. Mueller, T. J. Kim, R. A. Yetter, and F. L. Dryer, "Flow reactor studies and kinetic modeling of the H_2/O_2 reaction," *International Journal of Chemical Kinetics*, vol. 31, no. 2, pp. 113–125, 1999.

[41] J. Li, Z. Zhao, A. Kazakov, and F. L. Dryer, "An updated comprehensive kinetic model of hydrogen combustion," *International Journal of Chemical Kinetics*, vol. 36, no. 10, pp. 566–575, 2004.

[42] M. P. Burke, M. Chaos, Y. Ju, F. L. Dryer, and S. J. Klippenstein, "Comprehensive H_2O_2 kinetic model for high-pressure combustion," *International Journal of Chemical Kinetics*, vol. 44, no. 7, pp. 444–474, 2012.

[43] M. Ó Conaire, H. J. Curran, J. M. Simmie, W. J. Pitz, and C. K. Westbrook, "A comprehensive modeling study of hydrogen

oxidation," *International Journal of Chemical Kinetics*, vol. 36, no. 11, pp. 603–622, 2004.

[44] J. E. Kent, *B.E.thesis*, the University of Sydney, 2003.

[45] C. N. Markides, *Autoignition in turbulent flows [Ph.D. thesis]*, University of Cambridge, Cambridge, UK, 2005.

[46] C. N. Markides, G. De Paola, and E. Mastorakos, "Measurements and simulations of mixing and autoignition of an n-heptane plume in a turbulent flow of heated air," *Experimental Thermal and Fluid Science*, vol. 31, no. 5, pp. 393–401, 2007.

[47] C. S. Yoo, R. Sankaran, and J. H. Chen, "Three-dimensional direct numerical simulation of a turbulent lifted hydrogen jet flame in heated coflow: flame stabilization and structure," *Journal of Fluid Mechanics*, vol. 640, pp. 453–481, 2009.

[48] N. Peters, *Turbulent Combustion*, Cambridge University Press, Cambridge, UK, 2000.

[49] I. Stanković and B. Merci, "Analysis of auto-ignition of heated hydrogen-air mixtures with different detailed reaction mechanisms," *Combustion Theory and Modelling*, vol. 15, no. 3, pp. 409–436, 2011.

[50] C. B. Devaud, R. W. Bilger, and T. Liu, "A new method of modeling the conditional scalar dissipation rate," *Physics of Fluids*, vol. 16, no. 6, pp. 2004–2011, 2004.

[51] C. J. Cobos, H. Hippler, and J. Troe, "High-pressure falloff curves and specific rate constants for the reactions $H+O_2 \leftrightarrows HO_2 \leftrightarrows HO + O$," *Journal of Physical Chemistry*, vol. 89, no. 2, pp. 342–349, 1985.

[52] J. Troe, "Detailed modeling of the temperature and pressure dependence of the reaction $H+O_2(+M) \rightarrow HO_2(+M)$," *Proceedings of the Combustion Institute*, vol. 28, no. 2, pp. 1463–1469, 2000.

[53] M. A. Mueller, R. A. Yetter, and F. L. Dryer, "Measurement of the rate constant for $H+O_2 + M(m = N_2, ar)$ using kinetic modeling of the high-pressure $H_2/O_2/NO_x$ reaction," *Proceedings of the Combustion Institute*, vol. 27, no. 1, pp. 177–184, 1998.

[54] J. V. Michael, M.-C. Su, J. W. Sutherland, J. J. Carroll, and A. F. Wagner, "Rate constants for $H+O_2 + M \rightarrow HO_2+M$ in seven bath gases," *Journal of Physical Chemistry A*, vol. 106, no. 21, pp. 5297–5313, 2002.

[55] A. N. Pirraglia, J. V. Michael, J. W. Sutherland, and R. B. Klemm, "A flash photolysis-shock tube kinetic study of the H atom reaction with $H+O_2 \leftrightarrows OH+O$ (962 K ≤ T ≤ 1705 K) and $H+O_2 + Ar \rightarrow HO_2 + Ar$ (746 K ≤ T ≤ 987 K)," *Journal of Physical Chemistry*, vol. 93, no. 1, pp. 282–291, 1989.

[56] J. P. Hessler, "Calculation of reactive cross sections and micro-canonical rates from kinetic and thermochemical data," *Journal of Physical Chemistry A*, vol. 102, no. 24, pp. 4517–4526, 1998.

[57] Z. Hong, D. F. Davidson, E. A. Barbour, and R. K. Hanson, "A new shock tube study of the $H+O_2 \rightarrow OH + O$ reaction rate using tunable diode laser absorption of H_2O near 2. 5 μm," *Proceedings of the Combustion Institute*, vol. 33, no. 1, pp. 309–316, 2011.

12

Thermodynamic Model for Updraft Gasifier with External Recirculation of Pyrolysis Gas

Fajri Vidian,[1] Adi Surjosatyo,[2] and Yulianto Sulistyo Nugroho[2]

[1]*Department of Mechanical Engineering, Faculty of Engineering, Universitas Sriwijaya, Jalan Raya Palembang - Prabumulih km 32, Indralaya, Ogan Ilir, Sumatera Selatan 30662, Indonesia*
[2]*Department of Mechanical Engineering, Faculty of Engineering, Universitas Indonesia, Kampus Baru UI Depok, Jawa Barat 16424, Indonesia*

Correspondence should be addressed to Fajri Vidian; fajri.vidian@unsri.ac.id

Academic Editor: Ishwar K. Puri

Most of the thermodynamic modeling of gasification for updraft gasifier uses one process of decomposition (decomposition of fuel). In the present study, a thermodynamic model which uses two processes of decomposition (decomposition of fuel and char) is used. The model is implemented in modification of updraft gasifier with external recirculation of pyrolysis gas to the combustion zone and the gas flowing out from the side stream (reduction zone) in the updraft gasifier. The goal of the model obtains the influences of amount of recirculation pyrolysis gas fraction to combustion zone on combustible gas and tar. The significant results of modification updraft are that the increases amount of recirculation of pyrolysis gas will increase the composition of H_2 and reduce the composition of tar; then the composition of CO and CH_4 is dependent on equivalence ratio. The results of the model for combustible gas composition are compared with previous study.

1. Introduction

The types commonly used in gasification are the fixed bed (updraft and downdraft) and fluidized bed. The updraft gasifier has many advantages but, on the other hand, produced high level of tar [1]. Several modifications to gasifier with recirculation of pyrolysis gas have been conducted to reduce the tar [2–4]. In previous study, we have modified type of updraft gasifier with pyrolysis gas recirculated externally to the combustion zone. Furthermore, the produced gas exits at the reduction zone (side stream) to reduce the tar [4].

Modeling is needed in the gasification process as a tool to predict the results that will be obtained from an experiment. Some thermodynamic modeling using Aspen Plus has been done before on updraft gasifier. Chen et al. conducted a model using one process of decomposition of MSW as fuel to see the effect of flue gases of combustion on the LHV; the results showed the flue gas will increase the LHV [5]. He et al. applied a model on Lurgi gasification process using one process of decomposition of coal as fuel to see the effect of oxygen/coal ratio on exergetic efficiency of the process;

the results of modeling showed oxygen/coal ratio affecting exergetic efficiency of the process [6]. Modeling generally only uses one process of decomposition using ultimate analysis of fuel. In this study, we use a model using two processes of decomposition (decomposition of fuel and char) using the ultimate analysis of fuel and char. The model is implemented on updraft gasifier with external recirculation of pyrolysis gas [4].

Modeling aims to get the effect of the amount of external recirculations of pyrolysis gas flow to the combustion zone on the composition of the combustible gas (CO, H_2, and CH_4) and tar composition (C_7H_8O, $C_{10}H_8$).

2. Methodology

The process that occurs during the gasification includes drying, pyrolysis, reduction, and combustion; meanwhile each stage of the process results in a product for the next stage or directly results in a gasification process product. The processes of drying and pyrolysis are the processes that are not in an equilibrium state and may take place instantaneously [7]. The composition of the product in its process is calculated using the mass and energy balance principles. The process of gasification and combustion is a process that can take place in an thermodynamic equilibrium state [8]. The composition of the product in thermodynamic equilibrium conditions can be calculated using Gibbs free energy minimization [9].

The reactions of combustion and gasification taking place at equilibrium conditions are as follows:

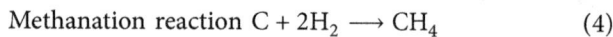

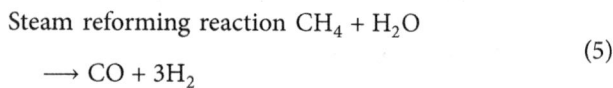

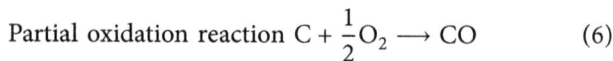

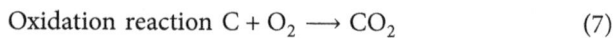

Boudouard reaction $C + CO_2 \longrightarrow 2CO$ (1)

Water gas reaction $C + H_2O \longrightarrow CO + H_2$ (2)

Water gas shift reaction $CO + H_2O \longrightarrow CO_2 + H_2$ (3)

Methanation reaction $C + 2H_2 \longrightarrow CH_4$ (4)

Steam reforming reaction $CH_4 + H_2O$

$\longrightarrow CO + 3H_2$ (5)

Partial oxidation reaction $C + \frac{1}{2}O_2 \longrightarrow CO$ (6)

Oxidation reaction $C + O_2 \longrightarrow CO_2$ (7)

In this simulation process, tar is modeled as cresol and naphthalene. Cresol is to represent nearly approaching heavy tar and naphthalene to represent light tar [10]. Tar will experience reforming when entering the combustion and gasification zones to form H_2 and CO through the reaction of steam reforming and dry reforming [11]:

$C_xH_yO_z + H_2O \longrightarrow CO + H_2 + tar$ (8)

$C_xH_yO_z + CO_2 \longrightarrow CO + H_2O + tar$ (9)

In equilibrium, the total Gibbs free energy (G_{total}) of the system is in a minimum state. The concentration of each compound (n_i) is obtained by minimizing the objective function G_{total}:

$$G_{total} = \sum_{i=1}^{N} n_i \Delta G_{f,i}^0 + \sum_{i=1}^{N} n_i RT \ln\left(\frac{n_i}{\sum n_i}\right).$$ (10)

Limits for completion of the object function are expressed in the mass balance of elements:

$$\sum_{i=1}^{N} a_{ij} n_i = A_j, \quad j = 1, 2, 3, \ldots, k.$$ (11)

G_{total} is total Gibbs free energy, G_f^0 is standard Gibbs free energy, R is universal gas constant, T is temperature, N is number of compounds, a_{ij} is number of atoms in element number j in mole of compound number i, and A_j is number of atoms in element number j in mixture of reaction.

The Gibbs minimization equation can be solved using Langrange multiplier:

$$L = G_{total} - \sum_{j=1}^{K} \lambda j \left[\sum_{i=1}^{N} a_{ij} n_i - A_j \right],$$ (12)

where λ is Langrange multiplier.

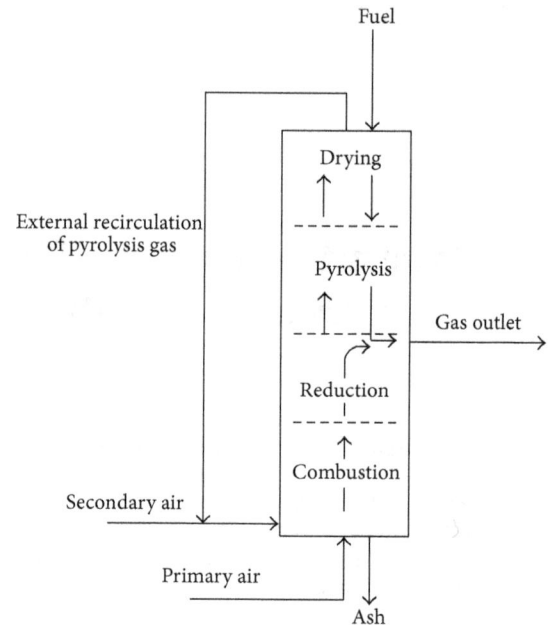

FIGURE 1: Updraft gasifier with external recirculation of pyrolysis gas [4].

Equilibrium conditions will be obtained if the partial derivative of the Langrange function is zero:

$$\left[\frac{\partial L}{\partial n_i} \right] = \frac{\Delta G_{f,i}^0}{RT} + \sum_{i=1}^{N} \ln\left(\frac{n_i}{\sum n_i} \right) + \frac{1}{RT} \sum_{j=1}^{K} \lambda j \left(\sum_{i=1}^{N} a_{ij} n_i \right) = 0.$$ (13)

Calorific value of gas resulting from the gasification process is calculated based on

$$LHV \left(kJ/Nm^3 \right) = y_{co} \cdot 12621 + y_{H_2} \cdot 10779 + y_{CH_4} \cdot 35874.$$ (14)

Modeling starts with the approach to the process of experiments that have been carried out by Surjosatyo et al. [4] as shown in Figure 1. Furthermore, the approaches are used to facilitate the simulation process. The approach is taken to the gas flows out from the gasifier at the end of the pyrolysis process, so that, in modeling, the fraction of the pyrolysis gas uncirculated does not enter the equilibrium zone and directly goes out of the reactor (Figure 2). Primary air and secondary air in the experiment are the total quantity of air for the combustion process, so that, in modeling, the total quantity of air only is used to enter the equilibrium zone.

Modeling is developed by applying the mass and energy balance as well as the Gibbs free energy minimization in the configuration of the updraft gasifier reactor with an external recirculation of pyrolysis gas as described in Figure 1. The mass and energy balance are used in the stages of drying, pyrolysis, reduction, and combustion processes. Gibbs free energy minimization is used during the combustion and

Thermodynamic Model for Updraft Gasifier with External Recirculation of Pyrolysis Gas

TABLE 1: Proximate and ultimate analysis of fuel.

	Unit	Value	Normalization
Proximate analysis			
Moisture	Mass fraction (%)	10.24	
Ash	Mass fraction (%)	2.71	
Volatile	Mass fraction (%)	71.80	
Fixed carbon	Mass fraction (%)	15.25	
Ultimate analysis (dry basis)			
Carbon	Mass fraction (%)	43.33	48.27
Hydrogen	Mass fraction (%)	5.11	5.69
Oxygen	Mass fraction (%)	38.61	43.01

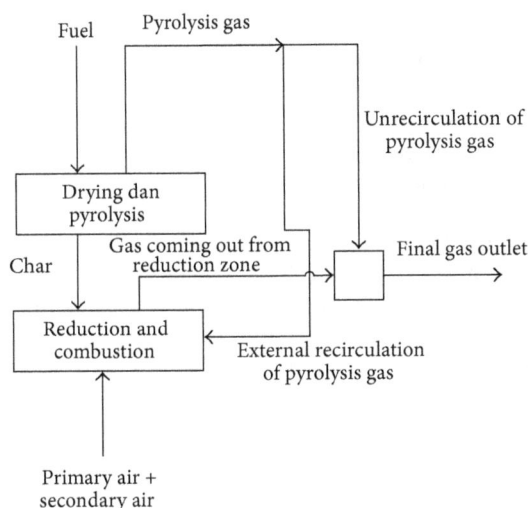

FIGURE 2: Configuration model updraft gasifier with external recirculation of pyrolysis gas.

TABLE 2: Pyrolysis product (fuel decomposition).

Compound	Mass (kg)	Mass fraction
CO	0.49	0.10
CH_4	0.09	0.02
H_2O	1.79	0.36
CO_2	0.51	0.10
C_7H_8O (tar)	0.76	0.15
$C_{10}H_8$ (tar)	0.36	0.07
Char	0.88	0.18
Ash	0.12	0.02

TABLE 3: Pyrolysis product of wood [13, 14].

Spices	Mass fraction (%) [13]	Mass fraction (%) [14]
Gas	64.8	51
Pyrolysis water		26
Tar (organic compound)	20.0	15
Char	15.2	21

reduction (gasification) processes to obtain the product composition.

Modeling of the pyrolysis zone is a process of decomposition of nonconventional fuel elements into the conventional compound [12]. The conversion process is facilitated by using proximate and ultimate analysis data available in Table 1 which is further decomposed into its conventional compounds as presented in Table 2. The conventional data could be compared to the data of literature obtained by Wang et al. [13] and Milhé et al. [14] as presented in Table 3. It shows tendency of the same distribution of the composition.

Modeling in the reduction and combustion zones is performed using proximate and ultimate analysis data of char. The proximate and ultimate analysis data of char is decomposed into its conventional compound. The product of decomposition is mixed with the pyrolysis products and combustion air. The mixture then enters the thermodynamic equilibrium reactions in the process of gasification and combustion.

The ratio of the amount of gas resulting from the process of pyrolysis recirculated to enter the combustion chamber to the total amount of gas resulting from the pyrolysis process

is called the fraction of the pyrolysis gas recirculating. The fraction of pyrolysis gas recirculation is a variable that is used to see the effect of pyrolysis gas recirculation on composition of combustible gas (CO, H_2, and CH_4) and tar (C_7H_8O and $C_{10}H_8$). Modeling carried out under operating parameters is presented in Table 4.

Modeling is performed using Aspen Plus Simulator. The process that occurs in the drying and the pyrolysis zone simulated using RYield reactor block [7, 15] and the processes in the zone of combustion and gasification are modeled using RGibbs [8, 16]. Thermodynamic properties are calculated using the state equation of Peng-Robinson. The enthalpy and density of nonconventional component are calculated using the models of HCOALGEN and DCOALIGT.

3. Results and Discussions

3.1. The Effect of the Fraction of Pyrolysis Gas Recirculation. Figure 3 presents the results for simulating the effect of the fraction of pyrolysis gas recirculation to the combustion zone

TABLE 4: Simulation parameters.

Number	Parameter	Value
1	Stoichiometric combustion air (kg/h)	25.6
2	Fuel flow rate (kg/h)	5
3	Equivalence ratio (ER)	0.2–0.4
4	Pyrolysis gas recirculation fraction	0-1

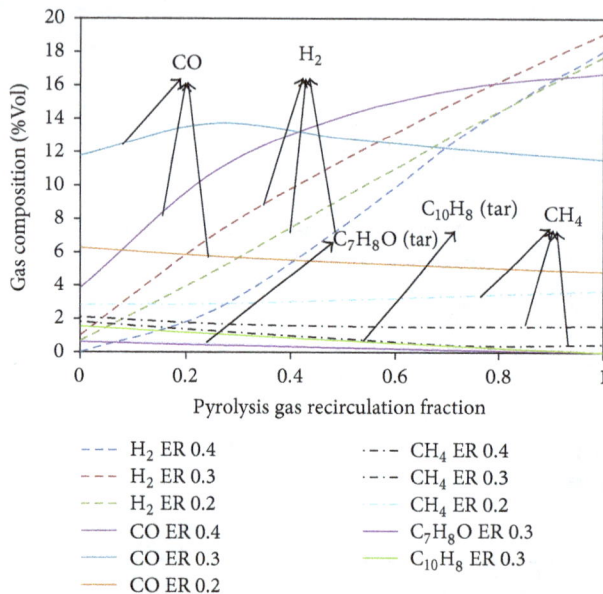

---	H_2 ER 0.4	-·-- CH_4 ER 0.4
---	H_2 ER 0.3	-·-- CH_4 ER 0.3
---	H_2 ER 0.2	---- CH_4 ER 0.2
——	CO ER 0.4	—— C_7H_8O ER 0.3
——	CO ER 0.3	—— $C_{10}H_8$ ER 0.3
——	CO ER 0.2	

FIGURE 3: Gasification product according to pyrolysis gas recirculation fraction and equivalence ratio.

on the combustible gas (CO, H_2, and CH_4) and tar (C_7H_8O, $C_{10}H_8$).

H_2 concentration increases by the addition of pyrolysis gas recirculation flow fraction for each equivalence ratio due to the greater amount of water vapor and tar carried away by the flow of pyrolysis gas. The reactions contributing to increasing H_2 are the water gas reaction, water gas shift reaction, and steam reforming tar reaction. The modeling results show the same trend as the experimental results [3, 4, 17, 18]. The increase in equivalence ratio from 0.2 to 0.3 tends to increase the concentration of H_2 but in the equivalence ratio above 0.3 will reduce the composition of H_2. This is because of a shift in reaction to the Boudouard reaction and reverse water gas shift due to the increase in heat of combustion. The maximum concentration of H_2 is obtained on the condition that the amount of pyrolysis gas recirculation flow is maximum (pyrolysis gas recirculation flow fraction of 1).

CO concentration tends to decrease by the increase in the fraction of pyrolysis gas recirculation flow until the equivalence ratio reaches 0.3 and then will increase. The modeling results show the same tendency as the experimental results [3, 4, 17, 18]. This is due to increasing Boudouard reaction which is influenced by the increase in the heat released by the combustion process and CO_2 carried by the pyrolysis gas.

CH_4 concentration tends to increase by the increase in the fraction of pyrolysis gas recirculation flow until the equivalence ratio reaches 0.2 and then will decrease. This is because, in the equivalence ratio below 0.2, the methanation reaction takes place better as it is an exothermic reaction. The addition of pyrolysis gas will increase the amount of H_2 for the methanation reaction. CH_4 carried by the fraction of pyrolysis gas recirculation flow also contributes to the increase of CH_4. In the equivalence ratio above 0.2, the methanation reaction will decrease by the increase of heat released on combustion reaction. CH_4 in the pyrolysis gas will react with H_2O to produce H_2 through steam reforming reaction.

The tar content in the producer gas decrease by the increase in fractions of pyrolysis gas recirculation flow to the combustion zone as presented in Figure 3. This is because the tar experiences decomposition through the reforming process into H_2 and CO as it receives heat resulting from combustion process.

The pyrolysis gas recirculation flow fractions of zero (0) and one (1) represent each of the updraft and downdraft operations. The difference in the operating condition leads to significant difference in composition of combustible gas and tar, especially the concentration of H_2. It shows the significance of the effect of H_2O (water vapor) and tar contents in the pyrolysis gas on H_2 concentration. Research by Ueki et al. [19] and Saravanakumar et al. [20] presents a direct comparison between the operation of the updraft and downdraft gasifier or crossdraft gasifier. The results show there is a large difference in concentration between H_2 and tar.

The simulation results show that the variation of pyrolysis gas recirculation fraction gives an influence on the composition of the combustible gas and tar. Therefore, its variables are very important to do the experiment until the fraction of the pyrolysis gas recirculation flow reaches maximum.

3.2. The Effect of Equivalence Ratio. In the condition where the recirculation flow fraction is maximum (pyrolysis gas recirculation fraction of 1) as shown in Figure 4, an increase in the equivalence ratio will increase the concentration of H_2 until equivalence ratio of 0.3 and CO until equivalence ratio of 0.4 and then it will decrease. This is because the process will move towards the stoichiometric combustion where more heat is released. The increase in heat causes the Boudouard and the water gas reactions to be better. However, under a certain condition, its reaction will decrease because of the decrease of char quantity to produce heat when it burns a lot by the addition of the air supply (an increase of equivalence ratio) [21]. An increase in the equivalence ratio reduces the concentration of CH_4. It is caused by a decrease in methanation reaction. Methanation reaction is a reaction that releases heat; therefore it will be dominantly in the lower equivalence ratio. The results of model have a similar tendency with model results reported by Doherty et al. [7], Puig-Arnavat et al. [21], and Reed and Das [22], especially for Reed and Das model in which H_2 seems to continuously decrease. This is because in this modeling there is a heat loss.

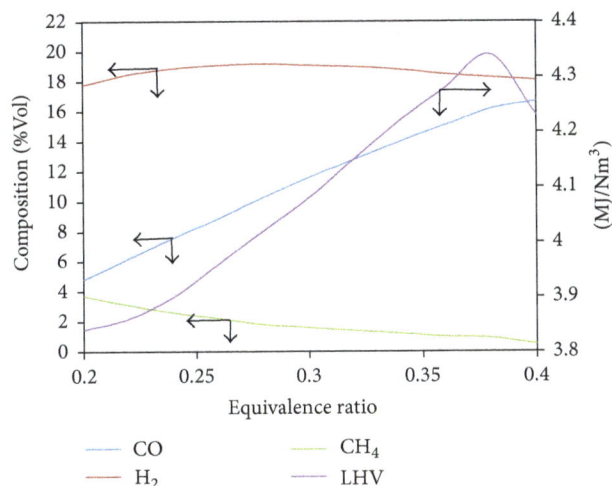

FIGURE 4: Combustible gas and LHV according to equivalence ratio at the maximum pyrolysis gas recirculation fraction.

TABLE 5: Comparison between models and experiments for gas composition in ER 0.3 and fraction of pyrolysis gas recirculation of 0.5.

Combustible gas	Model (% mol)	Experiment (% mol)	Error (%)
CO	12.8	14.4	11.1
H_2	11.5	14.93	22.9
CH_4	1.6	3.49	54.1

The simulation results in equivalence ratio of 0.2 to 0.4 at the maximum pyrolysis gas recirculation flow which results in a high LHV. In the experiment, it is necessary to select a mechanism that can discharge the pyrolysis gas recirculation flow maximum with total combustion air between 0.2 and 0.4 of stoichiometric air.

The LHV maximum of 4.34 MJ/Nm³ is in the equivalence ratio of 0.38. The simulation shows the best operating condition which occurs in the equivalence ratio of 0.38 with a fraction of the maximum pyrolysis gas flow. This condition would provide the highest value of LHV and the lowest of tar.

3.3. Comparison between Simulation and Experiment.
Comparison of the results of simulations and experiments which have been conducted by Surjosatyo et al. [4] is presented in Table 5. The results of the comparison show a trend of values which are almost equal to the concentration of CO and H_2. The CH_4 composition shows a bit far different value. The difference in the CH_4 value is almost equal to that obtained from thermodynamic modeling carried out by Kuo et al. [8], Jarungthammachote and Dutta [23], and Barratieri et al. [24]. This is because the CH_4 final product is affected by factors beyond the equilibrium (nonequilibrium). Thus, it cannot be predicted purely thermodynamically [24].

In general, the results of the simulation show lower values, because the parameters in the simulation are still different to the experimental condition.

4. Conclusions

The simulation results show that an increase in the fraction of the pyrolysis gas recirculation flow will affect the composition of the combustible gas and tar.

The concentration of H_2 increases in each equivalence ratio constant. The concentration of CO decreases until the equivalence ratio reaches 0.3 and increases above that. The concentration of CH_4 increases until the equivalence ratio reaches 0.2 and decreases above that.

The composition of C_7H_8O (tar) and $C_{10}H_8$ (tar) decreases by the increase in the fraction of pyrolysis gas recirculation flow into the combustion zone.

The best operating condition exists when the amount of recirculated flow is maximum (pyrolysis gas recirculation flow fraction of 1) and the equivalence ratio is 0.38.

Modeling and experimental results show a tendency to approach the same value, especially composition of CO and H_2.

Conflict of Interests

The authors declare that there is no conflict of interests regarding the publication of this paper.

Acknowledgments

The authors acknowledge Mr. Prof. Dr. Ir. Herri Susanto and Muflih Arisa Adnan, S.T. from the Department of Chemical Engineering, Institut Teknologi Bandung, for all the supports in the form of guidance, advice, inputs, and corrections in preparation and completion of the modeling process and the paper.

References

[1] P. Plis and R. K. Wilk, "Theoretical and experimental investigation of biomass gasification process in a fixed bed gasifier," *Energy*, vol. 36, no. 6, pp. 3838–3845, 2011.

[2] H. Susanto and A. A. C. M. Beenackers, "A moving-bed gasifier with internal recycle of pyrolysis gas," *Fuel*, vol. 75, no. 11, pp. 1339–1347, 1996.

[3] Y. Cao, Y. Wang, J. T. Riley, and W.-P. Pan, "A novel biomass air gasification process for producing tar-free higher heating value fuel gas," *Fuel Processing Technology*, vol. 87, no. 4, pp. 343–353, 2006.

[4] A. Surjosatyo, F. Vidian, and Y. S. Nugroho, "Experimental gasification of biomass in an updraft gasifier with external recirculation of pyrolysis gases," *Journal of Combustion*, vol. 2014, Article ID 832989, 6 pages, 2014.

[5] C. Chen, Y.-Q. Jin, J.-H. Yan, and Y. Chi, "Simulation of municipal solid waste gasification in two different types of fixed bed reactors," *Fuel*, vol. 103, pp. 58–63, 2013.

[6] C. He, X. Feng, and K. H. Chu, "Process modeling and thermodynamic analysis of Lurgi fixed-bed coal gasifier in an SNG plant," *Applied Energy*, vol. 111, pp. 742–757, 2013.

[7] W. Doherty, A. Reynolds, and D. Kennedy, "The effect of air preheating in a biomass CFB gasifier using ASPEN Plus simulation," *Biomass and Bioenergy*, vol. 33, no. 9, pp. 1158–1167, 2009.

[8] P.-C. Kuo, W. Wu, and W.-H. Chen, "Gasification performances of raw and torrefied biomass in a downdraft fixed bed gasifier using thermodynamic analysis," *Fuel*, vol. 117, part B, pp. 1231–1241, 2014.

[9] S. Jarungthammachote and A. Dutta, "Equilibrium modeling of gasification: gibbs free energy minimization approach and its application to spouted bed and spout-fluid bed gasifiers," *Energy Conversion and Management*, vol. 49, no. 6, pp. 1345–1356, 2008.

[10] C. Li and K. Suzuki, "Tar property, analysis, reforming mechanism and model for biomass gasification—an overview," *Renewable and Sustainable Energy Reviews*, vol. 13, no. 3, pp. 594–604, 2009.

[11] W.-G. Wu, Y.-H. Luo, Y. Chen et al., "Experimental investigation of tar conversion under inert and partial oxidation conditions in a continuous reactor," *Energy & Fuels*, vol. 25, no. 6, pp. 2721–2729, 2011.

[12] S. Begum, M. G. Rasul, D. Akbar, and N. Ramzan, "Performance analysis of an integrated fixed bed gasifier model for different biomass feedstocks," *Energies*, vol. 6, no. 12, pp. 6508–6524, 2013.

[13] Y. Wang, T. Namioka, and K. Yoshikawa, "Effects of the reforming reagents and fuel species on tar reforming reaction," *Bioresource Technology*, vol. 100, no. 24, pp. 6610–6614, 2009.

[14] M. Milhé, L. Van De Steene, M. Haube, J.-M. Commandré, W.-F. Fassinou, and G. Flamant, "Autothermal and allothermal pyrolysis in a continuous fixed bed reactor," *Journal of Analytical and Applied Pyrolysis*, vol. 103, pp. 102–111, 2013.

[15] M. B. Nikoo and N. Mahinpey, "Simulation of biomass gasification in fluidized bed reactor using ASPEN PLUS," *Biomass and Bioenergy*, vol. 32, no. 12, pp. 1245–1254, 2008.

[16] N. Ramzan, A. Ashraf, S. Naveed, and A. Malik, "Simulation of hybrid biomass gasification using Aspen plus: a comparative performance analysis for food, municipal solid and poultry waste," *Biomass and Bioenergy*, vol. 35, no. 9, pp. 3962–3969, 2011.

[17] W. Zou, C. Song, S. Xu, C. Lu, and Y. Tursun, "Biomass gasification in an external circulating countercurrent moving bed gasifier," *Fuel*, vol. 112, pp. 635–640, 2013.

[18] K. Jaojaruek, S. Jarungthammachote, M. K. B. Gratuito, H. Wongsuwan, and S. Homhual, "Experimental study of wood downdraft gasification for an improved producer gas quality through an innovative two-stage air and premixed air/gas supply approach," *Bioresource Technology*, vol. 102, no. 7, pp. 4834–4840, 2011.

[19] Y. Ueki, T. Torigoe, H. Ono, R. Yoshiie, J. H. Kihedu, and I. Naruse, "Gasification characteristics of woody biomass in the packed bed reactor," *Proceedings of the Combustion Institute*, vol. 33, no. 2, pp. 1795–1800, 2011.

[20] A. Saravanakumar, T. M. Haridasan, and T. B. Reed, "Flaming pyrolysis model of the fixed bed cross draft long-stick wood gasifier," *Fuel Processing Technology*, vol. 91, no. 6, pp. 669–675, 2010.

[21] M. Puig-Arnavat, J. C. Bruno, and A. Coronas, "Modified thermodynamic equilibrium model for biomass gasification: a study of the influence of operating conditions," *Energy & Fuels*, vol. 26, no. 2, pp. 1385–1394, 2012.

[22] T. B. Reed and A. Das, *Handbook of Biomass Downdraft Gasifier Engine Systems*, Solar Technical Information Program, Solar Energy Research Institute, Golden, Colo, USA, 1988.

[23] S. Jarungthammachote and A. Dutta, "Thermodynamic equilibrium model and second law analysis of a downdraft waste gasifier," *Energy*, vol. 32, no. 9, pp. 1660–1669, 2007.

[24] M. Barratieri, P. Baggio, L. Fiori, and M. Grigiante, "Biomass as an energy resource thermodynamic constrain on the performance of the conversion process," *Bioresources Technology*, vol. 99, no. 15, pp. 7063–7073, 2008.

13

Real Costs Assessment of Solar-Hydrogen and Some Fossil Fuels by means of a Combustion Analysis

Giovanni Nicoletti, Roberto Bruno, Natale Arcuri, and Gerardo Nicoletti

Mechanical, Energetic and Management Engineering Department, University of Calabria, V. P. Bucci 46/C, Arcavacata, 87036 Rende, Italy

Correspondence should be addressed to Giovanni Nicoletti; giovanni.nicoletti@unical.it

Academic Editor: Kazunori Kuwana

In order to compare solar-hydrogen and the most used fossil fuels, the evaluation of the "external" costs related to their use is required. These costs involve the environmental damage produced by the combustion reactions, the health problems caused by air pollution, the damage to land from fuel mining, and the environmental degradation linked to the global warming, the acid rains, and the water pollution. For each fuel, the global cost is determined as sum of the market price and of the correspondent external costs. In order to obtain a quantitative comparison, the quality of the different combustion reactions and the efficiency of the technologies employed in the specific application sector have to be considered adequately. At this purpose, an entropic index that considers the degree of irreversibility produced during the combustion process and the degradation of surroundings is introduced. Additionally, an environmental index that measures the pollutants released during the combustions is proposed. The combination of these indexes and the efficiency of the several technologies employed in four energy sectors have allowed the evaluation of the total costs, highlighting an economic scenario from which the real advantages concerning the exploitation of different energy carrier are determined.

1. Introduction

Nowadays energy saving applications and environmental protection could be achieved by the diffusion of newer and cleaner energy resources [1, 2]. The gradual and continuous decrement of the reserves of fossil fuels and the problems of environmental pollution concerning their use determines an unpostponable transition to renewable energy [3]. The replacement of traditional fossil fuels is not difficult to assess also in the field of energy carriers, where hydrogen could represent a satisfactory solution in every sector of society. Moreover, it can be considered completely renewable if produced by electrolysis process of the water supplied by photovoltaic technologies [4–6].

The exploitation of the electrolytic hydrogen certainly will lead to significant reductions of the major air pollutants, in particular CO_2, SO_x, and particulate dusts, by improving the city livability and the environmental quality [7–9]. Additionally, solar-hydrogen diffusion could lead to the improvement of the photovoltaic and fuel cells technologies [10].

The diffusion of electrolytic hydrogen is currently impeded by technical and economic difficulties: some problems concerning storage and transportation have to be solved [11–14], besides the current market price of solar-hydrogen that makes it strongly unattractive [15].

In this paper, an economic comparison to compare solar-hydrogen and some fossil fuels was carried out by means of the real economy approach. The competitiveness among the considered fuels is carried out by considering the total costs supported in a specific application sector. The total cost includes not only the market price of the fuel, but also the "external costs" related to the fuel exploitation. These external costs consider the environmental damage produced by the combustion reactions, the health problems caused by air pollution, the damage to land from fuel mining, and the environmental degradation linked to the global warming, the acid rains, and the water pollution. Usually, in energy sectors supplied by fossil fuels, the external costs are not considered, making the comparison with solar-hydrogen with no meaning [16].

The latter, in fact, reduces the harmful effects linked to the energy production with fossil fuels, allows the control of air pollution, and reduces some problems linked to water and soil utilization [1].

Several authors have investigated the hydrogen costs in different energy fields. Weinert et al. determined the costs required for the refueling of cars equipped with fuel cells technologies in Shangai [17]. Melaina, instead, investigated the costs concerning the hydrogen stations in US to evaluate the potential of hydrogen diffusion in the transportation sector [18]. Prince-Richard et al. have carried out a technoeconomic analysis in order to evaluate the costs concerning the hydrogen produced by electrolysis and destined to fuel cell vehicles [19]. Finally, Veziroğlu and Barbir proposed a particular approach to evaluate the real costs of different fuels, but by now these costs' result is outdated [20].

Unlike the latter document, in this paper a different approach to determine the total costs of solar-hydrogen and of traditional fossil fuels, based on thermodynamic evaluations, is introduced. These evaluations are required to consider the different effects linked to the combustion reactions; in function of the fuel exploitation, different quantity of released heat, emitted pollutants, and diverse effects on the external environment are produced. In order to take into account the mentioned aspects, in the economic analysis an appropriate weight factor that modifies the external costs and takes into account the aforementioned aspects is necessary. This weight factor is defined as product between two quality indexes: the first is an "entropic" index to measure the irreversibility degree of the combustion reaction and the quality of the same reaction and the second is an "environmental" index to determine the quantity of pollutants emitted in atmosphere. Additionally, since the quantity of fuel required in specific application sectors depends on the involved technology, the total cost is adjusted in function of the thermodynamic efficiency of the employed device. Thus, also considering different fuels with dissimilar thermodynamic and environmental properties, the employment of such parameters is normalized, allowing the obtainment of homogenous results. The comparison is carried out between solar-hydrogen and the currently diffuse fossil fuels in industrialized and emerging countries such as methane, coal, and gasoline/diesel.

The entropic impact index was investigated in thermodynamic open systems, whose state is defined by the values of enthalpy, Gibbs free energy, and chemical work [21, 22]. In function of the fuel properties, it measures the combustion quality and the environmental damage, since during the combustion process the lower the entropy production, the lower the impact on surroundings [23]. The environmental impact index, indeed, has been determined by considering the quantity of pollutants emitted and normalized in function of the thermal energy released during the combustion reaction. Finally, the product of the proposed indexes determines the

global weight factor "p" required for the total cost evaluation. In order to assess the economic advantages concerning the exploitation of a specific fuel in opposition to another one, a reference fuel has to be recognized. For instance, if the benefits consequent to the solar-hydrogen exploitation have to be weighted against the methane, the latter will be classified as reference fuel. Therefore, the total costs of the investigated fuels can be evaluated by the relation [24]:

$$S = \left\{ C + E \cdot \left(\frac{p_o}{p} \right) \right\} \cdot \left(\frac{\eta_o}{\eta} \right), \qquad (1)$$

where S is the total (or real) cost per unit of produced energy (€/kWh); C is the fuel price market per unit of energy delivered to the customer, including taxes (€/kWh); E is the external cost consequent to the fuel use per unit of energy, expressed in current money (€/kWh); η is the efficiency of the device supplied by a specific fuel in a specific application sector; η_o is the efficiency of the technology supplied by the reference fuel in the same application sector; p_o is the weight factor determined for the reference fuel.

Equation (1) allows the evaluation of the so-called "real costs," by moving the focus of the investigation to the end of the process of energy production and by extending the evaluation of the real costs by including the effects produced from the final use of the same fuels. The economic analyses have been carried out by considering the residential, the industrial, the power generation, and the transportation sectors as application areas.

2. The Exploitation of Solar-Hydrogen as Energy Carrier

A promising solution in the field of the clean fuels is represented by solar-hydrogen [1]. A hypothetical scheme of the complete cycle of energy production involving hydrogen as energy carrier starts from photovoltaic fields, and the possible exploitation in different energy sectors is shown in Figure 1. Every involved application provides liquid water, which could be reused to supply newly the hydrogen cycle, making the process more sustainable.

The technical difficulties to the exploitation of solar-hydrogen concern the hydrogen storage systems and the structure of the hydrogen transportation, but the recent research campaign in these fields seems to show promising results [12–14, 25].

A large emitter of macropollutants (CO_2, NO_x, CO, SO_x, and particulate) is represented by the sector of electric power generation that could be adequately supplied by hydrogen pipeline and can evolve towards the innovative use of new technologies as the fuel cells, realizing the tasks of high efficiency and contemporaneous production of thermal energy [26]. Alternatively, solar-hydrogen could be used also in appropriate burners, by using pure oxygen or air as oxidant substances, to produce water and a significant amount of thermal energy. The same fuel cells could be employed in the residential sector for cogenerative applications. Finally, solar-hydrogen could be used as new energy carrier in the transportation sector to supply the existing vehicles equipped

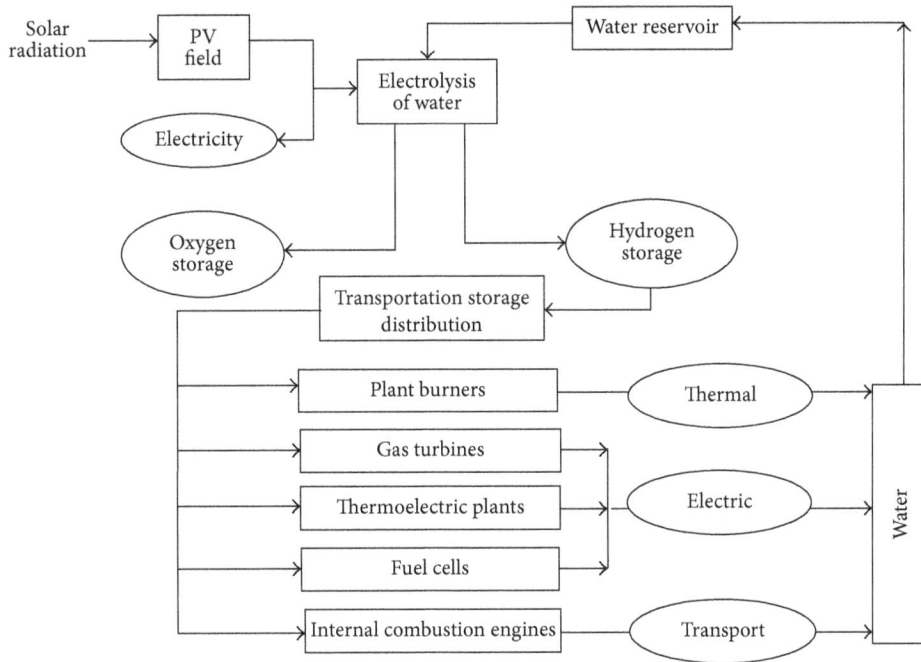

FIGURE 1: Possible scheme of exploitation of hydrogen produced by solar radiation.

with electric engines. The electrolysis process, beyond hydrogen, allows the production of oxygen; therefore a further advantage is linked to the absence of nitrogen oxides in the combustion products.

In the scheme of Figure 1, hydrogen primarily assumes the tasks of storing and conveying the solar radiation towards the final use. Consequently, it is appropriate to consider the hydrogen as an energy carrier, as well as the electricity. In Figure 2, a more detailed scheme of solar-hydrogen exploitation is reported, together with the correspondent energy fluxes and the efficiencies of the involved processes. Possible ways to exploit solar hydrogen at different temperature levels are reported. Low enthalpy heat could be employed in heat pumps for heating application in buildings.

3. The Entropic and Environmental Indexes in Combustion Processes

The evaluation of the total cost by (1) requires the analytical determination of the weight factor "p," defined in this paper as product between entropic and environmental indexes. The first measures the quality of the chemical reaction, the second quantifies the emitted pollutant substances, and both are normalized in function of the amount of heat released during the combustion.

The combustion reactions represent an open thermodynamic system in which the entropy variation is sum of two terms: the first intrinsic due to irreversibility (always positive) and the second extrinsic (positive or negative) linked to the heat exchange through the border of the thermodynamic system [27]. During a combustion reaction, the produced

chemical work plays a determining role in order to evaluate the entropy change, the enthalpy of formation, and the Gibbs potential [28].

The latter considers the energy contained in the initial components in a rate of available energy $\Delta \widetilde{h}$ and in another rate of constrained (not available) energy that can be written as

$$\Delta \left(T\widetilde{s} + \mu \right) \equiv \Delta \left(T\widetilde{s}^{*} \right). \tag{2}$$

In (2), the term $\widetilde{s}^{*}$ indicates a fictitious parameter that considers both the total entropy per unit of molar mass ($\widetilde{s}$) and the intrinsic work of the reaction (μ).

By considering conservatively the gasoline/diesel fuel as C_8H_{18} (isooctane), which represents the newest and cleaner member of the hydrocarbons family, and the coal as pure carbon, the combustion reactions of the considered fuels in their stoichiometric form can be written as

(i) hydrogen:

$$H_2O \leftrightarrows H_2 + \frac{1}{2}O_2 \tag{3}$$

(ii) coal (gaseous form):

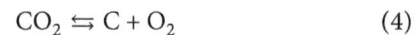

$$CO_2 \leftrightarrows C + O_2 \tag{4}$$

(iii) natural gas (methane):

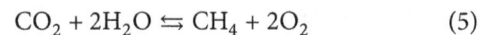

$$CO_2 + 2H_2O \leftrightarrows CH_4 + 2O_2 \tag{5}$$

(iv) gasoline:

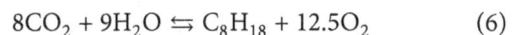

$$8CO_2 + 9H_2O \leftrightarrows C_8H_{18} + 12.5O_2 \tag{6}$$

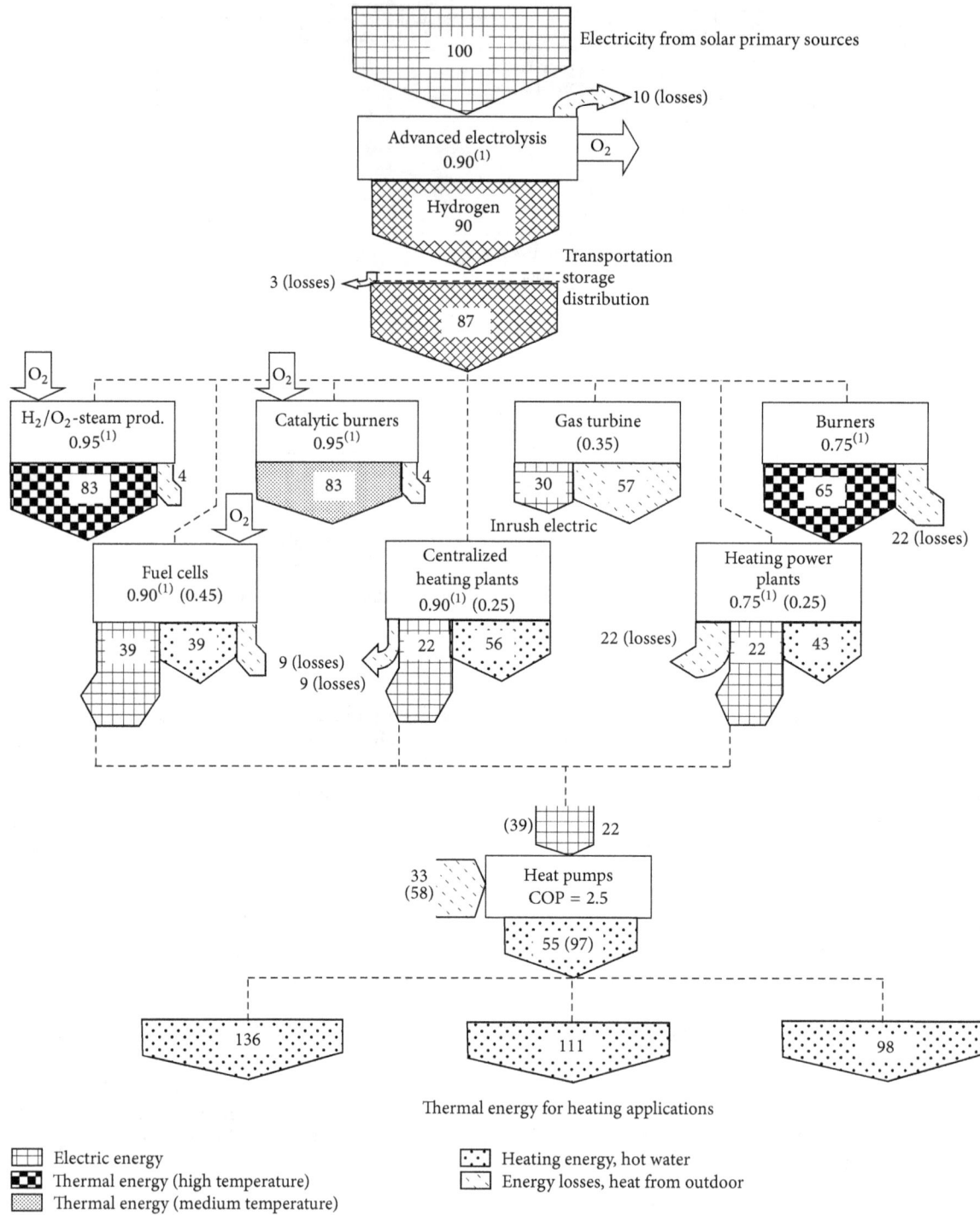

FIGURE 2: Hypothetic energy flux for the complete solar-hydrogen exploitation.

For each of these reactions, the values per mole unit of $\Delta\tilde{h}$, $\Delta\tilde{s}^*$, and $\Delta\tilde{g}$ can be found in [29–31], at reference values of pressure and temperature equal, respectively, to 1 atm, 298 K, and by fixing the reference entropy to the value calculated at a pressure of 1 atm. The values are listed in Table 1 together with the molar specific heat at constant pressure and a reference absolute temperature $\langle T \rangle$, defined as the mean value between the ambient temperature (298 K) and the combustion temperature. The specific heat at constant pressure is reported per mole unit for dimensional homogeneity.

TABLE 1: Some thermodynamic parameters of hydrogen, methane, coal, and gasoline/diesel evaluated during stoichiometric combustion.

Fuels	$\Delta \tilde{h}$ (J/mole)	$\Delta \tilde{s}^*$ (J/mole·K)	$\Delta \tilde{g}$ (J/mole)	$\tilde{c}_p$ (J/mole·K)	$\langle T \rangle$ (K)
H_2	−251 100	56.6	−0.39	14.304	1451
C	−394 500	57.4	−0.53	15.576	1285
CH_4	−870 500	90.9	−1.05	35.840	1155
C_8H_{18}	−5 486 500	1644.5	−11.69	254.790	2035

TABLE 2: Entropic impact indexes evaluated for hydrogen, methane, coal, and gasoline/diesel supposing stoichiometric reaction combustions.

Fuels	i_{js}
H_2	0.672
C	0.813
CH_4	0.879
C_8H_{18}	0.390

For each fuel, the entropic degradation during the combustions reaction is calculated by the proposed dimensionless index:

$$i_{js} = \left(1 - \left| \frac{\Delta \tilde{s}^* \cdot \langle T \rangle}{\Delta \tilde{h}} \right| \right) \quad (j = a \longrightarrow d), \qquad (7)$$

where j indicates the considered fuel (a = solar-hydrogen; b = coal; c = methane; d = gasoline/diesel); $\Delta \tilde{s}^*$ is the specific molar entropic generation of the combustion process; $\Delta \tilde{h}$ is the specific molar enthalpy of the chemical reaction, assuming a similar meaning to the lower heating power of the jth fuel.

The dimensionless entropic index i_{js} varies from the maximum value of 1, for the ideal combustion, where $\Delta \tilde{s}^* = 0$, to the minimum value of 0, for the extreme case of total energy degradation. In the latter case, the increment of the temperature ΔT due to combustion involves a total transformation of the thermal energy in technical work, required to eject the mass of the combustion products, with no advantage in terms of heat released from the combustion reaction. By using the values reported in Table 1, the value concerning the entropic index obtained for each fuel is listed in Table 2.

Considering exclusively the combustion quality, methane and coal assume a better score than hydrogen and gasoline/diesel; therefore the degradation of the external environment is smaller than the latter fuels. The hydrogen score is greater only than gasoline/diesel, but at this step the environmental properties are not considered yet. In order to quantify the effects linked to the pollutants emitted during the combustion reactions, another index, which considers exclusively the environmental properties of each fuel, is required. The latter is evaluated by (8), in function of the thermophysical properties of the jth fuel and of the kth pollutant content produced during standard combustion:

$$i_{jE} = \prod i_{jk}. \qquad (8)$$

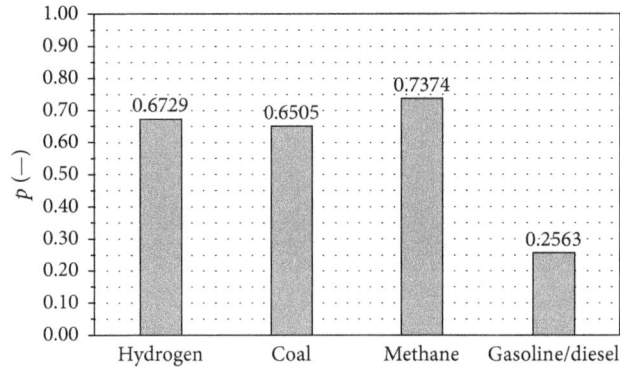

FIGURE 3: Weight factor for the evaluation of the external cost of the considered fuels.

The term i_{jk} that compares in (8) is determined by means of the following equation [32]:

$$i_{jk} = \left(1 - \frac{m_{jk}}{\left|\Delta \tilde{h}\right| / \left(\tilde{c}_p \cdot \langle T \rangle\right)} \right)_{k=1 \to 4}, \qquad (9)$$

where m_{jk} indicates the quantity of pollutant (kg) released per kg of burned fuel; these parameters are grouped in Table 3 considering as main emitted substances CO_2, SO_2, NO_x, and particulate dusts. Actually, the emission factors depend not only on the fuel thermodynamic characteristics, but also on the involved technology in the specific application sector. In simplified way, the values listed in Table 3 represent average emission coefficients indicated in [33] related to the European situation. For the gasoline, the emission factors have been modified taking into account also the pollutants produced by diesel combustion. The results obtained applying (9) are listed in Table 4, where the last column lists the environmental indexes determined by (8). The solar-hydrogen score becomes unitary because its exploitation with pure oxygen in fuel cells was supposed. Moreover, the water vapor was not considered as pollutant substance because the hypothesis of its complete recovery was adopted. Water vapor provided by fuel cells in urban areas, in fact, could be considered as a greenhouse gas.

In Figure 3 the weight factor "p" to apply in (1) for the economic analysis, evaluated by means of the proposed indexes, is shown.

Despite the worse emission factors with respect to solar-hydrogen, the better combustion characteristics of the methane allow the obtainment of the best score. Conversely, the better environmental characteristics of solar-hydrogen compensate the gap with coal in terms of combustion quality. Among the investigated fuels, the gasoline/diesel presents the worst score for effect of the more penalized values obtained for the entropic and the environmental indexes.

4. Evaluation of the Total Costs of the Fuels in Different Application Sectors

The evaluation of global cost of the investigated fuels requires the knowledge of the correspondent market prices and the

TABLE 3: Emission factor for the standard combustion per kg of burned fuel.

Fuels		1 CO_2 [kg/kg]	2 SO_2 [kg/kg]	3 NO_x [kg/kg]	4 Particulate [kg/kg]
a	H_2	0	0	0	0
b	C	3.196	0.0156	0.0469	0.0046
c	CH_4	2.745	0.0205	$2.1 \cdot 10^{-5}$	$1 \cdot 10^{-6}$
d	C_8H_{18}	3.197	0.0012	$6.3 \cdot 10^{-5}$	$2.4 \cdot 10^{-4}$

TABLE 4: Specific and global environmental indexes obtained for the investigated fuels.

Fuels		i_{j1}	i_{j2}	i_{j3}	i_{j4}	i_{jE}
a	H_2	1.0000	1.0000	1.0000	1.0000	**1**
b	C	0.8034	0.9990	0.9971	0.9997	**0.8001**
c	CH_4	0.8386	0.9999	0.9998	1.0000	**0.8385**
d	C_8H_{18}	0.6573	0.9998	0.9999	0.9999	**0.6572**

external costs. The market price of fossil fuels, in function of the application sector, can be found in different sources [34, 35]. The cost of the hydrogen produced by photovoltaic technologies, by excluding other methods of solar energy exploitation, is subjected to numerous investigations [6]. It is mainly formed by the cost of the electricity provided by the photovoltaic field (C_{PV}) and by the cost due to the electrolysis process (C_{El}) [36]. In the last few years, the cost of the photovoltaic electricity was strongly reduced in relation to the recent incentive campaign adopted from different countries [37]. The cost of the electrolysis process is largely dependent on the cost of electricity, on the efficiencies, and on the capital costs of the systems. Because the system efficiency can be limitedly increased (current value is 78%), the cost of the electrolysis process cannot be reduced by an efficiency increment as much as a significant reduction of the electricity price. For instance, by benefiting for forecourt systems of the same electricity prices for industries, the reduction of the cost of electrolysis process could be of 31% [38]. Moreover, if the hydrogen economy grows and the electrolyzer systems are mass produced, a substantial reduction in capital cost could be also achieved. Lastly, the cost related to the electrolysis process can be calculated in simplified way using the following relationship [36]:

$$C_{El} = \frac{I_0 (A + M)}{\eta \cdot E_{el}} = \frac{\text{Yearly global specific cost}}{\text{Yearly specific energy produced}}, \quad (10)$$

where I_0 is the investment cost sustained to acquire the electrolyzer, including installation cost (€); A is the actualization factor, usually set equal to the amortized cost and by providing a useful life of the plant equal to 20 years; M is the percentage factor concerning the annual cost of maintenance and management of the system, including the staff cost, set to 5%; η is the electrolyzer efficiency in order to transform electric energy in chemical energy stored as hydrogen (E_{H_2}).

By supposing large-scale production, a global cost of about 6.67 €/kg for solar-hydrogen has been determined

exploiting the technology of plane PV [39]. The lower heating value for hydrogen is equal to 34.86 kWh/kg; therefore a correspondent specific energy cost of 0.191 €/kWh has been determined. In [40], with reference to the European power generation sector, in 2013 an average levelised cost of electricity ranging from 0.059 €/kWh (coal) to 0.086 €/kWh (combined power plant fired by methane) is reported. Consequently, without considering the additional costs related to the retransformation of hydrogen into electrical energy, the obtained value is itself more 2–4 times higher than the cost obtained through exploitation of conventional fossil fuels. The economic competitiveness of solar-hydrogen could increase with a consistent reduction of the investment costs, both for the part concerning the photovoltaic system, by adopting more efficient technologies, and for the part concerning the electrolysis process. Regarding the reduction of the costs related to the PV technology, a considerable improvement of the scenario has been investigated in [41], where a specific energy cost to 0.09 €/kWh has been quantified by using concentrator PV fields, characterized by higher solar conversion efficiencies (37%). In this way, this specific cost becomes close to other hydrogen production methods, such as gas reformation, wind, and nuclear electrolysis. The determined gap cost with traditional fossil fuels becomes less severe if the social and environmental costs associated with their use are considered [42]. The "hidden" costs of fossil fuels can be quantified by the parameter E that compares in (1). These values, for each considered fossil fuel, have been obtained from [43] where a suitable analytical tool makes the external cost for different energy carrier available. This tool was developed starting from the data carried out for six big cities of emerging countries. The comparison between solar-hydrogen and the considered fossil fuels, in different fields of use (residential, industrial, electrical, and transportation), is shown in the following figures by separating the normalized fuel price ($C \cdot \eta_0/\eta$) from the normalized external cost ($E \cdot p_0 \cdot \eta_0/p \cdot \eta$). The obtained values have been determined hypothesizing fuel cells technologies for the hydrogen exploitation, with the following values of the efficiency η in function of the application sector:

(i) 0.8 for the residential and industrial sectors, by supposing to use fuel cells in cogenerative operation to produce heat and electricity at the same time;

(ii) 0.45 for the power generation and the transportation sectors, by supposing the employment of electric engines in the cars.

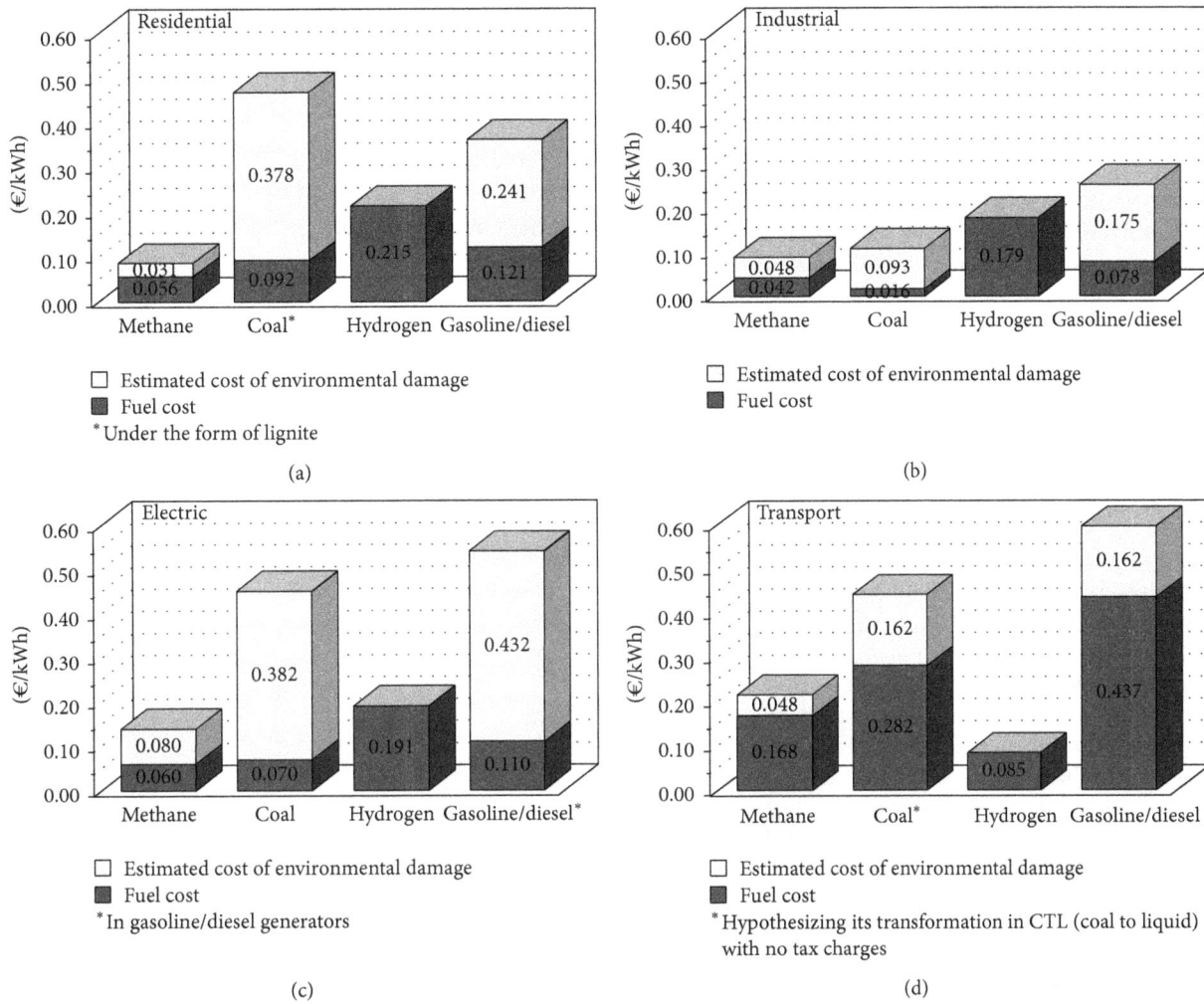

FIGURE 4: (a) Fuel prices and external costs in the residential sector. (b) Fuel prices and external costs in the industrial sector. (c) Fuel prices and external costs in the power generation sector. (d) Fuel prices and external costs in the transport sector.

Regarding the efficiency η of traditional technologies supplied by fossil fuels, the adopted value is

(i) 0.9 for residential sector, hypothesizing hybrid boilers for the production of heat and electricity (equipped, e.g., with Stirling engines [44]);

(ii) 0.75 for industrial sector in cogenerative operations;

(iii) 0.45 for the power generation sector, resulting as average value between the efficiency of combined cycles and conventional Rankine cycles;

(iv) 0.20 for the transportation sector.

The analytical results of the economic analysis are plotted in Figures 4(a)–4(d), in function of the application sector and for the considered fuels.

Currently, the real cost of solar-hydrogen becomes already profitable for the transportation sector, taking advantage from the better efficiency of the electric traction. Moreover, the absence of taxation costs makes the market price of solar-hydrogen lower than those of the fossil fuels. For the industrial sector, solar-hydrogen in fuel cells becomes profitable only if compared to technologies supplied by gasoline/diesel, because the lower market prices of methane and coal still compensate the external costs. A better situation has been detected for the power generation sector, where solar-hydrogen is more attractive than coal and gasoline, while the methane technologies are still profitable. Despite the lower market prices, the position of coal and gasoline/diesel in the ranking is strongly penalized from the elevated external costs. Finally, in residential sector fuel cells supplied by solar-hydrogen present a better position compared to coal and gasoline/diesel technologies, but methane remains still profitable in relation to the improvement of the boiler technologies that limit the external costs.

5. Conclusions

The real costs of energy system supplied by solar-hydrogen and some fossil fuels have been determined. The real cost includes the "external" costs related to the final use of the fuels, considering air pollution and environmental damage.

The investigated fuels present different thermodynamic characteristics and, usually, are employed in different devices characterized by dissimilar values of thermodynamic efficiency. In this paper, to obtain homogenized results that consider the mentioned aspects, a weight factor has been determined in order to assess an economic analysis based on the "real economy" approach. The weight factor was evaluated by means of a thermodynamic approach employed to quantify two quality indexes: an entropic index concerning the quality of the combustion reaction, and an environmental index to quantify the main emitted pollutant substances. Moreover, the proposed indexes have been normalized in function of the heat released during the combustion process. The thermodynamic approach described in this paper has allowed a direct comparison among four fuels usually employed as energy carriers: solar-hydrogen, methane, coal, and gasoline/diesel. Regarding the proposed weight factor, the results show a position of hydrogen secondary only to methane, which presents the better combination between combustion characteristics and emitted pollutants. From an economic point of view, considering four energy application sectors (residential, industrial, power generation, and transport), the real costs have highlighted that solar-hydrogen becomes already more attractive in the transportation sector, presenting the lowest value. In the industrial sector, the lower market prices of methane and coal make these fuels more attractive, since they compensate widely the external costs. Therefore, the position of solar-hydrogen will improve only with a large augment of the market price of fossil fuel in the next future, for effect of their inevitable depletion. The same evaluations can be done for the energy sector, where solar-hydrogen is already competitive than coal in relation to the large external costs of the latter. Nowadays, in the residential sector, solar-hydrogen is a good alternative for the replacement of old boilers supplied by gasoline/diesel or coal. The position of solar-hydrogen is still worse than methane, since the improvement and the diffusion of condensation boilers equipped with burner at low combustion temperatures have produced limited external costs. In reality, the evaluated real costs for solar-hydrogen become optimistic, since its market price is not suffered by tax charges. In the future, the replacement of the traditional fuels will cause an inevitable taxation also on the new energy carriers, including solar-hydrogen, to guarantee the financial revenue derived from fuel commerce. Therefore, the economic scenario above introduced will be surely different. However, the determined real cost gaps among the investigated fuels will become useful to evaluate an adequate taxation level for solar-hydrogen. On the other hand, the mass production of electrolyzer technologies, the reduction of the PV electricity costs, or the improvement of the PV technologies will determine a reduction of the solar-hydrogen cost that could compensate the tax charges.

In a short-term scenario, in absence of tax charge of solar-hydrogen, its diffusion becomes already profitable than gasoline/diesel technologies in every application sector. Therefore, the obtained results show that an economical persuasion to exploit solar-hydrogen as new energy carrier in different application sectors could already exist.

Nomenclature

A: Actualization factor [—]
$\tilde{c}$: Molar specific heat [J/mole·K]
C: Fuel market cost [€]
E: External cost [€]
$\tilde{g}$: Molar Gibbs energy [J/mole]
$\tilde{h}$: Molar enthalpy [J/mole]
I: Investment [€]
i: Impact index [—]
m: Pollutant emission factor [g/kg]
M: Management and maintenance rate [—]
p: Weight factor [—]
P: Pressure [Pa]
$\tilde{s}$: Molar entropy [J/mole·K]
S: Real cost [€]
T: Temperature [K].

Greek Symbols

η: Efficiency [—]
μ: Apparent work [J/mole]
$\tilde{v}$: Molar volume [m^3/mole].

Subscripts and Superscripts

0: At the year 0
el: Electricity
E: Environmental
El: Electrolyzer
H$_2$: Hydrogen
j: Fuel type
k: Pollutant type
p: Constant pressure
PV: Photovoltaic
o: Reference
S: Combustion.

Conflict of Interests

The authors declare that there is no conflict of interests regarding the publication of this paper.

References

[1] I. Dincer and C. Acar, "A review on clean energy solutions for better sustainability," *International Journal of Energy Research*, vol. 39, no. 5, pp. 585–606, 2015.

[2] P. D. Lund, "Clean energy systems as mainstream energy options," *International Journal of Energy Research*, vol. 40, no. 1, pp. 4–12, 2016.

[3] N. Armaroli and V. Balzani, *Energy for a Sustainable World: From the Oil Age to a Sun-Powered Future*, John Wiley & Sons, Hoboken, NJ, USA, 2010.

[4] G. Nicoletti, "Hydrogen as solar energy storage—technical and economical aspects," in *Proceedings of the International Congress Hypothesis*, Cassino, Italy, June 1995.

[5] J. R. White, "Comparing solar energy alternatives," *International Journal of Energy Research*, vol. 8, no. 1, pp. 39–52, 1984.

[6] J. Turner, G. Sverdrup, M. K. Mann et al., "Renewable hydrogen production," *International Journal of Energy Research*, vol. 32, no. 5, pp. 379–407, 2008.

[7] J. Andrews and B. Shabani, "Re-envisioning the role of hydrogen in a sustainable energy economy," *International Journal of Hydrogen Energy*, vol. 37, no. 2, pp. 1184–1203, 2012.

[8] Gi. Nicoletti and Ge. Nicoletti, "L'idrogeno come opzione energetica sostenibile," in *Proceedings of the 8th Italian Conference CIRIAF*, Perugia, Italy, April 2010.

[9] J. M. Ogden, *Alternative Fuels and Prosects—Overview*, John Wiley & Sons, Hoboken, NJ, USA, 2010.

[10] P. Kumar, K. Dutta, S. Das, and P. P. Kundu, "An overview of unsolved deficiencies of direct methanol fuel cell technology: factors and parameters affecting its widespread use," *International Journal of Energy Research*, vol. 38, no. 11, pp. 1367–1390, 2014.

[11] J.-Y. Hwang, S. Shi, X. Sun, Z. Zhang, and C. Wen, "Electric charge and hydrogen storage," *International Journal of Energy Research*, vol. 37, no. 7, pp. 741–745, 2013.

[12] K. Kendall, "Hydrogen and fuel cells in city transport," *International Journal of Energy Research*, vol. 40, no. 1, pp. 30–35, 2016.

[13] A. Züttel, "Hydrogen storage methods," *Naturwissenschaften*, vol. 91, no. 4, pp. 157–172, 2004.

[14] I. A. Gondal and M. H. Sahir, "Prospects of natural gas pipeline infrastructure in hydrogen transportation," *International Journal of Energy Research*, vol. 36, no. 15, pp. 1338–1345, 2012.

[15] S. Blanchette Jr., "A hydrogen economy and its impact on the world as we know it," *Energy Policy*, vol. 36, no. 2, pp. 522–530, 2008.

[16] L. I. González-Monroy and A. Córdoba, "Financial costs and environmental impact optimization of the energy supply systems," *International Journal of Energy Research*, vol. 26, no. 1, pp. 27–44, 2002.

[17] J. X. Weinert, L. Shaojun, J. M. Ogden, and M. Jianxin, "Hydrogen refueling station costs in Shanghai," *International Journal of Hydrogen Energy*, vol. 32, no. 16, pp. 4089–4100, 2007.

[18] M. W. Melaina, "Initiating hydrogen infrastructures: preliminary analysis of a sufficient number of initial hydrogen stations in the US," *International Journal of Hydrogen Energy*, vol. 28, no. 7, pp. 743–755, 2003.

[19] S. Prince-Richard, M. Whale, and N. Djilali, "A techno-economic analysis of decentralized electrolytic hydrogen production for fuel cell vehicles," *International Journal of Hydrogen Energy*, vol. 30, no. 11, pp. 1159–1179, 2005.

[20] T. N. Veziroğlu and F. Barbir, "Economic comparison of hydrogen and fossil fuel systems," in *Clean Utilization of Coal: Coal Structure and Reactivity, Cleaning and Environmental Aspects*, Y. Yurum, Ed., vol. 370 of *NATO ASI Series*, pp. 295–313, Springer, Dordrecht, The Netherlands, 1992.

[21] G. Nicoletti and R. Arienti, "Aspetti Entropici dell'Impatto Ambientale da combustione," in *Proceedings of the 4th Italian Conference "Analisi Ambientale in Italia"*, Milan, Italy, 1992.

[22] I. Dincer and M. A. Rosen, "Thermodynamic aspects of renewables and sustainable development," *Renewable and Sustainable Energy Reviews*, vol. 9, no. 2, pp. 169–189, 2005.

[23] N. Kousuke, T. Toshimi, and K. Shinichi, "Analysis of entropy generation and exergy loss during combustion," *Proceedings of the Combustion Institute*, vol. 29, no. 1, pp. 869–874, 2002.

[24] T. N. Veziroglu, "Economic comparison of solar hydrogen energy system with fossil fuel system," in *Solar-Wasserstoff-Versorgung: Internationales Symposium*, Zürich, Switzerland, November 1989.

[25] A. Fonseca, V. Sá, H. Bento, M. L. C. Tavares, G. Pinto, and L. A. C. N. Gomes, "Hydrogen distribution network optimization: a refinery case study," *Journal of Cleaner Production*, vol. 16, no. 16, pp. 1755–1763, 2008.

[26] I. Dincer and M. A. Rosen, "Sustainability aspects of hydrogen and fuel cell systems," *Energy for Sustainable Development*, vol. 15, no. 2, pp. 137–146, 2011.

[27] E. A. Guggenheim, *Thermodynamics*, North Holland, Amsterdam, The Netherlands, 5th edition, 1977.

[28] Hutte, *Theoretische Grundlagen*, W. Ernst & Sohn, Berlin, Germany, 28th edition, 1985.

[29] L. Bornstein, *Chemische Handbuch*, vol. II/4, Auer, 6th edition, 1959.

[30] P. Chilton, *Chemical Engineering Handbook*, McGraw-Hill, New York, NY, USA, 1973.

[31] R. C. Weast, *Handbook of Chemistry and Physics*, CRC Press, New York, NY, USA, 1977.

[32] G. Nicoletti, N. Arcuri, G. Nicoletti, and R. Bruno, "A technical and environmental comparison between hydrogen and some fossil fuels," *Energy Conversion and Management*, vol. 89, pp. 205–213, 2015.

[33] EEA, "EMEP/EEA air pollutant emission inventory guidebook 2013: technical guidance to prepare national emission inventories," European Environmental Agency Technical Report 12/2013, European Environment Agency, Copenhagen, Denmark, 2013.

[34] http://www.fuel-prices-europe.info/.

[35] VVAA, *DECC Fossil Fuel Price Projections—Gov.uk*, Department of Energy and Climate Change of UK, 2014.

[36] D. Coiante, "La produzione fotovoltaica dell'idrogeno," *Notiziario ENEA "Energia ed Innovazione"*, no. 11-12, pp. 24–41, 1990.

[37] VVAA, "Global Market Outlook for Photovoltaics 2014–2018," EPIA—European Photovoltaic Industry Association, http://www.cleanenergybusinesscouncil.com/site/resources/files/reports/EPIA_Global_Market_Outlook_for_Photovoltaics_2014-2018_-_Medium_Res.pdf.

[38] VVAA, *Technology Brief: Analysis of Current-Day Commercial Electrolyzers*, NREL—National Renewable Energy Laboratory—Department of Energy, Washington, DC, USA, 2004.

[39] P. E. Dodds and W. McDowall, "A review of hydrogen production technologies for energy system models," UKSHEC Working Paper 6, UCL Energy Institute, University College London, London, UK, 2012.

[40] C. Kost, J. N. Mayer, J. Thomsen et al., "Levelized cost of electricity renewable energy technologies," Report, Fraunhofer Institute for Solar Energy Systems ISE, Freiburg, Germany, 2013, http://www.ise.fraunhofer.de/en/publications/veroeffentlichungen-pdf-dateien-en/studien-und-konzeptpapiere/study-levelized-cost-of-electricity-renewable-energies.pdf.

[41] J. R. Thomson, R. D. McConnell, and M. Mosleh, "Cost analysis of a concentrator photovoltaic hydrogen production system," in *Proceedings of the International Conference on Solar Concentrators for the Generation of Electricity or Hydrogen*, Scottsdale, Ariz, USA, May 2005.

[42] National Research Council, *Hidden Costs of Energy: Unpriced Consequences of Energy Production and Use*, The National Academies Press, Washington, DC, USA, 2010.

[43] K. Lvovsky, G. Hughes, D. Maddison, B. Ostro, and D. Pearce, "Environmental costs of fossil fuels: a rapid assessment method with application to six cities," Environment Department Paper, Report for the World Bank, 2010.

[44] M. De Paepe, P. D'Herdt, and D. Mertens, "Micro-CHP systems for residential applications," *Energy Conversion and Management*, vol. 47, no. 18-19, pp. 3435–3446, 2006.

14

Review on Recent Advances in Pulse Detonation Engines

K. M. Pandey and Pinku Debnath

Department of Mechanical Engineering, National Institute of Technology, Silchar, Assam 788010, India

Correspondence should be addressed to K. M. Pandey; kmpandey2001@yahoo.com

Academic Editor: Sergey M. Frolov

Pulse detonation engines (PDEs) are new exciting propulsion technologies for future propulsion applications. The operating cycles of PDE consist of fuel-air mixture, combustion, blowdown, and purging. The combustion process in pulse detonation engine is the most important phenomenon as it produces reliable and repeatable detonation waves. The detonation wave initiation in detonation tube in practical system is a combination of multistage combustion phenomena. Detonation combustion causes rapid burning of fuel-air mixture, which is a thousand times faster than deflagration mode of combustion process. PDE utilizes repetitive detonation wave to produce propulsion thrust. In the present paper, detailed review of various experimental studies and computational analysis addressing the detonation mode of combustion in pulse detonation engines are discussed. The effect of different parameters on the improvement of propulsion performance of pulse detonation engine has been presented in detail in this research paper. It is observed that the design of detonation wave flow path in detonation tube, ejectors at exit section of detonation tube, and operating parameters such as Mach numbers are mainly responsible for improving the propulsion performance of PDE. In the present review work, further scope of research in this area has also been suggested.

1. Introduction

In present days the attention of researchers in propulsion field from all over the world has turned towards the pulse detonation engine historical background, thermodynamics analysis, detonation initiation, and deflagration to detonation wave transition device as their main subject in detonation combustion research area. Another review research on rotating detonation engine model and application on aerospace and turbomachinery and performance are also included in this area. These involve researches from the United States, Russia, Japan and China, Germany, and Malaysia. Numbers of research publications have increased significantly in the past few decades. The main attraction of detonation combustion was to generate shock wave that is followed by combustion wave [1]. Pratt and Whitney began to develop the pulse detonation engine in 1993. Their research approach was to study the deflagration to detonation transition through the pulse detonation engine [2]. The feasibility study of a reaction device operating on intermittent gaseous detonation wave is considered by Nicholls et al. [3]. They conducted a study to investigate the thrust, fuel flow, air flow, and temperature over the range of operating conditions. Recently many countries give much importance to research of multimode combined detonation engine in hypersonic aircrafts propulsion system [4]. Kailasanath [5] studied a review on practical implementation on pulse detonation engine and deflagration to detonation transition was also studied in obstacle geometry. Again Kailasanath studied the development of pulse detonation engine. The detonation combustion parameters such as Chapman velocity and pressure are well derived in this study [6]. Wilson and Lu [7] summarized integrated studies for both PDE and RDE based propulsion system. They focused detonation waves to hypersonic flow simulation and power generation. Smirnov et al. [8] studied numerical simulation of detonation engine fed by fuel-oxygen mixture. The advantage of a constant volume combustion cycle as compared to constant pressure combustion was in terms of thermodynamic efficiency focused for advanced propulsion on detonation engine.

2. Reviews on Experimental Analysis

Chen et al. [9] investigated experimentally the nozzle effects on thrust and inlet pressure of an air-breathing pulse detonation engine. Their results showed that thrust augmentation of converging-diverging nozzle, diverging nozzle, or straight nozzle is better than that of converging nozzle on whole operating conditions. Li et al. [10] conducted an experiment on PDRE model utilizing kerosene as the fuel, oxygen as oxidizer, and nitrogen as purge gas. The thrust and specific impulse were investigated experimentally. Their obtained results showed that the thrust of PDRE test model was approximately proportional to detonation frequency. The time average thrust was around 107 N. Yan et al. [11] studied the performance of pulse detonation engine with bell-shaped convergent-divergent nozzle. This experiment has been done using kerosene as liquid fuel, oxygen as oxidizer, and nitrogen as purge gas. Their tested result has shown that the maximum thrust augmentation is approximately 21%. Allgood et al. [12] experimentally measured the damped thrust of multicycle pulse detonation engine with exhaust nozzle. Their results showed that diverging nozzle increases the performance with increase in fill fraction. Peng et al. [13] studied the two-phase dual-tube air-breathing pulse detonation engine (APDE) experiments to improve the understanding of characteristics of valveless multitube APDE. From the experimental results it is seen that the comparison between single- and dual-tube firing and the operation pattern of single-tube firing is beneficial to reduce the disturbance in the common air inlet. Yan et al. [14] experimentally investigated pulse detonation rocket engine with injectors and nozzles. The injectors were tested for atomization and mixing of two phase reactants. They observed that nozzles are the critical component for improving the performance of PDE. From their tested results they observed that a nozzle with high contraction ratio and high expansion ratio generated the highest thrust augmentation of 27.3%. Kasahara et al. [15] tested "Todoroki" rocket system at different operating conditions. The maximum thrust was produced slightly above 70 N with a specific impulse up to 232 s. The frequency of the system, even for a constant supply rate of propellants, varied over the range of 40–160 Hz. Copper et al. [16] measured impulse by using a Ballistic pendulum arrangement for detonation and deflagration in a tube closed at one end. They also studied the effect of internal obstacles on the transition from deflagration to detonation (see Figure 6). Their experimental results and prediction from analytical model are agreed within 15%. Hinkey et al. [17] experimentally demonstrated a rotary valve pulse detonation engine combustor for high frequency operation. Their experiments series were conducted in rotary valve single combustor and rotary valve multicombustor pulse detonation engine. The main measuring parameters are thrust as well as combustor wall pressure histories, oxidizer, and fuel mass flow rate (see Figure 14). Their concept of system operation was successfully demonstrated in rotary valve multicombustor PDE. In Japan, recently a single-tube pulse detonation rocket system that can slide on rails was fabricated. In the sliding test, the system operated 13 cycles at the frequency of 6.67 Hz [18]. Li et al. [19] illustrated the detonation initiation area in a detonation chamber. This detonation tube was closed at one end and open at another, which was composed of thrust wall and ignition section. The three spiraling internal grooves like semicircle, square, and inversed-triangle grooves were used in this experimentation. The results showed that the spiraling

internal groove can effectively enhance DDT. Asato et al. [20] have studied experimentally the effects of rapid flame propagation, rotating velocity, and Shchelkin spiral dimension in the vortex flow on DDT characteristics. The vortex flow was generated and that was promoted by Shchelkin spiral dimensions and DDT distance in the vortex flow could be shortened by 50–57%. New et al. [21] experimentally investigated Shchelkin spiral effect on multicycle pulse detonation engine. The effectiveness of Shchelkin spiral parameters on DDT phenomenon was studied using propane-oxygen mixture at low energy ignition source. The various configurations like spiral blockage ratio and spiral length to diameter ratio were also studied. In these studies shorter length configurations and highest blockage ratio were successful and sustained DDT was achieved. Wang et al. [22] performed the numbers of experiments on spiral configuration in pulse detonation engine. Their analysis provided the design data for deflagration to detonation transition rule in curved detonation chamber. Some experiments have been conducted using nine tubes in resistance experiments and result indicates that there is no detonation wave formed in the straight tube, but fully developed detonation waves have been obtained at selected spiral tubes. Panicker et al. [23] studied specific techniques for deflagration to detonation transition that were considered including Shchelkin spirals, grooves, converging-diverging nozzles, and orifice plates. They observed that Shchelkin spiral is to be the best performer for deflagration to detonation wave transition among other DDT enhancement devices.

Valiev et al. [24] investigated the "Flame Acceleration in Channels with Obstacles in the Deflagration-to-Detonation Transition." They found that obstacle mechanism is much stronger to accelerate deflagration flame to detonation wave. The physical mechanism of deflagration flame acceleration in an obstacle channel is quite different from Shchelkin spiral mechanism. The obstacle mechanism is much stronger for deflagration to detonation wave transition and it depends on operating conditions. The mechanism of viscous heating was also identified with proper modifications of obstacle geometry. Gaathaug et al. [25] numerically studied the deflagration to detonation transition in a turbulent jet behind an obstacle. The spiral internal grooves and inversed-triangle grooves were tested to enhance DDT and results showed that the spiraling internal groove can effectively enhance DDT. Moen et al. [26] studied the influence of obstacles on propagation of cylindrical flame. The freely expanding cylindrical type flame speed depends on obstacle configuration and achieved appropriate turbulence in obstructed flame travel path.

Ogawa et al. [27] studied flame acceleration and DDT in square obstacle array by solving Navier-Stokes equations. The computational simulation shows that deflagration wave acceleration was effected by obstacle series. Johansen and Ciccarelli [28] investigated the effect of obstacle blockage ratio on development of unburnt gas flow field for varying obstacle height. The computational simulations show that turbulence production increases with increasing number of blockage. Gamezo et al. [29] studied numerically deflagration wave acceleration and deflagration to detonation transition in obstructed channels. From the simulation they observed that detonation is ignited when a Mach stem formed by the

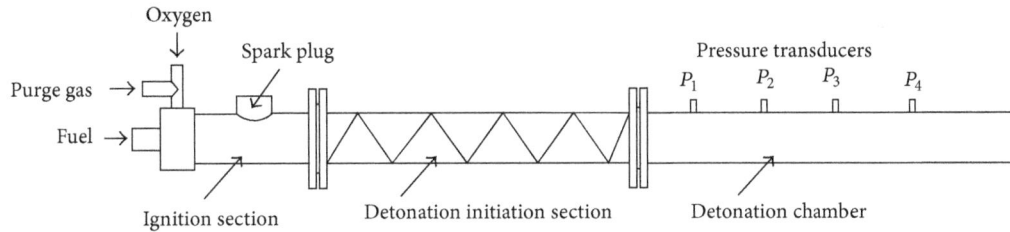

FIGURE 1: Schematic diagram of detonation tube with spiral [19].

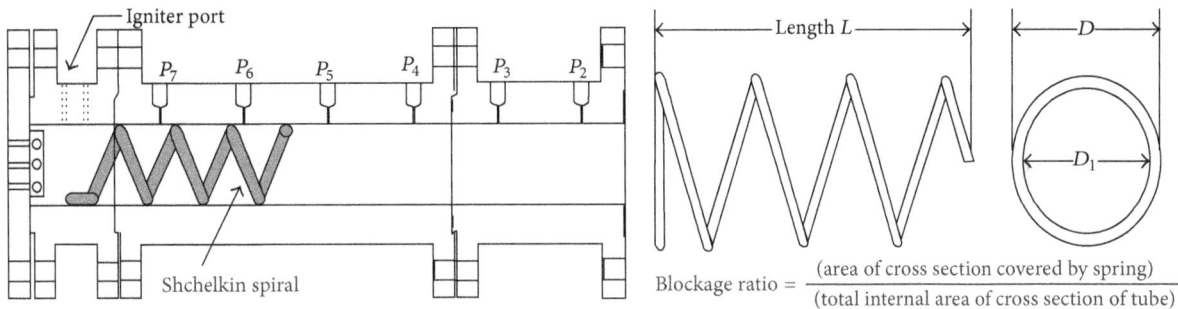

$$\text{Blockage ratio} = \frac{(\text{area of cross section covered by spring})}{(\text{total internal area of cross section of tube})}$$

FIGURE 2: Schematic view of Shchelkin spiral configuration [21].

diffracting shock reflecting from the side wall collides with an obstacle. Johansen and Ciccarelli [30] studied the effect of obstacle blockage ratio on the development of unburnt gas flow field ahead of flame front in an obstacle channel using large eddy simulation. The simulations indicate that turbulence production increases with increasing number of obstacles. The quasi-detonation regime is characterized by an average flame velocity that significantly depends on the geometry of detonation tube (see Figure 1) [31]. A series of high frequency detonation wave tests is conducted by Huang et al. [32] in small scale pulse detonation engine using kerosene-air as the fuel-oxidizer to seek DDT enhancement efficiency. They observed that DDT distance and transition time were reduced. Rudy et al. [33] studied that the flame acceleration in obstructed channel has important applications for supersonic propulsion technology. The mechanism of DDT in hydrogen-air mixtures was experimentally investigated in obstacle channel using pressure profiles, wave velocities, and numerical calculations. Their results also show that obstacle blockage ratio and spacing have strong influence on detonation wave velocity stability. A weak thrust experiment on pulse detonation engine was conducted by Baklanov et al. [34]. The test was done to perform the oxidant on engine operation mode influence of ring obstacle on deflagration to detonation transition. Air-hydrogen and air-hydrocarbon mixture was used in this test.

Ciccarelli et al. [42] studied experimentally the effect of obstacle blockage on the rate of flame acceleration and on the final quasi-steady flame-tip velocity. In a smooth tube detonation transition occurs when the flame acceleration eventually leads to a terminal velocity under 1000 m/s. The freely expanding cylindrical flame speed depends on obstacle configuration and achieved appropriate turbulence in obstacles flame flow path. Gamezo et al. [43] experimentally investigated

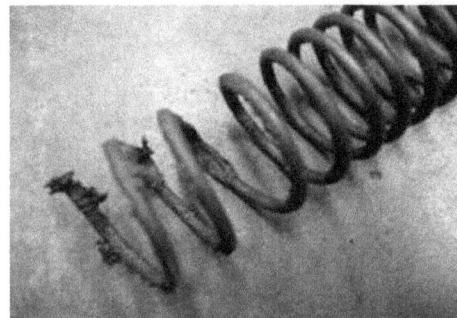

FIGURE 3: Photographs of experimental Shchelkin spiral [23].

the flame acceleration and DDT in hydrogen-air mixture in obstacle channel using 2D and 3D reactive Navier-Stokes numerical simulations and observed the regimes of supersonic turbulent flames, quasi-detonation, and detonation flame propagation behind the leading shock wave. Johansen and Ciccarelli [44] studied the effect of obstacle blockage ratio on development of unburnt gas flow field using varying obstacle height. The effect of blockage ratio on flame acceleration was investigated in an obstructed square cross section channel. Paxson [35] developed a simple computational code to access the impact of DDT (see Figure 8) enhancing obstacles on pulse detonation engine (see Figures 4 and 7). The simulation was to examine the relative contributions from drag and heat transfer. Pulse detonation engine observed that heat transfer is more significant than aerodynamic drag. Frolov [36] studied the deflagration to detonation transition in gas and drop air-fuel mixture. In this study reflecting elements could improve the fast deflagration to detonation transition of kerosene-air mixture. Teodorczyk [37] studied the flame

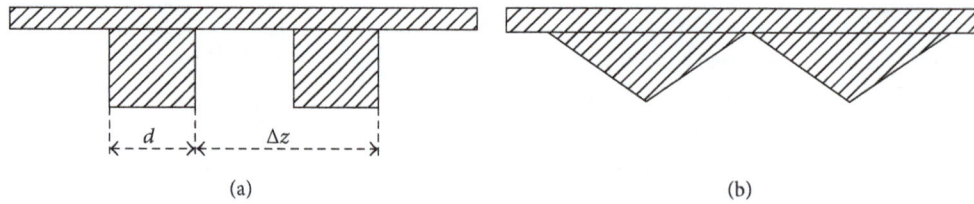

FIGURE 4: Schematic view of typical obstacle shape discussed in ICDERS-2007 [24].

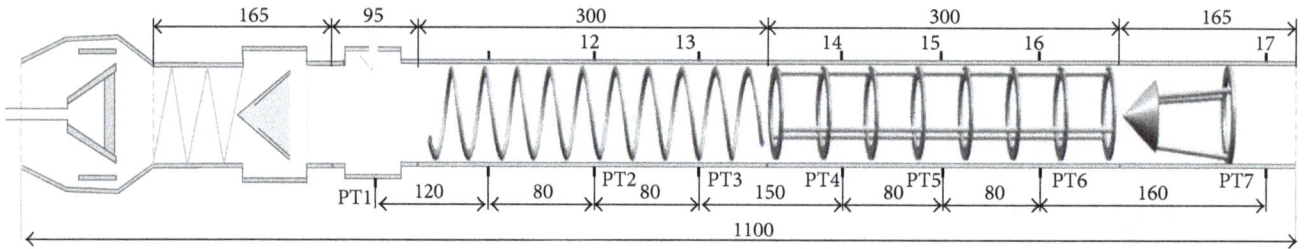

FIGURE 5: Schematic view of the small scale pulse detonation engine test setup with Shchelkin spiral and obstacle blockage [32].

FIGURE 6: Experimental setup of obstacles location along the detonation tube [33].

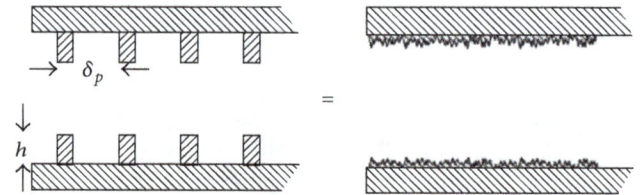

FIGURE 7: Schematic view of pulse detonation engine: 1 and 5: fuel and oxidant feed manifolds, 2: spark plug, 3: combustion chamber, 4: detonation wave, 6: ring obstacle, 7: prechamber, and 8: nozzle [34].

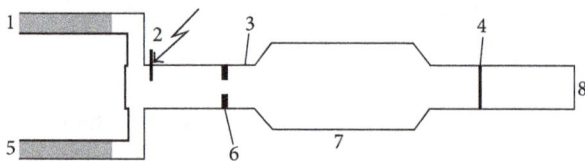

FIGURE 8: Schematic view of DDT obstacle layout [35].

propagation using channel heights of 0.01, 0.02, 0.04, and 0.08 m experimentally. The velocities of flame propagation in the obstructed channel were established in this study. The experimental result was found that deflagration to detonation transition and flame propagation regimes were established.

Frolo et al. [45] studied detonation characteristics in a U-bend tube to simulate DDT and their analysis concluded that U-bend is useful for quicker DDT. Frolov et al. [38] studied shock to detonation transition in U-bend tube experimentally and computationally both. The simulation results demonstrated a considerable effect of U-bend tube on detonation initiation. Semenov et al. [46] proposed the parabolic contraction and conical expansion for detonation initiation in a tube. The shine shaped wall is proposed for optimized geometry of conical expansion. They observed that the minimal incident shock wave velocity of 680 ± 20 m/s is approximately Mach number of 2. The U-bend is used to optimize the design of pulse detonation engine by Frolov et al. [39]. The numerical simulation of this optimization process reveals some features of deflagration to detonation transition in U-bend tubes.

Ejector is a device which is placed downstream of the pulse detonation combustor exit, coaxial the detonation tube, and it is used for propulsion performance implementation. Allgood and Gutmark [47] provided two-dimensional ejectors to pulse detonation engine for parametric study and performance was observed for inlet geometry and axial position relative to the exhaust section of PDE. Yan et al. [48] conducted an experiment in a small scale pulse detonation rocket engine (see Figure 5) which was used as a predetonator to initiate detonation in its ejectors. In this experiment they observed that flame propagation upstream at the entrance of the ejector was inevitable, which affected the detonation initiation process in the ejector. Another experimental study was performed by Bai and Weng [49] for investigating the ejector's effect on the performance of a pulse detonation engine. Their results indicated that thrust augmentation increases at high operating frequency. Canteins et al. [50] experimentally as well as numerically observed that performances of PDE

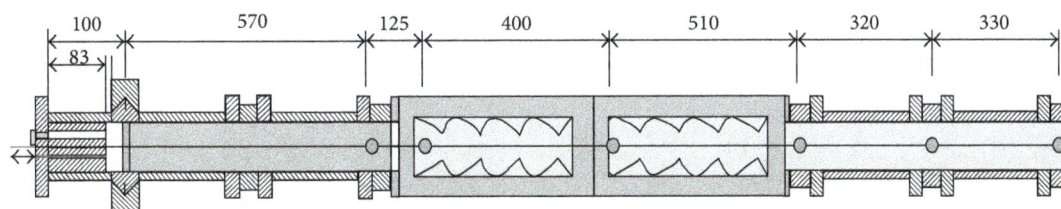

FIGURE 9: Scheme view of experimental unit with profiled regular obstacles [36].

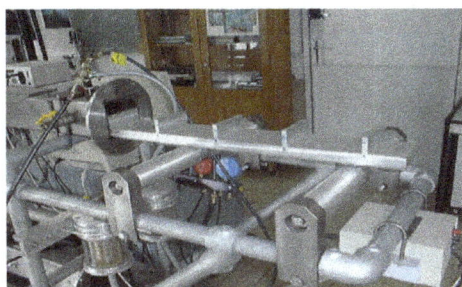

FIGURE 10: The photo of the channel with obstacles [37].

are varied with three geometry parameters of the ejector, that is, inner diameter, ejector length, and ejector position relative to the thrust wall of combustion chamber. For these configurations an ejector increases the specific impulse up to 60%. Cha et al. [51] suggested combined effect of bypass and ejector concept on the analyses of air-breathing pulse detonation engine. The calculated results showed that the APDE performance is determined by shock loss induced by throat and diameter of the nozzle. The experiments were also designed by Santoro et al. [52] to probe different aspects of PDE/ejector setup. The results indicate that, for the geometries studies, the maximum thrust augmentation of 24% is achieved. Linear array and compact box array of detonation tube with axisymmetric ejectors were examined by Hoke et al. [53]. In this study secondary flow was engaged with the ejector lip by linear detonation tube. Further the thrust augmentation was found as a function of ejector entrance to detonation tube exit distance. This thrust augmentation factor of 2.5 was observed using a tapered ejector and it was also observed that thrust augmentation depends on the distance between the detonation tube exit and the ejector entrance. The maximum thrust depends on both positive and negative values of this distance [54]. The tapered ejector and cylindrical ejector configurations were tested by Paxson et al. [55] for thrust augmentation. From this test desired results are achieved in tapered ejector configuration. Glaser et al. [56] conducted an experiment on the operation of pulse detonation engine-driven ejectors. The experimental studies employed an H_2-air mixture in PDE with ejectors to improve the performance augmenters driven. In this research straight and diverging ejectors were investigated. The optimum axial placement was found to be downstream of the pulse detonation engine. To improve the performance of ejector driven by an air-breathing pulse detonation engine the experiments were carried out with convergent nozzle

at different operating frequencies by Changxin et al. [57]. The maximum thrust augmentation was obtained in single-stage ejector for $L/D = 2$. Korobov and Golovastov [58] studied the ejector influence effectiveness of a detonation engine and their results showed that the use of an ejector can yield thrust augmentation by 17%. Huang et al. [59] experimentally studied the noise radiation characteristics of multicycle pulse detonation engine with and without ejectors. The results implied that the pulse sound pressure level increased with the increase in operating frequency. But the ejectors system showed that ejectors could decrease the peak sound pressure level of pulse detonation engine. Qiu et al. [60] conducted the operating characteristics of aerodynamic valves for pulse detonation engine for adaptive control of fuel. Their experimental results showed that residual fuel droplets moved upstream with backflow followed by flame and flame could propagate across the valve.

Matsuoka et al. [61] developed rotary valve for pulse detonation engine to analyze its basic characteristics and performance and they obtained maximum time averaged thrust of 71 N. Again Matsuoka et al. [62] developed the liquid-purge method as a novel approach to a pulse detonation combustor and they observed that detonations were successfully initiated by this method. Fan et al. [63] experimentally investigated the mixing and ignition characteristics of fully developed detonation in pulse detonation engine. The experimental results show that duty cycles should be kept as close as possible to attain effective mixing. Frolov et al. [64] conducted a low frequency demonstrator liquid fuel pulse detonation engine and their results reported that DDT occurs within a very short distance. An experimental study was investigated by Tangirala et al. [65] in multitube PDC-turbine hybrid system. The turbine component efficiency was found to be similar under PDC-fired operation. A series of experiments were carried out by Li et al. [66] on a pulse detonation engine using liquid kerosene-oxygen mixture to investigate the detonation wave initiation. The successful detonation wave was achieved when length of the spiral (see Figure 11) was increased six times of inner diameter of detonation tube. Stevens et al. [67] experimentally investigated that detonation produces shock strength and separation distance between shock and decoupled flame. Their observations indicated that higher Mach number is favored when attempting to reinitiate detonation on a reflecting ramp. Fan et al. [68] experimentally investigated on two-phase pulse detonation engine and their obtained results have demonstrated that the averaged thrust of PDE is approximately proportional to the volume of detonation chamber and detonation frequency.

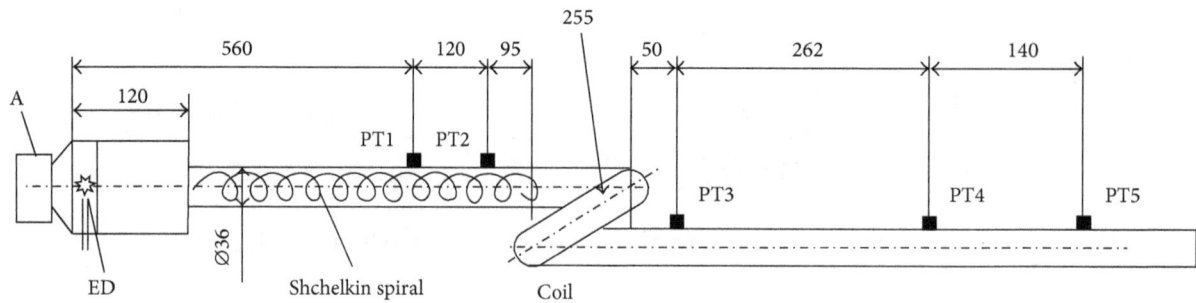

FIGURE 11: Schematic diagram of Shchelkin spiral and a coil inside the detonation tube [38].

The effects of small perturbations with varying blockage ratio on the critical tube diameter problems are investigated by Mehrjoo et al. [69]. They found that optimal blockage ratio is approximately 8% to 10% and suggested that it can be useful for design application of pulse detonation engines propulsion and power system. Huang et al. [70] conducted an experiment using liquid kerosene fuel and they observed heating liquid kerosene enhancing the engine performance, which was helpful for deflagration to detonation transition process. The experimental investigation was conducted by Deng et al. [71] to examine the operability of turbo machinery that a single stage radial flow turbine of a turbocharger in pulse detonation combustor operates. Their experimental results showed that the experimental rig can operate stably under the frequency up to 10 Hz. Cha et al. [72] developed pulse temperature and stream concentration test device to measure the temperature and steam concentration at the exhaust plum of a pulse detonation engine. Their experimental results demonstrate that the frequency of the temperature and pressure is the same and their device is able to capture every pulse detonation quickly and precisely. Thrust measurement experiment was carried out by Morozumi et al. [73] using a four-cylinder pulse detonation engine with rotary valve. Their experimental results achieved time averaged thrust of 258.5 N and specific impulse of 138.7 s. The detailed reactive detonation combustion flow features are experimentally investigated by Zitoun and Desbordes [74]. They observed that length to diameter ratio of combustion chamber is most important for thrust and specific impulse deficit. Fan et al. [75] conducted an experiment to investigate the effects of fuel preheating and adding additives on the detonation initiation time. From this analysis they achieved adding the additives to liquid kerosene, the detonation initiation time was reduced from 0.75 ms to 0.34 ms.

3. Reviews on Computational Analysis

Experimental analysis in pulse detonation engine is needed to have observations carefully on detonation combustion simulations; some researchers observed that at the same time numerical simulations are equally important to visualize the detonation combustion phenomena in PDE combustor. The PDE combustor with and without obstacles having hydrogen-air mixture has been simulated by Soni et al. [76] using a commercially available CFD code. They observed that obstacles are useful for PDE combustor design and development. Amin et al. [77] studied the effects of various nozzle geometries and operating conditions on the performance of a pulse detonation engine. The CFD results indicate that a diverging nozzle is more effective than a converging-diverging nozzle at low ambient pressure. Tangirala et al. [78] simulated the performance of PDE in subsonic and supersonic flight conditions. Their first parametric studies were employed in 2D CFD model. The results indicated that the exit nozzle enhances thrust generation, maintains operating pressure, and also controls operating frequency. Shao et al. [79] studied the effect of different types of nozzle on continuous detonation engine (CDE) using one step 3D numerical chemical reaction model. The analyzed four types of nozzles are, namely, the constant-area nozzle, Laval nozzle, diverging nozzle, and converging nozzle. The results indicated that Laval nozzle has great scope to improve the propulsion performance of this system. Ma et al. [80] studied the comparison between single-tube and multitube PDEs and influence on the nozzle flow field to improve propulsive performance. In this study thrust variation in transverse direction occurred due to reduction in axial-flow oscillations and offers a wider operation range in terms of valve timing. They showed that convergent nozzle helps to keep the original chamber pressure constant and consequently improves the engine net performance. Wintenberger and Shepherd [81] studied the flow path in a single-tube pulse detonation engine. They developed this analytical model for simulating the flow and for estimating the performance and they compared it with ideal ramjet engine. They observed that engine thrust depends on detonation tube impulse, momentum, and pressure terms. The result also indicated that the total pressure losses were caused due to unsteadiness of the flow. Kailasanath and Patnaik [82] presented a review of computational studies on pulse detonation engine. In this paper their objective was to evaluate the time dependent numerical simulations of PDE. They observed that the initial conditions in simulations are the significant effect on overall performance. Ma et al. [83] numerically studied internal flow dynamics of pulse detonation engine, which used ethylene as a combustible fuel. The detonation wave dynamics and flow evaluation were simulated. For the engine design optimization sensitivity study of operation time was also conducted. Frolov and Aksenov [84] demonstrated the deflagration to detonation transition in a tube with continuous flow of a prevaporized TS-1 jet kerosene-air mixture at

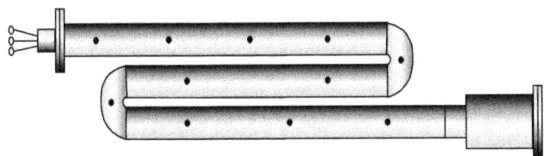

FIGURE 12: Experimental setup of U-bend tube indicates the pressure transducer location [39].

atmospheric pressure. In this analysis fuel combustion was observed in detonation combustion mode. Johansen and Ciccarelli [85] conducted large eddy simulation of initial flame acceleration in an obstructed channel (see Figure 10). The effect of obstacle blockage ratio on development of unburnt gas flow field ahead of flame front in an obstacle channel was investigated using large eddy simulation. The computational simulations indicate that turbulence production increases with increasing obstacle blockage ratio. The static thrust is easily achievable through a Lockwood-Hiller design, with a U-shaped configuration. This design needed to be tested for detonation wave simulation [40]. Again Frolov et al. [86] performed the detonation wave propagation in planar channel and cylindrical tube with two U-shaped bends of limiting curvature. Otsuka et al. [87] studied the influence of U-bend (see Figure 12) on detonation wave propagation with computational fluid dynamics analysis. Their results show that detonation waves disappear near the U-bend curvature inlet and restart after passing through it and also found that the U-bend with small channel width and curvature radius can induce fast DDT. The starting vortex was generated and performance was affected by the geometry of ejectors. On the other side ejector length is less important on overall performance compared to the ejector diameter [41]. The unsteady thrust augmentation was studied by Zheng et al. [88]. The optimum starting vortex was generated by optimum diameter of ejector and investigation also showed a minor effect of ejector length. The thrust augmentation was found to increase with ejector length. The ejector performance was observed to be strongly dependent on the operating fill fraction [89]. Zhang et al. [90] investigated the ejectors performance on thrust augmentation in pulse detonation engine. The numerical results were describing the primary detonation wave propagation processes of PDE and secondary detonation wave propagation in ejector system. The details of flow field of detonation wave propagation inside the detonation tube and injection into the ejectors were studied. The rotary wave ejector concept has highly significant potential for thrust augmentation relative to a basic pulse detonation engine [91]. Yi et al. [92] studied the effect of ejector in pulse detonation rocket propulsion system. They observed various features including detonation-shock interaction, detonation diffraction, and vortex formation in propagation of hydrogen-oxygen mixture detonation. Stoddard et al. [93] studied the CFD analysis on detonation wave propagation and resulting exhaust gas dynamics. The simulations are taken towards optimizing a static thrust on pulse detonation engine.

Tan et al. [94] studied the detonation wave on film dynamics and concluded that thermal protection is very important

and necessary for the design of a pulse detonation combustor. He and Karagozian [95] studied the transient, reactive compressible flow phenomena in pulse detonation wave engine computationally. The engine performance parameter and in addition engine noise were estimated within and external to detonation tube. Yungster et al. [96] studied numerically the formation of NO_x in hydrogen-fuelled pulse detonation engine. Their results indicated that NO_x formation in pulse detonation engine is very high for stoichiometric mixture. Ma et al. [83] studied numerically internal flow dynamics in a valveless air-breathing pulse detonation engine operating on ethylene fuel. Their calculated pressure histories and gross specific impulse of 1215 s show good agreement with experimental results.

4. Performance Analysis on Detonation Combustion

Historically, exergy analysis was developed to evaluate the process of power extraction from heat energy. The applications of exergy analysis for the performance assessment of power generation cycles have increased in recent years [97]. The exergy analysis gives a quantitative theoretical useful work that is obtained from different energy form in combustion process and is a function of the system and environment. The exergy of a fluid stream depends on reference environment [98]. Rouboa et al. [99] studied the exergy losses during the shock and rarefaction wave of hydrogen-air mixture. Their main objectives were the exergetic efficiency analysis using 1.5%, 2.5%, and 5% hydrogen mass fraction in combustible air. They obtained exergetic efficiency of 77.2%, 73.4%, and 69.7% for aforesaid hydrogen concentrations. Bellini and Lu [100] studied the efficiency in terms of availability of power device operated by detonation wave. The second law or exergetic efficiency was computed based on the fuel availability and results are then compared with the exergetic efficiency applied to deflagration system. The exergy analysis makes clear that this device is efficient for power generation. Hutchins and Metghalchi [101] have performed energy and exergy analysis of the pulse detonation engine. Different fuels such as methane and jet propulsion (JP10) (see Figure 13) have been used in this analysis. The exergetic efficiency has been calculated in pulse detonation engine and simple gas turbine engine. The compared result shows that for same pressure ratio pulse detonation engine has better efficiency and effectiveness than gas turbine system. Bellini and Lu [102] studied the exergy analysis in pulse detonation power device which is designed for power production using methane (CH_4) and propane (C_3H_8). The exergetic efficiency was studied for different design parameters like cycle frequency and detonation tube length. The theoretical analysis of pulse detonation engine over gas turbine has been carried out for electric power production. Safari et al. [103] studied that the transport equation of entropy is introduced in large eddy simulation to perform exergy analysis of turbulent combustion systems. In this methodology, the effects of chemical reaction and entropy generation contribution appear in closed forms. Fuel exergy of hydrogen was analyzed by Wu et al. [104] and computational results indicate that fuel

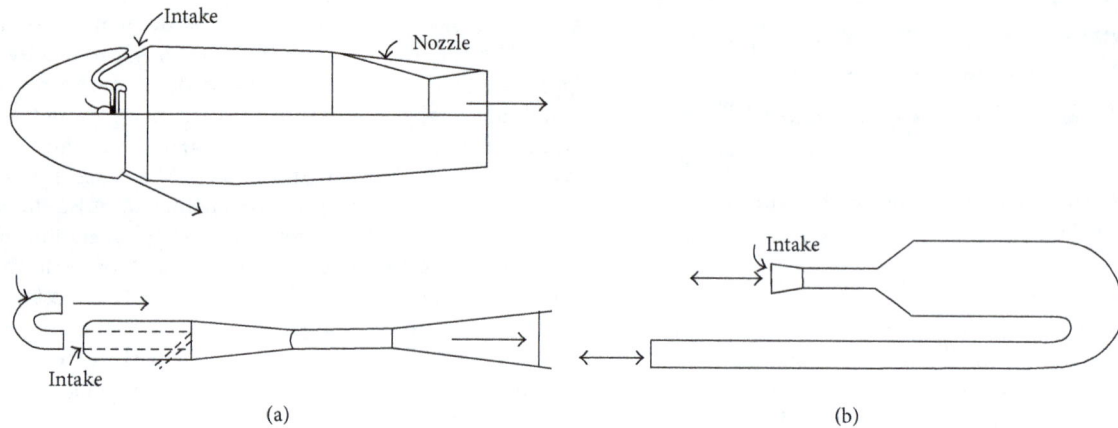

FIGURE 13: Non-PDE pulse jet designs and valveless pulsejet configurations [40].

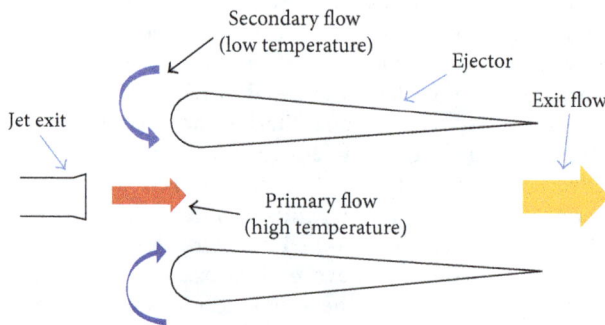

FIGURE 14: Combustion product mass flows through the ejectors [41].

of pulse detonation engine and their results show that below Mach 3 performance of PDE cycle is better than real Brayton cycle. Wu et al. [110] studied the propulsive efficiency of pulse detonation engine and system performance in terms of thrust and specific impulse and, also, compared with gas turbine and ramjet engines. Li and Kailasanath [111] studied that the flow field pressure wave developed in pulse detonation engine, which is responsible for thrust production. Cooper et al. [112] studied the specific impulse over the Mach number regime. Their analysis represents that the specific impulse varies with Mach number up to a certain limit; after that it does not change with Mach number increases. Schauer et al. [113] studied the research efforts on detonation initiation, propagation, heat transfer, and other parameters for the performance analysis.

5. Reviews on Detonation Combustion

In the 1970s a substantial effort was undertaken to investigate the feasibility of using detonative propulsion for thrusters in the dense or high-pressure atmospheres of solar system planets [114]. In subsonic flow condition the combustion of propellant in longitudinal pulsed and spin detonation waves is experimentally studied by Bykovskii and Vedernikov [115] and they observed that stability of detonation regimes depends on counter pressure of the medium. Detonation combustion produces higher rate of energy release compared to the deflagration combustion. The rate of energy release in detonation mode of combustion normally depends on the magnitudes of deflagration flame [116]. The fact that contributes to the research is more towards the gaseous fuel rather than liquid fuels which need an additional system for PDE, which evaporate the liquid fuel to gaseous state before it can get detonated. However liquid fuels have advantages over the gaseous fuel as it is best suitable for the volume limited aerospace system and at high altitude applications [117]. Radulescu and Hanson [118] studied the convective heat losses in reactive flow field in pulse detonation engine. They showed that nondimensional tube length L/D ratio, that is, length to the diameter ratio, governs the amount of heat losses. Shock

exergy of hydrogen exceeded the lower heating value (LHV) and was less than the higher heating value (HHV). The fuel-air mixture is ignited by high energy, high frequency source in combustion chamber. Hu et al. [105] studied that the fluid viscosity is a significant factor resulting in the energy loss in most fluid dynamical systems. The efficiency of PDE is affected by fuel viscosity due to exergy losses in pulse detonation engine. The energy loss in the Burgers model excited by periodic impulses is investigated based on the generalized multisymplectic method. Som and Sharma [106] developed the theoretical model of energy and exergy balance in a spray combustion process in a gas turbine combustor at various operating conditions. They observed that the velocity, temperature, and pressure field of the combustor are required for evaluating the flow availabilities and process irreversibility from a two-phase separated flow model of spray combustion.

Theoretical analysis of pulse detonation engine was estimated by Endo and Fujiwara [107]. The time average thrust was estimated from their simple theoretical model. Zel'dovich [108] studied the detonation combustion products calculation using Jouguet theory. The products of combustion are quite high in kinetic energy as well as heat energy and their statement is in good agreement with experimental analysis. Heiser and Pratt [109] studied thermodynamic cycle analysis

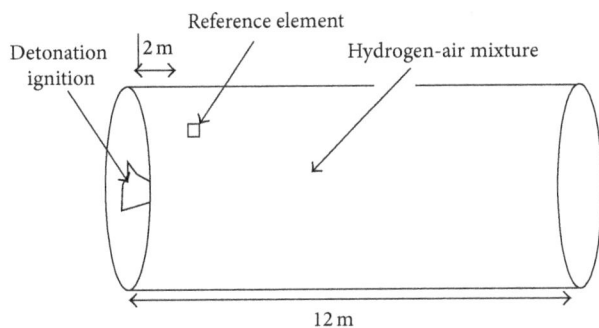

FIGURE 15: Schematic view of shock tube for exergy analysis [99].

wave propagation in a detonation tube of pulse detonation engine was investigated experimentally by Driscoll et al. [119]. Their results indicated that crossover length of detonation tube is directly related to strength of transfer shock in detonation tube. Li et al. [120] studied detonation structure generated by wedge-induced, oblique shocks in hydrogen-oxygen-nitrogen mixtures. Their simulations show a multidimensional detonation structure consisting of nonreactive oblique shock, induction zone, set of deflagration waves, and reactive shock (see Figure 15). The gaseous detonations in a channel with porous walls are investigated experimentally and numerically by Mazaheri et al. [121]. In this analysis both regular and irregular detonation waves were performed by Euler simulations. Wu and Lee [122] studied the stability of spinning detonation near the detonation limits. The detonation velocities as well as the wave structure are observed for long distance of propagation. They found that the fluctuation of local velocity, spinning detonation propagation, and fluctuations are increasing towards the limits. A two-dimensional structure of detonation wave in stoichiometric methane-air mixture in channel has been performed by Trotsyuk et al. [123]. From these calculations the irregular structure of detonation wave is obtained using a two-step kinetic model (see Figure 9).

6. Reviews on Rotating Detonation Engines

Voisekhovsky [124] performed experiments on continuously rotating detonations in 1960. The fuel blending technique is proposed by George et al. [125] to achieve detonation initiation of hydrocarbon-air mixtures in a rotating detonation engine. Wu et al. observed that the stabilizing mechanism of hydrogen addition may be physical rather than chemical for fuel plenum. The sudden change of the stagnation pressure has an immediate influence on the average axial velocity at head end of the rotating detonation engine [126]. The rotating detonation combustion process and kinetic properties of reactive flow within the combustion chamber were studied by Meng et al. [127]. They observed the inner structure of the flow field of detonation at 1110 microseconds after ignition. The experimentation on two kinds of geometry of rotating detonation engine was studied by Kindracki et al. [128]. The

thrust time profile as well as detonation velocity was calculated from pressure picks. Wang et al. [129] conducted experiments on continuous rotating detonation engine using hydrogen as a gaseous fuel. They observed that in annular combustor four different combustion patterns were observed as follows, that is, one wave in homorotating mode, two waves in homorotating mode, one couple in heterorotating mode, and two couples in heterorotating mode. Meng et al. [127] studied the three-dimensional numerical simulation investigating the combustion process and kinetic properties of reactive flow within the rotating detonation engine combustion chamber. Peng et al. [130] experimentally studied rotating detonation engine with the slot-orifice impinging injection method. They found that deflagration to detonation transition was observed in the annular combustion chamber and DDT time exhibited an obvious randomness and successful rate was up to 94%.

7. Some Results with CFD Analysis

7.1. Nozzle Effect on Detonation Wave Acceleration. The performance of single phase pulse detonation engine was studied using computational fluid dynamics as shown in Figure 16. In these analyses detonation wave dynamic pressure and velocity contour were analyzed in pulse detonation engine with divergent nozzle. From the contour plots analysis it was observed that strong detonation wave dynamic pressure and velocity appeared in detonation tube [131].

7.2. Effect of Obstacle Geometry on Detonation Wave Temperature. The temperature contour plots of detonation wave in obstructed detonation tube having blockage ratio of BR = 0.4, 0.5, 0.6, and 0.7 at obstacle spacing of $S = 4$ cm are shown in Figure 17. The strong detonation wave propagation temperature of 3185 K was found in obstacle spacing of $S = 4$ cm and detonation wave acceleration temperature greatly depends on blockage ratio; the strong propagation temperature increases the thermodynamic performance, which was found in blockage ratio of BR = 0.4 [132].

7.3. Effect of Shchelkin Spiral on Detonation Wave Acceleration. The strong detonation wave dynamic pressure of 30.1×10^5 Pa was obtained in detonation tube with Shchelkin spiral configuration (see Figure 2), which is near about C-J detonation wave as shown in Figure 18(a). The detonation wave dynamic pressure of 10.1×10^5 Pa was obtained in clean configuration as shown in Figure 18(b). From the detonation wave magnitude, it was observed that Shchelkin spiral (see Figure 3) accelerates the detonation wave at same operating Mach number [133].

8. Conclusions

In the above comprehensive review study it is observed that there is more needed research in detonation wave flow path design and detonation wave acceleration at high supersonic Mach number. Due to this research drawback future possible research is that the change in exhaust nozzle design in the exit

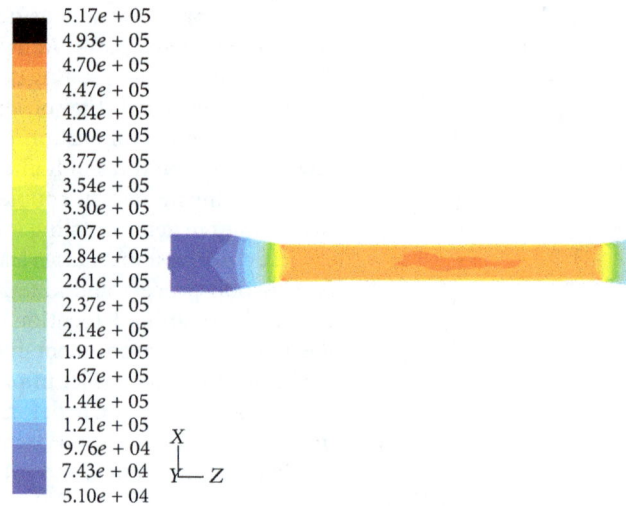

Contours of dynamic pressure (Pa) (time = 1.0000e + 00)

Fluent 6.3 (3d, dp, pbns, ske, unsteady)

(a)

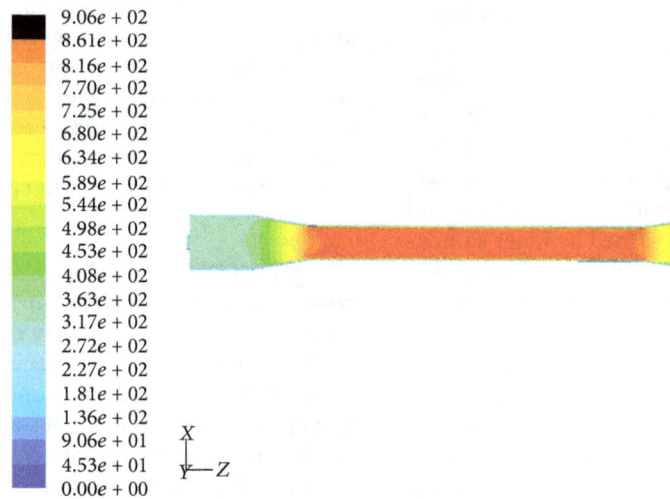

Contours of velocity magnitude (m/s) (time = 1.0000e + 00)

Fluent 6.3 (3d, dp, pbns, ske, unsteady)

(b)

FIGURE 16: Detonation wave (a) dynamic pressure and (b) velocity flow field contours analysis at Ma = 1 [131].

section of detonation tube can give better propulsion performance for detonation based pulse detonation engine. Further Shchelkin spiral improved the detonation wave initiation and acceleration inside the detonation tube of pulse detonation engine. Another successful detonation wave initiation as well as acceleration was observed inside the detonation tube with obstacles geometry. The U-bend shape detonation tube and ejectors at exit section of detonation tube can enhance deflagration flame to detonation wave transition. Exergetic efficiency of hydrogen-air detonation improves the thermodynamic performance of pulse detonation engine. The series of computational simulations performed the desired

objectives of detonation combustion phenomena. Some further research scopes are as follows:

(a) The detonation combustion wave can be simulated using orifice plates and multiple reflections in detonation tube.

(b) Internal grooves effect on detonation wave characteristics can be analyzed.

(c) The combined effect of blockage and Shchelkin spiral can enhance the deflagration to detonation wave transition.

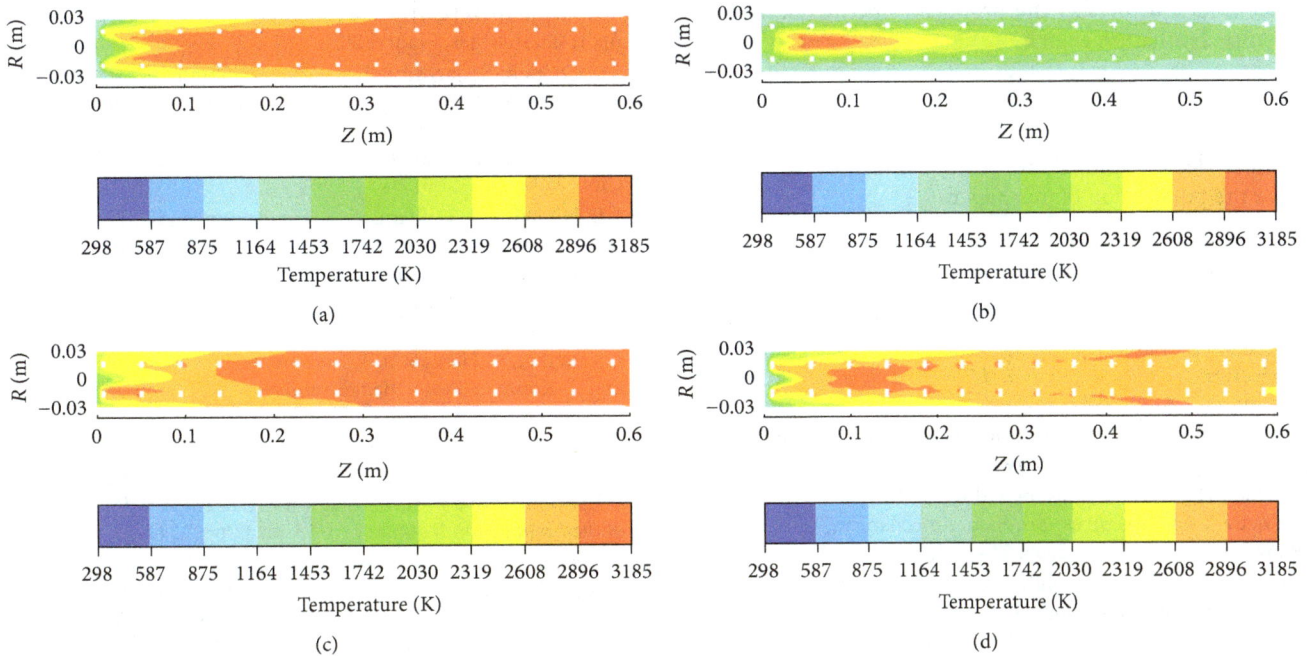

FIGURE 17: Contour of axial flame temperature distribution generated by an accelerating flame in obstacle landed detonation tube for (a) BR = 0.4, (b) BR = 0.5, (c) BR = 0.6, and (d) BR = 0.7 at Ma = 5 for spacing of $S = 4$ cm [132].

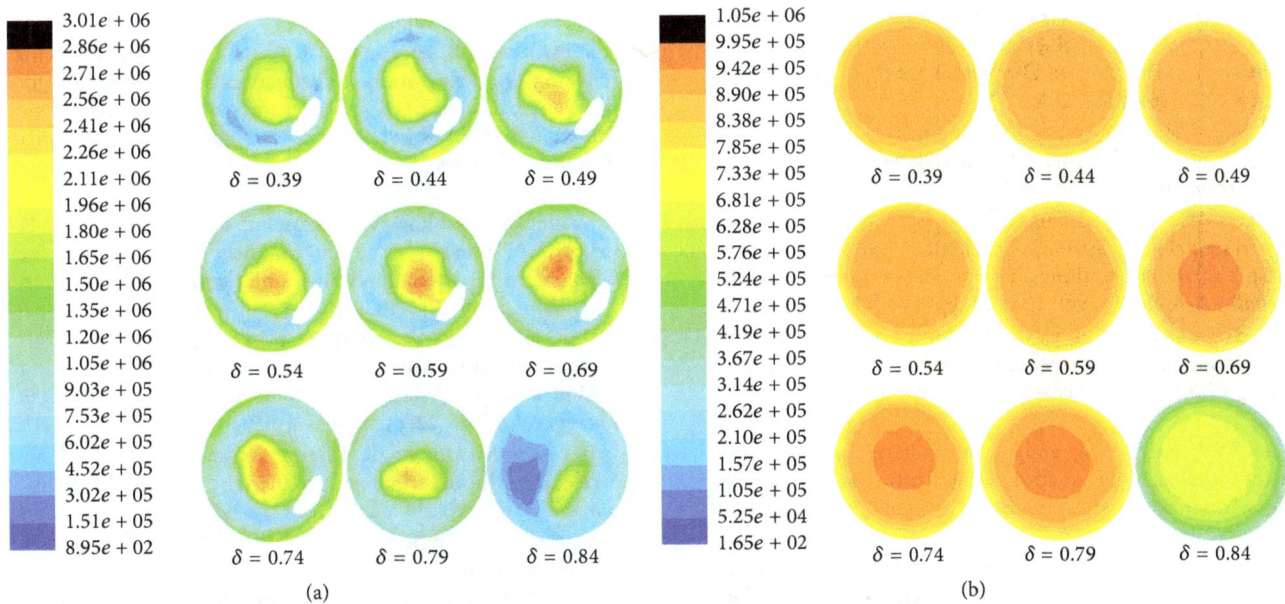

FIGURE 18: Contour plots evaluation of wave dynamic pressure (Pascal) field (a) in Shchelkin spiral and (b) without Shchelkin spiral in detonation tube at Ma = 1.4 [133].

(d) Detonation wave can be simulated using multi-injectors system.

(e) The detonation combustion can be simulated using different fuel utilization in pulse detonation engine combustor, like kerosene as a combustible fuel.

(f) Heat transfer can be analyzed in detonation tube at different frequencies.

(g) Optimize the detonation parameter, that is, detonation wave pressure, temperature, density, and species mass fraction, by using different optimization technique to predict the propulsion performance of pulse detonation engine.

Conflict of Interests

The authors declare that there is no conflict of interests regarding the publication of this paper.

References

[1] G. D. Roy, S. M. Frolov, A. A. Borisov, and D. W. Netzer, "Pulse detonation propulsion: challenges, current status, and future perspective," *Progress in Energy and Combustion Science*, vol. 30, no. 6, pp. 545–672, 2004.

[2] Pratt and Whitney, "Pratt & Whitney Acquires Propulsion technologyBusiness," 2003, http://www.pw.utc.com/pr_012301.asp.

[3] J. A. Nicholls, H. R. Wilkinson, and R. B. Morrison, "Intermittent detonation as a thrust-producing mechanism," *Journal of Jet Propulsion*, vol. 27, no. 5, pp. 534–541, 1956.

[4] L. Jin, W. Fan, K. Wang, and Z. Gao, "Review on the recent development of multi-mode combined detonation engine," *International Journal of Turbo & Jet Engines*, vol. 30, no. 3, pp. 303–312, 2013.

[5] K. Kailasanath, "Review of propulsion applications of detonation waves," *AIAA journal*, vol. 38, no. 9, pp. 1698–1708, 2000.

[6] K. Kailasanath, "Recent developments in the research on pulse detonation engines," *AIAA Journal*, vol. 41, no. 2, pp. 145–159, 2003.

[7] R. D. Wilson and F. K. Lu, "Summary of recent research on detonation wave engines at UTA," in *Proceedings of the International Workshop on Detonation for Propulsion*, Busan, South Korea, November 2011.

[8] N. N. Smirnov, V. B. Betelin, V. F. Nikitin, Y. G. Phylippov, and J. Koo, "Detonation engine fed by acetylene-oxygen mixture," *Acta Astronautica*, vol. 104, no. 1, pp. 134–146, 2014.

[9] W. Chen, W. Fan, Q. Zhang, C. Peng, C. Yuan, and C. Yan, "Experimental investigation of nozzle effects on thrust and inlet pressure of an air-breathing pulse detonation engine," *Chinese Journal of Aeronautics*, vol. 25, no. 3, pp. 381–387, 2012.

[10] Q. Li, W. Fan, C.-J. Yan, C.-Q. Hu, and B. Ye, "Experimental investigation on performance of pulse detonation rocket engine model," *Chinese Journal of Aeronautics*, vol. 20, no. 1, pp. 9–14, 2007.

[11] Y. Yan, W. Fan, K. Wang, and Y. Mu, "Experimental investigation of the effect of bell-shaped nozzles on the two-phase pulse detonation rocket engine performance," *Combustion, Explosion and Shock Waves*, vol. 47, no. 3, pp. 335–342, 2011.

[12] D. Allgood, E. Gutmark, J. Hoke, R. Bradley, and F. Schauer, "Performance measurements of multicycle pulse-detonation-engine exhaust nozzles," *Journal of Propulsion and Power*, vol. 22, no. 1, pp. 70–77, 2006.

[13] C. Peng, W. Fan, L. Zheng, Z. Wang, and C. Yuan, "Experimental investigation on valveless air-breathing dual-tube pulse detonation engines," *Applied Thermal Engineering*, vol. 51, no. 1-2, pp. 1116–1123, 2013.

[14] Y. Yan, W. Fan, K. Wang, X.-D. Zhu, and Y. Mu, "Experimental investigations on pulse detonation rocket engine with various injectors and nozzles," *Acta Astronautica*, vol. 69, no. 1-2, pp. 39–47, 2011.

[15] J. Kasahara, K. Marsuoka, T. Nakamichi, M. Esumi, A. Matsuo, and I. Funaki, "Study of high frequency rotary valve pulse detonation rocket engine," in *Proceedings of the DWP Workshop*, pp. 11–13, Bourges, France, 2011.

[16] M. Cooper, S. Jackson, J. Austin, E. Wintenberger, and J. E. Shepherd, "Direct experimental impulse measurements for detonations and deflagrations," in *Proceedings of the 37th Joint Propulsion Conference and Exhibit*, AIAA-2001-3812, Salt Lake City, Utah, USA, 2001.

[17] J. B. Hinkey, J. T. Williams, S. E. Henderson, and T. R. A. Bussing, "Rotary-valved, multiple-cycle, pulse detonation engine experimental demonstration," in *Proceedings of the 33rd Joint Propulsion Conference and Exhibit*, AIAA, pp. 27–46, Seattle, Wash, USA, 1997.

[18] J. Kasahara, A. Hasegawa, T. Nemoto, H. Yamaguchi, T. Yajima, and T. Kojima, "Performance validation of a single-tube pulse detonation rocket system," *Journal of Propulsion and Power*, vol. 25, no. 1, pp. 173–180, 2009.

[19] J.-L. Li, W. Fan, C.-J. Yan, H.-Y. Tu, and K.-C. Xie, "Performance enhancement of a pulse detonation rocket engine," *Proceedings of the Combustion Institute*, vol. 33, no. 2, pp. 2243–2254, 2011.

[20] K. Asato, T. Miyasaka, Y. Watanabe, and K. Tanabashi, "Combined effects of vortex flow and the Shchelkin spiral dimensions on characteristics of deflagration-to-detonation transition," *Shock Waves*, vol. 23, no. 4, pp. 325–335, 2013.

[21] T. H. New, P. K. Panicker, F. K. Lu, and H. M. Tsai, "Experimental investigations on DDT enhancements by shchelkin spirals in a PDE," AIAA 2006-552, AIAA, 2006.

[22] W. Wang, H. Qiu, W. Fan, and C. Xiong, "Experimental study on DDT characteristics in spiral configuration pulse detonation engines," *International Journal of Turbo & Jet Engines*, vol. 30, no. 3, pp. 261–270, 2013.

[23] P. K. Panicker, F. K. Lu, and D. R. Wilson, "Practical methods for reducing the deflagration-to-detonation transition length for pulse detonation engines," in *Proceedings of the 9th International Symposium on Experimental and Computational Aerothermodynamics of Internal Flows (ISAIF '09)*, Gyeongju, Republic of Korea, September 2009.

[24] D. Valiev, V. Bychkov, V. Akkerman, C. K. Law, and L.-E. Eriksson, "Flame acceleration in channels with obstacles in the deflagration-to-detonation transition," *Combustion and Flame*, vol. 157, no. 5, pp. 1012–1021, 2010.

[25] A. V. Gaathaug, K. Vaagsaether, and D. Bjerketvedt, "Experimental and numerical investigation of DDT in hydrogen-air behind a single obstacle," *International Journal of Hydrogen Energy*, vol. 37, no. 22, pp. 17606–17615, 2012.

[26] I. O. Moen, M. Donato, R. Knystautas, and J. H. Lee, "Flame acceleration due to turbulence produced by obstacles," *Combustion and Flame*, vol. 39, no. 1, pp. 21–32, 1980.

[27] T. Ogawa, V. N. Gamezo, and E. S. Oran, "Flame acceleration and transition to detonation in an array of square obstacles," *Journal of Loss Prevention in the Process Industries*, vol. 26, no. 2, pp. 355–362, 2013.

[28] C. Johansen and G. Ciccarelli, "Numerical simulations of the flow field ahead of an accelerating flame in an obstructed channel," *Combustion Theory and Modelling*, vol. 14, no. 2, pp. 235–255, 2010.

[29] V. N. Gamezo, T. Ogawa, and E. S. Oran, "Flame acceleration and DDT in channels with obstacles: effect of obstacle spacing," *Combustion and Flame*, vol. 155, no. 1-2, pp. 302–315, 2008.

[30] C. Johansen and G. Ciccarelli, "Numerical simulations of the flow field ahead of an accelerating flame in an obstructed channel," *Combustion Theory and Modelling*, vol. 14, no. 2, pp. 235–255, 2010.

[31] D. A. Kessler, V. N. Gamezo, E. S. Oran, and R. K. Zipf, "Simulation of deflagration-to-detonation transition in premixed CH_4-air in large-scale channels with obstacles," in *Proceedings of the 47th AIAA Aerospace Sciences Meeting Including the New Horizons Forum and Aerospace Exposition*, Orlando, Fla, USA, January 2009.

[32] Y. Huang, H. Tang, J. Li, and C. Zhang, "Studies of DDT enhancement approaches for kerosene-fueled small-scale pulse detonation engines applications," *Shock Waves*, vol. 22, no. 6, pp. 615–625, 2012.

[33] W. Rudy, R. Porowski, and A. Teodorczyk, "Propagation of hydrogen-air detonation in tube with obstacles," *Journal of Power Technologies*, vol. 91, no. 3, pp. 122–129, 2011.

[34] D. I. Baklanov, S. V. Golovastov, V. V. Golub, N. V. Semin, and V. V. Valodin, "Model of low-thrust pulse detonation device with valveless fuel feed," *Progress in Propulsion Physics*, vol. 1, pp. 341–352, 2009.

[35] D. E. Paxson, "Performance impact of deflagration to detonation transition enhancing obstacles," in *Proceedings of the 47th Aerospace Sciences Meeting*, AIAA-2009-502, AIAA, Orlando, Fla, USA, January 2009.

[36] S. M. Frolov, "Fast deflagration-to-detonation transition," *Russian Journal of Physical Chemistry B*, vol. 2, no. 3, pp. 442–455, 2008, Pleiades Publishing, 2008, Original Russian Text, S. M. Frolov, 2008, Published in *Khimicheskaya Fizika*, vol. 27, no. 6, pp. 32–46, 2008.

[37] A. Teodorczyk, "Scale effects on hydrogen–air fast deflagrations and detonations in small obstructed channels," *Journal of Loss Prevention in the Process Industries*, vol. 21, no. 2, pp. 147–153, 2008.

[38] S. M. Frolov, V. S. Aksenov, and I. O. Shamshin, "Detonation propagation through U-bends," in *Nonequilibrium Processes. Combustion and detonation*, G. Roy, S. Frolov, and A. Starik, Eds., vol. 1, pp. 348–364, Torus Press, Moscow, Russia, 2005.

[39] S. M. Frolov, V. S. Aksenov, and I. O. Shamshin, "Reactive shock and detonation propagation in U-bend tubes," *Journal of Loss Prevention in the Process Industries*, vol. 20, no. 4–6, pp. 501–508, 2007.

[40] C. Tharratt, "The propulsive duct," *Aircraft Engineering and Aerospace Technology*, vol. 37, no. 12, pp. 359–371, 1965.

[41] F. Zheng, A. V. Kuznetsov, W. L. Roberts, and D. E. Paxson, "Influence of geometry on starting vortex and ejector performance," *Journal of Fluids Engineering*, vol. 133, no. 5, Article ID 051204, 2011.

[42] G. Ciccarelli, C. T. Johansen, and M. Parravani, "The role of shock–flame interactions on flame acceleration in an obstacle laden channel," *Combustion and Flame*, vol. 157, no. 11, pp. 2125–2136, 2010.

[43] V. N. Gamezo, T. Ogawa, and E. S. Oran, "Numerical simulations of flame propagation and DDT in obstructed channels filled with hydrogen-air mixture," *Proceedings of the Combustion Institute*, vol. 31, no. 2, pp. 2463–2471, 2007.

[44] C. T. Johansen and G. Ciccarelli, "Visualization of the unburned gas flow field ahead of an accelerating flame in an obstructed square channel," *Combustion and Flame*, vol. 156, no. 2, pp. 405–416, 2009.

[45] S. M. Frolo, V. S. Aksenov, and I. O. Shamshin, "Shock wave and detonation propagation through U-bend tubes," *Proceedings of the Combustion Institute*, vol. 31, no. 2, pp. 2421–2428, 2007.

[46] I. V. Semenov, P. S. Utkin, V. V. Markov, S. M. Frolov, and V. S. Aksenov, "Numerical and experimental investigation of detonation initiation in profiled tubes," *Combustion Science and Technology*, vol. 182, no. 11-12, pp. 1735–1746, 2010.

[47] D. Allgood and E. Gutmark, "Experimental investigation of a pulse detonation engine with a 2D ejector," in *Proceedings of the 42nd AIAA Aerospace Sciences Meeting and Exhibit*, AIAA-2004-864, Reno, Nev, USA, January 2004.

[48] Y. Yan, W. Fan, and Y. Mu, "Preliminary experimental investigation on detonation initiation in the ejector of a pulse detonation rocket engine," *International Journal of Turbo & Jet Engines*, vol. 29, no. 4, pp. 299–307, 2012.

[49] Q. D. Bai and C. S. Weng, "Experimental study of ejectors' effect on the performance of pulse detonation rocket engine," *Applied Mechanics and Materials*, vol. 628, pp. 293–298, 2014.

[50] G. Canteins, F. Franzetti, E. Zocłońska, B. A. Khasainov, R. Zitoun, and D. Desbordes, "Experimental and numerical investigations on PDE performance augmentation by means of an ejector," *Shock Waves*, vol. 15, no. 2, pp. 103–112, 2006.

[51] X. Cha, Y. Chuan-Jun, and Q. Hua, "Analysis of an air-breathing pulsed detonation engine with bypass and ejector," *International Journal of Turbo & Jet-Engines*, vol. 25, no. 2, pp. 129–136, 2008.

[52] R. J. Santoro, P. Sibtosh, R. Shehadeh, S. Saretto, and S. Y. Lee, "Experimental study of a pulse detonation engine driven ejector," in *Proceedings of the 39th AIAA/ASME/SAE/ASEE Joint Propulsion Conference and Exhibit*, AIAA 2003-4972, Huntsville, Ala, USA, July 2003.

[53] J. L. Hoke, A. G. Naples, L. P. Goss, and F. R. Schauer, "Schlieren imaging of a single-ejector multi-tube pulsed detonation engine," in *Proceedings of the 47th AIAA Aerospace Sciences Meeting Including The New Horizons Forum and Aerospace Exposition*, Orlando, Fla, USA, January 2009.

[54] J. Wilson, A. Sgondea, D. E. Paxson, and B. N. Rosenthal, "Parametric investigation of thrust augmentation by ejectors on a pulsed detonation tube," *Journal of Propulsion and Power*, vol. 23, no. 1, pp. 108–115, 2007.

[55] D. E. Paxson, P. J. Litke, F. R. Schaue, R. P. Bradley, and J. L. Hoke, "Performance assessment of a large scale pulsejet-driven ejector system," in *Proceedings of the 44th AIAA Aerospace Sciences Meeting and Exhibit*, Reno, Nev, USA, January 2006.

[56] A. J. Glaser, N. Caldwell, E. Gutmark, J. Hoke, R. Bradley, and F. Schauer, "Study on the operation of pulse-detonation engine-driven ejectors," *Journal of Propulsion and Power*, vol. 24, no. 6, pp. 1324–1331, 2008.

[57] P. Changxin, F. Wei, Z. Qun, Y. Cheng, C. Wenjuan, and Y. Chuanjun, "Experimental study of an air-breathing pulse detonation engine ejector," *Experimental Thermal and Fluid Science*, vol. 35, no. 6, pp. 971–977, 2011.

[58] A. E. Korobov and S. V. Golovastov, "Numerical investigation of the ejector influence on the effectiveness of a jet nozzle of a detonation engine," *High Temperature*, vol. 53, no. 1, pp. 118–123, 2015.

[59] X. Huang, M. Li, Y. Yuan, Y. Xiong, and L. Zheng, "Experimental investigation on noise radiation characteristics of pulse detonation engine-driven ejector," *Advances in Mechanical Engineering*, vol. 7, no. 6, 2015.

[60] H. Qiu, C. Xiong, C.-J. Yan, and L.-X. Zheng, "Effect of aerodynamic valve on backflow in pulsed detonation tube," *Aerospace Science and Technology*, vol. 25, no. 1, pp. 1–15, 2013.

[61] K. Matsuoka, M. Esumi, K. B. Ikeguchi, J. Kasahara, A. Matsuo, and I. Funaki, "Optical and thrust measurement of a pulse detonation combustor with a coaxial rotary valve," *Combustion and Flame*, vol. 159, no. 3, pp. 1321–1338, 2012.

[62] K. Matsuoka, T. Mukai, and T. Endo, "Development of a liquid-purge method for high-frequency operation of pulse detonation combustor," *Combustion Science and Technology*, vol. 187, no. 5, pp. 747–764, 2015.

[63] W. Fan, C.-J. Yan, Q. Li, Y.-Q. Ding, and J.-H. Li, "Experimental investigations on detonability limits of pulse detonation rocket engine," *International Journal of Turbo & Jet Engines*, vol. 23, no. 4, pp. 277–282, 2006.

[64] S. M. Frolov, V. S. Aksenov, and V. S. Ivanov, "Experimental demonstration of the operation process of a pulse-detonation liquid rocket engine," *Russian Journal of Physical Chemistry B*, vol. 5, no. 4, pp. 664–667, 2011.

[65] V. E. Tangirala, A. Rasheed, and A. J. Dean, "Performance of a pulse detonation combustor-based hybrid engine," in *Proceedings of the ASME Turbo Expo: Power for Land, Sea and Air*, vol. 3 of *ASME Proceedings, Paper no. GT2007-28056*, pp. 403–414, Montreal, Canada, May 2007.

[66] J. Li, W. Fan, C. Yan, and Q. Li, "Experimental investigations on detonation initiation in a kerosene-oxygen pulse detonation rocket engine," *Combustion Science and Technology*, vol. 181, no. 3, pp. 417–432, 2009.

[67] C. A. Stevens, P. I. King, F. R. Schauer, and J. L. Hoke, "An experimental study of detonation diffraction and shock initiation," in *Proceedings of the 51st AIAA Aerospace Sciences Meeting Including the New Horizons Forum and Aerospace Exposition*, AIAA 2013-1027, Grapevine, Tex, USA, January 2013.

[68] W. Fan, C. Yan, X. Huang, Q. Zhang, and L. Zheng, "Experimental investigation on two-phase pulse detonation engine," *Combustion and Flame*, vol. 133, no. 4, pp. 441–450, 2003.

[69] N. Mehrjoo, R. Portaro, and H. D. Ng, "A technique for promoting detonation transmission from a confined tube into larger area for pulse detonation engine applications," *Propulsion and Power Research*, vol. 3, no. 1, pp. 9–14, 2014.

[70] X. Huang, Y. Yan, Y. Mu, L. Zheng, and L. Chen, "Experimental investigation on heating kerosene using thrust tube waste heat of pulse detonation engine," *International Journal of Turbo & Jet-Engines*, vol. 30, no. 4, pp. 339–345, 2013.

[71] J. Deng, L. Zheng, C. Yan, L. Jiang, C. Xiong, and N. Li, "Experimental investigation of a puls detonation combustor integrated with a turbine," *International Journal of Turbo & Jet-Engines*, vol. 27, no. 2, pp. 119–134, 2010.

[72] X. Cha, Y. Chuan-Jun, and Q. Hua, "Experiments and numerical simulation of exhaust temperature of pulsed detonation engine," *International Journal of Turbo and Jet Engines*, vol. 25, no. 2, pp. 137–144, 2008.

[73] T. Morozumi, K. Matsuoka, R. Sakamota, Y. Fujiwara, and J. Kasahara, "Study on a four-cylinder pulse detonation rocket engine with a coaxial high frequency rotary valve," in *Proceedings of the 51st AIAA Aerospace Sciences Meeting including the New Horizones Forum and Aerospace Exposition*, AIAA 2013-027, Dallas, Tex, USA, January 2013.

[74] R. Zitoun and D. Desbordes, "Propulsive performances of pulsed detonations," *Combustion Science and Technology*, vol. 144, no. 1, pp. 93–114, 1999.

[75] Z.-C. Fan, W. Fan, H.-Y. Tu, J.-L. Li, and C.-J. Yan, "The effect of fuel pretreatment on performance of pulse detonation rocket engines," *Experimental Thermal and Fluid Science*, vol. 41, pp. 130–142, 2012.

[76] S. K. Soni, A. Singh, M. Sandhu, A. Goel, and R. K. Sharma, "Numerical simulation to investigate the effect of obstacle on detonation wave propagation in a pulse detonation engine combustor," *International Journal of Emerging Technology and Advanced Engineering*, vol. 3, no. 3, pp. 458–464, 2013.

[77] M. R. Amin, H. Z. Rouf, and J.-L. Cambier, "Numerical investigation on the effects of nozzle geometry on the performance of a pulse detonation engine," *Journal of Mechanical Engineering*, vol. 51, no. 7-8, pp. 484–490, 2005.

[78] V. N. Tangirala, A. J. Dean, N. Tsuboi, and A. K. Hayashi, "Performance of a pulse detonation engine under subsonic and supersonic flight conditions," in *Proceedings of the 45th AIAA Aerospace Sciences Meeting and Exhibit*, AIAA 2007-1245, Reno, Nev, USA, January 2007.

[79] Y. Shao, M. Liu, and J. Wang, "Continuous detonation engine and effects of different types of nozzle on its propulsion performance," *Chinese Journal of Aeronautics*, vol. 23, no. 6, pp. 647–652, 2010.

[80] F. Ma, J.-Y. Choi, and V. Yang, "Thrust chamber dynamics and propulsive performance of multitude pulse detonation engines," *Journal of Propulsion and Power*, vol. 21, no. 4, pp. 681–691, 2005.

[81] E. Wintenberger and J. E. Shepherd, "Model for the performance of airbreathing pulse-detonation engines," *Journal of Propulsion and Power*, vol. 22, no. 3, pp. 593–603, 2006.

[82] K. Kailasanath and G. Patnaik, "Performance estimates of pulsed detonation engines," *Proceedings of the Combustion Institute*, vol. 28, no. 1, pp. 595–601, 2000.

[83] F. Ma, J.-Y. Choi, and V. Yang, "Internal flow dynamics in a valveless airbreathing pulse detonation engine," *Journal of Propulsion and Power*, vol. 24, no. 3, pp. 479–490, 2008.

[84] S. M. Frolov and V. S. Aksenov, "Deflagration-to-detonation transition in a kerosene-air mixture," *Doklady Physical Chemistry*, vol. 416, no. 1, pp. 261–264, 2007.

[85] C. Johansen and G. Ciccarelli, "Modeling the initial flame acceleration in an obstructed channel using large eddy simulation," *Journal of Loss Prevention in the Process Industries*, vol. 26, no. 4, pp. 571–585, 2013.

[86] S. M. Frolov, V. S. Aksenov, and I. O. Shamshin, "Propagation of shock and detonation waves in channels with U-shaped bends of limiting curvature," *Russian Journal of Physical Chemistry B*, vol. 2, no. 5, pp. 759–774, 2008.

[87] S. Otsuka, M. Suzuki, and M. Yamamoto, "Numerical investigation on detonation wave through U-bend," *Journal of Thermal Science*, vol. 19, no. 6, pp. 540–544, 2010.

[88] F. Zheng, A. V. Kuznetsov, W. L. Roberts, and D. E. Paxson, "Numerical study of a pulsejet-driven ejector," in *Proceedings of the 45th AIAA/ASME/SAE/ASEE Joint Propulsion Conference & Exhibit*, Denver, Colorado, August 2009.

[89] D. Allgood, E. Gutmark, J. Hoke, R. Bradley, and F. Schauer, "Performance studies of pulse detonation engine ejectors," *Journal of Propulsion and Power*, vol. 24, no. 6, pp. 1317–1323, 2008.

[90] H.-H. Zhang, Z.-H. Chen, X.-H. Sun, X.-H. Jiang, and B.-M. Li, "Numerical investigations on the thrust augmentation mechanisms of ejectors driven by pulse detonation engines," *Combustion Science and Technology*, vol. 183, no. 10, pp. 1069–1082, 2011.

[91] M. R. Nalim, Z. A. Izzy, and P. Akbari, "Rotary wave-ejector enhanced pulse detonation engine," *Shock Waves*, vol. 22, no. 1, pp. 23–38, 2012.

[92] T. H. Yi, D. R. Wilson, and F. K. Lu, "Detonation wave propagation in an ejector augmented pulse detonation rocket," in *Proceedings of the 44th AIAA Aerospace Sciences Meeting and Exhibit*, Reno, Nev, USA, January 2006.

[93] W. A. Stoddard, M. Mihaescu, and E. Gutmark, "Simulation of detonation based cycles in tubes with multiple openings," in *Proceedings of the 46th AIAA/ASME/SAE/ASEE Joint Propulsion Conference & Exhibit*, AIAA 2010-6745, Nashville,Tenn, USA, July 2010.

[94] X.-M. Tan, J.-Z. Zhang, and X.-T. Wang, "Effects of pulse detonation wave on film dynamics," *Engineering Applications of Computational Fluid Mechanics*, vol. 5, no. 4, pp. 499–505, 2011.

[95] X. He and A. R. Karagozian, "Numerical simulation of pulse detonation engine phenomena," *Journal of Scientific Computing*, vol. 19, no. 1–3, pp. 201–224, 2003.

[96] S. Yungster, K. Radhakrishnan, and K. Breisacher, "Computational study of NO_x formation in hydrogen-fuelled pulsed detonation engines," *Combustion Theory and Modelling*, vol. 10, no. 6, pp. 981–1002, 2013.

[97] H. Taniguchi, K. Mouri, T. Nakahara, and N. Arai, "Exergy analysis on combustion and energy conversion processes," *Energy*, vol. 30, no. 2–4, pp. 111–117, 2005.

[98] R. Petela, "Application of exergy analysis to the hydrodynamic theory of detonation in gases," *Fuel Processing Technology*, vol. 67, no. 2, pp. 131–145, 2000.

[99] A. Rouboa, V. Silva, and N. Couto, "Exergy analysis in hydrogen-air detonation," *Journal of Applied Mathematics*, vol. 2012, Article ID 502979, 16 pages, 2012.

[100] R. Bellini and F. K. Lu, "Exergy analysis of a pulse detonation power device," in *Proceedings of the 10th Brazilian Congress of Thermal Sciences and Engineering (ENCIT '04)*, Brazilian Society of Mechanical Sciences and Engineering, ABCM, Rio de Janeiro, Brazil, November 2004.

[101] T. E. Hutchins and M. Metghalchi, "Energy and exergy analyses of the pulse detonation engine," *Journal of Engineering for Gas Turbines and Power*, vol. 125, no. 4, pp. 1075–1080, 2003.

[102] R. Bellini and F. K. Lu, "Exergy analysis of a pulse detonation power device," *Journal of Propulsion and Power*, vol. 26, no. 4, pp. 875–877, 2010.

[103] M. Safari, M. R. H. Sheikhi, M. Janbozorgi, and H. Metghalchi, "Entropy transport equation in large eddy simulation for exergy analysis of turbulent combustion systems," *Entropy*, vol. 12, no. 3, pp. 434–444, 2010.

[104] Z. Wu, S. Zhou, and L. An, "The second law (exergy) analysis of hydrogen," *Journal of Sustainable Development*, vol. 4, no. 1, pp. 260–263, 2011.

[105] W. Hu, Z. Deng, and G. Xie, "Energy loss in pulse detonation engine due to fuel viscosity," *Mathematical Problems in Engineering*, vol. 2014, Article ID 735926, 5 pages, 2014.

[106] S. K. Som and N. Y. Sharma, "Energy and exergy balance in the process of spray combustion in a gas turbine combustor," *Journal of Heat Transfer*, vol. 124, no. 5, pp. 828–836, 2002.

[107] T. Endo and T. Fujiwara, "A simplified analysis on a pulse detonation engine model," *Transactions of the Japan Society for Aeronautical and Space Sciences*, vol. 44, no. 146, pp. 217–222, 2002.

[108] Y. B. Zel'dovich, "To the question of energy use of detonation combustion," *Journal of Propulsion and Power*, vol. 22, no. 3, pp. 588–592, 2006.

[109] W. H. Heiser and D. T. Pratt, "Thermodynamic cycle analysis of pulse detonation engines," *Journal of Propulsion and Power*, vol. 18, no. 1, pp. 68–76, 2002.

[110] Y. Wu, F. Ma, and V. Yang, "System performance and thermodynamic cycle analysis of airbreathing pulse detonation engines," *Journal of Propulsion and Power*, vol. 19, no. 4, pp. 556–567, 2003.

[111] C. Li and K. Kailasanath, "Partial fuel filling in pulse detonation engines," *Journal of Propulsion and Power*, vol. 19, no. 5, pp. 908–916, 2003.

[112] M. Cooper, J. E. Shepherd, and F. Schauer, "Impulse correlation for partially filled detonation tubes," *Journal of Propulsion and Power*, vol. 20, no. 5, pp. 947–950, 2004.

[113] F. Schauer, J. Stutrud, and R. Bradly, "Detonation initiation studies and performance results for pulsed detonation engine applications," in *Proceedings of the 39th AIAA Aerospace Sciences Meeting*, AIAA 2001-1129, Reno, Nev, USA, January 2001.

[114] K. Kim, G. Varsi, and L. H. Back, "Blast wave analysis for detonation propulsion," *AIAA Journal*, vol. 15, no. 10, pp. 1500–1502, 1977.

[115] F. A. Bykovskii and E. F. Vedernikov, "Continuous detonation of a subsonic flow of a propellant," *Combustion, Explosion and Shock Waves*, vol. 39, no. 3, pp. 323–334, 2003.

[116] N. Smirnov, "Pulse detonation engines: advantages and limitations," in *Advanced Combustion and Aerothermal Technologies*, NATO Science for Peace and Security Series C: Environmental Security, pp. 353–363, Springer, Dordrecht, The Netherlands, 2007.

[117] M. Z. Ahmad Faiz, M. A. Wahid, K. M. Saqr, and U. Haffis, "Single pulse detonation study of natural gas," in *Proceedings of the International Conference on Theoretical and Applied Mechanics, and International Conference on Fluid Mechanics and Heat & Mass Transfer (MECHANICS-HEAPFL '10)*, pp. 78–83, Stevens Point, Wis, USA, 2010.

[118] M. I. Radulescu and R. K. Hanson, "Effect of heat loss on pulse-detonation-engine flow fields and performance," *Journal of Propulsion and Power*, vol. 21, no. 2, pp. 274–285, 2005.

[119] R. Driscoll, W. Stoddard, A. St George, and E. Gutmark, "Shock transfer and shock-initiated detonation in a dual pulse detonation engine/crossover system," *AIAA Journal*, vol. 53, no. 1, pp. 132–139, 2015.

[120] C. Li, K. Kailasanath, and E. S. Oran, "Detonation structures behind oblique shocks," *Physics of Fluids*, vol. 6, no. 4, pp. 1600–1611, 1994.

[121] K. Mazaheri, Y. Mahmoudi, M. Sabzpooshani, and M. I. Radulescu, "Experimental and numerical investigation of propagation mechanism of gaseous detonations in channels with porous walls," *Combustion and Flame*, vol. 162, no. 6, pp. 2638–2659, 2015.

[122] Y. Wu and H. S. J. Lee, "Stability of spinning detonation waves," *Combustion and Flame*, vol. 162, no. 6, pp. 2660–2669, 2015.

[123] A. V. Trotsyuk, P. A. Fomin, and A. A. Vasil'ev, "Numerical study of cellular detonation structures of methane mixtures," *Journal of Loss Prevention in the Process Industries*, vol. 36, pp. 394–403, 2015.

[124] B. V. Voisekhovsky, "A spin stationary detonation," *Applied Mechanics and Technical Physics*, vol. 3, pp. 157–164, 1960.

[125] A. George, R. Driscoll, V. Anand, D. Munday, and E. J. Gutmark, "Fuel blending as a means to achieve initiation in a rotating detonation engine," in *Proceedings of the 53rd AIAA Aerospace Sciences Meeting*, AIAA SciTech, Kissimmee, Fla, USA, January 2015.

[126] D. Wu, Y. Liu, Y. Liu, and J. Wang, "Numerical investigations of the restabilization of hydrogen-air rotating detonation engines," *International Journal of Hydrogen Energy*, vol. 39, no. 28, pp. 15803–15809, 2014.

[127] L. Meng, Z. Shuang, W. Jianping, and C. Yifeng, "Parallel three-dimensional numerical simulation of rotating detonation

engine on graphics processing units," *Computers & Fluids*, vol. 110, pp. 36–42, 2015.

[128] J. Kindracki, P. Wolański, and Z. Gut, "Experimental research on the rotating detonation in gaseous fuels–oxygen mixtures," *Shock Waves*, vol. 21, no. 2, pp. 75–84, 2011.

[129] C. Wang, W. Liu, S. Liu, L. Jiang, and Z. Lin, "Experimental investigation on detonation combustion patterns of hydrogen/vitiated air within annular combustor," *Experimental Thermal and Fluid Science*, vol. 66, pp. 269–278, 2015.

[130] L. Peng, D. Wang, X. Wu, H. Ma, and C. Yang, "Ignition experiment with automotive spark on rotating detonation engine," *International Journal of Hydrogen Energy*, vol. 40, no. 26, pp. 8465–8474, 2015.

[131] P. Debnath and K. M. Pandey, "Performance investigation on single phase pulse detonation engine using computational fluid dynamics," in *Proceedings of the ASME International Mechanical Engineering Congress & Exposition (IMECE '13)*, IMECE2013-66274, San Diego, Calif, USA, November 2013.

[132] P. Debnath and K. M. Pandey, "Effect of blockage ratio on detonation flame acceleration in pulse detonation combustor using CFD," *Applied Mechanics and Materials*, vol. 656, pp. 64–71, 2014.

[133] P. Debnath and K. M. Pandey, "Computational study of deflagration to detonation transition in pulse detonation engine using shchelkin spiral," *Applied Mechanics and Materials*, vol. 772, pp. 136–140, 2015.

15

Review of Sensing Methodologies for Estimation of Combustion Metrics

Libin Jia, Jeffrey D. Naber, and Jason R. Blough

Michigan Technological University, 1400 Townsend Drive, Houghton, MI 49931, USA

Correspondence should be addressed to Libin Jia; libinj@mtu.edu

Academic Editor: Sergey M. Frolov

For reduction of engine-out emissions and improvement of fuel economy, closed-loop control of the combustion process has been explored and documented by many researchers. In the closed-loop control, the engine control parameters are optimized according to the estimated instantaneous combustion metrics provided by the combustion sensing process. Combustion sensing process is primarily composed of two aspects: combustion response signal acquisition and response signal processing. As a number of different signals have been employed as the response signal and the signal processing techniques can be different, this paper did a review work concerning the two aspects: combustion response signals and signal processing techniques. In-cylinder pressure signal was not investigated as one of the response signals in this paper since it has been studied and documented in many publications and also due to its high cost and inconvenience in the application.

1. Introduction

Determination of combustion metrics for an internal combustion engine has the potential of providing feedback for closed-loop combustion phasing control to meet current and upcoming emission and fuel consumption regulations. Closed-loop control of the combustion process has been a focus for engine research and development [1–3]. Open-loop operation based on calibration maps which are conservatively set based upon laboratory operation can give a quick response and is relatively easy to control. However, the open loop cannot adapt to the changes caused by the condition variations such as injector aging and fuel quality [2].

In comparison, closed-loop control considers the condition changes in the control mechanism and enables operation closer to the optimum fuel consumption and emissions target. Combustion metrics detected or estimated through a sensor, which is referred to as combustion sensing output, provides feedback information to control the combustion process. In this paper, the combustion metrics being addressed involves SOC (start of combustion), peak pressure location, peak apparent heat release rate location, CA50 (crank-angle location for 50% fuel burnt), and so forth. The feedback information provided by the combustion sensing

process also refers to the abnormal combustion phenomena including knock and misfire events. In this paper, a review was performed for combustion sensing methodologies concerning combustion metrics estimation with respect to combustion sensing response signals and signal processing techniques.

2. Response Signal for Combustion Event

In-cylinder pressure waveform is the most commonly used signal which provides the information for engine combustion control [4–6]. In-cylinder pressure signal has been historically used in the laboratory and more recently in series production to derive the combustion metrics and provide feedback for combustion phasing control [7–9]. However, the measurement of the in-cylinder pressure is typically obtained with intrusive sensors that require a special mounting process and engine structure modification. Also the in-cylinder pressure transducer has a high cost for mass production for diesel engines [10]. So the response signals discussed in this paper do not include the in-cylinder pressure signal.

2.1. Crank-Shaft Speed Fluctuation. Due to the variations of the in-cylinder pressure waveform during a combustion cycle, the crank-shaft speed fluctuation varies in a complex way which depends on the engine parameters. How the speed fluctuation varies with the engine in-cylinder pressure changes has been explored by many researchers so as to develop a good alternative to the direct intrusive in-cylinder pressure measurement [11–19].

Based on a model that relates the crank-shaft speed and the in-cylinder pressure, the in-cylinder pressure can be estimated with the input of instantaneous speed signal measured by a crank-shaft speed sensor. The sensor can be an optical encoder or a magnetic pickup transducer which are easy to mount and low in cost.

Moro et al. [11] proposed a linear dependency between the in-cylinder pressure and the engine speed signal and experimentally verified it for 38 different engine running conditions. The equation that can represent this linear dependency is given as [11]

$$Ar\left(p\left(t\right)-p_{\mathrm{misf}}\left(t\right)\right)f_{\mathrm{crank}}=J\ddot{\theta}_s\left(t\right)-\ddot{\theta}_{s\,\mathrm{misf}}\left(t\right)$$
$$=J\Delta\ddot{\theta}_s\left(t\right),\tag{1}$$

where A is piston area in m^2, r is the crank radius in m, p_{misf} is the in-cylinder pressure in case of misfire (bar), $\ddot{\theta}_s$ is the synthetic engine acceleration in $\mathrm{rad/s}^2$, $\ddot{\theta}_{s\,\mathrm{misf}}$ is the synthetic engine acceleration in case of misfire $\mathrm{rad/s}^2$, J is the moment of initial ($\mathrm{kg\cdot m}^2$), and f_{crank} is the crank-slider kinematics function.

A frequency response function between the in-cylinder pressure and the engine speed can be obtained by converting (1) into frequency domain. However, this FRF is sensitive to engine running conditions and the FRF obtained based on one condition does not lead to the estimated in-cylinder pressure with high accuracy when condition varies. So a FRF mapping was created in this paper based on 38 different steady-state engine conditions with the engine speed and the manifold pressure as the condition parameters to distinguish different test conditions. For the conditions falling into the FRF mapping, interpolation technique was used for both real and imaginary harmonic components to obtain the estimated FRF. The pressure recovery results for low speed low load, high speed low load, and low speed high load conditions were shown in this paper.

Connolly and Yagle modeled the cylinder combustion pressure via the crank-shaft velocity from a statistical point of view [12]. The model involves three sequent components. First, by replacing the time domain independent variables with crank-angle variables, a nonlinear differential model between the crank's shaft speed and the in-cylinder pressure signal can be simplified. Secondly, the in-cylinder pressure signal was parameterized by the sample modeling sequence based on a stochastic model which uses the sum of the deterministic waveform and an amplitude-modulated cosine window. Third, an estimation of the in-cylinder pressure based on the crank-shaft speed signal was achieved through a state-space deconvolution process which utilized a Kalman filter. Moreover, signal to noise ratio effects to the in-cylinder pressure estimation were also evaluated in this paper. Results showed that for low to moderate noise level conditions the reasonable deconvolution can be reached.

Shiao and Moskwa [13] employed a sliding observer to estimate the in-cylinder pressure and combustion heat release for an SI engine. To estimate the in-cylinder pressure with high accuracy, the error between the measured and the estimated crank-shaft speed was taken as the feedback to reduce the dynamic error of the estimated in-cylinder pressure. The unobservability problem arises for the pressure estimation around the top-dead center and thus introduces significant estimation error. This problem was partly solved by adapting the parameters of the observer. Then the estimated in-cylinder pressure was used to compute the cylinder heat release. Also, detection of misfire or abnormal combustion events was achieved through the estimated heat release.

Additional investigators examining combustion metrics analysis based on the crank-shaft speed fluctuation can be found in [14–16]. In addition to estimating the in-cylinder pressure waveform and the heat release, engine crank-shaft speed was also used to recover the engine torque [17–19]. For most cases, the crank-shaft speed was fed into an engine model which was simplified based on assumptions to estimate the engine torque.

2.2. In-Cylinder Ion Current. Ion current in the combustion chamber is measured via the spark plug. After the high-voltage discharge, the ion current across the spark plug gap is obtained by applying a DC voltage across the gap and measuring the resulting current. The ion current is affected by gas flow, geometry of flame, electric potential, ion density, and the angle between the flame and electrode [20]. The ionization of gases in the cylinder occurs in two phases. When the fuel reacts with the oxygen during combustion, the first phase ionization occurs which can be considered as chemical phase. The second phase, defined as thermal phase, occurs when the burnt gases are compressed by the increased in-cylinder pressure [21]. The most consistent dependency between the ion current and in-cylinder pressure occurs on the peaks of the two signals for both amplitude and the location perspectives. This has been verified by the researches in [20–25].

Martychenko et al. [22] detected the breakdown voltage across the spark plug gap and modeled the relationship between the peak of the voltage and the peak of the in-cylinder pressure based on second-order polynomial function. The coefficients of the second-order polynomial function for the conditions with varied engine speed are different. However, the coefficients can be curve-fitted by a linear function of engine speeds. Hellring and Holmberg [21] proposed least squares fit method to estimate the in-cylinder pressure peak position of spark ignited engines based on the ion current signal. This method was proved to have a better robustness and accuracy than multilayer perceptron and Gaussian curve fit methods for peak in-cylinder pressure estimation.

Gazis et al. [28] explored the possibility of estimating in-cylinder characteristics based on the ion current with one simple and computationally inexpensive neural network, adaptive linear type of network. Thirteen extracted characteristics of the ion current were taken as the input and four characteristics of the in-cylinder pressure (peak pressure position, peak pressure magnitude, the width of curve at half of its height, and the area of the curve between inlet valve closing (IVC) and exhaust valve opening (EVO)) as the output to train the network with the purpose of predicting the four characteristics of in-cylinder pressure. Also, based on the same neural network structure but with the whole ion current signal (time domain sampled between IVC and EVO) as the input and the whole in-cylinder pressure signal (time synced with ion current signal) as the output, the in-cylinder pressure curve rather than just some characteristics of the in-cylinder pressure signal can be estimated. The peak pressure location, as one of the most important in-cylinder pressure characteristics, was predicted with the mean error at 0.062 degrees and standard deviation at 2.55 degrees.

Ion current was also used to detect engine knock, misfire, or incomplete combustion [24, 29–31]. Kumar et al. [24] applied a band-pass filter on the ion current signal and the filtered output indicates the engine knock. Danne et al. [29] compared the ion current based knock detection with the conventional methods of pressure based and accelerometer-based knock detection on a large-displacement, air-cooled, V-twin motorcycle engine. It was found that the ion current based method can detect the inaudible knock more accurately and reject the mechanical noise more effectively than the other two conventional methods. Zhu et al. [30] found that the in-cylinder ion current can detect misfire or incomplete combustion. Also, the ion current signal can be used to compute minimum spark advance for best torque (MBT) to measure the combustion stability [30, 32]. However, the results are only limited to a fixed load over a narrow speed range.

2.3. Accelerometer Signal. Accelerometers are mounted externally on the engine block or the engine head to detect the combustion events by measuring the vibration signals which are transmitted from the in-cylinder oscillation to the engine outer surface. However, as the accelerometer detects the vibrations from the sources in addition to the cylinder oscillation including the valve dynamics and piston slaps, the signal may vary from cylinder to cylinder and over operating conditions. So the utilization of the accelerometer signal for combustion metrics detection relies on the signal processing technique which can eliminate the effects from other sources.

Naber et al. [33] evaluated the effectiveness and accuracy of accelerometer-based knock detection. The distributions of the accelerometer-based knock intensity metrics for various operation conditions including varied speeds, loads, cam timings, and knock levels were measured and fitted by a log-norm distribution. The log-norm model was verified to provide a good fit of the distributions and the distribution characteristics including skewness and peakedness. In addition, a good correlation can be seen between the cylinder pressure based knock intensity metrics and the accelerometer-based knock intensity metrics. Guillemin et al. [34] estimated the instantaneous engine knock by fitting the accelerometer signals with Gaussian function

on a 2.2 L HCCI engine and measure the start of combustion where the knock level is out of the user-defined threshold.

Characteristics of the accelerometer signal which are related to the characteristics of in-cylinder pressure or apparent heat release rate were investigated and extracted [35–37]. Some characteristics of the in-cylinder pressure or apparent heat release including start of combustion, CA50, and peak pressure crank-angle location are closely related to the combustion process and thus can be used as the feedback to control the combustion process. Arnone et al. [35] band-pass filtered the in-cylinder pressure signal and the accelerometer signal within 650–1000 Hz and found that the accelerometer signal can locate the sudden rise of the in-cylinder pressure signal (so as to denote the start of combustion), diffusive combustion process, and the peak of the in-cylinder pressure on a water cooled Lombardini LDW442CRS direct injection common rail diesel engine. Chiavola et al. [36] computed the cumulative heat release based on the measured in-cylinder pressure and investigated the relationship between the accelerometer signal and the cumulative heat release on a two-cylinder diesel engine equipped with a common rail injection system. By superimposing the filtered accelerometer signal to the cumulative heat release, it was found that the filtered accelerometer signal can locate the start of combustion, the beginning of main combustion, and MFB50 (50% of the burnt fuel mass). Taglialatela et al. [37] investigated the correlation between the in-cylinder pressure signal and the features derived from the accelerometer signal on a 4 L single cylinder SI engine. Time-frequency spectrogram method was utilized to analyze the accelerometer signal to present more features of the accelerometer signal than the analysis in time domain. The result indicates a direct correlation between the peak pressure location and the maximum amplitude of accelerometer signal in time-frequency domain for all the engine operating conditions conducted in this paper. So the maximum amplitude location of the accelerometer signal can be used as the feedback for a closed-loop control system of spark advance.

Polonowski et al. [38] explored the potential of accelerometers to recover the in-cylinder pressure curve on a 1.9 L four-cylinder, turbocharged, HPCR, direct injection diesel engine. In this paper, standard signal processing techniques including Fast Fourier Transform (FFT) and coherence were employed and results showed that a strong coherence presented between the in-cylinder pressure signal and the accelerometer signal within frequency band of 0.5 kHz to 4 kHz with the coherence value over 0.9. Also, this research found that the accelerometer location did result in a varied coherence value between the in-cylinder pressure signal and the accelerometer signal. The optimal locations for the accelerometer placement were determined based on both offline and online coherence analysis. In his later work [39], frequency response function (FRF) was used to quantify the relationship between accelerometer response and AHR and the relationship between accelerometer response and in-cylinder pressure. A technique termed as spectrum weighting was utilized to combine FRFs from all conducted test conditions into a single FRF by weighting the

FRF magnitude and phase information for each frequency by the coherent output power at that specific frequency.

This obtained FRF was used to estimate in-cylinder pressure and AHR on a cycle-cycle basis. The maximum pressure gradient determination was shown to have a root mean square error (RMSE) accuracy of 15% of actual maximum pressure gradient. The location based metrics had the RMSE as small as 0.29° and more than 80% of the estimated peak apparent heat locations were within 1° crank-angle.

3. Signal Processing Techniques to Correlate the Combustion Metrics and the Response Signal

3.1. Frequency Response Function (FRF). The pressure curve recovery depends on the transfer path modeling between the acquired source signal (in-cylinder pressure signal) and the response signal (e.g., vibration signal). Frequency response function which represents the frequency domain relationship between the in-cylinder pressure signal, $P(\omega)$, and the accelerometer signal, $A(\omega)$, can be presented by

$$A(\omega) = P(\omega) H(\omega). \tag{2}$$

$H(\omega)$ is the transfer path in frequency domain and is defined as the frequency response function (FRF). With the FRF obtained by measuring both $P(\omega)$ and $A(\omega)$ based on the representative operating conditions, the in-cylinder pressure signal of any other condition can be recovered with the measured $A(\omega)$ and the obtained $H(\omega)$ by

$$P(\omega) = A(\omega) H^{-1}(\omega). \tag{3}$$

However, as the response signal is sensitive to not only the in-cylinder pressure oscillation but the rotating crank-slider and vibration from other mechanical parts including piston slaps and valve dynamics which varies with the engine operating conditions, the FRF with assumption of linear dependency in frequency domain between the source signal and the response signal does not have a good robustness over engine operating conditions. This conclusion has been confirmed by researchers [39–42]. Gao and Randall [40] applied the FRF computed from 2400 rpm full load condition to reconstruct the in-cylinder pressure with the accelerometer signal measured at 3600 rpm and full load condition as the input. The significant recovery error for the in-cylinder pressure waveform indicated that the transfer path modeled by FRF cannot be considered consistent over engine operating conditions. Morello et al. [41] attempted to overcome the drawback of FRF application for the heat release recovery based on the accelerometer signal by optimizing the time domain window applied to the accelerometer signal. Also, a Vold-Kalman order tracking filter was employed to eliminate the abnormal harmonics of the singular value decomposition results of both the accelerometer signal and in-cylinder pressure signal. However, no significant heat release estimation accuracy improvement can be seen for start of injection (SOI) sweep test conditions. Polonowski [39] also tried to improve the FRF performance by adding a weighting function to the FRF. However, the robustness improvement for FRF is still limited.

Gao and Randall [40] explained why the variation of the FRF results in an error of source estimation with the Laplace transform. It can be seen that the variation of FRF introduces incomplete cancellation of the poles and zeros in the dominator with the accelerometer signal as the numerator. The incomplete cancellation of non-minimum-phase zeros will make the extra poles or zeros of FRF present and make the inverse filtering unstable. As a result errors will be introduced to the estimation result.

3.2. Cepstrum Analysis. Complex cepstral analysis is a non-linear homomorphic signal process which is being utilized in many areas including machine diagnostics, image processing, speech, and radar signal processing. A cepstrum is reached by taking the inverse Fourier transform of the logarithm of a signal spectrum. The complex cepstrum $X_c(t)$ can be expressed as

$$X_c(t) = F^{-1}\left\{\log\left(F\left\{x(t)\right\}\right)\right\}. \tag{4}$$

$x(t)$ is the signal in time domain, F represents the Fourier transform algorithm, and F^{-1} denotes inverse Fourier transform form.

Equation (5) can be obtained by applying logarithm to (2):

$$\log(A) = \log(P) + \log(H). \tag{5}$$

After computing the inverses Fourier transform

$$F^{-1}\left(\log(A)\right) = F^{-1}\left(\log(P)\right) + F^{-1}\left(\log(H)\right). \tag{6}$$

According to (4), the cepstrum of FRF can be obtained as

$$H_c(t) = A_c(t) - P_c(t). \tag{7}$$

The advantage of this method is that the convolution process is converted to an addition process in cepstrum domain. El-Ghamry et al. [42] applied the complex cepstrum analysis on the root mean square acoustic emission signal. The complex cepstrum of FRF, $H_c(t)$, was evaluated for four complete combustion cycles at 1280 rpm 30 Nm condition. By inserting $H_c(t)$ and $A_c(t)$ into (7), the cylinder pressure signal can be identified as

$$P_c(t) = A_c(t) - H_c(t). \tag{8}$$

However, this method only gave good estimation of the in-cylinder pressure signal for the same engine operating conditions based on which $H_c(t)$ was computed.

Another important application of complex cepstral analysis is for signal smoothing. The complex cepstrum was utilized to improve the robustness of the transfer path by smoothing both the source signal and the response signal [34, 40, 43]. Smoothing the FRF actually reduces the variations of the FRF associated with different operating conditions. The schematic illustration for smoothing the FRF is shown in Figure 1.

In the process described in Figure 1, the smoothing of magnitude and phase was realized by applying a low-pass filter in quefrency domain to lifter the corresponding

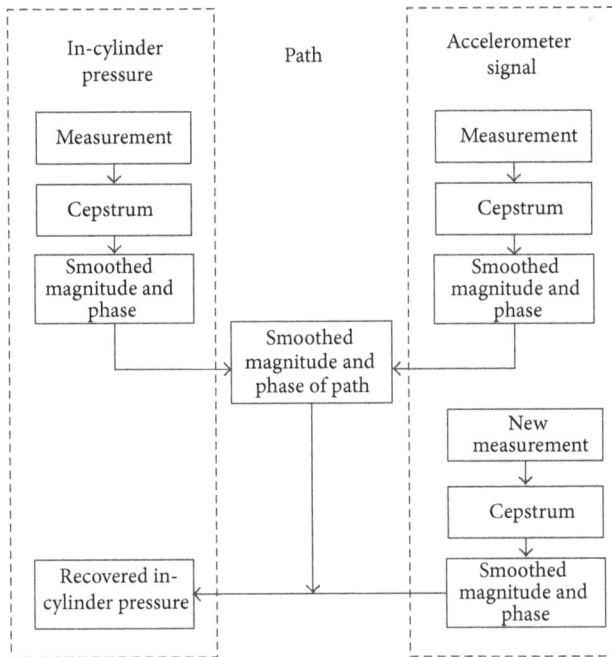

FIGURE 1: In-cylinder pressure estimation based on the cepstral smoothing technique.

content of complex cepstrum. In other words, the liftering process was achieved by applying a window around zero quefrency. Kim and Lyon [43] also discussed the effects of the window length in the signal smoothing and results showed that the shorter the window, the smoother the log spectrum. The smoothed amplitude and phase of FRF were obtained through (7) by smoothing the phase and amplitude of both the accelerometer signal and the premeasured in-cylinder pressure under the given conditions. By inserting the smoothed FRF into (8), the in-cylinder pressure signal was recovered by inputting the smoothed accelerometer signal measured from the same conditions on another engine structure of the same type.

Gao and Randall [40] applied the complex cepstral smoothing technique to obtain a smoothed FRF for in-cylinder pressure recovery. They compared the results based on four different in-cylinder pressure recovery methods, including two inverse filtering procedures (see (3) and (8)), the cepstral smoothed FRF, and the time domain smoothed FRF. It showed that the pressure waveforms recovered from the two smoothing operations can better match the measured ones than the pressure waveforms recovered from the two inverse filtering operations.

3.3. System Identification.
FRF method assumed a linear dependency in frequency domain between the in-cylinder pressure and the vibration signal. However, the low robustness of the FRF with respect to the engine operating condition variations proved that the linear dependency needs to be adapted. System identification approach modeled the transfer path between the in-cylinder pressure and the accelerometer signal with a nonlinear hypothesis. Villarino and Böhme [44] modeled the transfer path as a filter which was applied to the

in-cylinder pressure signal to output the accelerometer signal. Also, it was assumed that the accelerometer signal consists of a superposition of K components with each component for one cylinder. The model was expressed as

$$a_n = \left(1 - B\left(q^{-1}\right)\right) a_n + \sum_{k=1}^{K} H_k \left(q^{-1}, n\right) p_{k,n} + \omega_n. \quad (9)$$

a_n as the measured accelerometer signal is the sum of the in-cylinder pressure p_n filtered by a time-variant filter $H_k(q^{-1}, n)$, past accelerometer samples termed by $B(q^{-1}) = 1 + \sum_{m=1}^{M} a_m q^{-m}$, and the noise termed by ω_n. q^{-1} is the left shift operator and works as $q^{-m} a_n = a_{n-m}$.

The optimal filter coefficients $\widehat{B}(\cdot)$ and $\widehat{H}_k(\cdot)$ were estimated by minimizing the error ϵ_n between the estimated in-cylinder accelerometer signal $\widehat{a}_n$ and the measured one a_n:

$$\left[\widehat{B}, \widehat{H}_1, \ldots, \widehat{H}_k\right] = \arg\min \left(\sum_{n=1}^{N} \epsilon_n^2\right). \quad (10)$$

For the ease of reconstruction of the in-cylinder pressure, in-cylinder pressure trace was decomposed into three parts which are associated with the same dependent parameters. Expectation maximization algorithm was employed to recover the dependent parameters. The results showed that the peak pressure location estimation yields a mean error of $0.04°$ with the standard deviation at $4.78°$. However, the high estimation accuracy was limited to the same engine operating conditions. No results were reported when this method was applied to a varied engine operating condition.

Wagner et al. [45] built a physical model which denoted the speed dependence of the transfer path between the in-cylinder pressure and the accelerometer signal. SGN algorithm was used to identify the parameters of the transfer function speed-independently. Each pressure in this paper was considered to be composed of two parts with the first one introduced by the compression due to the piston movement and the second one generated by the pressure rise due to the combustion event. As the parameters were identified speed-independently, only one set of transfer path parameters needs to be stored for the in-cylinder pressure estimation.

Other than recovering the in-cylinder pressure signal, system identification approach was also used for misfire detection [46–48]. A function was developed to interpret the ratio between the energy of the signal and the energy of the noise, termed as signal energy-to-noise ratio in Villarino's work. A higher load can cause the signal energy-to-noise ratio to increase. A threshold value was determined with the function value lower than the threshold value indicating the occurrence of the misfire.

3.4. Neural Network.
The modeled transfer paths described in the previous sections only work for limited engine operating conditions. The transfer behavior between the accelerometer and the combustion metrics (with the in-cylinder pressure as the example) is a nonlinear dynamic path highly depending on the input and the engine operating condition. For this reason, another nonlinear modeling approach, neural network, was employed to investigate the relationship

between the combustion metrics and the response signal, including crank speed fluctuation [12, 15], the vibration signal [49, 50], and hybrid of crank speed fluctuation and vibration signal [26].

Gu et al. [15] modeled the relationship between the cranks shaft speed and the in-cylinder pressure with a radial basis function (RBF) neural network on a four-cylinder DI diesel engine. With network trained with the selected data, the in-cylinder pressure can be expressed as

$$y_{kj}(\theta) = \sum_{m=1}^{M} h_{km}(\theta) w_{mj}, \quad (11)$$

where $h_{km}(\cdot)$ is the radial basis function and w_{mj} are the weighting vectors. The RBF is composed of a linear layer represented by (11) and a nonlinear layer with radial basis functions as the components. The radial basis function is expressed as

$$h_{km} = \exp\left(-\frac{\|x_k - c_m\|}{r_m}\right), \quad (12)$$

where c_m is the hidden unit center, x_k represents the crank speed input, and r_m is the radius of the Gaussian function. $\|\cdot\|$ represents the Euclidean distance between the vectors. Results showed that the recovered pressure waveform matches well with the measured one for nine engine conditions with varied engine speeds and loads for all the phases: compression, peak pressure, and rise and fall of the combustion. Indicated mean effective pressure (IMEP) was also computed based on the recovered in-cylinder pressure signal and the IMEP from the recovered in-cylinder pressure can follow the respective measured values closely.

Taglialatela et al. [10] utilized the multilayer perceptron neural network to model the relationship between the crank-shaft speed and parameters extracted from the in-cylinder pressure, including peak pressure value and peak pressure angular location, instead of the pressure waveform. With the trained neural network, the peak pressure amplitude can be estimated with minimum error of 2.31 bar and maximum error of 6.97 bar which are 4.1% and 8.0%, respectively, in relative percentage scale. The peak pressure location can be estimated with minimum error of 1.38 crank-angle degrees and maximum of 5.20 crank-angle degrees.

Bizon et al. [49] reconstructed the in-cylinder pressure signal on a single cylinder 0.5 L diesel engine with the engine block vibration as the input signal to a trained RBF neural network. This paper focused on the RBF neural network parameters optimization with respect to the number of neurons and the spread parameter. 50 centers and spread parameter of 3.2 were finally determined and the RBF network structured with the optimized parameters was evaluated based on the peak pressure amplitude, peak pressure location, and the MBF50 which are derived from the recovered in-cylinder pressure. The peak pressure value estimation error was under 3% in relative RMSE and the peak location and MBF50 were both below 1.5 crank-angle degrees.

Johnsson [26] also employed the RBF neural network for the in-cylinder pressure recovery but with the hybrid of

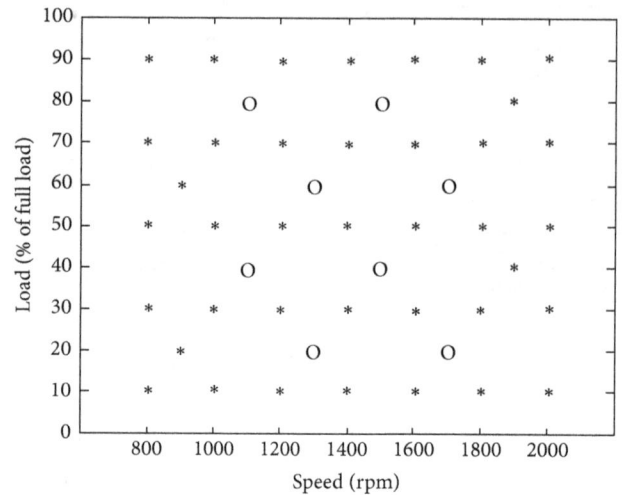

FIGURE 2: Engine operating conditions (o: validation data; *: training data) [26].

vibration signal and the crank-shaft speed as the input on a 9-litre, 6-cylinder, and inline four-stroke diesel engine. Because the coherence analysis indicated that the crank-shaft angular speed has the highest coherence with the in-cylinder pressure for the lowest frequency while the vibration signal has the highest coherence with the in-cylinder pressure for higher frequency, a recursive hybrid learning procedure was applied to train the neural network. K-means clustering algorithm found the k centers and used the regularization to determine the weights. Fourier transform of vibration signal and the crank-shaft signal are taken as the input to the RBF neural network, so both the input and output are complex values. The training data and the validation data are presented in Figure 2.

The results showed that the RMS error of maximum pressure based on the validation data was 3.5 bar, the location for the maximum pressure was 1.5 degrees, and RMS error for IMEP was 0.7 bar.

3.5. Wavelet Method. Although the frequency domain signal processing techniques including the Fourier transform and the FRF have advantages in analyzing the raw signal and building the transfer path, the Fourier transform result is inefficient for nonstationary problems such as the engine vibration signal with the nonstationary effects introduced by the combustion events [51].

Also, the Fourier-transform-based technique is not capable of detecting the temporal variations of the periodicities due to its pure frequency domain dependency [52]. Wavelet transform as a popular time-frequency transform decomposes the signal into different frequency bands and allows the feature analysis associated with these frequency bands. This property makes wavelet transform a useful tool to analyze the signals with time discontinuities and sharp spikes, such as the vibration signal or the radiated sound of an engine during combustion.

Kim and Min [27] applied the Meyer wavelet transform to the engine block vibration signal obtained on a controlled

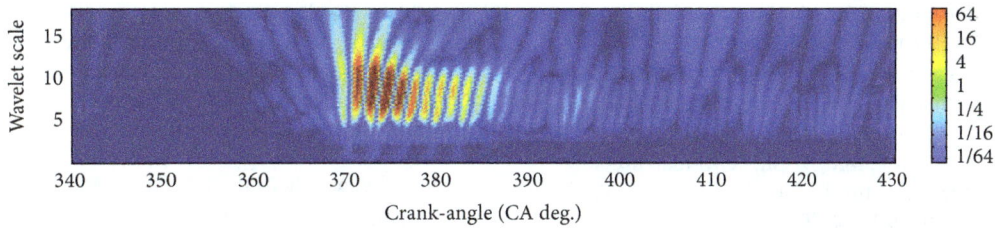

FIGURE 3: Wavelet transform result in a specific cycle for 1500 rpm [27].

autoignition engine to detect the start of combustion which was defined as 2% mass fraction burned. The engine block vibration signal within 500 Hz to 7 kHz was converted into wavelet scale with an interval of 100 Hz. A threshold value was determined to correspond to the start of combustion and the wavelet scale which grows greater than this threshold value was used to locate the start of combustion for each frequency band. Then the averaged start of combustion for each frequency band was the final determined result. Figure 3 indicates the engine block vibration wavelet transform results and the start of combustion was determined at 363 crank-angle degrees (3 degrees after top-dead center).

Hariyanto et al. [53] defined the pressure based start of combustion based on derivatives of the pressure trace and took it as reference. The technique for the start of combustion determination based on the vibration signal wavelet analysis used the trial and error method. Results showed that the averaged difference of the start of combustion between the two determination methods is below 1 crank-angle degree. The correlation coefficient of the start of combustion derived from in-cylinder pressure and the accelerometer signal is higher than 0.95.

As the wavelet transform presents more details about the signal in time-frequency domain, features extraction can be achieved based on the wavelet transform analysis. Then the correlation between the features extracted from the vibration or sound signal and the features extracted from the in-cylinder pressure trace was investigated. However, no transfer path based on the wavelet transform has been developed and applied for the pressure waveform recovery. More researches concerning application of wavelet transform for engine combustion or engine radiated noise can be found in [51, 52, 54].

4. Summary

Among the forementioned three signals for combustion metrics estimation or correlation, accelerometer signal was most utilized because the accelerometer has the advantage of low price and easy mounting as well as high reliability and durability. Also, utilization of multiple accelerometers which are placed at multiple locations on engine block can supply more than one input channel which have the potential of improving the combustion metrics estimation accuracy with the assistance of signal processing [55]. The limitation of usage of crank-shaft signal is that the instantaneous output torque near TDC where CI engine combustion typically starts is zero [56]. Also, the dynamics of the system limit the

dynamic content of the signal. Ion signals are dependent on engine conditions including speed, load, boost, air/fuel ratio, fuel additives, and spark plug condition [32]. So the accuracy of the combustion metrics estimation will be affected by the changes of these dependent conditions. Also, as deposit accumulation on the ion probe electrodes will decrease the ion current signal, a self-cleaning mechanism must be considered in its application [32].

Many of the methodologies used to quantify the relationship between the accelerometer signal and the in-cylinder pressure signal show promising results. However, the proposed methods used to model the transfer path, including FRF method and structural identification method, are only applicable for test conditions with limited variations of speed, load, and SOI. Neural network needs large amount of acquired data to train the network. The robustness of the network trained based on single engine is the biggest concern for its application. Also, a majority of the researches were performed on one- or two-cylinder engines which are relatively small in size with low power output.

Current researches for combustion sensing methodologies other than the in-cylinder pressure sensor approach are just confined in laboratory due to their limited effectiveness. High robustness, high efficiency, and good reliability are being pursued for each methodology so that the alternatives of in-cylinder pressure sensor can be used in actual engine management system.

Conflict of Interests

The authors declare that there is no conflict of interests regarding the publication of this paper.

References

[1] T. Schnorbus, S. Pischinger, T. Körfer, M. Lamping, D. Tomazic, and M. Tatu, "Diesel combustion control with closed-loop control of the injection strategy," SAE Technical Paper 2008-01-0651, 2008.

[2] S. Lee, J. Lee, S. Lee et al., "Study on reduction of diesel engine out emission through closed loop control based on the in-cylinder pressure with EGR model," SAE Technical Paper 2013-01-0322, SAE International, 2013.

[3] J. Franz, F. Schwarz, M. Guenthner et al., "Closed loop control of an HCCI multi-cylinder engine and corresponding adaptation strategies," SAE Technical Paper 2009-24-0079, SAE, 2009.

[4] J. Yanowitz, R. L. Mccormick, and M. S. Graboski, "In-use emissions from heavy-duty diesel vehicles," Environmental Science & Technology, vol. 34, no. 5, pp. 729–740, 2000.

[5] C. Amann, "Cylinder-pressure measurement and its use in engine research," SAE Paper no. 852067, SAE, 1985.

[6] B. K. Powell, G. P. Lawson, and G. Hogh, "Advanced real time powertrain system analysis," ASME Paper no. 87-ICE-46, ASME, 1987.

[7] M. Yoon, K. Lee, M. Sunwoo, and B. Oh, "Cylinder pressure based combustion phasing control of a CRDI diesel engine," SAE Technical Paper 2007-01-0772, 2007.

[8] Z. Yang, R. Stobart, and E. Winward, "Online adjustment of start of injection and fuel rail pressure based on combustion process parameters of diesel engine," SAE Technical Paper 2013-01-0315, SAE, 2013.

[9] Y. Huang, F. Yang, M. Ouyang, L. Chen, and X. Yang, "Optimal feedback control with in-cylinder pressure sensor under engine start conditions," SAE Technical Paper 2011-01-1422, 2011.

[10] F. Taglialatela, M. Lavorgna, E. Mancaruso, and B. M. Vaglieco, "Determination of combustion parameters using engine crankshaft speed," *Mechanical Systems and Signal Processing*, vol. 38, no. 2, pp. 628–633, 2013.

[11] D. Moro, N. Cavina, and F. Ponti, "In-cylinder pressure reconstruction based on instantaneous engine speed signal," *Journal of Engineering for Gas Turbines and Power*, vol. 124, no. 1, pp. 220–225, 2002.

[12] F. T. Connolly and A. E. Yagle, "Modeling and identification of the combustion pressure process in internal combustion engines," *Mechanical Systems and Signal Processing*, vol. 8, no. 1, pp. 1–19, 1994.

[13] Y. Shiao and J. J. Moskwa, "Cylinder pressure and combustion heat release estimation for SI engine diagnostics using nonlinear sliding observers," *IEEE Transactions on Control Systems Technology*, vol. 3, no. 1, pp. 70–78, 1995.

[14] S. Saraswati and S. Chand, "Reconstruction of cylinder pressure for SI engine using recurrent neural network," *Neural Computing and Applications*, vol. 19, no. 6, pp. 935–944, 2010.

[15] F. Gu, P. J. Jacob, and A. D. Ball, "A RBF neural network model for cylinder pressure reconstruction in internal combustion engines," in *Proceedings of the IEE Colloquium on Modeling and Signal Processing for Fault Diagnosis*, Digest no. 1996/260, Leicester, UK, September 1996.

[16] F. Liu, G. A. J. Amaratunga, N. Collings, A. Soliman, and F. Liu, "An experimental study on engine dynamics model based in-cylinder pressure estimation," SAE Paper 2012-01-0896, SAE International, 2012.

[17] T. Dinu, N. A. Henein, and W. Bryzik, "Determination of the gas-pressure torque of a multicylinder engine from measurements of the crankshaft's speed variation," *SAE Transactions*, vol. 107, no. 3, pp. 294–302, 1998.

[18] G. Rizzoni, "Estimate of indicated torque from crankshaft speed fluctuations: a model for the dynamics of the IC engine," *IEEE Transactions on Vehicular Technology*, vol. 38, no. 3, pp. 168–179, 1989.

[19] F. Liu, G. Amaratunga, N. Collings, and A. Soliman, "An experimental study on engine dynamics model based in-cylinder pressure estimation," SAE Technical Paper 2012-01-0896, SAE International, 2012.

[20] S. Yoshiyama, E. Tomita, and Y. Hamamoto, "Fundamental study on combustion diagnostics using a spark plug an ion probe," SAE Technical Paper 2000-01-2828, 2000.

[21] M. Hellring and U. Holmberg, "An ion current based peak-finding algorithm for pressure peak position estimation," SAE Technical Paper 2000-01-2829, 2000.

[22] A. A. Martychenko, J. K. Park, Y. S. Ko, A. A. Balin, J. W. Hwang, and J. O. Chae, "A study on the possibility of estimation of in-cylinder pressure by means of measurement of spark gap breakdown voltage," SAE Paper 01-1115, SAE International, 1999.

[23] M. Hellring and U. Holmberg, "An ion current based peak-finding algorithm for pressure peak position estimation," SAE Technical Paper 2000-01-2829, SAE, 2000.

[24] D. Kumar, A. Ramesh, M. K. Gajendra Babu, and P. V. Manivannan, "An ionization current based cylinder gas pressure estimation for knock detection and control in a single cylinder SI engine," Internal Combustion Engines 2009-32-0118, 2009.

[25] S. Yoshiyama, E. Tomita, and Y. Hamamoto, "Fundamental study on combustion diagnostics using a spark plug as ion probe," *SAE Transactions*, vol. 109, no. 3, pp. 1990–2002, 2000.

[26] R. Johnsson, "Cylinder pressure reconstruction based on complex radial basis function networks from vibration and speed signals," *Mechanical Systems and Signal Processing*, vol. 20, no. 8, pp. 1923–1940, 2006.

[27] S. Kim and K. Min, "Detection of combustion start in the controlled auto ignition engine by wavelet transform of the engine block vibration signal," *Measurement Science and Technology*, vol. 19, no. 8, Article ID 085407, 2008.

[28] A. Gazis, D. Panousakis, R. Chen, and W.-H. Chen, "Computationally inexpensive methods of ion current signal manipulation for predicting the characteristics of engine in-cylinder pressure," *International Journal of Engine Research*, vol. 7, no. 3, pp. 271–282, 2006.

[29] N. M. Danne, D. L. S. Hung, G. G. Zhu, and J. McKoskey, "Knock detection for a large displacement air-cooled V-twin motorcycle engine using in-cylinder ionization signals," SAE Technical Paper 2008-32-0028, 2008.

[30] G. Zhu, D. Hung, and J. Winkelman, "Combustion characteristics detection for low pressure direct injection engine using ionization signal," SAE Technical Paper 2006-01-3317, SAE, 2006.

[31] M. Bellenoue, S. Labuda, S. Julien, A. P. Chernukho, and A. N. Migoun, "Application of ionization probes for diagnostics of knocking combustion," in *Proceedings of the 7th Mediterranean Combustion Symposium*, Cagliari, Italy, September 2011.

[32] G. G. Zhu, C. F. Daniels, and J. Winkelman, "MBT timing detection and its closed-loop control using in-cylinder ionization signal," SAE Technical Paper 2004-01-2976, SAE, 2004.

[33] J. Naber, J. R. Blough, D. Frankowski, M. Goble, and J. E. Szpytman, "Analysis of combustion knock metrics in spark-ignition engines," SAE Technical Paper 2006-01 400, 2006.

[34] F. Guillemin, O. Grondin, J. Chauvin, and E. Nguyen, "Combustion parameters estimation based on knock sensor for control purpose using dedicated signal processing platform," SAE Technical Paper 2008-01-0790, 2008.

[35] L. Arnone, S. Manelli, G. Chiatti, and O. Chiavola, "In-cylinder pressure analysis through accelerometer signal processing for diesel engine combustion optimization," SAE Technical Paper 2009-01-2079, SAE, 2009.

[36] O. Chiavola, G. Chiatti, and E. Recco, "Accelerometer measurements to optimize the injection strategy," SAE Technical Paper 2012-01-1341, SAE International, 2012.

[37] F. Taglialatela, N. Cesario, M. Porto et al., "Use of accelerometers for spark advance control of SI engine," *SAE International Journal of Engines*, vol. 2, no. 1, pp. 971–981, 2009.

[38] C. J. Polonowski, V. K. Mather, and J. D. Naer, "Accelerometer based sensing of combustion in a high speed HPCR diesel engine," SAE Paper 01-0972, SAE International, 2007.

[39] C. Polonowski, *Accelerometer based measurements of combustion in an automotive turbocharged diesel engine [PhD dissertation]*, Michigan Technological University, Houghton, Mich, USA, 2009.

[40] Y. Gao and R. B. Randall, "Reconstruction of diesel engine cylinder pressure using a time domain smoothing technique," *Mechanical Systems and Signal Processing*, vol. 13, no. 5, pp. 709–722, 1999.

[41] A. J. Morello, J. R. Blough, J. Naber, and L. Jia, "Signal processing parameters for estimation of the diesel engine combustion signature," *SAE International Journal of Passenger Cars—Mechanical Systems*, vol. 4, no. 2, pp. 1201–1215, 2011.

[42] M. El-Ghamry, J. A. Steel, R. L. Reuben, and T. L. Fog, "Indirect measurement of cylinder pressure from diesel engines using acoustic emission," *Mechanical Systems and Signal Processing*, vol. 19, no. 4, pp. 751–765, 2005.

[43] J. T. Kim and R. H. Lyon, "Cepstral analysis as a tool for robust processing, deverberation and detection of transients," *Mechanical Systems and Signal Processing*, vol. 6, no. 1, pp. 1–15, 1992.

[44] R. Villarino and J. F. Böhme, "Fast in-cylinder pressure reconstruction from structure-borne sound using the EM algorithm," in *Proceedings of the IEEE International Conference on Acoustics, Speech, and Signal Processing (ICASSP '03)*, vol. 6, pp. 597–600, IEEE, Hong Kong, China, April 2003.

[45] M. Wagner, J. Böhme, and J. Förster, "In-cylinder pressure estimation from structure-borne sound," SAE Technical Paper 2000-01-0930, 2000.

[46] R. Villarino and J. F. Bohme, "Pressure reconstruction and misfire detection from multichannel structure-borne sound," in *Proceedings of the IEEE International Conference on Acoustics, Speech, and Signal Processing (ICASSP '04)*, vol. 2, IEEE, Montreal, Canada, May 2004.

[47] R. Villarino and J. F. Böhme, "Pressure reconstruction and misfire detection from multichannel structure-borne sound," in *Proceedings of the IEEE International Conference on Acoustics, Speech, and Signal Processing (ICASSP '04)*, vol. 2, pp. ii-141–ii-144, IEEE, Montreal, Canada, May 2004.

[48] R. Villarino and J. Böhme, "Misfire detection in spark-ignition engines with the EM algorithm," in *Proceedings of the 3rd IEEE International Symposium on Signal Processing and Information Technology (ISSPIT '03)*, pp. 142–145, Darmstadt, Germany, December 2003.

[49] K. Bizon, G. Continillo, E. Mancaruso, and B. M. Vaglieco, "Reconstruction of in-cylinder pressure in a diesel engine from vibration signal using a RBF neural network model," SAE Technical Paper 2011-24-0161, 2011.

[50] L. Jia, J. Naber, J. Blough, and S. A. Zekavat, "Accelerometer-based combustion metrics reconstruction with radial basis function neural network for a 9 L diesel engine," *Journal of Engineering for Gas Turbines and Power*, vol. 136, no. 3, Article ID 031507, 2014.

[51] J. M. Desantes and A. J. Torregrosa, "Wavelet transform applied to combustion noise analysis in high-speed DI diesel engines," SAE Technical Paper 2001-01-1545, SAE International, 2001.

[52] A. K. Sen, G. Litak, R. Taccani, and R. Radu, "Wavelet analysis of cycle-to-cycle pressure variations in an internal combustion engine," *Chaos, Solitons & Fractals*, vol. 38, no. 3, pp. 886–893, 2008.

[53] A. Hariyanto, K. Bagiasna, I. Asharimurti, A. Wijaya, I. K. Reksowardoyo, and W. Arismunandar, "Application of wavelet analysis to determine the start of combustion of diesel engines," SAE Technical Paper 2007-01-3556, 2007.

[54] J. Borg, G. Saikalis, S. Oho, and K. Cheok, "Knock signal analysis using the discrete wavelet transform," SAE Technical Paper 2006-01-0226, SAE International, 2006.

[55] L. Jia, J. Naber, and J. Blough, "Application of FRF with SISO and MISO model for accelerometer-based in-cylinder pressure reconstruction on a 9-L diesel engine," *Proceedings of the Institution of Mechanical Engineers C: Journal of Mechanical Engineering Science*, vol. 229, no. 4, pp. 629–643, 2015.

[56] J. B. Heywood, *Internal Combustion Engine Fundamentals*, McGraw-Hill, New York, NY, USA, 1988.

16

Augmenting the Structures in a Swirling Flame via Diffusive Injection

Jonathan Lewis, Agustin Valera-Medina, Richard Marsh, and Steven Morris

Cardiff School of Engineering, Queen's Buildings, Cardiff CF24 3AA, UK

Correspondence should be addressed to Jonathan Lewis; lewisj19@cf.ac.uk

Academic Editor: Kalyan Annamalai

Small scale experimentation using particle image velocimetry investigated the effect of the diffusive injection of methane, air, and carbon dioxide on the coherent structures in a swirling flame. The interaction between the high momentum flow region (HMFR) and central recirculation zone (CRZ) of the flame is a potential cause of combustion induced vortex breakdown (CIVB) and occurs when the HMFR squeezes the CRZ, resulting in upstream propagation. The diffusive introduction of methane or carbon dioxide through a central injector increased the size and velocity of the CRZ relative to the HMFR whilst maintaining flame stability, reducing the likelihood of CIVB occurring. The diffusive injection of air had an opposing effect, reducing the size and velocity of the CRZ prior to eradicating it completely. This would also prevent combustion induced vortex breakdown CIVB occurring as a CRZ is fundamental to the process; however, without recirculation it would create an inherently unstable flame.

1. Introduction

The depletion of fossil fuels and concern about the climate change have led to the development of new technologies to meet power generation demand, whilst maintaining security of supply and decreasing the environmental impact. The use of gas turbines, a well-developed technology, fired on nontraditional fuels, is an increasingly viable method for producing energy in the short to medium term. Fuels that can be used for this purpose range from those based on highly enriched hydrogenated blends to those that are produced from biomaterials [1–4]. Therefore, gas turbine technologies are evolving to cope with the use of these new fuels. However, operators are still finding problems with fuels that vary in composition, posing a new challenge to manufacturers to produce equipment with less stringent fuel requirements [5].

In order to reduce the emission of nitrogen oxides (NO_x), gas turbines operate using lean premixed combustion, utilising swirl stabilisation and resulting in the central recirculation zone (CRZ) of the flame becoming crucial. The CRZ provides heat to fresh reactants and anchors the flame. However, unless its size and shape are controlled, stability problems can arise. The CRZ can for instance readily extend back into the burner surrounding the fuel injector and facilitating early flashback (low stability limit) [6–8]. Flashback can be caused by (i) boundary layer flame propagation, (ii) turbulent flame propagation in the core flow, (iii) thermoacoustics, and (iv) upstream flame propagation of coherent vortical structures [7, 9–11].

Two of these mechanisms, that is, boundary layer flame propagation and upstream propagation of coherent structures, have been studied by several groups using natural gas. However, the use of unconventional fuels can be extremely detrimental to the control of these phenomena, and very little literature is available on this subject. High turbulence levels, one of the very useful features of swirling flow because of mixing potential, affect flashback limits detrimentally due to effects on turbulent flame speed (S_t) and it has been found [12, 13] that the current theoretical approximations of S_t do not agree with experimental values. Literature on this topic becomes more complex in terms of numerical modelling, but experiments tend to be different from numerical findings especially when complex flows are added to the field [13]. For instance, very little has been documented in terms of boundary layer propagation using atmospheric conditions [3, 14, 15], and these findings only show the evolution of 2D structures without swirl and under atmospheric conditions. Thus, the current knowledge on these mechanisms cannot adequately describe the flashback propensity of most practical combustor designs. Therefore, the recognition of the real pattern of this phenomenon is crucial in order to have systems and models capable of utilising new alternative fuels.

Some authors [16–21] have observed that the CRZ has a close connection to the stability of the system, with its shape, strength, and curvature being of high importance to its resistance to flashback and blow-off [20, 22]. Regular precession occurs in the CRZ, with its appearance dependent on the heat transfer regime, the mode of injection, and the increase in the interaction of the hot products and fresh reactants when confinement is imposed. The CRZ behaves as an intermittent structure that will propagate downstream in order to release some internal pressure as a product of the confinement, and intense recirculation at moderate to high swirl numbers [20]. This intermittency can also be detrimental to the phenomenon of flashback as the CRZ will evolve into combustion induced vortex breakdown(CIVB) [23], boundary layer propagation [15], or the production of turbulent burning along the vortical axis [24], all of which can be damaging to the system.

Dam et al. [12] demonstrated the combustion induced breakdown of vortices as being a result of the high velocity zones of a flame squeezing the recirculation zone, causing the CRZ to propagate upstream, ultimately inhibiting the recirculation from occurring. I could be argued that CIVB also plays its part in boundary layer and turbulent flame propagation by augmenting the position of the flame in a way that makes them happen more readily [25].

In terms of unconventional fuels, the primary goal of introducing CO_2 into the gas turbine combustor is to reduce the emissions of NO_x. This is achieved by cooling the flame; thus, the Zeldovich mechanism can be reduced [26]. Previous experimental and numerical studies have investigated the effect of dilution of Syngas fuels with various additives, including carbon dioxide, nitrogen, and steam [27–31]. In the majority, these studies focus on fundamental characteristics of the combustion process. However, the work by Lee et al. [27] and Khalil et al. [1] actually investigated the effect of diluting the premix fuel had on the emission of NO_x and CO from a model gas turbine.

Lee et al. [27] showed that reduction in ppmv NO_x per unit power is logarithmically related to the heat capacity of the total diluent added. Since carbon dioxide has a higher heat capacity than steam or nitrogen, a smaller mass flow rate is required for a comparable reduction in NO_x. Moreover, the use of CO_2 from carbon capture and storage facilities could reduce costs as well as capture equipment further downstream the combustion zone.

This study focuses on, but is certainly not limited to, flames that require a pilot to maintain stability. The injection of CO_2 is expected to augment the size and intensity of coherent structures whilst increasing mass flow rate through the exit nozzle, in much the same way methane pilot does, in such a way that CIVB is inhibited. The effect that the diffusive injection of air on stability will also be investigated and compared.

2. Experimental Facilities

A swirl burner constructed from stainless steel was used to examine the flame structure at atmospheric conditions (1bar, 293 K) at Cardiff University's Gas Turbine Research Centre (GTRC). External and sectioned views of the generic burner are presented in Figure 1. Secondary, full scale tests were performed using the high pressure combustion rig (HPCR), fitted with a proprietary gas turbine combustor, also at the GTRC. The rig is capable of delivering $5 \, kg \cdot s^{-1}$ of air at 900 K and 16 barA, thus allowing combustors to be operated at conditions applicable to use in a power generation derivative gas turbine engine.

2.1. Generic Swirl Burner. A single tangential inlet (a) feeds the premixed air and fuel to an outer plenum chamber (b) which uniformly distributes the gas to the slot type radial tangential inlets (c) which impart the swirling momentum on the premixed flow. Swirling premixed air and fuel then pass into the swirl chamber (d) and then into the exit nozzle (e). The central diffusion fuel injector (f), through which nonpremixed gases are introduced to the combustion zone, extends centrally through the combustor body to the exhaust.

The geometric swirl number (S_g) describes inlet conditions and burner geometry, allowing variations in flow pressure to be neglected [32]. It is defined in (1) for isothermal conditions, where density is constant, in radial type burner:

$$S_g = \frac{A_e \cdot r_t}{A_t \cdot r_e} \cdot \left(\frac{Q_{ta}}{Q_{to}} \right)^2, \tag{1}$$

where A_e is the area of burner nozzle exit (m^2), r_t is the effective radius of tangential inlets (c) (m), A_t is the area of tangential inlet (m^2), r_e is the radius of burner nozzle exit (m), Q_{ta} is the tangential flow rate (m^3/s), and Q_{to} is the total flow rate (m^3/s).

During these trials the geometry was kept constant, so the only change in geometric swirl number was caused by the injection of gases through the pilot injector.

The system was fed using compressed air through flexible hoses and a Coriolis meter for flow rate metering. Bottled methane was used as main fuel at a constant flow rate during the trials, fed through flexible hoses passing through another Coriolis meter. A Dantec PIV (particle image velocimetry) system consisting of a dual cavity Nd: YAG Litron Laser of 532 nm capable of operating at 15 Hz, a Dantec Dynamics laser sheet optics (9080X0651), was used to convert the laser beam into a 1 mm thick sheet. The laser sheet and thus the plane of measurement intersected the central axis of the burner exit nozzle. To record the images a Hi Sense MkII Camera model C8484-52-05CP was used, with 1.3 MPixel resolution at 8 bits. A 60 mm Nikon lens was used for resolution purposes, which allowed a field of view of approximately 75×75 mm, with a resolution of 5.35 pixels per mm and a depth of view of 1.5 mm. The inlet air was seeded with aluminum oxide (Al_2O_3) by an accumulator positioned 2.0 m upstream of the burner inlets.

After acquisition of the PIV data, a frame-to-frame adaptive correlation technique was then carried out with

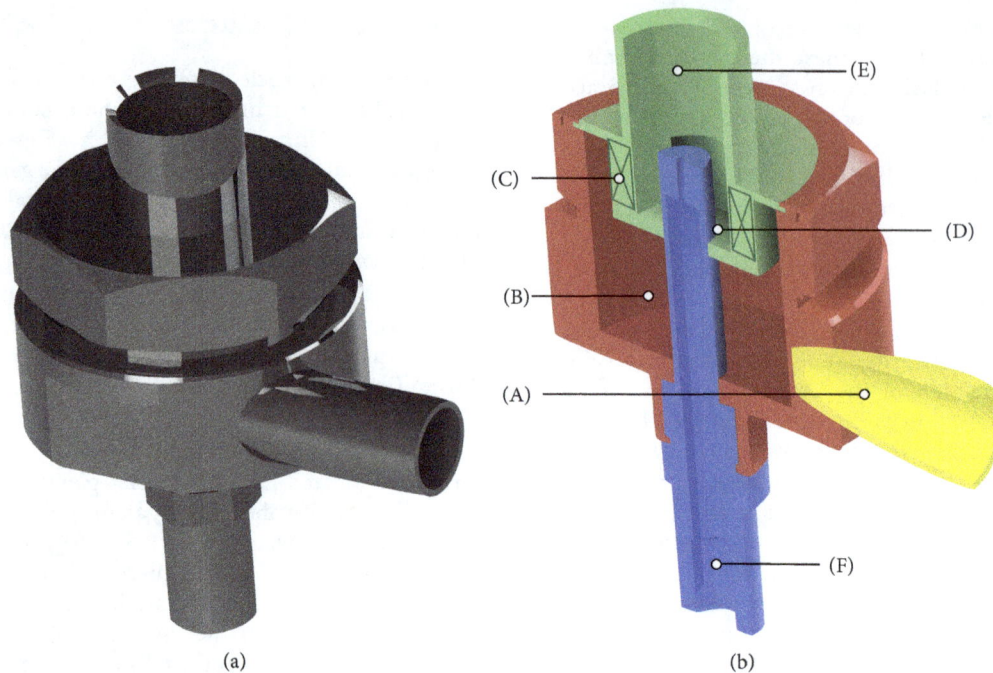

FIGURE 1: (a) External view and (b) sectioned schematic of the generic burner.

a minimum interrogation area of 32 × 32 pixels and a maximum of 64 × 64, with an adaptivity to particle density and velocity gradients. 150 pairs of frames per plane were used to create an average velocity map. Adaptivity in the analysis allowed very coherent images, with just a mask refinement of ~15–20% of the entire field of view. In order to reduce the parallax error, the line of view of the camera was positioned exactly in the middle of the nozzle using a calibration grid provided by the system manufacturer. The grid was used to correct, via software, any positioning issue. The field of view was calibrated with the central line of the burner in the centre of the grid, thus ensuring that the position of the system would not affect the results.

The cross sectional area and velocities (U) of the structures within the flame were assessed using the exported, numeric PIV data, with different components of the flame designated based on the axial velocity component, as indicated in Figure 2. These designations were assigned during previous experimentation with the generic burner [25]. A maximum value of U_a (axial velocity) = −0.3 m/s was used to define the CRZ, removing any outlying areas of recirculation. A boundary of U_t (total velocity) = 3 m/s was used to define the HMFR, and this was the maximum value within of U_r within the CRZ, as such flow with a velocity exceeding 3 m/s had a momentum higher than that in CRZ.

2.2. Method of Initial Experimentation. Varying amounts of carbon dioxide, air, and methane were injected through the diffusive pilot of the AGSB to assess the effect on flame structure. The three gases were selected for their differing combustion properties:

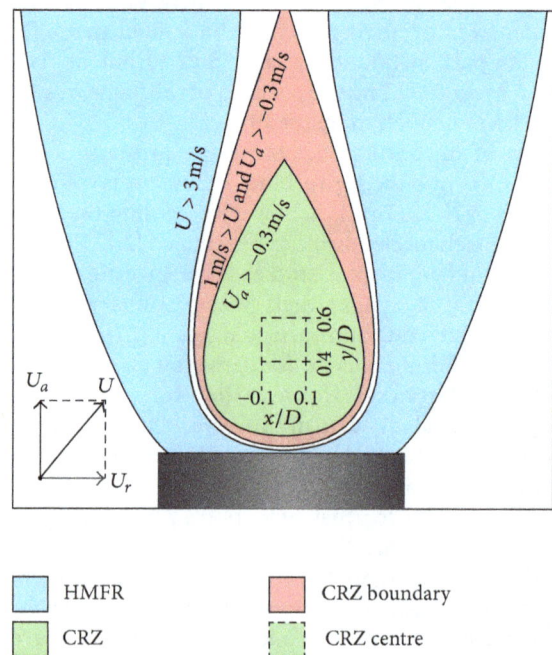

FIGURE 2: Designation of areas within the flame.

(i) methane which is often used as a pilot fuel in gas turbines and will increase the global equivalence ratio of the flame, global equivalence ratio being that of the premixed air and methane, and the

(ii) carbon dioxide which does not affect oxidant-to-fuel ratio but has been shown in previous studies to alter flame conditions [27–31],

TABLE 1: Details of AGSB test points 1–26.

| | Test point | Flow rates (g/s) | | Φ | Total power (kW) | Isothermal S_g | Bulk exit velocity (m/s) | Re | AFT (C) |
		Pilot (through f)	total						
Methane	1	0.000	3.640	1.00	10.0	1.05	4.88	4361	1953
	2	0.015	3.655	1.07	10.7	1.04	4.90	4379	1928
	3	0.030	3.670	1.15	11.5	1.02	4.92	4397	1875
	4	0.045	3.685	1.22	12.2	1.01	4.94	4415	1816
	5	0.059	3.699	1.30	13.0	0.99	4.96	4433	1760
	6	0.074	3.714	1.37	13.7	0.98	4.98	4451	1702
	7	0.089	3.729	1.45	14.5	0.97	5.00	4468	1645
	8	0.104	3.744	1.52	15.2	0.95	5.02	4486	1589
Air	9	0.000	3.640	1.00	10.0	1.05	4.88	4361	1953
	10	0.100	3.740	0.97	10.0	1.00	5.02	4482	1924
	11	0.201	3.841	0.94	10.0	0.95	5.15	4602	1902
	12	0.301	3.941	0.92	10.0	0.90	5.29	4722	1875
	13	0.402	4.042	0.90	10.0	0.86	5.42	4843	1846
	14	0.502	4.142	0.87	10.0	0.82	5.56	4963	1817
Carbon dioxide	15	0.000	3.640	1.00	10.0	1.05	4.88	4361	1953
	16	0.025	3.665	1.00	10.0	1.04	4.92	4391	1931
	17	0.050	3.690	1.00	10.0	1.03	4.95	4421	1923
	18	0.074	3.714	1.00	10.0	1.03	4.98	4451	1914
	19	0.099	3.739	1.00	10.0	1.02	5.02	4481	1906
	20	0.124	3.764	1.00	10.0	1.01	5.05	4510	1897
	21	0.149	3.789	1.00	10.0	1.00	5.08	4540	1889
	22	0.174	3.814	1.00	10.0	0.99	5.12	4570	1881

(iii) air which will reduce the flames global equivalence ratio.

In order to allow comparison between results, premix flow rates of 0.2 g/s methane and 3.44 g/s air were kept constant; these equate to a 10 kW, stoichiometric mixture. Details of the test points are displayed in Table 1. The flames are unconfined, as such ambient air will be entrained, and they will actually be leaner than the defined equivalence ratio.

The experimental method was the same for all three gases: the flame was lit and 150 images were recorded with the PIV system, and the flow rate of diffusion gases was then increased in regular intervals, with a further 150 images recorded at each interval, until the flame reached a point of quasistability. In order to maintain inlet and outlet conditions between test points, the flame was left alight.

3. Results

The PIV results in Figure 3 demonstrate the effect of introducing methane through the pilot on the flame structure. The left hand side of the PIV images shows the axial velocity (U_a) of the flows on scale of −5 to 8 meters per second. Negative velocities indicate that the flow downstream, towards the burner exit. On the right hand side of images total velocity (U_t) is shown, on a scale of 0 to 8 meters per second. The total velocity is a measure of magnitude, regardless of direction, so no negative velocities are possible. The combination of image masking and the neighbourhood validation causes the data shown to underestimate the velocity of the fluid as it leaves the exit nozzle. This is why the fluid appears to have exceptionally low velocities in the region where $0 < y/D < 0.1$.

The flow in test point (TP) 1 has the highest velocities at 0.25 y/D and $\pm 0.55\ x/D$, indicated by the red regions in both scalar plots, and there is also a subregion where axial velocity exceeds the 8 m/s on the scale shown. The unburned, premixed reactants are at their greatest velocity near the outside of the exit nozzle, where the tangential momentum imparted by the radial swirl passages has forced the flow away from the burner axis but boundary effects are negligible. The centrifugal force on the fluid also means that the reactants are the densest at the inner wall of the exit nozzle. Combustion occurs when the reactants leaving the exit nozzle, a significant increase in axial velocity results from hot gas expansion. An increase in radial velocity also occurs. The combined tangential and axial momentum of the products results in flow spreading in the x and y directions on the plane shown. As spreading occurs and the combustion process is complete, velocities decrease. Tangential momentum results in predominant expansion in the x direction being away from the burner axis. This creates a low pressure zone along the vertical burner axis, with a pressure differential that causes combustion products to flow towards the central axis

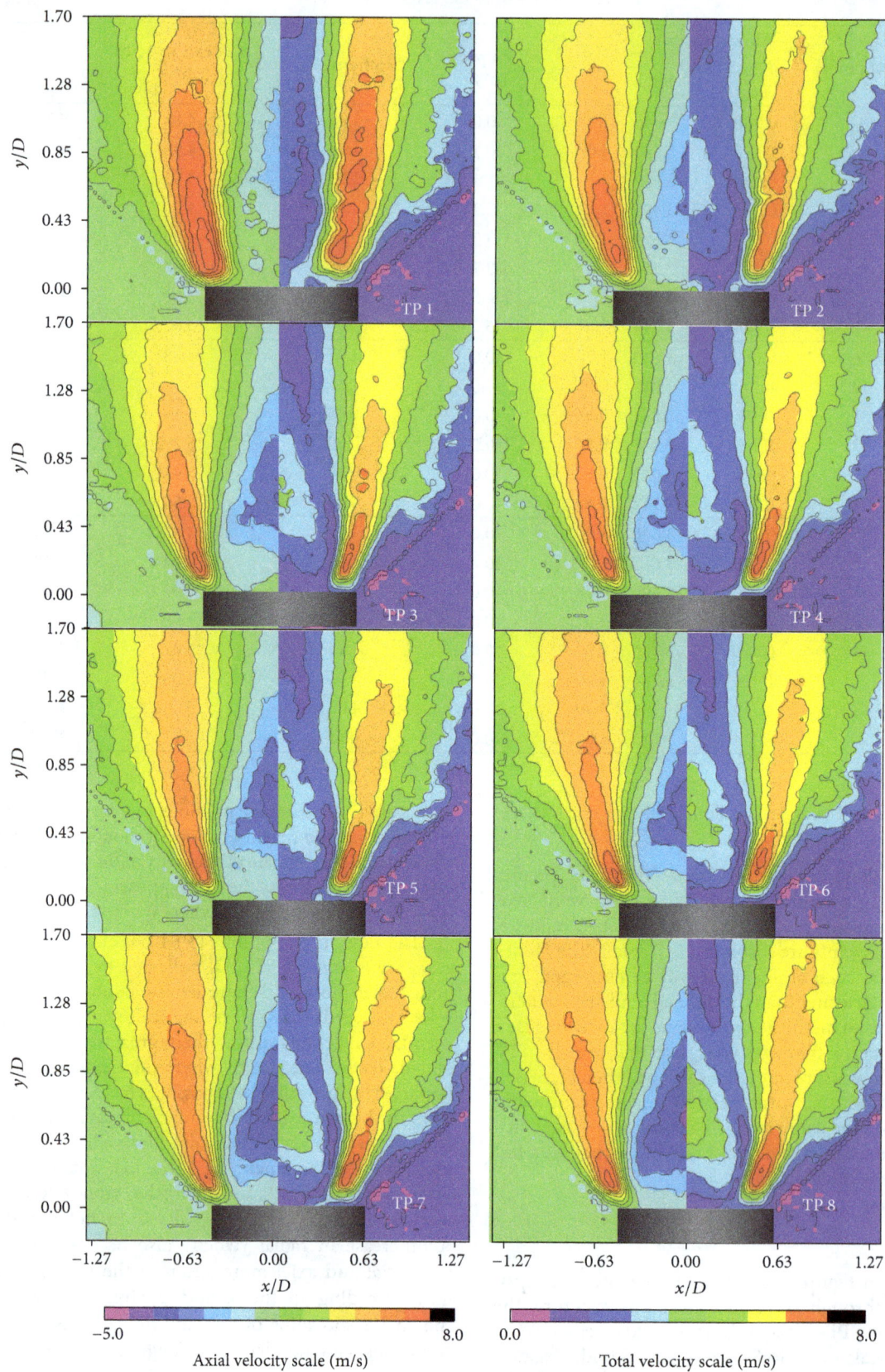

FIGURE 3: Scalar plots of TP 1 to TP 8 in the AGSB as detailed in Table 1, the left half of each image indicates axial velocity, and the right half total velocity.

and burner exit. The stagnation of the flow prior to being recirculated is indicated by the turquoise region in the axial velocity profile.

TP 1 has a very narrow, low velocity, CRZ. Within the recirculation zone the maximum value of U_t is 2 m/s and the minimum value of U_a is -1 m/s. The axial velocity of the recirculated gases becomes positive at $y/D = 0.4$. The introduction of methane through the pilot results in an increase in size and velocity of the CRZ, which can be seen in TP 2. The shape of recirculation zone also changes, with the width increasing disproportionally at its base. The negative axial velocity of the recirculated gases at the centre of the CRZ increases significantly, exceeding 2 m/s. Due to the increase in velocity of the recirculated gases, the region of negative axial velocity extends toward the burner exit, as a result the CRZ becomes elongated. The velocity in the HMFR in TP 2 is significantly lower than that in TP 1, in terms of both axial and total velocity; it appears to have become narrower as a result of the CRZ expanding.

The alteration in shape becomes further exaggerated as the flow rate of pilot methane increases through test points 3 to 8. With each increase in the flow rate of diffusive methane the width of the reverse flow region increases, as do the negative axial and total velocities of the gases within them. The injection of methane reduces the peak velocity of the reacting flows in the HMFR; however, the shape and overall velocity profiles in these regions remain largely unaffected.

The diffusive injection of air has a markedly different effect on the flow structure than was observed with methane. PIV images in Figure 4 effectively demonstrate how the air causes the size and mean velocity of the CRZ to reduce, eventually leading to its complete destruction. Test points 9 through 12 are displayed as in Figure 3, the left hand side of the images showing axial velocity (U_a) and the right hand side of images total velocity (U_t). The full, axial, and total velocity profiles of test points 13 and 14 are shown due to the high level of asymmetry in the flow structure.

The flow structure and velocities of TP 9 are very similar to those observed in TP 1; this is to be expected as the pre-mixed flow rates are the same. The structure of TP 10 however is very different to that of TP 2. Rather than increasing with diffusive flow rate, the width of the recirculation zone is unaltered from TP 9. Although total velocity is also unaltered, negative axial velocity is actually reduced.

The total and axial velocities in the HMFR of TP 11 are considerably lower than in TPs 9 and 10. The flame is also displaying significant asymmetry, with the low velocity recirculation zone positioned left of of the central axis of the burner. The increased addition of air continues to reduce the velocity of the flow within the CRZ and the HMFR in TP 12, whilst the asymmetry becomes more exagerated. The diffusive flow rate in TP 13 has caused an almost complete reduction of the CRZ, although a region exists where axial velocities are just below 0. In TP 14 the CRZ no longer exists, with no flow in the negative axial direction. Instead the CRZ is replaced by a central region of high positive axial velocity, on the right of the central burner axis as shown. The likely cause of the CRZ destruction is combustion inside of the CRZ, causing an expansion of gases along the central axis

of the burner. This gradually reduces the pressure differential that is required in order to induce recirculation.

The effect of reduced swirl number may also contribute to the results seen in test points 13 and 14, where the S_g is lower than that observed with CO_2 or CH_4. The delivery system restricted the flow rate of CO_2 and CH_4 to the values shown in Table 1. However, since the relationships between diffusive flow rate, for all three gases, and structure augmentation were established for geometric swirl numbers greater than 1, the effects of swirl number are considered secondary.

Carbon dioxide has a very similar effect of flow structure to methane; as a result, the progression caused by its diffusive injection between TP 15 to TP 22, which is shown in Figure 5, resembles the progression in Figure 3.

Test points 15 through 22 are displayed in Figure 3, the left hand side of the images showing axial velocity (U_a) and the right hand side of images total velocity (U_t). The flow structure of TP 15 resembles those of TP 1 and TP 9 where a narrow and low velocity recirculation zone exists. However, over the measured plane, the velocities of the flows in TP 15 are lower than those observed in TPs 1 and 9. This is likely the result of slightly different initial flow rates, which can be caused by changes in ambient conditions or compressor pressure. As the proceeding alterations in flow structure are of interest, this inconsistency in initial condition is acceptable.

In TP 16 it can be seen that the CRZ increases in width as carbon dioxide is diffusively injected, and the total and axial velocities within the CRZ are also increased. With methane this increase in velocity coincided with an elongation of the recirculation, with the region of reverse flow approaching the burner exit; with carbon dioxide this was not observed. A reduction in velocity in the HMFR is evident between TP 15 and TP 16.

Between TP 16 and TP 22, the increase in diffusive CO_2 flow rate results in an increase in the size of the CRZ, and axial and total velocities within it also increase, with the reverse flow region propagating slightly upstream. This reduction is expected as the dilution effect of the CO_2 will reduce flame temperatures and the expansion of the gas. The observed reduction in axial velocity is less pronounced than the reduction in total velocity, which suggests that radial velocity is reduced significantly.

3.1. Effect on Turbulence. When CO_2 is diffusively injected, the low velocity boundary that exists between the CRZ and the HMFR appears to be wider when CH_4 is injected. This is particularly evident when comparing vector plots of TP 5, with methane injection and TP 19, with carbon dioxide injection, which are shown in Figure 6. The CRZ is also visibly less defined in TP 19 than in TP 5; this is the result of higher levels of turbulence within the recirculation zone.

When all gases entering the burner are introduced tangentially, they are considered to be fully mixed prior to entering the combustion zone. With diffusive injection, mixing is forced to occur in the region of the burner exit, resulting in unequal mixtures within the combustion zone.

Simulation with CHEMKIN using the GRIMech 3.0 mechanism [33] predicts that under laminar conditions the

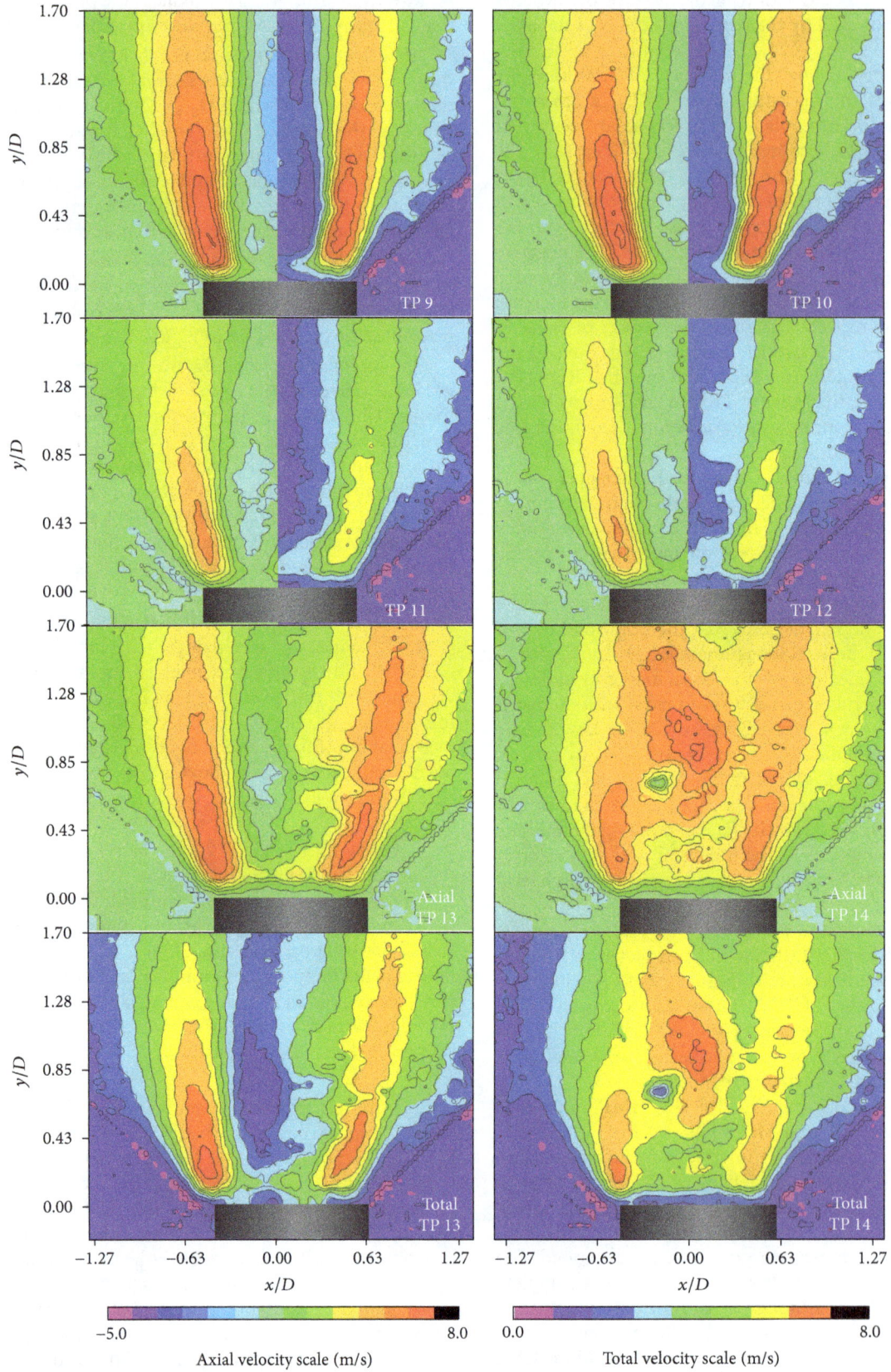

FIGURE 4: Scalar plots of TP 9 to TP 14 in the AGSB as detailed in Table 1, TP 9 to TP 12 the left half of each image indicates axial velocity, and the right half total velocity. TP 13 and TP 14 show full frames of axial and total velocity.

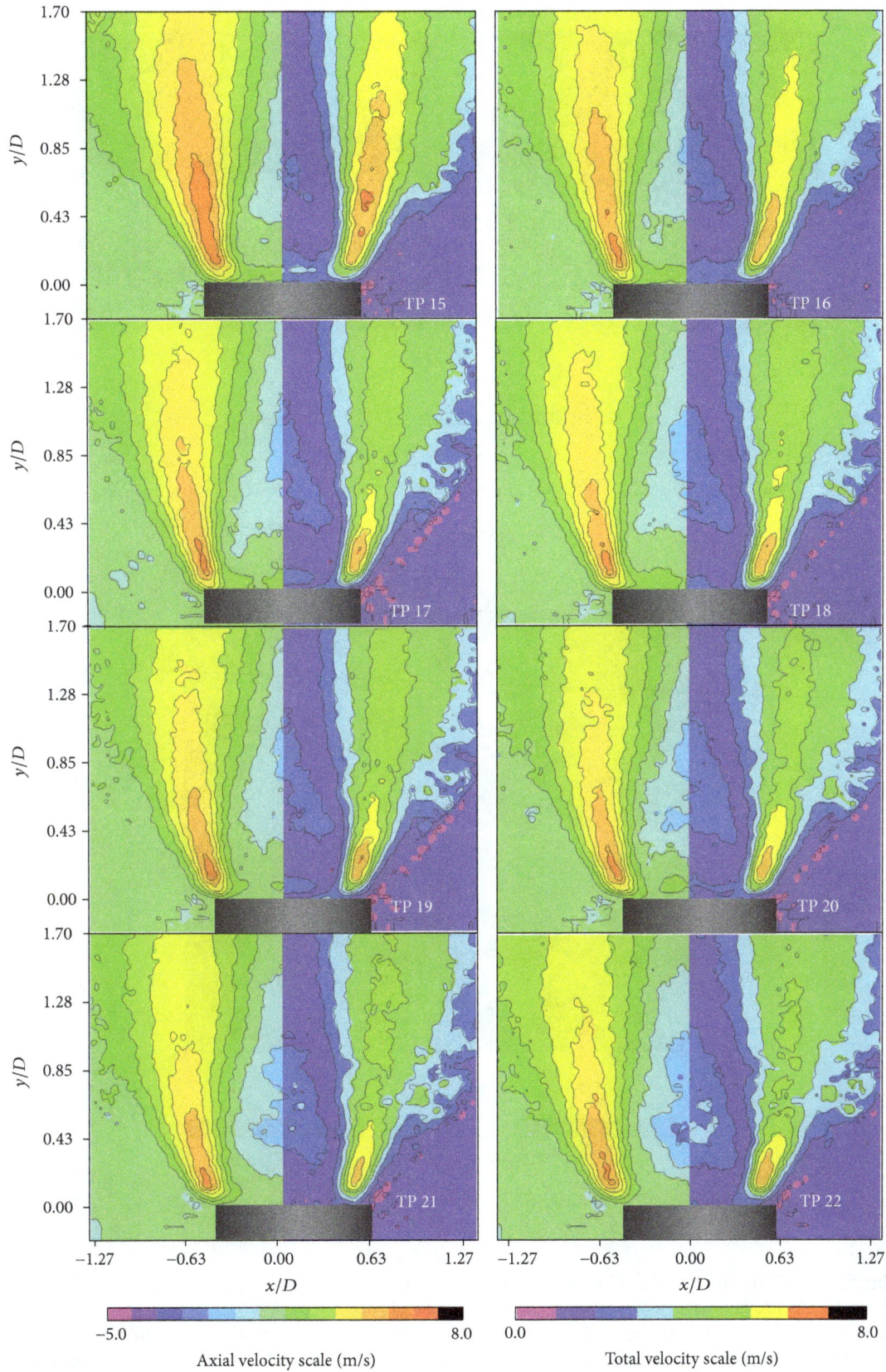

FIGURE 5: Scalar plots of TP 15 to TP 22 in the AGSB as detailed in Table 1, the left half of each image indicates axial velocity, and the right half total velocity.

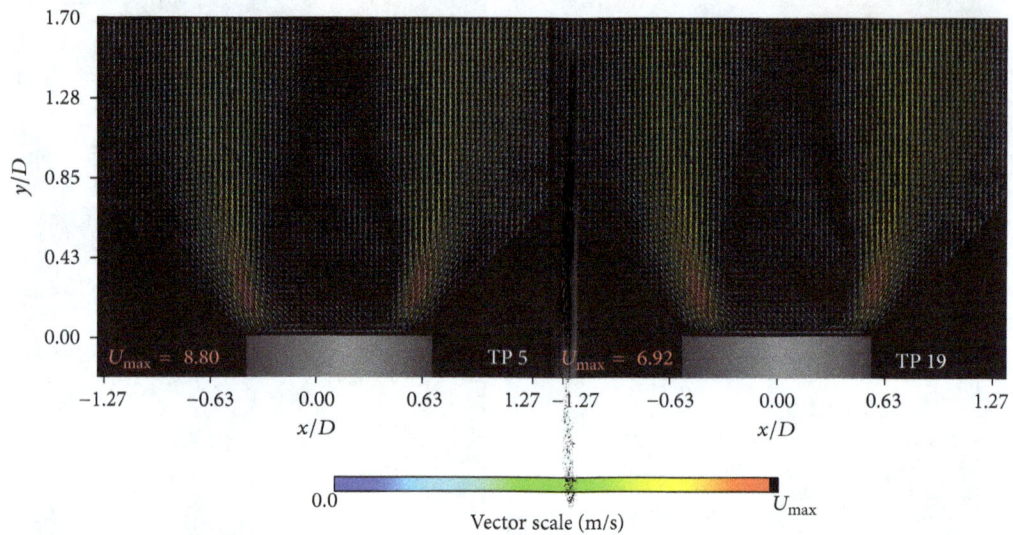

FIGURE 6: Vector plots of TP 5 and TP 19 in the AGSB with altered premix air flow rate and pilot CO_2 flow rate as detailed in Table 1: vectors indicate velocity and direction.

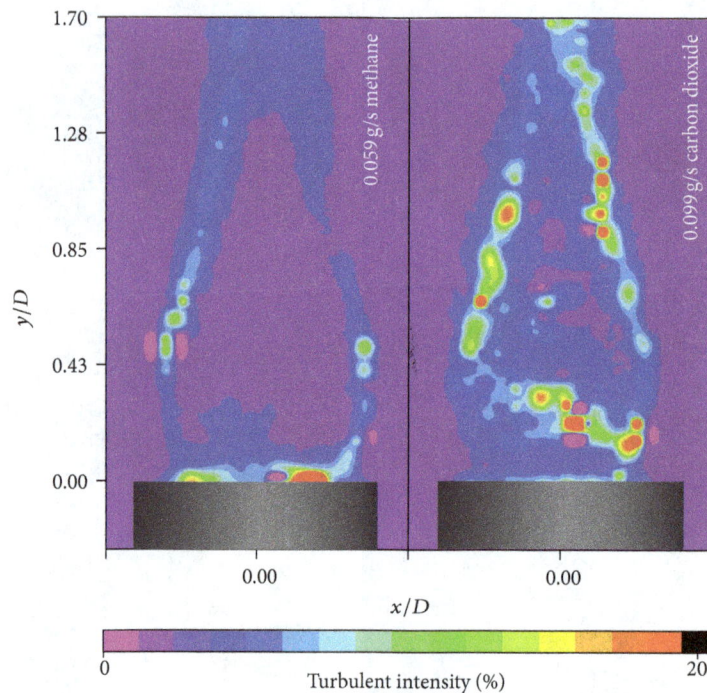

FIGURE 7: Comparison of turbulent intensity between flames in TP 5 and TP 19.

peak methane burning rate occurs at a rich equivalence ratio of approximately 1.1. Previous work has also utilised the combination of CHEMKIN and GRIMech 3.0 to show how CO_2 dilution reduces laminar burning rates [34]. With premixed gaseous fuel, combustion normally occurs at the boundary between the CRZ and the HMFR, where fresh reactants and hot products combine. The amalgamation of delayed mixing and unequal alterations in burning rates affect

the ability of the flame to propagate toward CRZ when CO_2 and CH_4 are injected.

The increase is in both size and mean velocity caused by CO_2 injection are not as prominent as with methane, and the CRZ has a less well-defined shape as an irregular flame front develops at the boundary between it and the HMFR, as is reflected in the level of the turbulent intensity. Figure 7 represents the turbulent intensity of the flames in

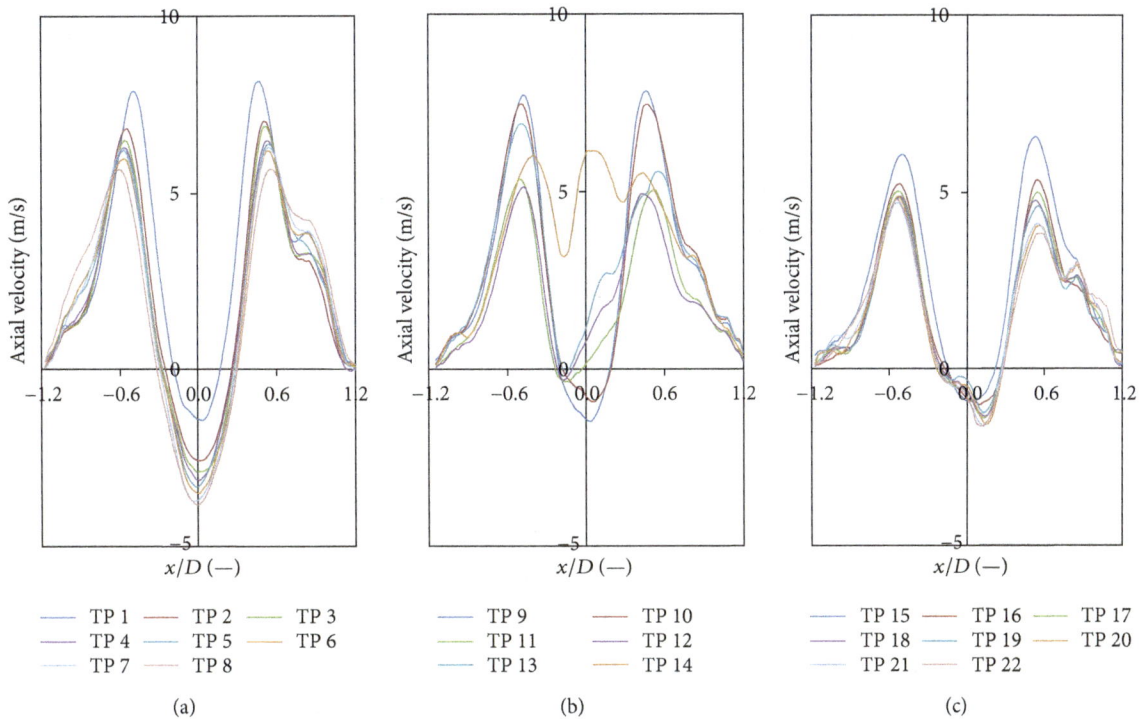

FIGURE 8: Axial velocities at $y/D = 0.5$ for test points detailed in Table 1 with the diffusive injection of (a) methane, (b) air, and (c) carbon dioxide.

TP 5 and TP 19 in the axial-radial plane. The method for calculating turbulent intensity in the measured plane (I), at each measurement point in the PIV frame, is shown in (2) and (3):

$$I = \frac{\sqrt{ke}}{\sqrt{U_a^2 + U_r^2}}, \tag{2}$$

where ke is the turbulent kinetic energy (m^2/s^2)

$$ke = \frac{1}{2} \cdot \left(u_a'^2 + u_r'^2 \right). \tag{3}$$

And u_a' is the root-mean-square of turbulent velocity fluctuations in the axial direction (m/s), u_r' is the root-mean-square of turbulent velocity fluctuations in the radial direction (m/s), U_a is the axial velocity (m/s), and U_r is the radial velocity (m/s).

The flame with diffusive methane injection has a very low turbulent intensity CRZ ($\overline{I} = 1.12\%$) and HMFR ($\overline{I} = 0.32\%$), with the intermediary boundary layer standing out due to the increased turbulent intensity ($\overline{I} = 3.07\%$). The HMFR of the flame with diffusive carbon dioxide injection also has a very low turbulence level ($\overline{I} = 0.44\%$). However, the turbulent intensity of the CRZ ($\overline{I} = 2.93\%$) and the intermediate boundary layer ($\overline{I} = 4.73\%$) are significantly higher.

The velocities in the methane injected flame are greater than those of the carbon dioxide injected flame, which has to be considered when comparing turbulent intensity, which is dependent on local velocity. When comparing the root-mean-square of the turbulent velocity fluctuations in the axial-radial plane, the mean turbulent kinetic energy levels in the CRZ boundaries of the methane and carbon dioxide flames are actually very similar, $\overline{k} = 48.32\,m^2/s^2$ and $\overline{k} = 44.70\,m^2/s^2$, respectively, whereas the CRZ of the carbon dioxide injected flame displays significantly more turbulence ($\overline{k} = 15.22\,m^2/s^2$) than that of the methane flame ($\overline{k} = 0.93\,m^2/s^2$).

3.2. Effect on Momentum. The axial velocity profiles at $y/D = 0.5$ for all test points are plotted in Figure 8, and a moving average of 5 data points is applied to smooth out local fluctuations from the global trend, for easier visual interpretation. In Figure 8(a), the profile of test point 1 demonstrates the negative axial velocities of the CRZ on the central axis of the burner, at $x/D = 0$. The velocities increase either side of the burner axis, peaking at the centre of the HMFR before reducing again as the unconfined flow dissipates. The progressive introduction of methane produces incremental increases in the negative velocities in recirculation zone, whilst incrementally reducing the velocity within the high momentum flow region. The point at which the lines cross the axis also diverges from $x/D = 0$, indicating that the CRZ is increasing in width whilst the HMFR is getting narrower. The profiles in Figure 8(b) demonstrate clearly how the introduction of air causes an asymmetrical distortion, and ultimate destruction of the CRZ. Between test point 9 and TP 10, the changes in velocity resemble those seen due to the injection of methane. Between TP 10 and TP 13, there is a significant reduction in the peak velocity in the right hand side of the HMFR. The CRZ in the positive radial

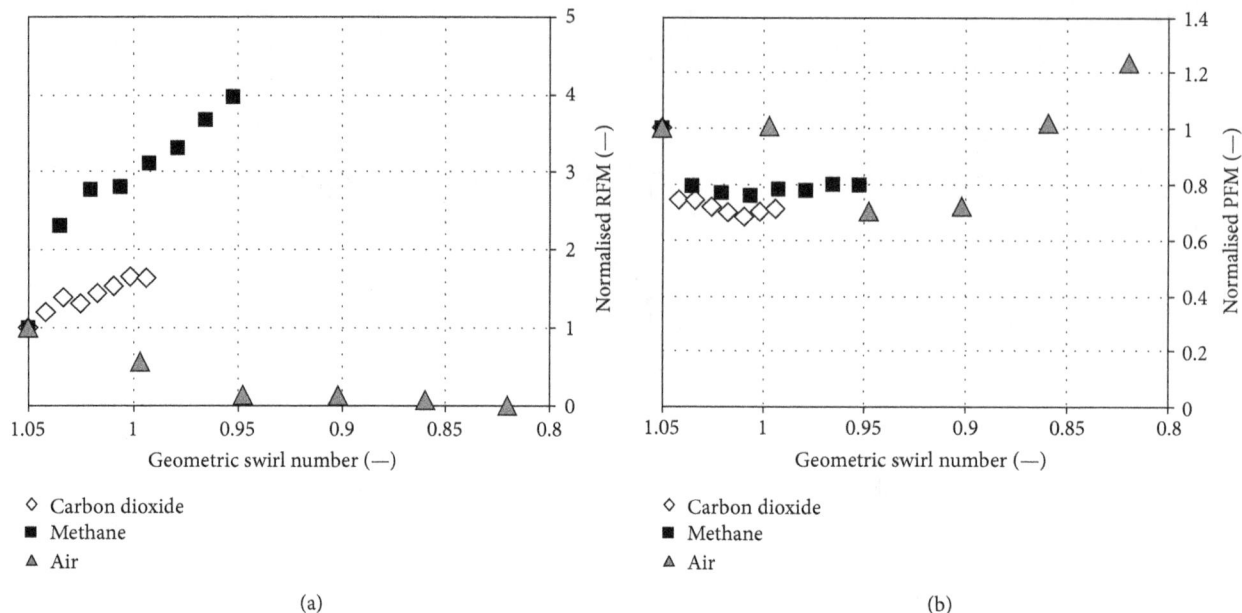

(a)

(b)

FIGURE 9: The (a) reverse and (b) positive flow momentum for all test points at $y/D = 0.5$.

direction of the burner is also subject to a greater change than in the negative radial direction, with a large incremental increase in the axial velocity seen. Between test points 13 and 14 a dramatic change occurs, with the entire range of axial velocities becoming positive, and the peak velocity occurring at $x/D = 0$.

However, the axial velocity profiles of the flames with CO_2 injection, test points 15 to 22, display the same behaviour as observed with the injection of CH_4. The diffusive injection causes the positive axial velocities in the HMFR to decrease, whilst the negative axial velocities in the CRZ increase. There is also evidence of the CRZ becoming wider as the value of x/D when axial velocity which equals 0 increases with diffusive flow rate, although the expansion is not as prominent as observed with methane injection.

The amount of stabilisation provided by the CRZ can be assessed by the reverse flow momentum (RFM) of the gases within it. Momentum is a product of velocity and mass, as the temperature, and as a result, density of the gases within the CRZ and the reverse flow momentum are unknown. The volume of a particular measurement point is fixed; therefore, if variations in density are ignored, it is possible to compare the flow momentum by comparing velocity. As such, changes in reverse flow momentum (RFM) at $y/D = 0$ are assessed by determining the areas below the x-axis, which are bound by the curves plotted in Figure 8.

The RFM of each test point is plotted against the isothermal geometric swirl number in Figure 9(a), with the results of TPs 1 to 8 normalised against the fully premixed flame of TP 1. Similarly, TPs 9 to 14 are normalised against TP 9, and TPs 15 to 22 are normalised against TP 15.

The increased diffusive injection of methane, which reduces S_g in accordance with (1), results in an increase in the RFM of the CRZ. Initially there is a very large increase,

with the RFM of TP 2 2.3 times greater than TP 1, and RFM then increases linearly with diffusive flow rate. The injection of diffusive CO_2 also increases the RFM of the CRZ; however, the rate of increase is significantly less than the one observed with CH_4. Injection of air has an opposing effect, with RFM reducing as the amount of air injected increases. The effects on positive flow momentum (PFM), which are calculated in a similar fashion to RFM at $y/D = 0$, with the area bound by the curve being positive of the x-axis, are shown in Figure 9(b). The injection of both CO_2 and CH_4, with initial diffusive flow rates causing 21% and 26% reductions in PFM, respectively. The increase in diffusive flow rate then has very little effect on the PFM, with the range of reduction of CH_4 being 21–25% and the range of reduction of CO_2 being 26–32%.

The complete alteration in structures caused by the introduction of air means that there are two effects on the PFM. The gradual destruction of the CRZ means that the main reaction zone expands, and the velocity in the HMFR reduces. Conversely, the destruction of the CRZ means that the regions of positive flow increases; the velocity in these regions will also increase due to the axial momentum of the injected air.

The radial velocity profiles at $y/D = 0.5$ for all test points are plotted in Figure 10, again a moving average of 5 data points is applied. In Figure 10(a), the radial velocities of TPs 1 to 8 are shown. Rotational symmetry about the origin is expected with regard to radial velocity; however, due to the slightly asymmetry in the burner this is not seen, with TP 1 displaying entirely negative radial velocities within the CRZ. The increased injection of CH_4 causes a general, incremental increase in radial velocity. TP 8 displays peak positive and negative values that are approximately as twice as those seen in TP 1.

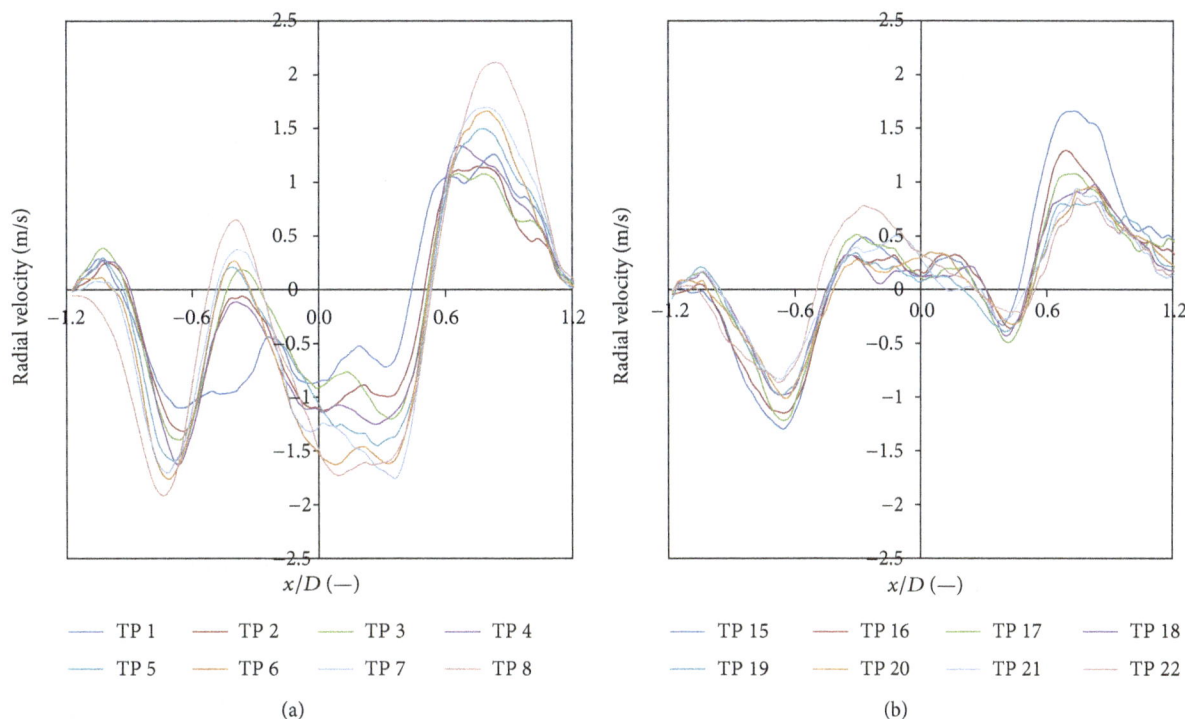

FIGURE 10: Radial velocities at $y/D = 0.5$ for test points detailed in Table 1 with the diffusive injection of (a) methane and (b) carbon dioxide.

Coupled with the decrease in axial velocities these results suggest that the level of swirl has increased, which is why the RFM has increased relative to the PFM. The different effect is observed with the diffusive injection of CO_2; the radial plots for test points 15 to 22 are shown in Figure 10(b). A slight change in flame structure means that velocities within the CRZ are all positive. The injection of CO_2 has very little effect on the radial velocity inside the recirculation zone; however, the radial velocity in the HMFR decreases.

3.3. Effect on the CRZ.

Figure 11 shows the effect on the size and velocity of the central recirculation zone, based on the corresponding geometric swirl number. As in Figure 8, the results are normalised against the fully premixed case to allow easier comparison. Figure 11(a) shows how both CO_2 and CH_4 injection cause the size of the CRZ to increase dramatically until reaching $Sg = 1.03$ for CO_2 and 1.02 for CH_4; the size then increases at a steady linear rate. The injection of CH_4 has a greater effect on the size of the CRZ than CO_2 does, with the increased velocity of the recirculated flow resulting in the wider CRZ base seen in Figure 3 than in Figure 5.

This increased velocity is demonstrated in Figures 11(b) and 11(c), where the mean velocity in the CRZ is in its entirety, and its central regions are calculated. Over the region as a whole, a steady linear increase, of a similar order, is observed with both CO_2 and CH_4, with all but one point falling on the same line. However, as the position and size of CRZ are calculated based on a component of the velocity, the mean velocity over the entire CRZ is influenced by the constraints applied. The central region, which is determined

geometrically, is a better reflection on changes in velocity; a linear, sixfold increase is observed with methane, whereas CO_2 only causes a threefold, nonlinear, increase.

3.4. Effect on the HMFR.

Another structure in a swirling flame is the high momentum flow region; this is high velocity region which contains the main reaction zone of the flame. In this study, it is defined as being the region where axial velocity exceeds 3 m/s.

The effects of pilot injection on the size and mean velocity in this region are shown in Figure 12; it is observed that the effects of CO_2 and CH_4 injection differ. As increasing diffusive flow rates of both CO_2 and CH_4 cause S_g to approach 1.02, a reduction in cross sectional area of the HMFR is observed, below values of $S_g = 1.02$, and CO_2 continues to cause a reduction in HMFR area, whilst CH_4 results in a linear increase. A similar pattern is observed with regard to mean velocity, with both gases causing a reduction as S_g approaches 1.02, after which CO_2 continues to cause a reduction whilst CH_4 displays a trend of increasing velocity.

Between $S_g = 1.05$ and 1.02, methane and carbon dioxide both cause a reduction in size and velocity of HMFR as the CRZ is increasing in size. Although the premixed air and fuel mixture was stoichiometric, the flame was unconfined, and in an environment with a large excess of oxidant. This facilitated a large increase in the volume of the structures when methane was injected, whereas the cooling effect of the carbon dioxide actually reduced structure volume.

There are four mechanisms that may result in flame flashback [7, 9–11]; upstream flame propagation of coherent structures and boundary layer flame propagation in

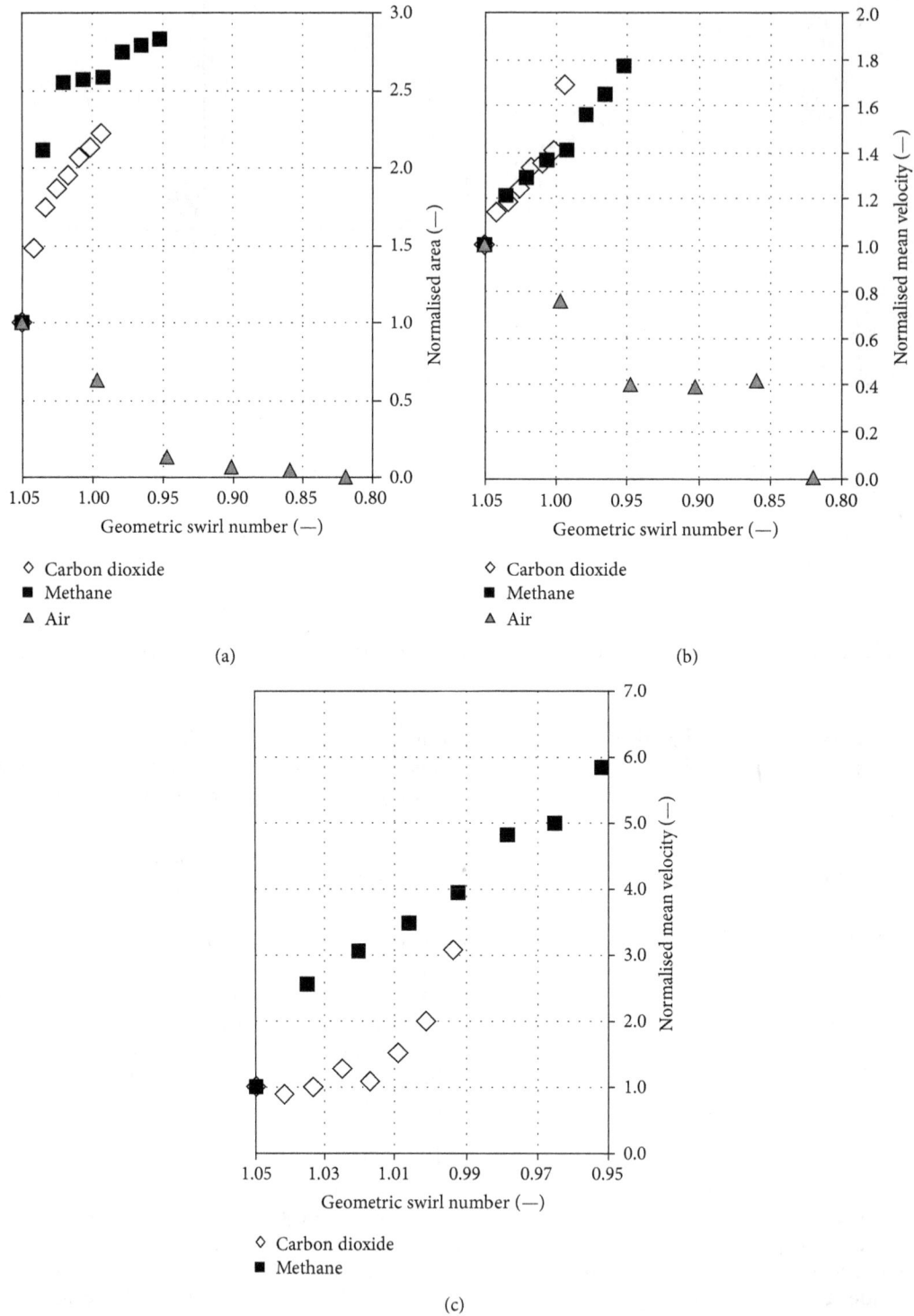

(a)

(b)

(c)

FIGURE 11: The normalised effect of the injection of carbon dioxide, methane, and air on (a) cross sectional area and (b) mean velocity of the entire CRZ and (c) mean velocity of the CRZ centre.

particular are symptomatic of combustion induced vortex breakdown.

Previous studies have demonstrated that CIVB can be the result of the CRZ being "squeezed" by the HMFR, which causes the CRZ to be displaced, upstream, ultimately

surrounding a central body in the combustor [12]. An increase in equivalence ratio caused this process to occur. However, depending on burner design, and critically the level of turbulence of the flow, it may also be caused by a reduction in equivalence ratio [25].

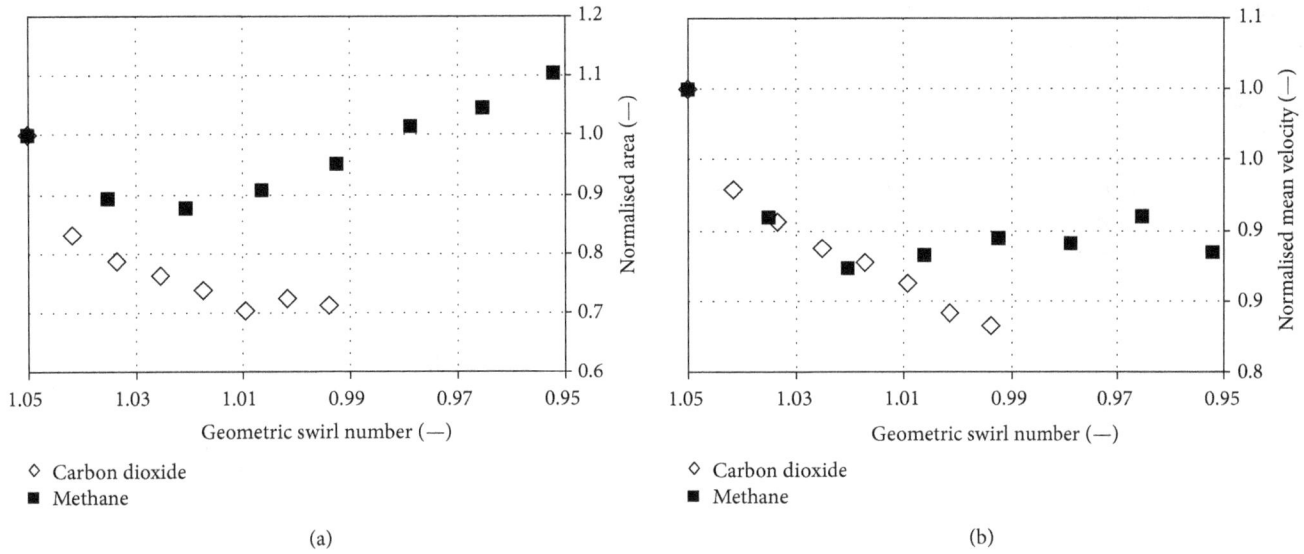

FIGURE 12: The normalised effect of the injection of carbon dioxide and methane on (a) the cross sectional area and (b) mean velocity of the HMFR.

Regardless of equivalence ratio changes, the interactions between the CRZ and HMFR flow structures are important factors relating to the initiation of CIVB, and this interaction may be altered by the diffusive injection of gases.

For the results taken in these trials the interaction between the CRZ and HMFR is assessed as described in (4), where R_M refers to the ratio of total RFM in the CRZ and total PFM in the HMFR; A_{CRZ}, A_{HMFR}, $\overline{U}_{CRZ}$, and $\overline{U}_{HMFR}$ refer to the cross sectional area and mean axial velocities of the CRZ and HMFR, respectively:

$$R_M = \frac{A_{CRZ} \cdot \overline{U}_{CRZ}}{A_{HMFR} \cdot \overline{U}_{HMFR}}. \qquad (4)$$

The full size of the HMFR is cropped by the measurement area. However, local changes in the measured area are indicative of global changes. The increasing value of R_M as more diffusive fuel is added indicates that the RFM of the CRZ is increasing compared to the PFM of the HMFR, and as a result reducing the propensity of the flame to flashback due to combustion induced vortex breakdown. The normalised effects of carbon dioxide and methane on R_M are displayed in Figure 13.

The differing effects of CO_2 and CH_4 on the CRZ, and in particular the HMFR, compensate for each other to produce a very similar response on the flame structure as a whole. The introduction of methane causes a large initial rise in R_M as S_g approaches 1.02, at which point the rate of change reduces dramatically. When $1.05 < S_g < 1.02$, there is a strong change in R_M as the interaction between the CRZ and HMFR is altered, and with $S_g < 1.02$, the change in R_M becomes less significant as the increased flow rate of methane causes an increase in global flame structure size and velocity.

Carbon dioxide produces a near linear response over the range of results, with the highest value of R_M being greater than that of methane.

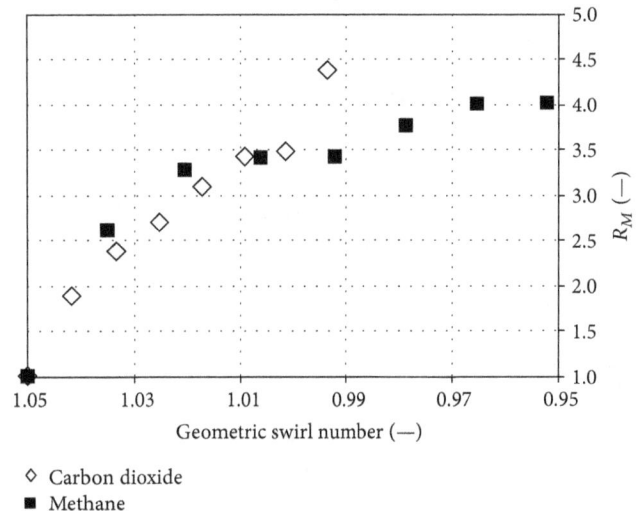

FIGURE 13: Normalised effect of the injection of carbon dioxide and methane on the ratio between RFM and PFM.

4. Summary and Conclusions

Experimentation using a generic swirl burner and PIV system demonstrated how the diffusive injection of different gases into a premixed methane flame altered its coherent structures.

The injection of both methane and carbon dioxide resulted in an increase in cross sectional area and velocity of gases within the central recirculation zone whilst maintaining flame stability. For the unconfined flame, diffusive injection of methane increased the burning rates and temperatures within the swirling flows, effecting flame propagation and hot gas expansion. Radial momentum increased whilst axial momentum decreased, increasing the volume of the recirculated gases when methane was injected. This leads to an expansion of the recirculation zone near the burner exit and a stable, well-defined boundary between the CRZ and high momentum flow region. The injection of carbon dioxide reduced the burning rate and temperature of gases within the flame, inhibiting flame propagation and reducing the hot gas expansion. Both axial and radial momentums were proportionally reduced, so the increased mass flow rate increased the size and velocity of the recirculation zone.

A differing effect on the high momentum flow region of the flame was seen. When flow rates of the diffusively injected gases produced isothermal geometric swirl numbers above 1.02, the size and velocity of the HMFR decreased, due to CH_4 increasing the level of swirl and CO_2 decreasing flame temperatures. When methane flow rates resulted in $S_g < 1.02$, this the cumulative effect of increasing swirl and flame temperature resulted in an increase in HMFR size and velocity.

Despite differing effects of the level of swirl and flame temperature, the actual effect on the RFM of the CRZ compared to the PFM of the HMFR was the same. The expansion of the CRZ compared to the HMFR opposes the changes in flame structure that result in combustion induced vortex breakdown; therefore, the diffusive injection of either CH_4 or CO_2 could be used as a measure to mitigate CIVB.

Tests were also performed with the diffusive injection of air. Its effect on the flow structures heavily contrasted those of CO_2 and CH_4, reducing the size and velocity of the CRZ before completely preventing recirculation occurring. Although the overall result would almost certainly reduce the likelihood of CIVB occurring, the destruction of the CRZ would result in an inherently unstable flame that would be highly susceptible to blow off.

Conflict of Interests

The authors declare that there is no conflict of interests regarding the publication of this paper.

Acknowledgments

This work was partly funded and supported via the European Union FP7 Project H_2IGCC, and the ERDF funded Low Carbon Research Institute.

References

[1] A. E. E. Khalil, V. K. Arghode, A. K. Gupta, and S. C. Lee, "Low calorific value fuelled distributed combustion with swirl for gas turbine applications," *Applied Energy*, vol. 98, pp. 69–78, 2012.

[2] A. Doherty, E. E. Walsh, and K. McDonnell, "The direct use of post-processing wood dust in gas turbines," *Journal of Sustainable Bioenergy Systems*, vol. 2, no. 3, pp. 60–64, 2012.

[3] C. Eichler, G. Baumgartner, and T. Sattelmayer, "Experimental investigation of turbulent boundary layer flashback limits for premixed hydrogen-air flames confined in ducts," *Journal of Engineering for Gas Turbines and Power*, vol. 134, no. 1, Article ID 011502, 2012.

[4] N. Syred, M. Abdulsada, A. Griffiths, T. O'Doherty, and P. Bowen, "The effect of hydrogen containing fuel blends upon flashback in swirl burners," *Applied Energy*, vol. 89, no. 1, pp. 106–110, 2012.

[5] J. Bower, D. Abbott, and S. James, "The impact of natural gas composition variations on the operation of gas turbines for power generation," in *Proceedings of the 6th International Conference on Future Gas Turbine Technology*, Brussels, Belguim, 2012.

[6] M. Subramanya and A. Choudhuri, "Investigation of combustion instability effects on the flame characteristics of fuel blends," in *Proceedings of the 5th International Energy Conversion Engineering Conference (IECEC '07)*, pp. 817–830, June 2007.

[7] T. Lieuwen, V. McDonell, D. Santavicca, and T. Sattelmayer, "Burner development and operability issues associated with steady flowing syngas fired combustors," *Combustion Science and Technology*, vol. 180, no. 6, pp. 1169–1192, 2008.

[8] J. D. Thornton, B. T. Chorpening, T. G. Sidwell, P. A. Strakey, E. D. Huckaby, and K. J. Benson, "Flashback detection sensor for hydrogen augmented natural gas combustion," in *Proceedings of the ASME Conference*, pp. 739–746, May 2007.

[9] J. Fritz, M. Kröner, and T. Sattelmayer, "Flashback in a swirl burner with cylindrical premixing zone," *Journal of Engineering for Gas Turbines and Power*, vol. 126, no. 2, pp. 276–283, 2004.

[10] M. Kroner, "Flashback limits for combustion induced vortex breakdown in a swirl burner," ASME Paper GT-2002-30075, ASME Turbo Expo, Amsterdam, The Netherlands, 2002.

[11] M. Kröner, J. Fritz, and T. Sattelmayer, "Flashback limits for combustion induced vortex breakdown in a swirl burner," *Journal of Engineering for Gas Turbines and Power*, vol. 125, no. 3, pp. 693–700, 2003.

[12] B. Dam, G. Corona, M. Hayder, and A. Choudhuri, "Effects of syngas composition on combustion induced vortex breakdown (CIVB) flashback in a swirl stabilized combustor," *Fuel*, vol. 90, no. 11, pp. 3274–3284, 2011.

[13] B. Dam, N. Love, and A. Choudhuri, "Flashback propensity of syngas fuels," *Fuel*, vol. 90, no. 2, pp. 618–625, 2011.

[14] C. Eichler and T. Sattelmayer, "Experiments on flame flashback in a quasi-2D turbulent wall boundary layer for premixed methane-hydrogen-air mixtures," *Journal of Engineering for Gas Turbines and Power*, vol. 133, no. 1, Article ID 011503, 2011.

[15] C. Eichler and T. Sattelmayer, "Premixed flame flashback in wall boundary layers studied by long-distance micro-PIV," *Experiments in Fluids*, vol. 52, no. 2, pp. 347–360, 2012.

[16] A. Valera-Medina, *Coherent structures and their effects on processes occurring in swirl combustors [Ph.D. thesis]*, Cardiff University, 2009.

[17] A. Valera-Medina, N. Syred, and A. Griffiths, "Visualisation of isothermal large coherent structures in a swirl burner," *Combustion and Flame*, vol. 156, no. 9, pp. 1723–1734, 2009.

[18] A. Valera-Medina, N. Syred, P. Bowen, and A. Crayford, "Studies of swirl burner characteristics, flame lengths and relative pressure amplitudes," *Journal of Fluids Engineering*, vol. 133, no. 10, Article ID 101302, 2011.

[19] A. Valera-Medina, N. Syred, P. Kay, and A. Griffiths, "Central recirculation zone analysis in an unconfined tangential swirl burner with varying degrees of premixing," *Experiments in Fluids*, vol. 50, no. 6, pp. 1611–1623, 2011.

[20] A. Valera Medina, N. Syred, and P. Bowen, "Central recirculation zone analysis using a confined swirl burner for terrestrial energy," *Journal AIAA Propulsion and Power*, vol. 29, no. 1, pp. 195–204, 2013.

[21] M. Stöhr, I. Boxx, C. D. Carter, and W. Meier, "Experimental study of vortex-flame interaction in a gas turbine model combustor," *Combustion and Flame*, vol. 159, no. 8, pp. 2636–2649, 2012.

[22] K. S. Kedia and A. F. Ghoniem, "Mechanisms of stabilization and blowoff of a premixed flame downstream of a heat-conducting perforated plate," *Combustion and Flame*, vol. 159, no. 3, pp. 1055–1069, 2012.

[23] M. Kröner, T. Sattelmayer, J. Fritz, F. Kiesewetter, and C. Hirsch, "Flame propagation in swirling flows: effect of local extinction on the combustion induced vortex breakdown," *Combustion Science and Technology*, vol. 179, no. 7, pp. 1385–1416, 2007.

[24] G. Blesinger, R. Koch, and H. J. Bauer, "Influence of flow field scaling on flashback of swirl flames," *Experimental Thermal and Fluid Science*, vol. 34, no. 3, pp. 290–298, 2010.

[25] J. Lewis, R. Marsh, A. Valera-Medina, S. Morris, and H. Baej, "The use of CO_2 to improve stability and emissions of an IGCC combustor," in *ASME Turbo Expo 2014: Turbine Technical Conference and Exposition*, Dusseldorf, Germany, June 2014.

[26] J. Warnatz, U. Maas, and R. Dibble, *Combustion*, Springer, Berlin, Germany, 1999.

[27] M. C. Lee, S. B. Seo, J. Yoon, M. Kim, and Y. Yoon, "Experimental study on the effect of N_2, CO_2, and steam dilution on the combustion performance of H_2 and CO synthetic gas in an industrial gas turbine," *Fuel*, vol. 102, pp. 431–438, 2012.

[28] J. Park, D. S. Bae, M. S. Cha et al., "Flame characteristics in H_2/CO synthetic gas diffusion flames diluted with CO_2: effects of radiative heat loss and mixture composition," *International Journal of Hydrogen Energy*, vol. 33, no. 23, pp. 7256–7264, 2008.

[29] A. Konnov, I. Dyakov, and J. Ruyck, "Nitric oxide formation in premixed flames of H_2 + CO + CO_2 and air," *Proceedings of the Combustion Institute*, vol. 295, pp. 2171–2177, 2002.

[30] J. Natarajan, T. Lieuwen, and J. Seitzman, "Laminar flame speeds of H_2/CO mixtures: Effect of CO_2 dilution, preheat temperature, and pressure," *Combustion and Flame*, vol. 151, no. 1-2, pp. 104–119, 2007.

[31] H. J. Burbano, J. Pareja, and A. A. Amell, "Laminar burning velocities and flame stability analysis of H_2/CO/air mixtures with dilution of N_2 and CO_2," *International Journal of Hydrogen Energy*, vol. 36, no. 4, pp. 3232–3242, 2011.

[32] N. Syred and J. M. Beér, "Combustion in swirling flows: a review," *Combustion and Flame*, vol. 23, no. 2, pp. 143–201, 1974.

[33] C. T. Bowman, M. Frenklach, W. C. Gardiner, and G. P. Smith, "The "GRIMech 3.0" chemical kinetic mechanism," 1999, http://www.me.berkeley.edu/gri-mech/.

[34] S. de Persis, F. Foucher, L. Pillier, V. Osorio, and I. Gökalp, "Effects of O_2 enrichment and CO_2 dilution on laminar methane flames," *Energy*, vol. 55, pp. 1055–1066, 2013.

17

Nonpremixed Counterflow Flames: Scaling Rules for Batch Simulations

Thomas Fiala and Thomas Sattelmayer

Lehrstuhls für Thermodynamik, Technische Universität München, Boltzmannstraße 15, 85747 Garching, Germany

Correspondence should be addressed to Thomas Fiala; fiala@td.mw.tum.de

Academic Editor: Constantine D. Rakopoulos

A method is presented to significantly improve the convergence behavior of batch nonpremixed counterflow flame simulations with finite-rate chemistry. The method is applicable to simulations with varying pressure or strain rate, as it is, for example, necessary for the creation of flamelet tables or the computation of the extinction point. The improvement is achieved by estimating the solution beforehand. The underlying scaling rules are derived from theory, literature, and empirical observations. The estimate is used as an initialization for the actual solver. This enhancement leads to a significantly improved robustness and acceleration of batch simulations. The extinction point can be simulated without cumbersome code extensions. The method is demonstrated on two test cases and the impact is discussed.

1. Introduction

Laminar steady nonpremixed counterflow flames have been investigated in numerous studies in the past [1–3]. Due to their quasi-one-dimensional nature, they are simulated on a computationally very inexpensive grid. Extensive chemical mechanisms can be applied, tested, and investigated. The results can be validated well against state-of-the-art experiments [1].

From counterflow flame simulations, detailed insight is gained into the structure of nonpremixed flames, including temperature and species profiles [3], heat release [3, 4], and flame radiation [5]. This setup is often used as a test bed for novel or reduced combustion mechanisms. Another purpose is the tabulation of flames in the form of flamelet libraries for use in turbulent combustion simulation [6].

Even though counterflow flame simulations usually run on a very small grid, the computation is time consuming because of the stiff equation system resulting from the detailed chemical mechanism. If the initialization is too far from the steady state solution, the solvers are prone to diverge.

A major aim of counterflow flame simulations is to compute extinction strain rates [7]. In this case, a second problem arises. At the extinction point, the numerical solution of a regular counterflow flame simulation becomes singular [2, 7]. The exact extinction flamelet therefore cannot be computed. To circumvent this problem, two sophisticated methods have been developed and applied on counterflow flames: the arc-length continuation method by Keller [8] and Giovangigli and Smooke [9] and the flame controlling continuation method by Nishioka et al. [10]. The idea behind both methods is to compute the unstable burning branch, that is, to calculate pseudostable flames which show maximum temperatures and strain rates below the extinction values. The extinction point can thus distinctly be identified. To realize this, the computational procedure has to be altered severely. A recent implementation is well documented by Wang et al. [7].

For both the computation of extinction strain rates and the creation of flamelet libraries, multiple simulations at varying strain rates are necessary. The operating pressure is often an investigated parameter as well. This can result in a large matrix of input parameters for which separate simulations are required. While the computational effort might still be acceptable for single simulations, it can be very large for the overall simulation campaign. A reduction of the computational effort is therefore favorable.

This paper develops a method to stabilize and accelerate batch counterflow flame simulations. These goals are achieved by improving the initialization for the iterative steps. Depending on the pressure and the strain rate changes, estimations for the change of the velocity, species, and temperature profiles are derived from the literature, theoretical considerations, and empirical observations.

2. Method

To properly deduce the scaling rules, the setup of the counterflow flame and its definitions are briefly described first. The approach to improve the batch computations is then derived theoretically and finally put into an applicable form.

2.1. Counterflow Flame Setup. A schematic view of the considered counterflow flame configuration is shown in Figure 1. Two axis-symmetric jets of fuel and oxidizer impinge on each other from opposed nozzles, thus creating an axis-symmetric flow field which can be described by the axial coordinate z and the radial coordinate r. Only the axial coordinate z has to be resolved to specify the flame [2].

The well-known governing partial differential equations are described by Kee et al. [2]. The problem can be solved numerically by evaluating the continuity equation, the radial momentum equation, a species conservation equation for each species considered, and the energy equation. Species source terms are calculated from detailed chemistry reaction mechanisms. Pressure, density, and temperature are linked by the ideal gas equation of state. Real gas effects were shown to have little influence on the structure of counterflow flames [7, 14].

The flow field is fully characterized by the axial coordinate z, the axial velocity u, the radial velocity divided by the radial coordinate $V = v/r$, the radial pressure curvature $\Lambda = (1/r)(\partial p/\partial r)$ (which is an eigenvalue and constant throughout the flame), the temperature T, and the species mass fractions Y_i for all considered species.

So-called plug-flow boundary conditions are applied on both the oxidizer and the fuel inlet [2, 3]. The mass flow rate area densities $\dot{m}_f = u_f \rho_f$ and $\dot{m}_o = u_o \rho_o$, the inlet temperatures T_f and T_o, and the inlet compositions $Y_{i,f}$ and $Y_{i,o}$ are specified on the fuel and oxidizer boundaries. The distance between fuel and oxidizer inlets is defined as the grid size d.

The counterflow flame is defined by setting the boundary conditions, the grid size, the operating pressure, and a chemical reaction mechanism. Counterflow flames are commonly characterized by their strain rate a. However, different definitions of the strain rate are used in the literature: a can be determined from the mean axial velocity gradient [3], the maximum axial velocity gradient [15], the axial velocity gradient at the stoichiometric point [16], or the

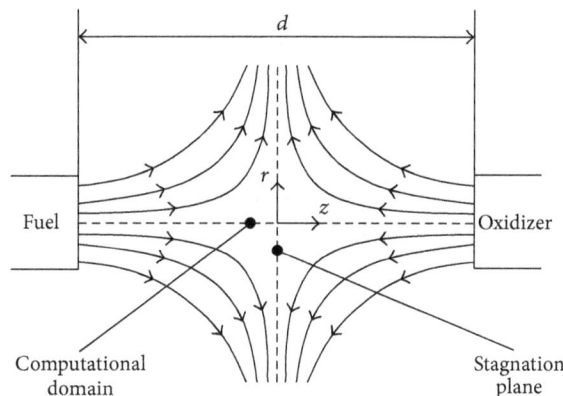

FIGURE 1: Schematic view of the counterflow flame configuration.

axial velocity gradient "before the flame" [3, 4, 6]. The latter corresponds to the boundary conditions in the potential flow setup where fuel and oxidizer are assumed to be infinitely far away. Sometimes, radial velocity gradients are used instead [2]. If oxidizer and fuel do not have equal density, the strain rates on either side of the flame are different [3]. All of these definitions are approximately directly proportional to each other. Nevertheless, attention has to be paid when comparing absolute results. In this paper, the maximum axial velocity gradient is used to characterize the strain rate.

In the plug-flow configuration, the potential flow boundary conditions can be approximated by

$$a_{f/o} = \sqrt{\frac{\Lambda}{\rho_{f/o}}} \tag{1}$$

for the fuel and for the oxidizer sides, respectively (this result can be obtained by reshaping (3.35) from [17] or (6.32) from [2]).

With all parameters defined, counterflow flames can be computed with standard software like OPPDIFF [18], Cosilab [19], FlameMaster [20], or Cantera [21]. These programs solve the underlying partial differential equations of the one-dimensional problem. To fully resolve the flame structure, an automatic grid refinement is often necessary. The solution procedure can take several minutes because of the stiff system of equations to be solved. The further away the final solution is from the initial solution, the larger the computation time is. If the initial solution is too far from the actual solution, the solver can easily diverge.

2.2. Approach. As pointed out in the introduction, for the calculation of extinction strain rates or the generation of flamelet tables, a large number of counterflow flames have to be simulated. This is generally done by performing a batch simulation.

The idea of improving the convergence behavior of batch simulations is to estimate the solution beforehand. This guess is used as the initialization for the actual computation. The estimation is based on the assumption that the structure of flames with finite-rate chemistry behaves similarly to the structure of flames with infinitely fast chemistry. In nonpremixed flames, the structure is mostly influenced by diffusion which is equally represented in flames with finite-rate and infinitely fast chemistry.

Counterflow diffusion flames with infinitely fast chemistry are known to have a self-similar structure [17]. According to Law [3], the spatial profiles in nonpremixed counterflow flames scale with the term

$$\sqrt{\frac{\lambda}{c_p \rho a}}. \qquad (2)$$

If the inlet temperatures are considered to be constant, the density ρ is directly proportional to the pressure p. Assuming that the thermal conductivity λ and the heat capacity at constant pressure c_p are independent of pressure and strain rate, the spatial profiles are directly proportional to the term $(ap)^{-1/2}$.

This formulation is equivalent to the postulation that the diffusion flame thickness (here, full width at half maximum (FWHM) of the temperature profile [3]) δ, which corresponds to the thickness of the mixing layer, scales with [16, 17]

$$\delta \sim \sqrt{\frac{D_{st}}{a}}. \qquad (3)$$

If the diffusion coefficient D is assumed to be inversely proportional to the pressure, the above relation can again be transformed to

$$\delta \sim (ap)^{-1/2}. \qquad (4)$$

This relation has been confirmed numerically to hold for a wide range of counterflow flames with detailed chemistry [4, 22].

2.3. *Application to Batch Computation.* To apply this on batch simulations, an existing solution with the strain rate a_{old} and the pressure p_{old} is considered. This flame has the known thickness δ_{old}. The thickness of a new flame δ_{new} to be calculated at a_{new} and p_{new} can be estimated according to (4):

$$\delta_{new} = \delta_{old} \cdot \left(\frac{a_{new}}{a_{old}}\right)^{-1/2} \left(\frac{p_{new}}{p_{old}}\right)^{-1/2}. \qquad (5)$$

Since all species profiles and the temperature profile are assumed similar with respect to the flame thickness, they can be scaled accordingly.

However, instead of scaling the flame on a fixed grid, the grid size d can be modified:

$$d_{new} = d_{old} \cdot \left(\frac{a_{new}}{a_{old}}\right)^{-1/2} \left(\frac{p_{new}}{p_{old}}\right)^{-1/2}. \qquad (6)$$

This procedure has the advantage that the temperature and the species mass fractions remain constant with respect to the grid points and do not have to be scaled separately. The magnitudes of both the temperature and the species mass fractions are supposed to change little for a batch step.

Therefore, the absolute values of these variables can remain unaltered. Additionally, a potential refinement of the grid is adapted as well automatically.

In contrast to the temperature and species profiles, the magnitudes of the velocity profiles, the pressure curvature, and the boundary conditions have to be updated. Using the definition of the mean strain rate

$$a_{mean} = \frac{|u_f - u_o|}{d} \qquad (7)$$

and the fact that all strain rate definitions are approximately directly proportional to each other, the axial velocity profile can be assumed proportional to

$$u \sim ad \sim a^{1/2} p^{-1/2}. \qquad (8)$$

By substituting this relation into the continuity equation [21], a scaling for V can be derived:

$$V \sim ap^0. \qquad (9)$$

This leads to

$$u_{new} = u_{old} \cdot \left(\frac{a_{new}}{a_{old}}\right)^{1/2} \left(\frac{p_{new}}{p_{old}}\right)^{-1/2},$$

$$V_{new} = V_{old} \cdot \left(\frac{a_{new}}{a_{old}}\right). \qquad (10)$$

If the ideal gas densities are assumed to scale with the pressure, the mass flux densities at the boundaries adapt accordingly:

$$\dot{m}_{f/o} = \rho u_{f/o} \sim a^{1/2} p^{1/2},$$

$$\dot{m}_{f/o,new} = \dot{m}_{f/o,old} \cdot \left(\frac{a_{new}}{a_{old}}\right)^{1/2} \left(\frac{p_{new}}{p_{old}}\right)^{1/2}. \qquad (11)$$

Following (1), the pressure curvature Λ can be estimated:

$$\Lambda_{new} = \Lambda_{old} \cdot \left(\frac{a_{new}}{a_{old}}\right)^2 \frac{p_{new}}{p_{old}}. \qquad (12)$$

The equations derived above are summarized in Table 1. In addition to the general case, the scaling factors at constant pressure and $a \sim p^{3/2}$ are attached. The constant pressure case is important for the computation of isobaric flames, for example, for the computation of the extinction strain rate at a specific pressure or the creation of a flamelet table. The last column is added for the case that different pressures should be calculated. With rising pressure, the extinction strain rate increases. If a batch of counterflow flames up to extinction is to be computed for various pressures, starting each pressure batch at the same initial strain rate would lead to an increasing number of simulations with pressure. To compute approximately the same number of simulations for all pressures, the initial strain rate has to be modified. The relation $a \sim p^{3/2}$ was found empirically to be a good scaling for the initial strain rates. This is based on the analysis of hydrogen-oxygen and ethane-air flames. The exact quantity probably depends on the fuel and oxidizer compositions, but $a \sim p^{3/2}$ is recommended as a good starting guess for such a scenario.

TABLE 1: Overview over the magnitude scaling parameters.

		$p = \text{const.}$	$a \sim p^{3/2}$
$T \sim$	$a^0 p^0$	a^0	p^0
$Y_i \sim$	$a^0 p^0$	a^0	p^0
$d \sim$	$a^{-1/2} p^{-1/2}$	$a^{-1/2}$	$p^{-5/4}$
$u \sim$	$a^{1/2} p^{-1/2}$	$a^{1/2}$	$p^{1/4}$
$V \sim$	$a^1 p^0$	a	$p^{3/2}$
$\dot{m}_{f/o} \sim$	$a^{1/2} p^{1/2}$	$a^{1/2}$	$p^{5/4}$
$\Lambda \sim$	$a^2 p$	a^2	p^4

TABLE 2: The initial parameters of the example counterflow flame simulations.

	Speed-up H_2	Speed-up C_2H_6	Extinction point
Fuel	H_2	C_2H_6	H_2
Oxidizer	O_2	Air	O_2
Reaction mechanism	Ó Conaire et al. [11]	GRI-MECH 3.0 [12]	Li et al. [13]
p [bar]	1	1	1
$T_{f/o}$ [K]	300	300	300
d [mm]	3.6	20	18
$\dot{m}_f$ [kg/(m^2 s)]	0.50	0.24	0.10
$\dot{m}_o$ [kg/(m^2 s)]	3.00	0.72	0.60

TABLE 3: Simulation times for the pressure loops and the average strain rate loops with and without the scaling rules applied.

	With scaling rules (accelerated)	Without scaling rules (benchmark)
Pressure loop H_2-O_2 (64 bar)	t: 00:00:43.20	t: 00:14:11.52
	s: 00:00:00.55	s: 00:00:10.92
Pressure loop C_2H_6-air	t: 00:15:15.48	t: 02:47:35.05
	s: 00:00:08.03	s: 00:01:28.20
Strain rate loop H_2-O_2	t: 00:00:04.92	t: 00:01:48.66
	s: 00:00:00.30	s: 00:00:06.21
Strain rate loop C_2H_6-air	t: 00:04:07.81	t: 03:55:28.61
	s: 00:00:06.27	s: 00:06:00.73

t: total (accumulated) time per loop, s: time per average simulation.

3. Results and Discussion

The method described above is implemented on top of the Cantera environment (version 2.2a, revision r2777 [21]). The simulation is set up using the counterflow flame class of the Cython-based interface. The batch procedure is formulated as a Python script. In between the simulation runs, the new initial condition is estimated using the relations derived above.

The two benefits of the proposed method, the speed-up and the robustness, are demonstrated using two test cases described in the following.

3.1. Speed-Up Test Case. To indicate the speed-up achieved by the proposed method, a series of counterflow flames at various pressures and strain rates is simulated. This could be the basis for the creation of a flamelet table. Since the computation of the underlying counterflow flames is the most computationally expensive step in the creation of a flamelet library, such a task is assumed to speed up accordingly. The acceleration is shown by comparing a batch simulation which incorporates the proposed method with a state-of-the-art benchmark simulation. This is demonstrated for a hydrogen-oxygen flame and an ethane-air flame. The overview over the simulation parameters is given in Table 2.

Starting from initial counterflow flames at a pressure of 1 bar, flames up to 100 bar should be calculated. The pressure is increased by approximately 15% in each step. Additionally, at 13 of the previously computed pressure levels, isobaric flames should be calculated at increasing strain rates. The strain rates are raised by 25% in each step until the flame is extinguished.

For each fuel-oxidizer configuration, the test case is run two times. The accelerated run makes use of the procedure described in this paper. In the corresponding benchmark run, the initialization procedure is not applied. The pressure and strain rate are set in this case by changing only the operating pressure and the inlet mass fluxes, respectively. Both cases are run sequentially on a standard desktop PC (housing an Intel i5 CPU at 3.2 GHz) under standard load conditions.

For the hydrogen-oxygen flame, the benchmark computation fails to converge for pressures above 64 bar. At this point, the cumulated simulation time already amounts to 14:12 minutes. The accelerated run runs smoothly for all pressure levels; at 64 bar the cumulated simulation time is only 43 seconds. For the ethane-air flame, both variants are able to compute all pressure levels. The simulation time for the entire loop is accelerated from 2:47:35 hours to 15:15 minutes using the proposed method. The values are compared in Table 3.

The loops through the increasing strain rates profit even more from the preconditioning step. For the hydrogen-oxygen flame, the average time for an isobaric loop decreases from 1:49 minutes to only 5 seconds. For the ethane-air flame, the corresponding times are 3:55:29 hours and 4:08 minutes. At pressures above 10 bar for the hydrogen-oxygen flame and above 30 bar for the ethane-air flame, the solver diverges in the strain rate loop in the benchmark run. This problem does not occur for the accelerated run. This fact also indicates the improvement in robustness achieved by using the derived scaling rules.

Figures 2 and 3 show the axial velocity and temperature profiles over the normalized grid for the accelerated case of the hydrogen-oxygen flame. It can be seen that the flame thickness with respect to the grid size indeed remains constant. The axial velocity profiles also remain similar. This proves that the scaling rules derived from infinitely fast chemistry in fact apply to finite-rate chemistry counterflow flames.

3.2. Simulation of the Extinction Point. A second independent test case is evaluated to demonstrate the robustness increase

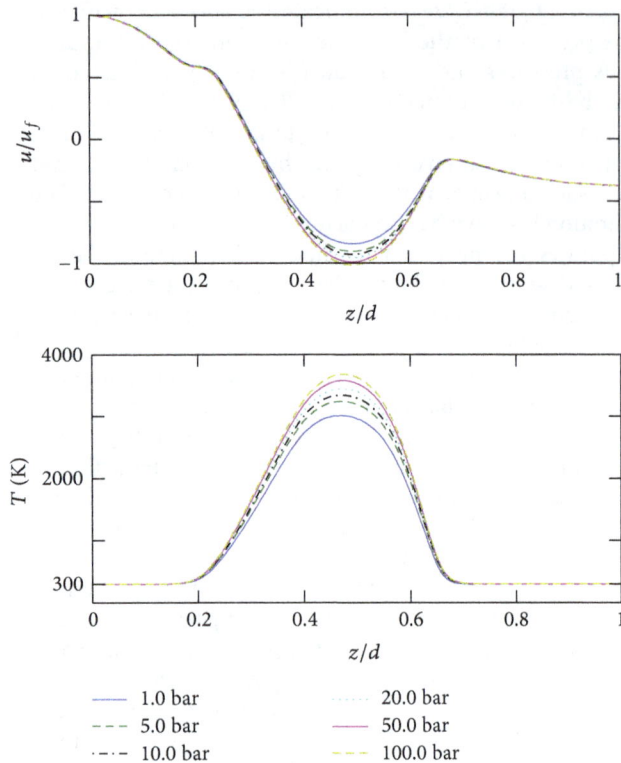

FIGURE 2: Profiles of the normalized axial velocity u/u_f and the temperature T over the normalized grid z/d for selected pressures of the hydrogen-oxygen flame.

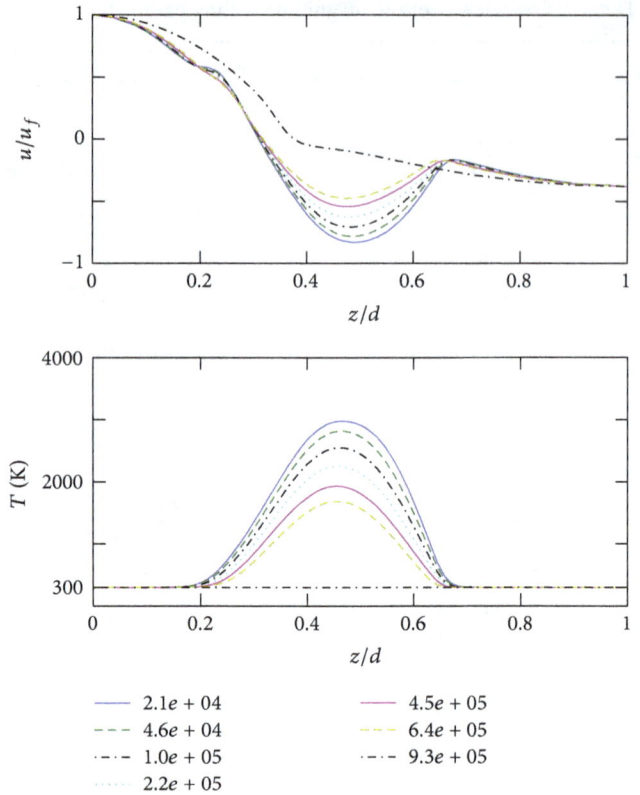

FIGURE 3: Profiles of the normalized axial velocity u/u_f and the temperature T over the normalized grid z/d for selected maximum strain rates [1/s] of the hydrogen-oxygen flame. The operating pressure is 1 bar.

achieved by this method. This makes it possible to compute the extinction point of counterflow flames very accurately without having to apply a cumbersome method to circumvent the singularity at this point.

For multiple flames at constant pressure levels, the strain rate is increased gently until the flame is extinguished. From the last burning flame, the strain rate is increased in smaller steps to get closer to the extinction point. This iteration is repeated until the temperature change is below a threshold value. The strain rate and the maximum temperature of the last burning solutions are taken as the extinction strain rate and extinction temperature.

This is demonstrated for a hydrogen-oxygen flame. The boundary and operating conditions of the initial simulation are given in Table 2. The output strain rates and temperatures at the extinction point are shown in Figure 4. The data are compared to the data by Wang et al. [7]. The same reaction mechanism and the boundary conditions were used as in the reference paper. While the extinction strain rate matches the reference values almost perfectly, the maximum flame temperature at extinction is slightly overpredicted. This is due to the fact that the reference data were created using a sophisticated two-point flame controlling continuation method. The method presented in this paper approaches the extinction point from higher temperatures.

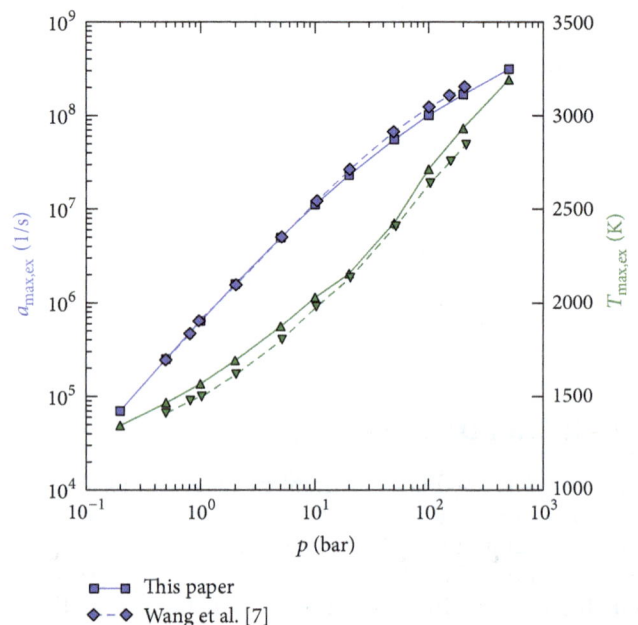

FIGURE 4: Maximum strain rates and maximum temperatures for H_2-O_2 counterflow flames just before the extinction point as a function of pressure. The reference data is taken from Wang et al. [7].

4. Conclusions

In this paper, a method is described which significantly improves the convergence behavior of batch counterflow flame simulations with varying pressure and strain rate. This is achieved by estimating the initial solutions from previously computed simulations. The scaling parameters for the solution vectors with respect to the pressure and the strain rate are derived from existing relations for flames with infinitely fast chemistry, dimensional analysis, and empirical observations.

The method is demonstrated using the Cantera software package [21] on a nonpremixed hydrogen-oxygen and an ethane-air counterflow flame. By computing batch simulations with and without applying the proposed routine, the calculation speed is observed to increase by a factor of 10–20 for a variation in pressure and a factor of 20–60 for variations in strain rate. Additionally, using the improved procedure, several operating points can be computed that fail to converge in the reference run.

The method is also used to compute the extinction point of a hydrogen-oxygen counterflow flame. The result very accurately matches data from the literature. Using the described simple procedure, the extinction point can be calculated without having to use cumbersome continuation methods.

Nomenclature

D: Diffusion coefficient [m^2/s]
T: Temperature [K]
V: Radial velocity potential [1/s]
Y: Mass fraction [—]
a: Strain rate [1/s]
c_p: Isobaric heat capacity [J/(kg K)]
d: Grid size [m]
$\dot{m}$: Mass flux density [kg/(s m^2)]
p: Pressure [Pa]
u: Axial velocity [m/s]
z: Axial coordinate [m]
Λ: Pressure curvature [N/m^4]
δ: Flame thickness [m]
λ: Heat conductivity [W/(m K)]
ρ: Density [m^3/kg]
f: Fuel inlet
o: Oxidizer inlet

Conflict of Interests

The authors declare that there is no conflict of interests regarding the publication of this paper.

Acknowledgments

The Deutsche Forschungsgemeinschaft (DFG) in the framework of Sonderforschungsbereich Transregio 40 provided financial support for this project. This work was supported by the German Research Foundation (DFG) and Technische Universität München within the funding programme Open Access Publishing.

References

[1] H. Tsuji, "Counterflow diffusion flames," *Progress in Energy and Combustion Science*, vol. 8, no. 2, pp. 93–119, 1982.

[2] R. J. Kee, M. E. Coltrin, and P. Glarborg, *Chemically Reacting Flow*, Wiley-Interscience, New York, NY, USA, 2003.

[3] C. K. Law, *Combustion Physics*, Cambridge University Press, New York, NY, USA, 2006.

[4] G. Ribert, N. Zong, V. Yang, L. Pons, N. Darabiha, and S. Candel, "Counterflow diffusion flames of general fluids: oxygen/hydrogen mixtures," *Combustion and Flame*, vol. 154, no. 3, pp. 319–330, 2008.

[5] T. Fiala and T. Sattelmayer, "On the use of OH* radiation as a marker for the heat release rate in high-pressure hydrogen-oxygen liquid rocket combustion," in *Proceedings of the 49th AIAA/ASME/SAE/ASEE Joint Propulsion Conference and Exhibit*, 2013-3780, San Jose, Calif, USA, July 2013.

[6] N. Peters, "Laminar diffusion flamelet models in non-premixed turbulent combustion," *Progress in Energy and Combustion Science*, vol. 10, no. 3, pp. 319–339, 1984.

[7] X. Wang, H. Huo, and V. Yang, "Supercritical combustion of general fluids in laminar counterflows," in *Proceedings of the 51st AIAA Aerospace Sciences Meeting Including the New Horizons Forum and Aerospace Exposition*, AIAA 2013-1165, American Institute of Aeronautics and Astronautics, Grapevine, Tex, USA, January 2013.

[8] H. B. Keller, *Applications of Bifurcation Theory*, Academic Press, New York, NY, USA, 1977.

[9] V. Giovangigli and M. D. Smooke, "Extinction of strained premixed laminar flames with complex chemistry," *Combustion Science and Technology*, vol. 53, no. 1, pp. 23–49, 1987.

[10] M. Nishioka, C. K. Law, and T. Takeno, "A flame-controlling continuation method for generating S-curve responses with detailed chemistry," *Combustion and Flame*, vol. 104, no. 3, pp. 328–342, 1996.

[11] M. Ó Conaire, H. J. Curran, J. M. Simmie, W. J. Pitz, and C. K. Westbrook, "A comprehensive modeling study of hydrogen oxidation," *International Journal of Chemical Kinetics*, vol. 36, no. 11, pp. 603–622, 2004.

[12] G. P. Smith, D. M. Golden, M. Frenklach et al., GRI-Mech 3.0.

[13] J. Li, Z. Zhao, A. Kazakov, and F. L. Dryer, "An updated comprehensive kinetic model of hydrogen combustion," *International Journal of Chemical Kinetics*, vol. 36, no. 10, pp. 566–575, 2004.

[14] S. Pohl, M.-M. Jarczyk, and M. Pfitzner, "A real gas laminar flamelet combustion model for the CFD-simulation of LOX/GH2 combustion," in *Proceedings of the 5th European Combustion Meeting (ECM '11)*, Number 86, Cardiff, UK, June-July 2011.

[15] S. R. Turns, *An Introduction to Combustion: Concepts and Applications*, Mechanical Engineering Series, McGraw-Hill, New York, NY, USA, 2000.

[16] T. Poinsot and D. Veynante, *Theoretical and Numerical Combustion*, R.T. Edwards, Philadelphia, Pa, USA, 2nd edition, 2005.

[17] N. Peters, *Turbulent Combustion*, Cambridge Monographs on Mechanics, Cambridge University Press, Cambridge, UK, 2000.

[18] A. E. Lutz, R. J. Kee, J. F. Grcar, and F. M. Rupley, *OPPDIFF: A Fortran Program for Computing Opposed-Flow Diffusion Flames*, Sandia National Laboratories, Eubank, Ky, USA, 1996.

[19] *COSILAB User Manual*, Rotexo, 2010.

[20] H. Pitsch, *Entwicklung eines Programmpaketes zur Berechnung eindimensionaler Flammen am Beispiel einer Gegenstromdiffu-sionsflamme [Diplomarbeit]*, Rheinisch-Westfälische Technis-che Hochschule Aachen, Aachen, Germany, 1993.

[21] D. Goodwin, N. Malaya, H. Moffat, and R. Speth, "Cantera: an object-oriented software toolkit for chemical kinetics, thermo-dynamics, and transport processes," 2013, https://code.google.com/p/cantera/.

[22] L. Pons, N. Darabiha, and S. Candel, "Pressure effects on non-premixed strained fames," in *Proceedings of the European Combustion Meeting*, Chania, Greece, April 2007.

Permissions

List of Contributors

George Vourliotakis, Dionysios I. Kolaitis and Maria A. Founti
Laboratory of Heterogeneous Mixtures and Combustion Systems, Thermal Engineering Section, School of Mechanical Engineering, National Technical University of Athens, 9 Heroon Polytechniou Street, Polytechnioupoli Zografou, 15780 Athens, Greece

Mohamed S. Shehata
Mechanical Engineering Department, Faculty of Engineering, King Khalid University, Abha 61413, Saudi Arabia

Mohamed M. ElKotb and Hindawi Salem
Mechanical Power Engineering Department, Faculty of Engineering, Cairo University, University Street, Giza 12316, Egypt

Florian Ettner, Klaus G. Vollmer and Thomas Sattelmayer
Lehrstuhl für Thermodynamik, Technische Universität München, 85748 Garching, Germany

Y. Gao and N. Chakraborty
School of Mechanical and Systems Engineering, Newcastle University, Claremont Road, Newcastle-Upon-Tyne NE1 7RU, UK

N. Swaminathan
Cambridge University Engineering Department, Trumpington Street, Cambridge CB2 1PZ, UK

G. Kats and J. B. Greenberg
Faculty of Aerospace Engineering, Technion-Israel Institute of Technology, 32000 Haifa, Israel

André Vagner Gaathaug, Dag Bjerketvedt and Knut Vaagsaether
Telemark University College, Faculty of Technology, 3918 Porsgrunn, Norway

Sandra Hennie Nilsen
Research, Development and Innovation, Section for Health, Safety and Water Management, Statoil ASA, 3905 Porsgrunn, Norway

Adi Surjosatyo, Fajri Vidian and Yulianto Sulistyo Nugroho
Department of Mechanical Engineering, Faculty of Engineering, Universitas Indonesia, UI Campus, Depok 16242, Indonesia

A. Andreini, C. Bianchini and A. Innocenti
Department of Industrial Engineering, University of Florence, Via di Santa Marta 3, 50139 Florence, Italy

Claude Valery Ngayihi Abbe and Robert Nzengwa
National Advanced School of Engineering, University of Yaounde, P.O. Box 337, Yaounde, Cameroon
Faculty of Industrial Engineering, University of Douala, P.O. Box 2701, Douala, Cameroon

Raidandi Danwe
National Advanced School of Engineering, University of Yaounde, P.O. Box 337, Yaounde, Cameroon
Higher Institute of the Sahel, University of Maroua, P.O. Box 46, Maroua, Cameroon

Zorica Kauf and Andreas Fangmeier
Institute of Landscape and Plant Ecology, University of Hohenheim, August-von-Hartmann Straße 3, 70599 Stuttgart, Germany

Roman Rosavec and Celjko Španjol
Department of Forest Ecology and Silviculture, Faculty of Forestry, University of Zagreb, Svetošimunska 25, 10002 Zagreb, Croatia

Ahmad El Sayed and Roydon A. Fraser
Department of Mechanical and Mechatronics Engineering, University of Waterloo, 200 University Avenue West, Waterloo, ON, Canada N2L 3G1

Fajri Vidian
Department of Mechanical Engineering, Faculty of Engineering, Universitas Sriwijaya, Jalan Raya Palembang - Prabumulih km 32, Indralaya, Ogan Ilir, Sumatera Selatan 30662, Indonesia

Adi Surjosatyo and Yulianto Sulistyo Nugroho
Department of Mechanical Engineering, Faculty of Engineering, Universitas Indonesia, Kampus Baru UI Depok, Jawa Barat 16424, Indonesia

Giovanni Nicoletti, Roberto Bruno, Natale Arcuri and Gerardo Nicoletti
Mechanical, Energetic and Management Engineering Department, University of Calabria, V. P. Bucci 46/C, Arcavacata, 87036 Rende, Italy

Giovanni Nicoletti, Roberto Bruno, Natale Arcuri, and Gerardo Nicoletti
Mechanical, Energetic and Management Engineering Department, University of Calabria, V. P. Bucci 46/C, Arcavacata, 87036 Rende, Italy

Libin Jia, Jeffrey D. Naber and Jason R. Blough
Michigan Technological University, 1400 Townsend Drive, Houghton, MI 49931, USA

Jonathan Lewis, Agustin Valera-Medina, Richard Marsh and Steven Morris
Cardiff School of Engineering, Queen's Buildings, Cardiff CF24 3AA, UK

Thomas Fiala and Thomas Sattelmayer
Lehrstuhls für Thermodynamik, Technische Universität München, Boltzmannstraße 15, 85747 Garching, Germany